SEMICONDUCTORS AND SEMIMETALS

VOLUME 12

Infrared Detectors II

SEMICONDUCTORS AND SEMIMETALS

Edited by *R. K. WILLARDSON*

COMINCO AMERICAN INCORPORATED
ELECTRONIC MATERIALS DIVISION
SPOKANE, WASHINGTON

ALBERT C. BEER

BATTELLE MEMORIAL INSTITUTE
COLUMBUS LABORATORIES
COLUMBUS, OHIO

VOLUME 12

Infrared Detectors II

1977

ACADEMIC PRESS New York San Francisco London
A Subsidiary of Harcourt Brace Jovanovich, Publishers

COPYRIGHT © 1977, BY ACADEMIC PRESS, INC.
ALL RIGHTS RESERVED.
NO PART OF THIS PUBLICATION MAY BE REPRODUCED OR
TRANSMITTED IN ANY FORM OR BY ANY MEANS, ELECTRONIC
OR MECHANICAL, INCLUDING PHOTOCOPY, RECORDING, OR ANY
INFORMATION STORAGE AND RETRIEVAL SYSTEM, WITHOUT
PERMISSION IN WRITING FROM THE PUBLISHER.

ACADEMIC PRESS, INC.
111 Fifth Avenue, New York, New York 10003

United Kingdom Edition published by
ACADEMIC PRESS, INC. (LONDON) LTD.
24/28 Oval Road, London NW1

Library of Congress Cataloging in Publication Data

Willardson, Robert K   ed.
    Semiconductors and semimetals.

    CONTENTS: v. Physics of III-V compounds.–v. 3.
Optic properties of III-V compounds.–v. 5.   In-
frared detectors.   [etc.]
    1.   Semiconductors.   2.   Semimetals.   I.   Beer,
Albert C., joint ed.   II.   Title.
QC610.92.W54   537.6'22   65-26048
ISBN: 0–12–752112–7 (v.12)

PRINTED IN THE UNITED STATES OF AMERICA

Contents

LIST OF CONTRIBUTORS ix
PREFACE xi
CONTENTS OF PREVIOUS VOLUMES xiii

Chapter 1   Operational Characteristics of Infrared Photodetectors
W. L. Eisenman, J. D. Merriam, and R. F. Potter

I. Introduction 1
II. Detector Parameters 2
III. Figures of Merit 5
IV. Measurement of Detector Parameters 8
V. Calculations of Performance for Specific Conditions 16
VI. Current State-of-the-Art Photodetectors 23
VII. Detectors in Advance of the State-of-the-Art 34
List of Symbols 38

Chapter 2   Impurity Germanium and Silicon Infrared Detectors
Peter R. Bratt

I. Introduction 39
II. Impurities in Germanium and Silicon 42
III. Impurity Photoconductivity 53
IV. Device Fabrication 77
V. Operating Characteristics 95
VI. Conclusions and Anticipated Future Developments 140

Chapter 3   InSb Submillimeter Photoconductive Detectors
E. H. Putley

I. Physical Principles 143
II. Description of Detector 147

III. Performance of Detector 150
IV. Applications and Future Developments 163
    Appendix. Amplifiers for Use with the InSb Submillimeter Detector . . . 164
    Addendum 166

Chapter 4 Far-Infrared Photoconductivity in High Purity GaAs

G. E. Stillman, C. M. Wolfe, and J. O. Dimmock

I. General Introduction 169
II. Material 176
III. Photoconductivity 208
IV. Detector Performance 262
V. Summary 288

Chapter 5 Avalanche Photodiodes

G. E. Stillman and C. M. Wolfe

I. Introduction 291
II. Avalanche Gain Mechanism 300
III. Multiplication Noise 314
IV. Electron and Hole Ionization Coefficients 325
V. Avalanche Photodiode Detectors 350
VI. Electroabsorption Avalanche Photodiode Detectors 380
VII. Summary and Conclusions 391

Chapter 6 The Josephson Junction as a Detector of Microwave and Far-Infrared Radiation

P. L. Richards

I. Introduction 395
II. The Alternating Current Josephson Effect 396
III. Real Junctions and Equivalent Circuits 399
IV. The Junction Impedance 405
V. Optimum Junctions for Detector Applications 408
VI. Noise 410
VII. Bolometer 414
VIII. Video Detector 417
IX. Heterodyne Detector with External Local Oscillator 421
X. Parametric Amplifier 429
XI. Other Devices of Interest 435
XII. Conclusions 439

Chapter 7 **The Pyroelectric Detector—An Update** 441
  E. H. Putley

AUTHOR INDEX 451
SUBJECT INDEX 464

List of Contributors

Numbers in parentheses indicate the page on which the authors' contributions begin.

PETER R. BRATT, *Santa Barbara Research Center, Goleta, California* (39)

J. O. DIMMOCK, *Office of Naval Research, Arlington, Virginia* (169)

W. L. EISENMAN, *U. S. Naval Electronics Laboratory Center, Infrared Technology Division, San Diego, California* (1)

J. D. MERRIAM, *U. S. Naval Electronics Laboratory Center, Infrared Technology Division, San Diego, California* (1)

R. F. POTTER, *U. S. Naval Electronics Laboratory Center, Infrared Technology Division, San Diego, California* (1)

E. H. PUTLEY, *Royal Signals and Radar Establishment, Great Malvern, Worcestershire, England* (143, 441)

P. L. RICHARDS, *Department of Physics, University of California, Berkeley, Berkeley, California* (395)

G. E. STILLMAN, *Department of Electrical Engineering and Materials Research Laboratory, University of Illinois, Urbana, Illinois* (169, 291)

C. M. WOLFE, *Department of Electrical Engineering, and Laboratory for Applied Electronic Sciences, Washington University, St. Louis, Missouri* (169, 291)

Preface

A highly important area where semiconductors and semimetals technology plays a major role is that of optical sensing devices. Receiving initial impetus because of infrared detector applications, these devices have spurred the development of modern opto-electronics, which promises to create a revolution in communication technology. A key factor responsible for such progress has been the advances in materials development and in device fabrication.

Because the improvements in materials-processing sophistication are reflected in the design and operational capabilities of a variety of optical detectors, it seems important to devote another volume of *Semiconductors and Semimetals* to the subject of infrared detectors in order to review important features not included in the earlier treatise (Volume 5) that was concerned with this subject. For example, the production of high-purity GaAs has permitted the development of fast photoconductors with good sensitivity in the far infrared (see Chapter 4); introduction of new materials and refinement of manufacturing techniques has led to the increased use of pyroelectric detectors for wider ranges of applications (as is discussed in Chapter 7). Optical communication requirements have caused renewed interest in avalanche photodiode detectors and use of GaAs in these devices seems particularly promising for low-noise wide-bandwidth applications at wavelengths where high quantum efficiency is possible (as discussed in Chapter 5); for longer wavelengths, beyond the normal absorption edge of GaAs, the GaAs electroabsorption avalanche photodiode is expected to be of considerable importance in applications involving integrated optical circuits where fast response is required at long wavelengths.

Over the years, the most popular types of laser detector have utilized specific impurities in germanium and silicon. Operating characteristics, fabrication procedures, and a discussion of a number of special features of these detectors are given (see Chapter 2). The characteristics of the InSb submillimeter photoconductive detectors based on the physical mechanism of hot electron photosensitivity are reviewed—including discussion of effects of a magnetic field, characteristics of optimum preamplifiers, and use of the devices in radio astronomy and in rocket-borne applications. An introductory chapter is provided, which gives rather detailed information on operational characteristics of infrared detectors, measurement of pertinent parameters, and calculations of performance for specific conditions. Finally, in the interest of completeness, the editors take some license with the scope of *Semiconductors and Semimetals* to include a review of detectors based on superconducting phenomena, namely the use of the Josephson junction as a detector of microwave and far-infrared radiation (Chapter 6). Besides providing a good theoretical background, the article gives a quite detailed discussion of some of the more promising applications, such as the heterodyne detector.

The editors are indebted to the many contributors and their employers who make this series possible. They wish to express their appreciation to Cominco American Incorporated and Battelle Memorial Institute for providing the facilities and the environment necessary for such an endeavor. Special thanks are also due to the editors' wives for their patience and understanding.

R. K. WILLARDSON
ALBERT C. BEER

Semiconductors and Semimetals

Volume 1 Physics of III–V Compounds

C. Hilsum, Some Key Features of III–V Compounds
Franco Bassani, Methods of Band Calculations Applicable to III–V Compounds
E. O. Kane, The $k \cdot p$ method
V. L. Bonch-Bruevich, Effect of Heavy Doping on the Semiconductor Band Structure
Donald Long, Energy Band Structures of Mixed Crystals of III–V Compounds
Laura M. Roth and Petros N. Argyres, Magnetic Quantum Effects
S. M. Puri and T. H. Geballe, Thermomagnetic Effects in the Quantum Region
W. M. Becker, Band Characteristics near Principal Minima from Magnetoresistance
E. H. Putley, Freeze-Out Effects, Hot Electron Effects, and Submillimeter Photoconductivity in InSb
H. Weiss, Magnetoresistance
Betsy Ancker-Johnson, Plasmas in Semiconductors and Semimetals

Volume 2 Physics of III–V Compounds

M. G. Holland, Thermal Conductivity
S. I. Novikova, Thermal Expansion
U. Piesbergen, Heat Capacity and Debye Temperatures
G. Giesecke, Lattice Constants
J. R. Drabble, Elastic Properties
A. U. Mac Rae and G. W. Gobeli, Low Energy Electron Diffraction Studies
Robert Lee Mieher, Nuclear Magnetic Resonance
Bernard Goldstein, Electron Paramagnetic Resonance
T. S. Moss, Photoconduction in III–V Compounds
E. Antončik and J. Tauc, Quantum Efficiency of the Internal Photoelectric Effect in InSb
G. W. Gobeli and F. G. Allen, Photoelectric Threshold and Work Function
P. S. Persham, Nonlinear Optics in III–V Compounds
M. Gershenzon, Radiative Recombination in the III–V Compounds
Frank Stern, Stimulated Emission in Semiconductors

Volume 3 Optical Properties of III–V Compounds

Marvin Hass, Lattice Reflection
William G. Spitzer, Multiphonon Lattice Absorption
D. L. Stierwalt and R. F. Potter, Emittance Studies
H. R. Philipp and H. Ehrenreich, Ultraviolet Optical Properties
Manuel Cardona, Optical Absorption above the Fundamental Edge
Earnest J. Johnson, Absorption near the Fundamental Edge
John O. Dimmock, Introduction to the Theory of Exciton States in Semiconductors
B. Lax and J. G. Mavroides, Interband Magnetooptical Effects

H. Y. Fan, Effects of Free Carriers on the Optical Properties
Edward D. Palik and George B. Wright, Free-Carrier Magnetooptical Effects
Richard H. Bube, Photoelectronic Analysis
B. O. Seraphin and H. E. Bennett, Optical Constants

Volume 4 Physics of III–V Compounds

N. A. Goryunova, A. S. Borschevskii, and D. N. Tretiakov, Hardness
N. N. Sirota, Heats of Formation and Temperatures and Heats of Fusion of Compounds $A^{III}B^{V}$
Don L. Kendall, Diffusion
A. G. Chynoweth, Charge Multiplication Phenomena
Robert W. Keyes, The Effects of Hydrostatic Pressure on the Properties of III–V Semiconductors
L. W. Aukerman, Radiation Effects
N. A. Goryunova, F. P. Kesamanly, and D. N. Nasledov, Phenomena in Solid Solutions
R. T. Bate, Electrical Properties of Nonuniform Crystals

Volume 5 Infrared Detectors

Henry Levinstein, Characterization of Infrared Detectors
Paul W. Kruse, Indium Antimonide Photoconductive and Photoelectromagnetic Detectors
M. B. Prince, Narrowband Self-Filtering Detectors
Ivars Melngailis and T. C. Harmon, Single-Crystal Lead–Tin Chalcogenides
Donald Long and Joseph L. Schmit, Mercury–Cadmium Telluride and Closely Related Alloys
E. H. Putley, The Pyroelectric Detector
Norman B. Stevens, Radiation Thermopiles
R. J. Keyes and T. M. Quist, Low Level Coherent and Incoherent Detection in the Infrared
M. C. Teich, Coherent Detection in the Infrared
F. R. Arams, E. W. Sard, B. J. Peyton, and F. P. Pace, Infrared Heterodyne Detection with Gigahertz IF Response
H. S. Sommers, Jr., Microwave-Biased Photoconductive Detector
Robert Sehr and Rainer Zuleeg, Imaging and Display

Volume 6 Injection Phenomena

Murray A. Lampert and Ronald B. Schilling, Current Injection in Solids: The Regional Approximation Method
Richard Williams, Injection by Internal Photoemission
Allen M. Barnett, Current Filament Formation
R. Baron and J. W. Mayer, Double Injection in Semiconductors
W. Ruppel, The Photoconductor–Metal Contact

Volume 7 Applications and Devices: Part A

John A. Copeland and Stephen Knight, Applications Utilizing Bulk Negative Resistance
F. A. Padovani, The Voltage–Current Characteristics of Metal–Semiconductor Contacts
P. L. Hower, W. W. Hooper, B. R. Cairns, R. D. Fairman, and D. A. Tremere, The GaAs Field-Effect Transistor

Marvin H. White, MOS Transistors
G. R. Antell, Gallium Arsenide Transistors
T. L. Tansley, Heterojunction Properties

Volume 7 Applications and Devices: Part B

T. Misawa, IMPATT Diodes
H. C. Okean, Tunnel Diodes
Robert B. Campbell and Hung-Chi Chang, Silicon Carbide Junction Devices
R. E. Enstrom, H. Kressel, and L. Krassner, High-Temperature Power Rectifiers of $GaAs_{1-x}P_x$

Volume 8 Transport and Optical Phenomena

Richard J. Stirn, Band Structure and Galvanomagnetic Effects in III–V Compounds with Indirect Band Gaps
Roland W. Ure, Jr., Thermoelectric Effects in III–V Compounds
Herbert Piller, Faraday Rotation
H. Barry Bebb and E. W. Williams, Photoluminescence I: Theory
E. W. Williams and H. Barry Bebb, Photoluminescence II: Gallium Arsenide

Volume 9 Modulation Techniques

B. O. Seraphin, Electroreflectance
R. L. Aggarwal, Modulated Interband Magnetooptics
Daniel F. Blossey and Paul Handler, Electroabsorption
Bruno Batz, Thermal and Wavelength Modulation Spectroscopy
Ivar Balslev, Piezooptical Effects
D. E. Aspnes and N. Bottka, Electric-Field Effects on the Dielectric Function of Semiconductors and Insulators

Volume 10 Transport Phenomena

R. L. Rode, Low-Field Electron Transport
J. D. Wiley, Mobility of Holes in III–V Compounds
C. M. Wolfe and G. E. Stillman, Apparent Mobility Enhancement in Inhomogeneous Crystals
Robert L. Peterson, The Magnetophonon Effect

Volume 11 Solar Cells

Harold J. Hovel, Introduction; Carrier Collection, Spectral Response, and Photocurrent; Solar Cell Electrical Characteristics; Efficiency; Thickness; Other Solar Cell Devices; Radiation Effects; Temperature and Intensity; Solar Cell Technology; Recent Results

CHAPTER 1

Operational Characteristics of Infrared Photodetectors

W. L. Eisenman, J. D. Merriam, and R. F. Potter

I.	INTRODUCTION	1
II.	DETECTOR PARAMETERS	2
	1. Detector Bias	3
	2. Radiation Signal Power	3
	3. Detector Signal	3
	4. Detector Noise	4
	5. Detector Temperature and Background Conditions	5
	6. Miscellaneous Physical and Electronic Properties	5
III.	FIGURES OF MERIT	5
	7. Responsivity	5
	8. Noise Equivalent Irradiance	7
	9. Noise Equivalent Power	7
	10. Detectivity (D-star)	7
	11. Responsive Time Constant	7
IV.	MEASUREMENT OF DETECTOR PARAMETERS	8
	12. Equipment	8
	13. Measurement Procedures	12
V.	CALCULATIONS OF PERFORMANCE FOR SPECIFIC CONDITIONS	16
	14. Responsivity	16
	15. Detectivity (D^*)	17
	16. Illustrative Sample Calculation	18
VI.	CURRENT STATE-OF-THE-ART PHOTODETECTORS	23
	17. Detectors Operating at Intermediate Cryogenic Temperatures	24
	18. Alloyed Detectors	25
	19. Criteria for Comparison of Detectors	27
	20. Thermal Detectors	32
VII.	DETECTORS IN ADVANCE OF THE STATE-OF-THE-ART	34
	21. Metal–Oxide–Metal Point Contact Diodes	34
	22. Josephson Junction Detectors	34
	23. Schottky Barrier Photodiode	35
	24. MIS InSb Photovaltaic Detector	35
	LIST OF SYMBOLS	38

I. Introduction

The developments in modern semiconductor materials technology have been both intensive and extensive. They have had such a special impact on the field of infrared detectors that two volumes of this treatise are devoted to the description of solid-state electrooptical signal transducers.

In this chapter we attempt to describe and compare the operating characteristics of modern infrared detectors. Parts II–V give a detailed description of the pertinent parameters, definitions, and procedures for measurement. These parts are limited to describing only those detectors having electrical outputs directly proportional to the input radiant signal power. (At the end of Part V we give an example of typical data for a PbSe photoconductor along with a set of sample calculations. These demonstrate how one is able to interpolate and extrapolate from standard report data to specific engineering information about the expected detector performance.)

The sixth part reviews some aspects of detectors characterized as being of the current state-of-the-art. Special emphasis is placed upon operation at intermediate cryogenic temperatures ($\sim 200°K$) which lend themselves to utilization with thermoelectric coolers. Also given in that part is a short review of detectors made up from the pseudobinary alloys of mercury–cadmium telluride and the lead–tin chalcogenides.

However convenient a figure of performance such as D^* is as a number for comparison among detectors, it does not tell a complete story. We have tried to convey that a more complete comparison can be made if other parameters such as noise and response as functions of frequency and bias are examined. Part VI concludes with a brief discussion of thermal detectors.

This introductory chapter closes (Part VII) by indicating four novel detectors that have promise for future applications, but would not be classed as current state-of-the-art.

II. Detector Parameters

The conditions under which a detector may be measured and used are almost infinite in number. To allow a standard measurement and reporting procedure obviously only a small number of parameters can be considered.[1-3] However, a limited number of parameters can afford a reasonable description of the detector characteristics under a wide variety of operating conditions. The only detectors considered here are those *whose output consists of an electrical signal that is proportional to the radiant signal power.* The parameters to be considered are the detector signal, detector noise, radiation signal power, detector bias, detector temperature, physical and electrical properties of the detector, and the background conditions under

---

[1] R. C. Jones, *J. Opt. Soc. Amer.* **39**, 327 (1949).
[2] R. C. Jones, *J. Opt. Soc. Amer.* **39**, 344 (1949).
[3] R. C. Jones, *Advan. Electron.* **5** (1953).

1. OPERATIONAL CHARACTERISTICS OF INFRARED PHOTODETECTORS 3

which the detector is operated. A list of the symbols used in this chapter will be found at the end of the chapter.

1. Detector Bias

Many detectors require an external bias of some kind and in the case of a photoconductive detector the bias is an applied electrical voltage. In general, the signal and noise characteristics of a detector are measured over the entire useful range of bias values. Since both the detector signal and noise are a function of the bias and modulation frequency, the optimum bias value reported is the value that maximizes the signal-to-noise ratio at a stated modulation frequency.

2. Radiation Signal Power

The radiation signal power incident on the detector must be described to allow a meaningful interpretation of the detector signal output. The spatial and spectral power density distributions on the detector should be reported. A blackbody source,[4] arranged to produce a uniform spatial power distribution on the detector, is generally used because this source readily provides a known spectral power density distribution. A blackbody source, operated at a temperature of 500°K, is commonly used in detector testing laboratories.

The radiation power incident on the detector is modulated and since the waveshape of the modulation may be different from laboratory to laboratory, the radiation signal power is defined as the power in the fundamental frequency component of a Fourier expansion of the modulation waveshape. The modulation frequency of the radiation is defined as the frequency of this fundamental component. The root-mean-square (rms) amplitude of the fundamental component is defined as the peak-to-peak amplitude of this component divided by $2^{3/2}$. A sinusoidal modulation is often used for the measurement of detector parameters.

3. Detector Signal

The signal is that voltage or current output from the detector that is coherent with the incident radiation signal power. In general, the magnitude of the signal is a function of the bias b applied to the detector, the modulation frequency f, the wavelength λ and power of the incident radiation J, and the detector area A. In functional notation, this can be written

$$V = V(b, f, \lambda, J, A).$$

For most usable detectors, the signal is a linear function of the radiation power density. Since only those detectors that have this property will be

---
[4] M. Holter, S. Nudelman, G. H. Suits, W. L. Wolfe, and G. J. Zissis, "Fundamentals of Infrared Technology." Macmillan, New York, 1962.

considered, V/JA is a constant for fixed b, f, and λ. In many detectors, the dependence of the signal upon the modulation frequency and the wavelength of the incident radiation signal power is separable. Detectors are said to have the factorability property in those wavelength regions for which this is true. Only detectors having this property are considered. We may now describe the detector signal as

$$V(b, J, A)v(\lambda)v(f).$$

In addition, the signal dependence upon the applied bias might be separable also. There are useful detectors that do not have this property, however, so caution must be used in making this assumption.

The dependence of the detector signal upon the applied bias is measured at a specified modulation frequency and with specified incident radiation. These data are reported as a part of a plot generally labeled "determination of optimum bias."

The dependence of a detector signal upon the frequency at which the incident radiation is modulated, measures the detectors temporal response. The measurements are usually made with constant radiation signal power and bias value. The results can be reported in a plot of relative signal versus modulation frequency. Because such a plot represents the frequency response of the detector signal, a time constant can be determined which is called the responsive time constant. The peak detective frequency often reported is the modulation frequency that maximizes the signal-to-noise ratio.

The signal dependence of a detector upon the wavelength of the radiation signal power is a measure of the detectors ability to respond to radiation of different wavelengths. These spectral responsivity measurements are made at constant bias value and modulation frequency. Such data are reported on a plot of relative signal, normalized to constant radiation signal power at each wavelength versus wavelength. Such a plot is usually labeled "relative spectral response."

If only a portion of the detector area is exposed to incident signal radiation, the signal of the detector may be a function of the position on the detector area where the radiation falls. A plot of the relative signal as a function of the position on the detector area that is illuminated with signal radiation is commonly labeled "sensitivity contour plot." Such measurements are not often made due to the experimental difficulty in obtaining the data.

4. DETECTOR NOISE

The detector noise is the electrical voltage or current output from the detector that is not coherent with the radiation signal power. The detector noise is a function of the bias, modulation frequency, and the area of the detector. It can be shown, under certain assumptions, that the detector signal-to-noise ratio varies directly as the square root of the detector area when the modulation frequency and bias are constant.[5] The noise is measured with the detector shielded from the radiation signal source. The detector noise, as reported, is referred to the output terminals of the detector and normalized to a 1 Hz effective noise bandpass.

The detector noise may be reported as a family of curves representing the different noise values obtained at several bias values plotted against frequency and labeled "noise spectrum."

5. DETECTOR TEMPERATURE AND BACKGROUND CONDITIONS

Since the signal and noise of a detector are greatly influenced by the operating temperature of the detector and the power radiated by the background, these quantities must be reported. The background radiation is often described by reporting the field of view of the detector and the spectral range through which the detector is illuminated by the background.

6. MISCELLANEOUS PHYSICAL AND ELECTRONIC PROPERTIES

To describe a detector adequately, it is necessary to report the detector area, the distance of the detector area from the aperture of the detector package, the detector resistance or impedance, and the over-all dimensions of the detector package. In the case of a cooled detector, the holding time of the Dewar will be useful information.

III. Figures of Merit

The measured data are sufficient to describe a detector. However, to provide ease of comparison between detectors, certain figures of merit, computed from the measured data, have been defined.[1-6] The more commonly used figures of merit are discussed below.

7. RESPONSIVITY

The responsivity is defined as

$$R = V/JA \quad (V/W). \tag{1}$$

Responsivity, the ratio of the rms value of the fundamental component of the signal voltage to the rms value of the fundamental component of the

---

[5] P. W. Kruse, L. D. McGlauchlin, and R. B. McQuistan, "Elements of Infrared Technology." Wiley, New York, 1962.
[6] R. C. Jones, D. Goodwin, and G. Pullan, Standard Procedure for Testing Infrared Detectors and for Describing Their Performance. Office of the Director of Defense Res. and Eng., Washington, D.C., 1960.

incident radiation power, is written in functional notation as

$$R(b,f,\lambda) = V(b,f,\lambda)/(P_\lambda(\text{ss})\Delta\lambda)_\lambda = \frac{V(b,f_0,\lambda_0)}{(P_\lambda(\text{ss})\Delta\lambda)_{\lambda_0}} v(\lambda)v(f)$$

$$= R_{\text{rms}}(b,f_0,\lambda_0)v(\lambda)v(f).$$

The absolute responsivity to a blackbody at temperature T is given by:

$$R_{\text{bb}}(b,f_r,T) = \int_0^\infty R(b,f_r,\lambda)(P_\lambda/P_{\text{rms}}(T))\,d\lambda$$

$$= R(b,f_0,\lambda_0)v(f_r)\int_0^\infty v(\lambda)(P_\lambda/P_{\text{rms}}(T))\,d\lambda.$$

Thus

$$R_{\text{bb}}(b,f_r,T) = [R(b,f,\lambda)v(f_r)/v(\lambda)v(f)]\int_0^\infty v(\zeta)(P_\zeta(T)/P_{\text{rms}}(T))\,d\zeta \quad (2)$$

where we have changed the integration variable to ζ, and also where $(P_\lambda(\text{ss})\Delta\lambda)_\lambda$ is quasi-monochromatic power from a spectral source at λ, incident on detector: λ_0 is *often* the wavelength at which $R(b,f,\lambda)$ is a maximum, i.e., $v(\lambda_0) = 1$; f_r is the frequency at which R_{bb} is measured; $P_{\lambda\text{rms}}(T)$ is the spectral radiant power of the blackbody source at temperature T; and $P_{\text{rms}}(T) = \int_0^\infty P_{\lambda\text{rms}}(T)\,d\lambda$. The term $v(f)/v(f_r)$ gives the frequency characteristics of the detector, and the term

$$\gamma_p(T) = \int_0^\infty v(\lambda)P_{\lambda\text{rms}}(T)\,d\lambda/P_{\text{rms}}(T) \quad (3)$$

gives the spectral power efficiency; that is, it indicates how effective the detector is compared to a "black" or "gray" detector (detectors with constant response at all wavelengths).[7]

The wavelength dependence of the responsivity can now be written

$$R(\lambda) = (R_{\text{bb}}/\gamma_p)v(\lambda). \quad (4)$$

Frequently, $R(\lambda_{\max})/R_{\text{bb}}$ is reported for $1/\gamma_p$; the term R_{bb}/γ_p gives the response of the detector at λ_0, which is often the spectral peak wavelength. Hence, to calculate the absolute signal voltage for a given radiation power spectrum, the blackbody response for the desired bias value, the relative modulation frequency response, and relative spectral response must be determined.

---

[7] R. A. Smith, F. E. Jones, and R. P. Chasmar, "The Detection and Measurement of Infrared Radiation." Oxford Univ. Press (Clarendon), London and New York, 1957.

8. Noise Equivalent Irradiance

The noise equivalent irradiance in functional notation is defined as

$$H_N(b,f,\lambda) = N(b,f)/A(\Delta f)^{1/2} R(b,f,\lambda). \tag{5}$$

H_N is the minimum radiant flux density necessary to produce a signal-to-noise ratio of 1 when the noise is normalized to a unit bandwidth. The units in common practice are $W/cm^2\ Hz^{1/2}$. Since the area of the detector is not taken into account in H_N, the figure describes the performance of a detector of a specific area.

9. Noise Equivalent Power

The noise equivalent power in functional radiation is defined as

$$P_N(b,f,\lambda) = AH_N = N(b,f)/(\Delta f)^{1/2} R(b,f,\lambda). \tag{6}$$

Here P_N is the minimum radiant flux necessary to produce a signal-to-noise ratio of 1 when the noise is normalized to unit bandwidth in units of $W/Hz^{1/2}$.

10. Detectivity (D-Star)

D-star or D^*, in functional notation, is defined[1-5] as

$$D^*(b,f,\lambda) = A^{1/2}/P_N = R(b,f,\lambda)(A\ \Delta f)^{1/2}/N(b,f). \tag{7}$$

D^* is the detectivity normalized to unit area and unit bandwidth, with usual units of $(cm\ Hz)^{1/2}/W$. Detectivity is the signal-to-noise ratio produced with unit radiant flux incident on the detector. Since the area dependence of the signal-to-noise ratio has been taken into account, D^* describes the general detector type rather than a detector of some particular area. The importance of D^* is that this figure of merit permits comparison of detectors of the same type, but having different areas.

The following relationships exist between D^* and the other figures of merit:

$$\text{responsivity} \quad R = D^* N (\Delta f A)^{1/2} \tag{8}$$

$$\text{noise equivalent irradiance} \quad H_N = 1/D^* A^{1/2} \tag{9}$$

$$\text{noise equivalent power} \quad P_N = A^{1/2}/D^*. \tag{10}$$

11. Responsive Time Constant

The "frequency response" curve of many detectors may be analytically described by the equation

$$v(f) = [1 + 2\pi f \tau]^{-1/2}.$$

If this equation is true for a detector, there is common agreement that the factor τ is the responsive time constant of the detector. If the above relationship does not hold true for a detector, then there is no common agreement upon the definition of the responsive time constant, and any use of this figure must be accompanied by an explicit definition.

IV. Measurement of Detector Parameters

12. EQUIPMENT

The equipment used to measure the necessary detector parameters consists of basic electrical and optical instruments. For the most part, these instruments can be obtained from various commercial suppliers and assembled to meet specific measurement requirements. Equipment costs for a modest laboratory may approach $100,000 and a large facility may easily have five times this amount invested in equipment.[5] While a substantial investment is necessary to equip a detector measurement laboratory properly, it should be stressed that experienced personnel are also absolutely essential for a successful detector measurement facility.

a. Electronics

The electronics generally used to measure detector signal and noise characteristics consists of a preamplifier, amplifier, and a spectrum analyzer. Auxiliary equipment consists of an oscillator and calibrated attenuator that are used for electrical calibration and a variable voltage supply suitable for providing detector bias. Figure 1 is a block diagram of a typical arrangement of equipment suitable for signal and noise measurements in the 1 kHz to 1 MHz frequency range. If a variety of detectors is to be measured, it

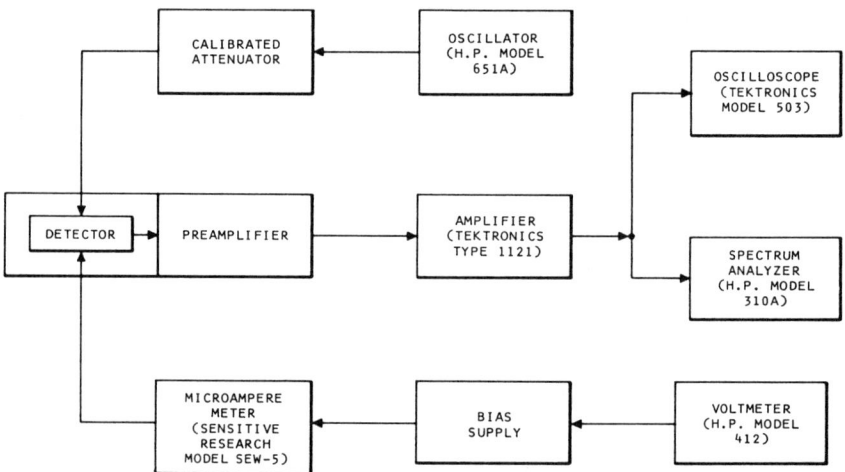

FIG. 1. Block diagram of typical arrangement of equipment suitable for response and noise voltage measurements in the range from 1 kHz to 1 MHz.

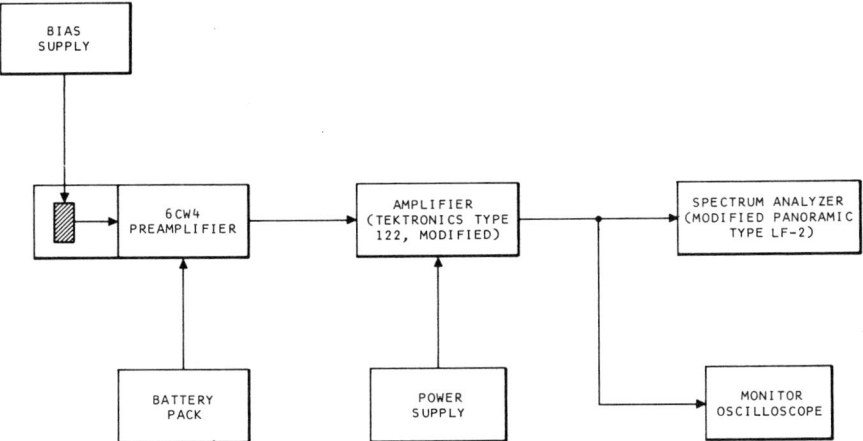

FIG. 2. Block diagram of arrangement for response and noise voltage measurements in the frequency range between 1 and 100 Hz.

will be found convenient to have several similar arrangements of instruments to cover the necessarily wide frequency range. A number of preamplifiers will also be necessary to cover a wide range of detector impedances.[8] When possible, the preamplifier should be placed together with the detector in a shielded enclosure to minimize input capacity. Apparatus suitable for measuring signal and noise of thermal-type detectors over a frequency range of 1 to 100 Hz is shown in Fig. 2.[9]

b. Blackbody Source

The primary requirement for detector responsivity measurements is a stable modulated source of radiant energy of known spectral irradiance. A blackbody source is a convenient means of meeting this requirement. The blackbody must be provided with a suitable modulator, which may be of a fixed frequency. However, if a fixed frequency is used, it will be found convenient to have at least three modulators available operating at frequencies of approximately 10, 100 and 1000 Hz. The source must produce an accurately known irradiance in the plane of the detector, and the irradiance must be uniform over the sensitive area of the detector. The spectral irradiance used is the rms amplitude of the fundamental component of the modulation frequency. In determining this value of rms spectral irradiance, the radiation from the modulator must also be taken into account. In order to compare measured data easily, many detector laboratories operate their blackbody sources at a temperature of 500°K. Root-mean-square irradiance

---

[8] J. A. Jamieson, *Proc. IRE* **47**, 1522 (1959).
[9] P. C. Caringella and W. L. Eisenman, *Rev. Sci. Instrum.* **6**, 659 (1962).

values of the order of microwatts per centimeter squared are appropriate for many types of detectors.

c. Variable Frequency Source

A variable frequency source is required to measure the dependence of detector signal on modulation frequency. Basically, any stable source having a suitable spectral output and equipped with a variable speed modulator may be used. The irradiance produced by the source must be uniform over the sensitive surface of the detector. At audio and subaudio frequencies, mechanical modulators are easily fabricated and perform in a satisfactory manner.[10] However, at frequencies above 50 kHz, mechanical modulators become awkward, blade diameters become large, aperture sizes are reduced to mil dimensions, and the blades must be driven at quite awesome speeds.

Electroluminescent diodes also can serve as variable frequency sources for frequencies into the megahertz region. The apparatus is quite simple, and one arrangement of instruments is shown in Fig. 3. The system utilizes the beat-frequency-oscillator output voltage available on several models of audio frequency wave analyzers. This BFO voltage has a frequency that is always equal to the input frequency of the analyzer. Thus, when the input frequency of the analyzer is varied over the frequency range of the equipment, the frequency of the BFO voltage automatically follows. The BFO voltage from the analyzer is amplified and applied to an electroluminescent diode.

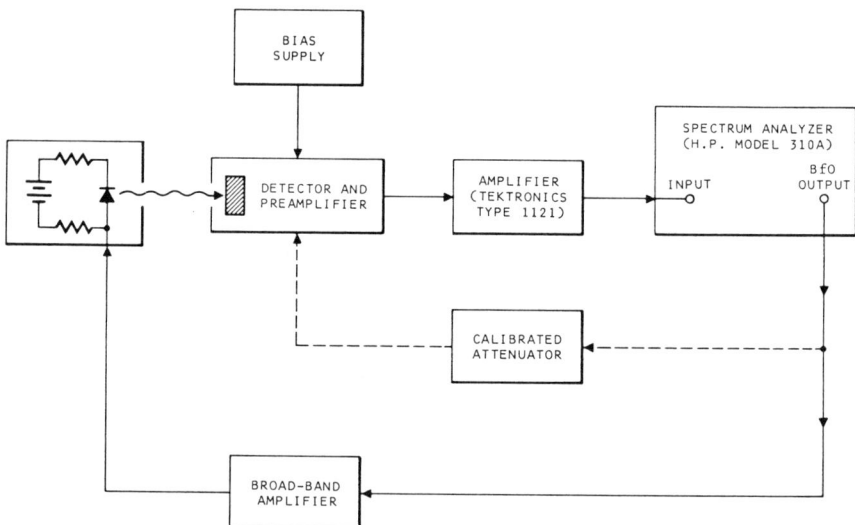

FIG. 3. Block diagram of equipment arrangement for frequency response measurement using an electroluminescent diode at frequencies up to 1.5 MHz.

[10] R. F. Potter, J. M. Pernett, and A. B. Naugle, *Proc. IRE* **47**, 1503 (1959).

The modulated radiation from the diode is focused upon the detector being measured. The electrical signal from the detector is amplified and then applied to the input of the analyzer.

Besides its simplicity, this arrangement has two distinct additional advantages. First the measurement of detector frequency response with this apparatus is quite rapid since only one control on the wave analyzer need be adjusted. Second, the narrow bandpass of the wave analyzer produces a large detector signal-to-noise ratio. Thus, a relatively small amount of input radiation is necessary.

The major disadvantage of this measurement method is the narrow spectral distribution of the radiated energy. Gallium arsenide and indium arsenide diodes, emitting at 0.8 and 3.2 μm, respectively, are commercially available. Hopefully, diodes of other materials will become available in the near future that will provide emission at longer wavelengths.

d. Monochromatic Source

The monochromatic source consists of a stable source of radiation, a modulator, a monochromator, and an optical system that illuminates the detectors. Tungsten filament lamps, Nernst glowers and glow bars are commonly used sources depending upon the wavelength region of the interest. The radiant output of both the glower and glow bar will be affected by air currents and these sources should be provided with a suitable housing or chimney. The modulator may be a fixed frequency device and is normally placed at the entrance slit of the monochromator. The monochromator should be capable of providing a wavelength band of radiation that is not wider than the order of 1/25th of the center wavelength. Scattered radiation may be a problem particularly at the longer wavelength and a double monochromator is preferable for this application; however, satisfactory performance may be obtained from a single monochromator by fitting the instrument with suitable rejection filters.[11] The detector being measured and the reference detector are alternately illuminated by an optical system placed at the exit slit of the monochromator. A typical arrangement is shown in Fig. 4.

e. Reference Detector

The spectral sensitivity of a detector is obtained by comparing the signal from the detector to the signal from a reference detector as a function of wavelength of the incident radiation. Obviously, the spectral sensitivity of the reference detector must be known. Detectors utilizing a cavity as the radiation receiver have been used as reference standards[12,13]; however,

---

[11] A. E. Martin, "Infrared Instrumentation and Techniques." Elsevier, Amsterdam, 1966.
[12] W. L. Eisenman, R. L. Bates, and J. D. Merriam, *J. Opt. Soc. Amer.* **53**, 729 (1963).
[13] W. L. Eisenman and R. L. Bates, *J. Opt. Soc. Amer.* **54**, 1280 (1964).

detectors of this type are difficult to obtain and hard to use because of their low sensitivity and slow speed of response. A radiation thermocouple is a convenient detector for use as a reference standard provided its spectral sensitivity has been determined (by comparison to a cavity-type detector). The relative spectral sensitivity of a typical radiation thermocouple compared to such a cavity is shown in Fig. 5. The decline in sensitivity is relatively smooth and there is no difficulty in correcting the data. Since the sensitivity of a thermal-electric type detector may not be uniform across the radiation receiver,[14] the optical system used at the exit slit of the monochromator should be adjusted so as to just completely flood the thermocouple receiver. Further, the monochromator should provide a reasonably uniform exit pupil.

13. Measurement Procedures

Test procedures may be divided into two independent groups. In the first group are the measurements that are necessary to determine the detector responsivity and in the second are the measurements that yield the root-power-spectrum of the noise. The arrangement of the electrical equipment is the same for both groups of measurements. Figure 1 shows the arrangement for the common case of a photoconductive detector that requires a single bias voltage.

a. Responsivity Measurements

The determination of detector responsivity involves three separate measurements using the three radiation sources. In all of these measurements

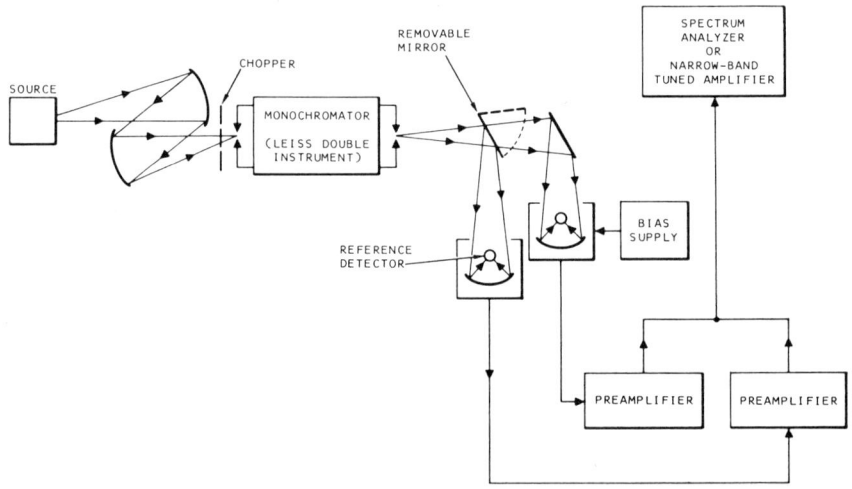

Fig. 4. Block diagram of instrument arrangement for measuring spectral response.

[14] R. Stair, W. E. Schneider, W. R. Walters, and J. K. Jackson, *Appl. Opt.* **4**, 703 (1965).

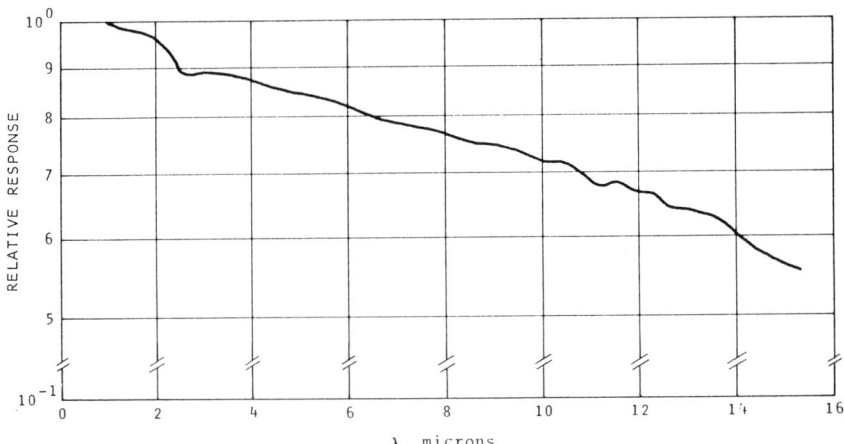

FIG. 5. Relative spectral response of a typical radiation thermocouple as compared to a special conical cavity reference detector.

the signal radiation must be normally incident on the detector and the amount of signal radiation must be confined to a range in which the output signal from the detector is proportional to the incident power. Confirmation of this linearity may be necessary in some cases.

The measurement of the responsivity involves the use of the factorability property. For those detectors where this property does not hold over the entire wavelength range of interest, the blackbody source and the variable frequency source must be equipped with a filter that limits the radiation to a wavelength band within which the factorability property holds.[13]

(1) The first step is to establish the range of bias values to be used with the detector being measured. Experience with similar detectors will usually indicate the approximate range of bias values. The range will normally cover at least one decade of bias voltage or current. The highest value of bias is normally known as the "manufacturer's maximum bias." Considerable care should be exercised if measurements are to be made at biases greater than the manufacturer's maximum value. The detector noise should be carefully monitored and the bias should be increased in small steps. Operating some types of detectors in this region of high bias is very risky, and experience must be relied upon to a great extent.

(2) The blackbody source equipped with a modulator of frequency f is used to irradiate the detector. The center frequency of the spectrum analyzer is set at f and the signal generator is set to zero output. The reading E_s of the output meter is noted. Then the radiation is removed, the signal generator is set to frequency f and the attenuator is adjusted to a value that gives the same reading E_s on the meter. The open circuit detector signal voltage V is the voltage across the calibrating resistor. These readings are

then repeated for several values of bias until the complete range of bias values has been covered.

The radiant power P_{rms} incident on the detector is obtained from the known irradiance J_{rms} upon multiplication by the detector area A:

$$P_{rms} = AJ_{rms}.$$

The corresponding detector responsivity is given by:

$$R_{bb}(b,f) = V_{rms}/P_{rms}.$$

(3) With the spectrum analyzer set at frequency f and the bias applied, the detector is irradiated by the monochromatic source. The center wavelength of the monochromator is varied over the wavelength range of interest and the relative signal voltage E_s of the detector is recorded as a function of wavelength. The detector under measurement is then replaced by the reference detector and the relative signal voltage E_{ref} of the reference detector is recorded as a function of wavelength. (Some detectors may exhibit changes in spectral response as a function of bias. If these changes are significant, then several spectral response curves must be obtained for different bias values.)

The relative response $v(\lambda)$ as a function of the wavelength is then calculated by

$$v(\lambda) = E_s \varepsilon(\lambda)/E_{ref},$$

where $\varepsilon(\lambda)$ is the relative spectral responsivity of the reference detector.

(4) The detector is now irradiated with the variable frequency source. As the modulation frequency of the source is varied, the center frequency of the spectrum analyzer is continuously adjusted to the modulation frequency. The detector signal voltage E_s, read on the meter, is recorded as a function of frequency. The source is then removed and the signal generator with a fixed attenuator setting is varied over the same range of frequencies. As the frequency of the signal generator is varied, the center frequency of the spectrum analyzer is continuously adjusted to the same frequency. The voltage E_c, read on the meter, is recorded as a function of frequency. The relative response $v(f)$ as a function of modulation frequency is then computed by

$$v(f) = E_s(f)E_c(f_0)/E_c(f)E_s(f_0),$$

where f_0 is such that $v(f_0) = 1$.

The frequency response of some detectors may vary as a function of applied bias. If the change in frequency response is significant, then several frequency response curves must be measured over the entire range of bias values.

b. Noise Power Measurements

The measurement of the noise characteristics of a radiation detector requires good judgment and experience to insure that the only noise recorded is noise generated in the detector, load resistor, and in the amplifier. Constant attention is required to prevent external sources of noise from influencing the results. It may be found convenient to place a wide-band oscilloscope in the electronic system ahead of the bandpass filter. The appearance of the noise trace on the oscilloscope is helpful in determining the presence of any extraneous noise.

In particular, the bias supply must not contribute appreciable noise. The bias source can be checked for internal noise by substituting a wire-wound resistor in place of the detector in the input circuit. The resistance of the wire-wound resistor should be approximately equal to the detector resistance. Bias is then applied to the circuit and the noise noted on the output meter. The noise generated in the wire-wound resistor should be independent of the current flowing through the resistor.

(1) Bias is applied to the detector, all radiation sources except ambient background are removed, and with the signal generator producing zero signal, the rms noise voltage indicated by the output meter is recorded as a function of frequency over the entire frequency range of interest. The voltage read is then denoted E_0.

(2) The detector and load resistor are replaced by a wire-wound resistor having approximately the same resistance as the parallel combination of the detector and load resistor. The temperature of this wire-wound resistor is maintained such that the thermal noise generated in the resistor is small compared to the noise generated in the amplifier. The rms noise voltage indicated by the output meter is again recorded as a function of frequency. This voltage is denoted E_a.

(3) The signal generator is now adjusted to produce a calibration signal E_c across the calibrating resistor. This calibration signal is made approximately 100 times larger than the detector noise. The spectrum analyzer is tuned to the frequency of the calibration signal and the voltage indicated on the output meter is recorded. This procedure is repeated over the entire frequency range of interest. The system gain $g(f)$ is thus determined as a function of frequency.

The root-power-spectrum per unit hertz bandpass, N, referred to the terminals of the detector, referred to an infinite load impedance, and corrected for amplifier noise, is calculated from the following formula:

$$N(f, b) = \left[\frac{E_o^2(f, b) - E_a^2(f)}{g^2(f)} - N_L^2 \left(\frac{R}{R_L} \right)^2 \right]^{1/2} \bigg/ (\Delta f)^{1/2}$$

where R is the resistance of the detector. The thermal noise voltage N_L, generated by the load resistor R_L in the noise bandwidth Δf is given by:

$$N_L^2 = 4kTR_L \Delta f.$$

Note that Δf is the effective noise bandwidth of the measurement equipment and is defined as[6]

$$\Delta f = \int_0^\infty [G^2(f)/G_m^2] \, df.$$

$G(f)$ is the gain of the system as a function of frequency f and G_m is the maximum value of the gain. The fixed center frequency f_m is the frequency corresponding to the maximum gain G_m.

V. Calculations of Performance for Specific Conditions

Since it is not possible to measure a detector under all possible conditions, a limited set of conditions must be selected for the tests. However, if the measurement conditions are properly chosen, it is feasible to predict how a detector will respond under a variety of operating conditions. As D^* is so closely related to H_N and P_N, let us consider the D^* of a particular detector. We shall, therefore, be concerned only with the parameters $b, f,$ and λ, which have been identified as the bias value, the modulation frequency, and the radiation wavelength, respectively.

14. Responsivity

Responsivity is written

$$R = R(b, f, \lambda).$$

The parameters (b, f, λ) enter into R only through the signal, since J and A are independent of them. Therefore, our concern here is only with

$$V = V(b, f, \lambda).$$

Because the parameters $b, f,$ and λ are separable in any usable detector, we may write

$$V = V(b)v(f)v(\lambda). \tag{11}$$

The parameters $V(b)$, $v(f)$, and $v(\lambda)$ can be measured separately and reported graphically in the following charts:

$V(b)$ determination of optimum bias,

$v(f)$ frequency response,

$v(\lambda)$ spectral response.

The subscript r will be used to designate the reported value and the subscript 1 to designate a desired value. Then the responsivity at bias b_1, modulation frequency f_1, and radiation wavelength λ_1 is given by the relation

$$R(b_1,f_1,\lambda_1) = R(b_r,f_r,\lambda_r)V(b_1)v(f_1)v(\lambda_1)/V(b_r)v(f_r)v(\lambda_r). \qquad (12)$$

Since the responsivity is measured with a blackbody usually at 500°K, the following relation is used:

$$R(b_r,f_r,\lambda_{max}) = R_{bb}(b_r,f_r,T)/\gamma_p(T), \qquad (13)$$

and since $v(\lambda)$ is normalized to the peak wavelength, we take $\lambda_r = \lambda_{max}$ and $v(\lambda_{max}) = 1$. We have

$$R(b_1,f_1,\lambda_1) = \frac{R_{bb}(b_r,f_r,T)}{\gamma_p(T)} \frac{V(b_1)}{V(b_r)} \frac{v(f_1)}{v(f_r)} \frac{v(\lambda_1)}{1}, \qquad (14)$$

where $R_{bb}(b_r,f_r,T)$ and γ_p^{-1} (sometimes reported as the ratio $R_{\lambda_{max}}/R_{bb}$) are reported parameters.

In general, a detector is used in applications where it is exposed to spectral source other than true blackbodies. These sources vary from monochromatic sources, line spectra, lasers, narrow-band filtered sources, wide-band filtered sources, etc. Thus if one wants the responsivity to a special spectral source one needs a parameter γ_{ss} comparable to the spectral power efficiency γ_p [Eq. (3)].

Given a spectral source with a $P_\lambda(ss)$ power density in W/μm, then

$$\gamma_{ss} = \left(\int_0^\infty v(\lambda)P_\lambda(ss)\,d\lambda\right) \Big/ \left(\int_0^\infty P_\lambda(ss)\,d\lambda\right). \qquad (15)$$

The responsivity of a detector for this spectral source is given by

$$R[b,f,P(ss)] = R_{bb}(b_r,f_r,T)\frac{\gamma_{ss}}{\gamma_p}\frac{V(b_1)}{V(b_r)}\frac{v(f_1)}{v(f_r)}, \qquad (16)$$

$$P(ss) = \int_0^\infty P_\lambda(ss)\,d\lambda.$$

Of course, if $P_\lambda(ss)$ has values only within the wavelengths λ_a and λ_b and zero everywhere else, these become the integration limits for calculating γ_{ss}.

15. Detectivity (D^*)

As stated earlier, D^* is a measure of the signal-to-noise ratio normalized to unit noise bandwidth and unit detector area. The parameters b,f,λ enter into D^* only through the signal and noise terms, being independent of A, J, and Δf. Since the signal has been treated in the previous section, we shall here examine the noise (which is independent of λ) as a function of b and f:

$$N = N(b,f).$$

Unfortunately, noise is not factorable into independent functions of the parameters b and f. In reporting the noise as a function of bias and frequency, a parameter family of curves is given for the different biases as a function of frequency.

At bias b_1 and frequency f_1, it is necessary to interpolate a frequency f_1 between curves of reported biases to find the noise $N(b_1,f_1)$.

There is a set of bias, frequency, and wavelength values (b_p, f_p, λ_p) that give a maximum D^* denoted as D^*_{mm}. Note that while $v(\lambda_p)$ would be unity, the same does not necessarily hold for $v(f_p)$, since f_p is the frequency that maximizes the signal-to-noise ratio rather than the signal. The values of b_p and f_p can be obtained from a plot of $D^*(b)$ versus frequency, where b_p and f_p correspond to the maximum value attained.

Given D^*_{mm}, we can find D^* at other given bias, frequency, and wavelength values with the formula

$$D^*(b_1,f_1,\lambda_1) = D^*_{mm}(b_p,f_p,\lambda_p) \frac{N(b_p,f_p)}{N(b_1,f_1)} \frac{V(b_1)}{V(b_p)} \frac{v(f_1)}{v(f_p)} \frac{v(\lambda_1)}{v(\lambda_p)}. \tag{17}$$

16. Illustrative Sample Calculation

A typical detector test result is presented here in a standardized format. Sample calculations using these data are also presented.

We shall calculate R, D^*, P_N, and H_N for a detector operated at different conditions than those given in the reported data. For this example, a photoconductive PbSe detector operated at liquid nitrogen temperature will be used. Its characteristics are shown in Figs. 6 and 7.

Let us suppose that we want to find the responsivity and signal-to-noise ratio where

$$f_1 = 10^3 \text{ Hz}, \quad b_1 = 75 \text{ } \mu\text{A}, \quad \lambda_1 = 6.0 \text{ } \mu\text{m}.$$

First we will find the responsivity. From the frequency response plot, we find

$$v(90 \text{ Hz}) = 1.0, \quad v(10^3 \text{ Hz}) = 0.78.$$

From the determination of optimum bias plot, we find

$$V(50 \text{ } \mu\text{A}) = 9.0 \times 10^{-3} \text{ V}, \quad V(75 \text{ } \mu\text{A}) = 1.3 \times 10^{-2} \text{ V}.$$

From the spectral response plot, we find

$$v(6.0 \text{ } \mu\text{m}) = 1.5 \times 10^{-1}.$$

1. OPERATIONAL CHARACTERISTICS OF INFRARED PHOTODETECTORS 19

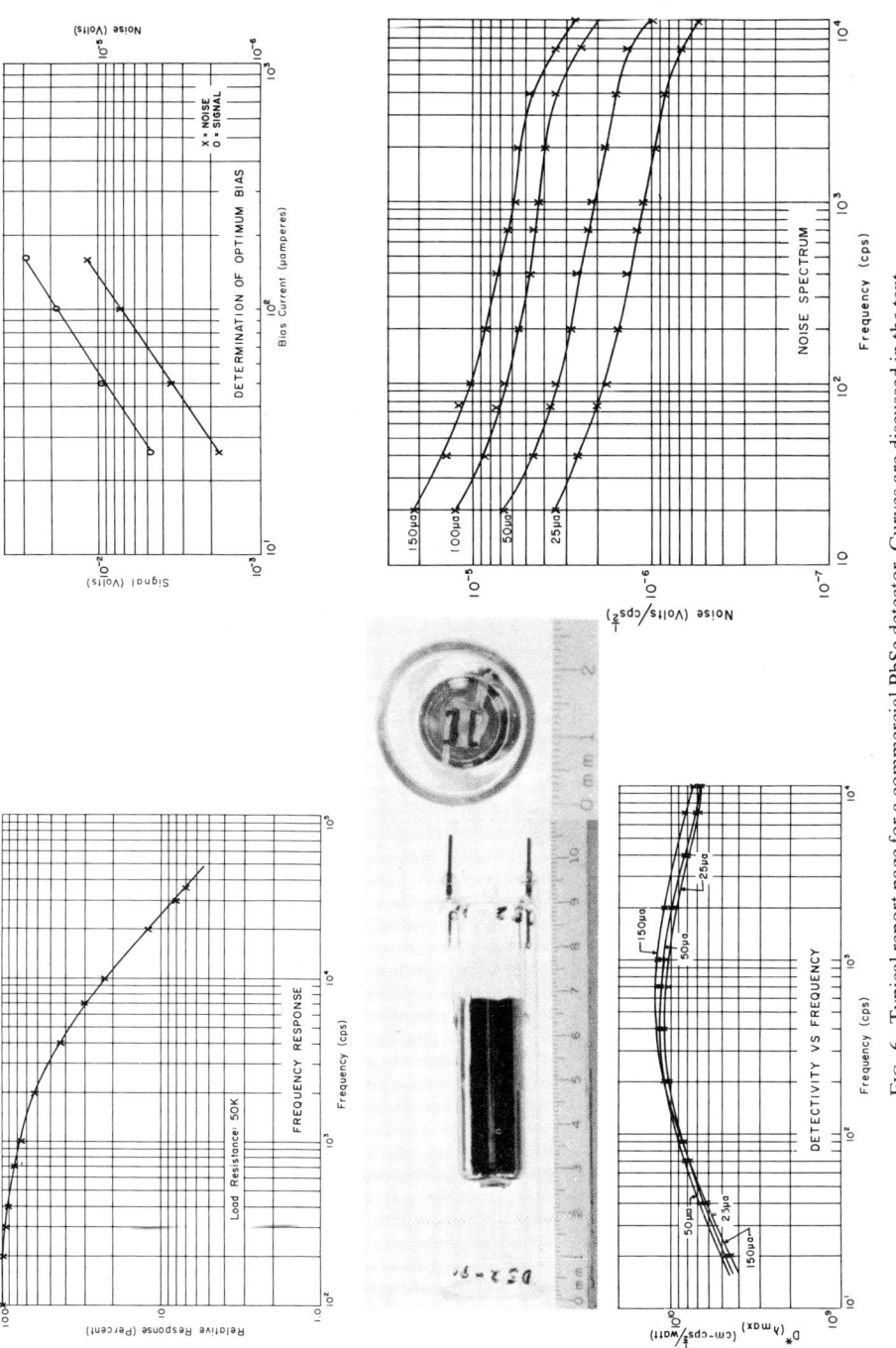

FIG. 6. Typical report page for a commercial PbSe detector. Curves are discussed in the text.

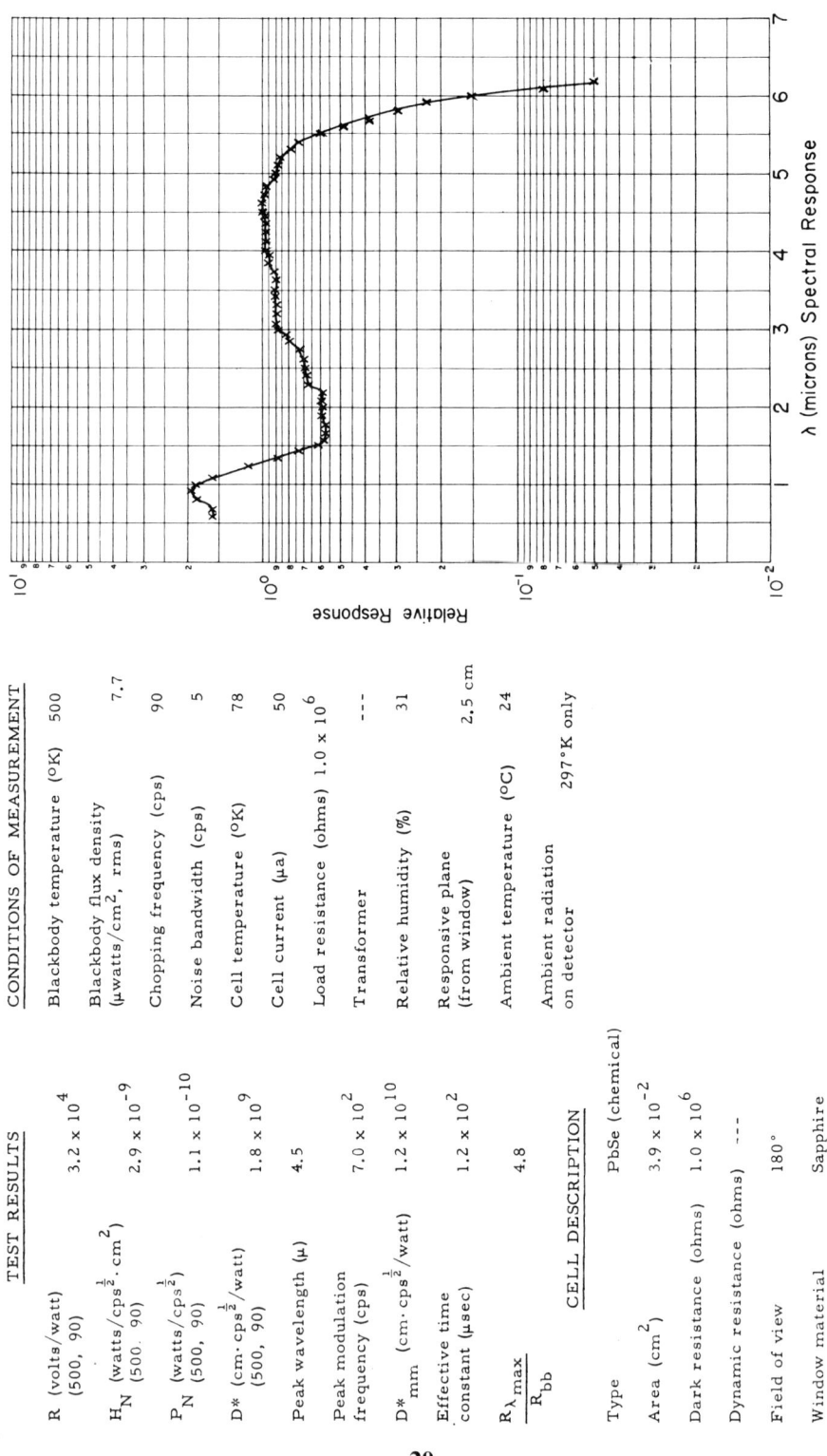

TEST RESULTS

R (volts/watt) (500, 90)		3.2×10^4
H_N (watts/cps$^{\frac{1}{2}}$ · cm^2) (500, 90)		2.9×10^{-9}
P_N (watts/cps$^{\frac{1}{2}}$) (500, 90)		1.1×10^{-10}
D^* (cm · cps$^{\frac{1}{2}}$/watt) (500, 90)		1.8×10^9
Peak wavelength (μ)		4.5
Peak modulation frequency (cps)		7.0×10^2
D^*_{mm} (cm · cps$^{\frac{1}{2}}$/watt)		1.2×10^{10}
Effective time constant (μsec)		1.2×10^2
$\dfrac{R_{\lambda max}}{R_{bb}}$		4.8

CELL DESCRIPTION

Type		PbSe (chemical)
Area (cm^2)		3.9×10^{-2}
Dark resistance (ohms)		1.0×10^6
Dynamic resistance (ohms)		---
Field of view		180°
Window material		Sapphire

CONDITIONS OF MEASUREMENT

Blackbody temperature (°K)	500
Blackbody flux density (μwatts/cm^2, rms)	7.7
Chopping frequency (cps)	90
Noise bandwidth (cps)	5
Cell temperature (°K)	78
Cell current (μa)	50
Load resistance (ohms)	1.0×10^6
Transformer	---
Relative humidity (%)	31
Responsive plane (from window)	2.5 cm
Ambient temperature (°C)	24
Ambient radiation on detector	297°K only

FIG. 7. Typical report page for the same commercial PbSe detectors as Fig. 6 giving the test results, cell description, the condition of measurement, and the relative response. The response curve is normalized to unity at the useful wavelength 4.5 μm.

1. OPERATIONAL CHARACTERISTICS OF INFRARED PHOTODETECTORS

and from the test results, we find

$$R_{bb}(50 \ \mu A, 90 \ Hz, 500°K) = 3.2 \times 10^4 \quad V/W,$$

$$1/\gamma_p = R_{\lambda_{max}}/R_{bb} = 4.8.$$

By use of Eq. (14) and substitution, we have

$$R(75 \ \mu A, 10^3 \ Hz, 6.0 \ \mu m) = (3.2 \times 10^4) \frac{1.3 \times 10^{-2}}{9.0 \times 10^{-3}} \frac{0.78}{1.0} (1.5 \times 10^{-1}) 4.8$$
$$= 2.6 \times 10^4 \quad V/W.$$

Therefore, if this detector were of unit area and we had a radiation source with a flux density of $1.0 \ \mu W/cm^2$, we would expect a signal level of 2.6×10^4 μV from the detector at $b = 75 \ \mu A, f = 10^3 \ Hz$, and $\lambda = 6.0 \ \mu m$.

To find D^* at f_1, b_1, λ_1 we must find the noise level at f_1, b_1. From the noise spectrum plot, we interpolate at $f = 10^3$ to find the noise level for $b = 75 \ \mu A$. This gives, for a 1-Hz system bandwidth,

$$N(75 \ \mu A, 10^3 \ Hz) = 3.1 \times 10^{-6} \quad V/Hz^{1/2},$$

$$N(150 \ \mu A, 7 \times 10^2 \ Hz) = 6.2 \times 10^{-6} \quad V/Hz^{1/2}.$$

To find D^* at b_1, f_1, λ_1, we use Eq. (17). Since the peak frequency is reported as 7.0×10^2 and the peak bias is shown in the detectivity versus frequency plot at $150 \ \mu A$, we find

$$D^*(b_1, f_1, \lambda_1) = D^*_{mm}(b_p, f_p, \lambda_p) \frac{N(150, 7.0 \times 10^2)}{N(75, 1.0 \times 10^3)} \frac{V(75 \ \mu A)}{V(150 \ \mu A)}$$
$$\times \frac{v(1.0 \times 10^3 \ Hz)}{v(7.0 \times 10^2 \ Hz)} v(6.0 \ \mu m),$$

$$D^*(b_1, f_1, \lambda_1) = (1.2 \times 10^{10}) \frac{6.2 \times 10^{-6}}{3.1 \times 10^{-6}} \frac{1.3 \times 10^{-2}}{2.9 \times 10^{-2}} \frac{0.78}{0.83} (1.5 \times 10^{-1}),$$

$$D^*(b_1, f_1, \lambda_1) = 1.5 \times 10^9 \quad Hz^{1/2} \ cm/W.$$

If the detector had 1 cm² area, the system of unit bandwidth, and the radiation flux density at $1.0 \times 10^{-6} \ W/cm^2$, we would expect a signal-to-noise ratio at b_1, f_1, λ_1 of 1.5×10^3.

The noise equivalent power at b_1, f_1, λ_1 is

$$P_N = A^{1/2}/D^*(b_1, f_1, \lambda_1) = (3.9 \times 10^{-2})^{1/2}/1.5 \times 10^9 - 1.3 \times 10^{-10} \quad W/Hz^{1/2}$$

and the noise equivalent irradiance at b_1, f_1, λ_1 is

$$H_N = P_N/A = 1.3 \times 10^{-10}/3.9 \times 10^{-2} = 3.4 \times 10^{-9} \quad W/cm^2\text{-}Hz^{1/2}$$

for unit system bandwidth.

Let us now consider that the detector will be exposed to radiation over a spectral range from 3.0 to 4.0 μm. For this example, we shall consider that the irradiance distribution from the radiation source peaks at 3.6 μm and is distributed as shown in Table I. For $\lambda < 3.0$ μm and for $\lambda > 4.0$ μm, $P_\lambda = 0$.

We wish to find the signal, the responsivity, and the D^* values when the detector is operated under the conditions stated below. To find the signal, Eq. (15) is used:

$$R(75 \ \mu A, 10^3 \ Hz, P(ss)) = \left[R(b_r, f_r, \lambda) \frac{V(75 \ \mu A)}{V(50 \ \mu A)} \frac{v(10^3 \ Hz)}{v(90 \ Hz)} \frac{\gamma_{ss}}{\gamma_p} \right].$$

Note that the first term gives the responsivity of the detector where $b = 75$ μA, $f = 1000$ Hz, and the λ peak = 4.5 μm.

We shall first consider only γ_{ss}, as the other quantities are easily calculated. The integration will be numerically performed by using the evenly spaced increment $\Delta\lambda = 0.2$ μm. A smaller increment would provide greater accuracy. Note that

$$J = \int_{3.0 \mu m}^{4.0 \mu m} H_{\lambda_i} \, d\lambda.$$

From the spectral response plot we obtain the values of $v(\lambda)$ over the spectral region of interest and consider the power falling on the detector (Table II):

$$\sum_i v(\lambda_i) H_{\lambda_i} \Delta\lambda = 2.06 \times 10^{-6} \ \text{W/cm}^2,$$

$$J = \sum_i H_{\lambda_i} \Delta\lambda = 2.25 \times 10^{-6} \ \text{W/cm}^2,$$

$$\gamma_{ss} = .916.$$

From the test results in Fig. 7, we obtain the values for $R(b_r, f_r, T)$ and $R_{\lambda_{max}}/R_{bb} = 1/\gamma_p$, and from detectivity versus frequency and frequency

TABLE I

λ_i	$H_{\lambda_i} \Delta\lambda$ (W/cm²)
3.0	5.0×10^{-8}
3.2	1.0×10^{-7}
3.4	5.0×10^{-7}
3.6	1.0×10^{-6}
3.8	5.0×10^{-7}
4.0	1.0×10^{-7}

TABLE II

λ_i (μm)	$v(\lambda_i)$	$H_{\lambda_i} \Delta\lambda$ (W/cm²)	$v(\lambda_i) H_{\lambda_i} \Delta\lambda$ (W/cm²)
3.0	0.87	5.0×10^{-8}	4.35×10^{-8}
3.2	0.90	1.0×10^{-7}	9.00×10^{-8}
3.4	0.90	5.0×10^{-7}	4.50×10^{-7}
3.6	0.90	1.0×10^{-6}	9.00×10^{-7}
3.8	0.95	5.0×10^{-7}	4.75×10^{-7}
4.0	0.97	1.0×10^{-7}	9.70×10^{-8}
		2.25×10^{-6} W/cm²	2.06×10^{-6} W/cm²

response (Fig. 6) the values for

$$V(75\ \mu A)/V(50\ \mu A) \quad \text{and} \quad v(10^3\ \text{Hz})/v(90\ \text{Hz}).$$

$$R(75\ \mu A, 10^3\ \text{Hz}, P_\lambda(\text{ss})) = (3.2 \times 10^4) \frac{1.3 \times 10^{-2}}{9.0 \times 10^{-3}} \frac{0.78}{1.0} 4.8\ (0.916)$$

$$= 1.58 \times 10^5 \quad \text{V/W}.$$

Since $J = \int_{3.0\mu m}^{4.0\mu m} H_{\lambda i}(\text{ss})\,d\lambda = 2.25 \times 10^{-6}$ W/cm² and the area (from cell description; Fig. 7) is 3.9×10^{-2} cm², we can calculate the signal from $V = RAJ$:

$$V(75\ \mu A, 10^3\ \text{Hz}, P(\text{ss})) = 1.3 \times 10^{-2} \quad \text{V}.$$

To find D^* at 75 μA, 10^3 Hz, $P_\lambda(\text{ss})$, we take the D^* (75 μA, 10^3 Hz, 6.0 μm) computed previously and use the relation

$$D^*(75, 10^3, P(\text{ss})) = D^*(75, 10^3, 6.0)\lambda_{\text{ss}}/v(6.0)$$

$$= 1.5 \times 10^9 \times 0.916/(1.5 \times 10^{-1})$$

$$= 9.16 \times 10^9 \quad (\text{Hz}^{1/2}\ \text{cm})/\text{W}$$

We can calculate the signal-to-noise ratio from the relation $D^*J\sqrt{A} = S/N$

$$(S/N)(75\ \mu A, 10^3\ \text{Hz}, P(\text{ss})) = 4.06 \times 10^3$$

for a unit system bandwidth and a unit detector area.

It should be clearly understood that the above values of signal, responsivity, and D^* apply only when this detector is exposed to a source having the definite spectral characteristics as given. A source having a different spectral distribution could yield considerably different values for these parameters. Therefore, when this type of calculation is made, the spectral distribution of the energy from the source, as well as the total power from the source, must be known.

VI. Current State-of-the-Art Photodetectors

The introductory chapter of Volume 5 of this treatise gives a resume of the well-known and fully characterized photodetectors, both intrinsic and extrinsic, in terms of the performance figure, D_λ^*.[15] (See also Kruse et al., Chapter 10.[5]) Similar curves have also been shown in an earlier review.[16] Such presentations are limited in their utility for the electrooptical system

---

[15] H. Levinstein, in "Semiconductors and Semimetals" (R. K. Willardson and A. C. Beer, eds.), Vol. 5, Chap. 1. Academic Press, New York, 1970.
[16] R. F. Potter and W. L. Eisenman, *Appl. Opt.* **1**, 567 (1962).

design engineer who seeks that detector that can optimize a sensor's performance. It can, however, readily be seen that the detectivities of many photodetectors are within factors of 2 or less from the ideal detector in a 300°K radiation ambient.

17. DETECTORS OPERATING AT INTERMEDIATE CRYOGENIC TEMPERATURES

Such curves also provide convincing evidence that maximum sensitivity at longer wavelengths can be achieved only at lower operating temperatures.

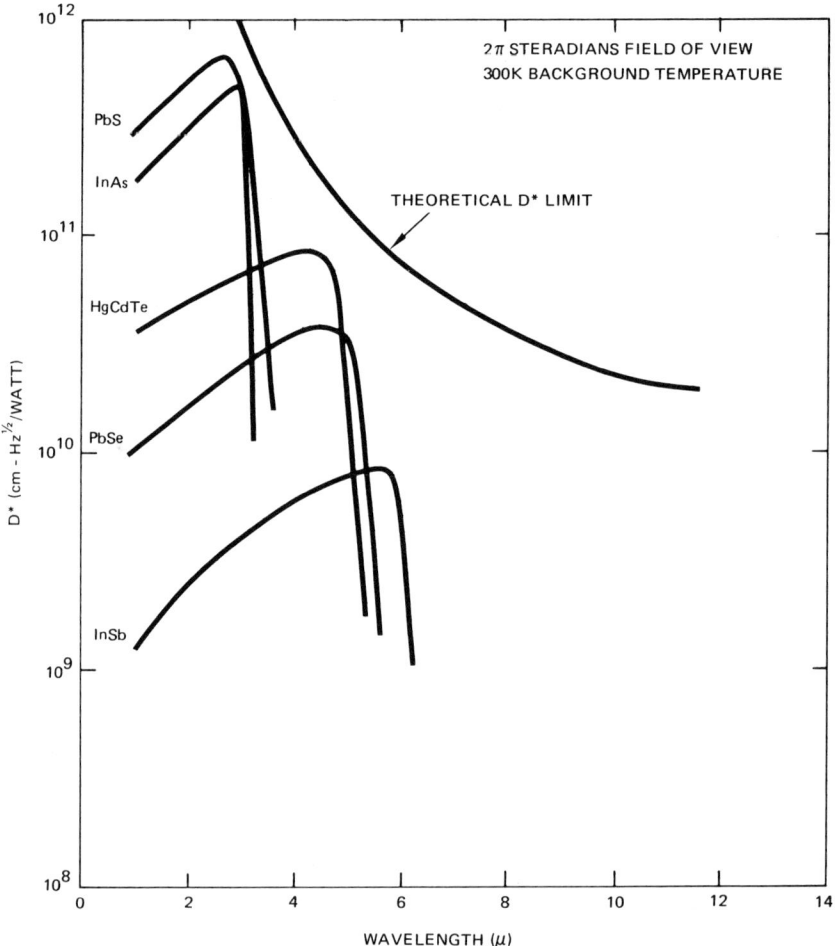

FIG. 8. Representative curves of D^* for four commercial photodetectors operated at 200°K. The theoretical curve gives the limit for D^* with a 300°K background radiation.

1. OPERATIONAL CHARACTERISTICS OF INFRARED PHOTODETECTORS 25

Cryogenic liquids such as He, N_2, and Ne are the most commonly used low temperature refrigerants. However, if the sensor requirements can permit detector operating temperatures of 195°K and higher, thermoelectric cooling should be considered. This is not the place to discuss thermoelectric coolers, but among the advantages are reliability, ease of operation, and long operational lifetime. Among the disadvantages one notes low cooling capacities, requirements for low voltage, and high current sources resulting in low efficiencies.

As thermoelectric coolers have a practical lower limit of 200°K representative D_λ^* curves for several intrinsic photoconductors are shown in Fig. 8.

Lead sulfide increases in D^* about an order of magnitude upon cooling to 200°K and approaches the theoretical limit. Unfortunately, the responsive time constant also increases by at least an order of magnitude. The temperature of 200°K appears to be optimum in maximizing the D^* of PbS. Some high performance cooled PbS detectors have exhibited reduced D^* values when the detectors were exposed to relatively low levels of visible and ultraviolet radiation.

Lead selenide detectors also increase about an order of magnitude in D^* when cooled to 200°K. This temperature also appears optimum for PbSe and the specific detectivity approaches the theoretical limit. PbSe does not operate well over long periods of time in a hard vacuum and this pecularity has presented some problems when this detector material is operated cooled. Recently, however, a change in Dewar design has largely solved this problem. Recent efforts with both PbS and PbSe have resulted in the development of cooled arrays using cold aperture stops, cold spectral filters, and thermoelectric coolers.

Indium arsenide increases in specific detectivity about an order of magnitude when cooled to 200°K. The responsive time constant of this detector material is in the range of 1 μsec. The temperature range of 200°K is quite suitable for the operation of InAs and this material can provide fast, sensitive, and reliable detectors in the 1–3 μm spectral region.

Although, the specific detectivity of indium antimonide increases from one to two orders of magnitude when cooled to 200°K, this temperature range is not low enough to achieve maximum D^* values for InSb detectors. InSb has been used to some extent at this temperature however, since it can provide D^* values approaching 10^{10} cm-Hz$^{1/2}$/W at wavelengths out to 6 μm and it has a short responsive time constant.

18. ALLOYED DETECTORS

Two new families of photodetector materials have been developed in recent years made up of the pseudobinary alloys of compound semiconductors. Combinations of HgTe and CdTe are now available (see the review

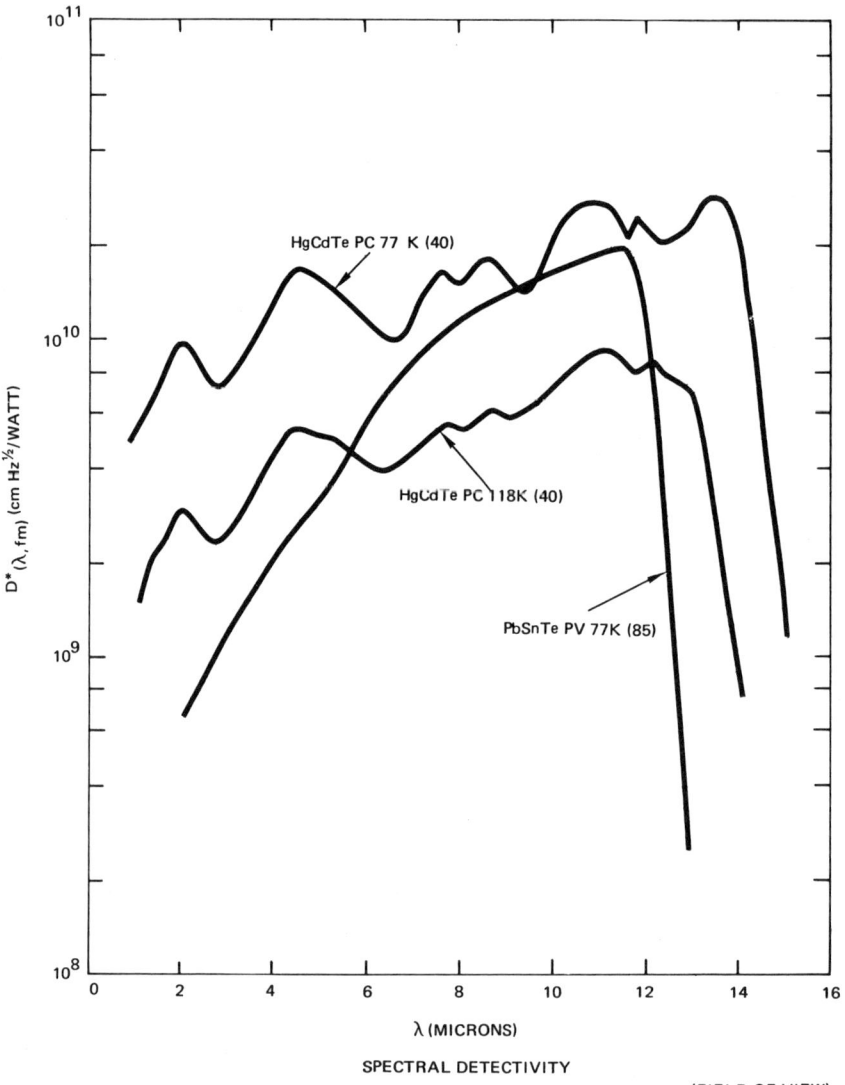

FIG. 9. Spectral D^* curves for a HgCdTe photoconductor and a PbSnTe photovoltaic detector are shown. The operating temperatures are given. The field of view in degrees is given in the parentheses.

of Long and Schmit, Chapter 5, Volume 5 of this treatise[17]) as $Hg_{1-x}Cd_xTe$ where $x \sim 0.2$ for operation in the 8–14 μm spectral region of the atmos-

[17] D. Long and J. L. Schmit, in "Semiconductors and Semimetals" (R. K. Willardson and A. C. Beer, eds.), Vol. 5, Chap. 5. Academic Press, New York, 1970.

pheric window. This family of detectors is usually referred to as "merc–cad telluride" (HgCdTe) as a generic term with the subscript suppressed. The HgCdTe detector of Fig. 8 is obviously made from an alloy where x differs considerably from 0.2.

A similar story applies to the lead–tin chalcogenides of which lead–tin–telluride is currently important. (Again the reader is referred to Volume 5 of this treatise to the review by Melngailis and Harman in Chapter 4.[18]) For operation in the 8–14 μm region the alloy $Pb_{1-x}Sn_xTe$ has been used as a photodiode with $x \sim 0.2$ also.

Curves for a PbSnTe photovoltaic detector operated at 77°K and a HgCdTe photoconductive detector at two different temperatures, but designed for 8–14 μm are shown in Fig. 9.

19. Criteria for Comparison of Detectors

Although the detectivity is important as a measure of the signal-to-noise ratio in the sensing element, one should examine other characteristics in

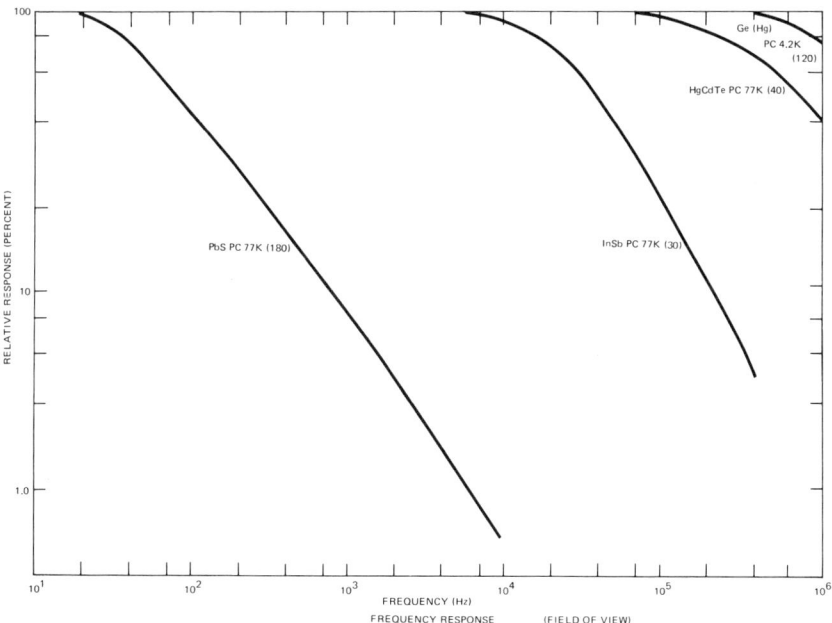

FIG. 10. Relative response versus frequency curves for four different photoconductor detectors are shown. Operating temperatures are given at the field of view (degrees) as given in the respective parentheses.

[18] I. Melngailis and T. C. Harman, in "Semiconductors and Semimetals" (R. K. Willardson and A. C. Beer, eds.), Vol. 5, Chap. 4. Academic Press, New York, 1970.

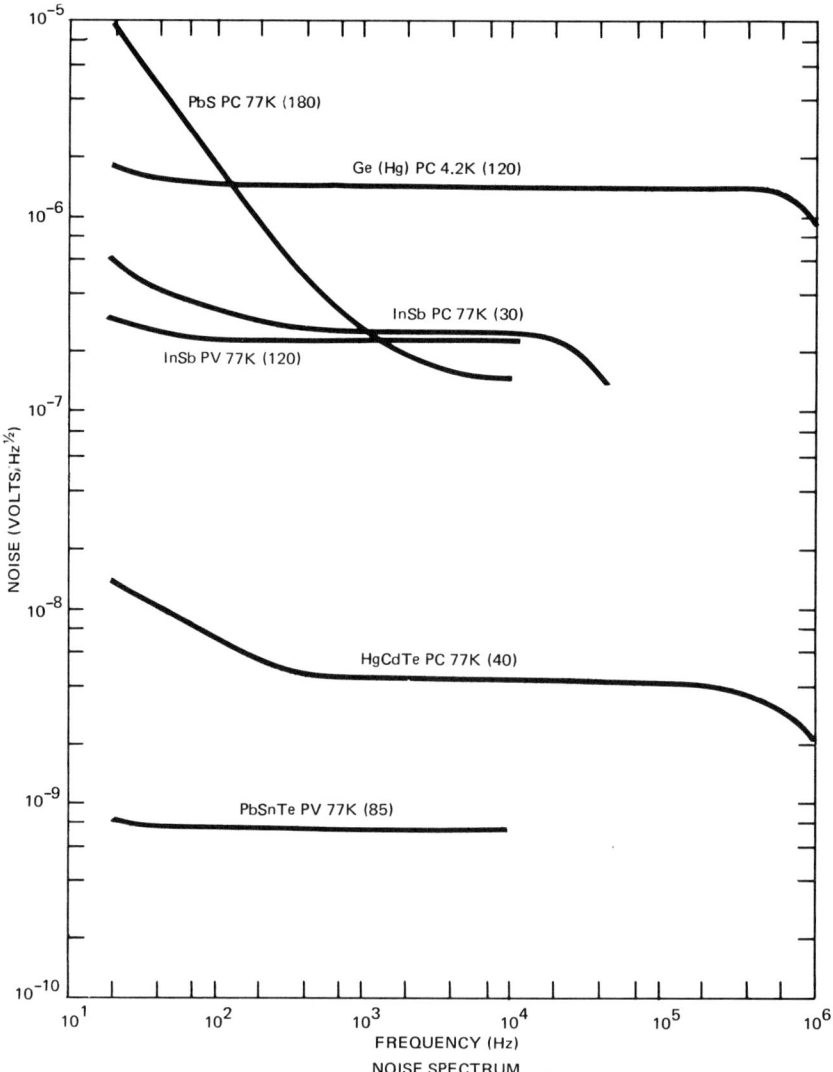

FIG. 11. The noise voltage per unit bandwidth for the four photoconductors (PC) of Fig. 10 plus two photovoltaic detectors (PV).

making a definitive comparison between detectors. The relative response as a function of modulation frequency and the root-noise-power spectrum, usually expressed as the noise voltage, are two such parameters. Representative curves for comparison of several detectors are shown in Figs. 10 and 11. When the data are combined and presented in the form of $D^*_{\lambda_{\max}}$ versus

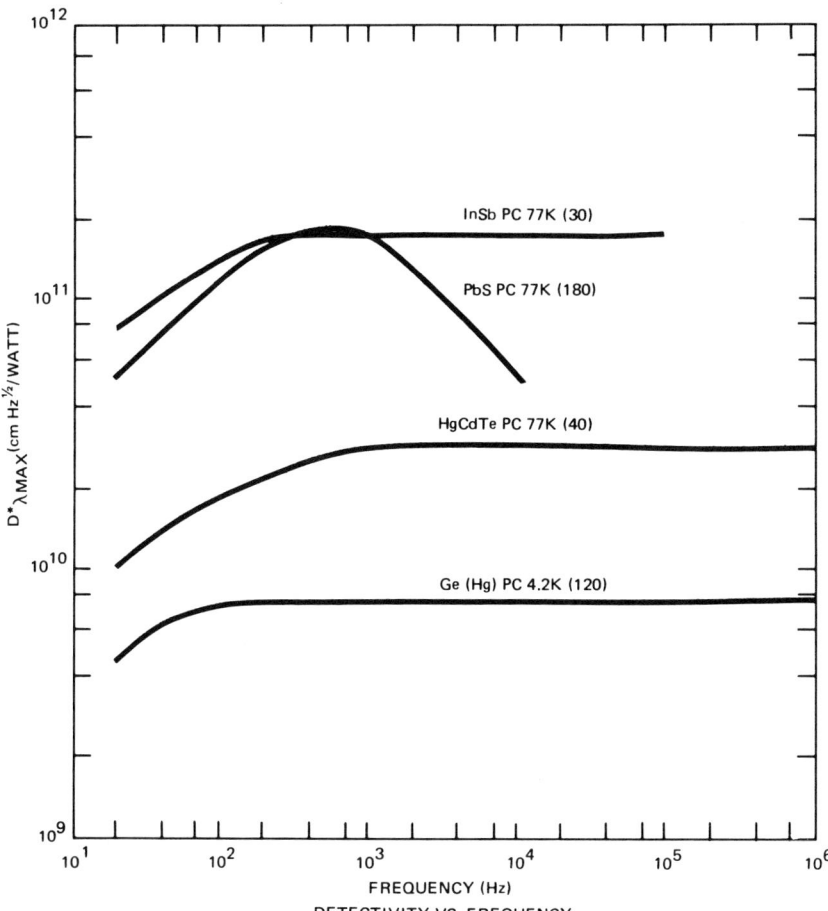

FIG. 12. $D^*_{\lambda\,\text{max}}$ versus frequency for the four photoconductors of Fig. 10. The PbS detector is not background limited and has a short detective time constant.

modulation frequency (Fig. 12), the three background limited photoconductors stand out from the PbS detector. As can be seen the detectivity frequency response extends to high frequencies.

Signal responsivity and noise voltages versus bias should also be measured and consulted for operating characteristics. The data sheet for the PbSe detector in Fig. 6 is an example of the value of these measurements. Both signal response and noise voltage increase at rates similar to that of the bias. The result is that detectivity is changed very little with bias current and the user has a much less severe noise figure requirement on his preamplifier circuits if he operates at the higher biases. A more subtle example is

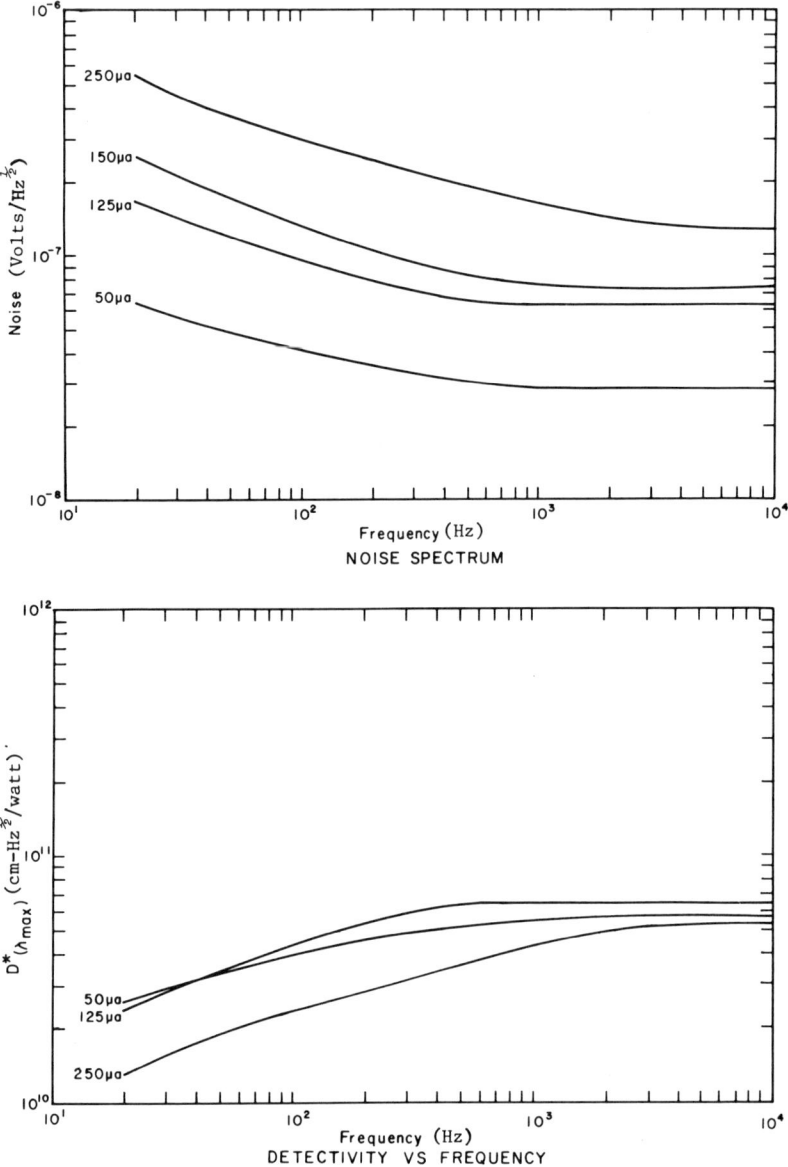

FIG. 13. Upper graph gives the frequency spectra of the noise voltage for an InSb photoconductor at different bias currents. The lower graph gives the corresponding $D^*_{\lambda_{max}}$ curves. As discussed in the text operating at the higher bias reduces the requirements of the noise figure for the preamplifier without an undue sacrifice in D^*.

shown in Fig. 13 for an InSb photoconductor. The noise voltages versus frequency for several bias currents are shown along with $D^*_{\lambda_{max}}$ for the corresponding biases. The result is that although 250 μA is not the optimum bias, operation of the detector at that bias current at the higher modulation frequencies would again relieve severe restrictions on the preamplifier noise figure and yet the system would suffer a minimal loss of detectivity.

Figure 14 shows representative noise voltage versus frequency for two detectors representative of their respective types; an intrinsic photoconductor (HgCdTe) and an extrinsic photoconductor Ge(Hg), both of which appear to be background limited in their characteristics.

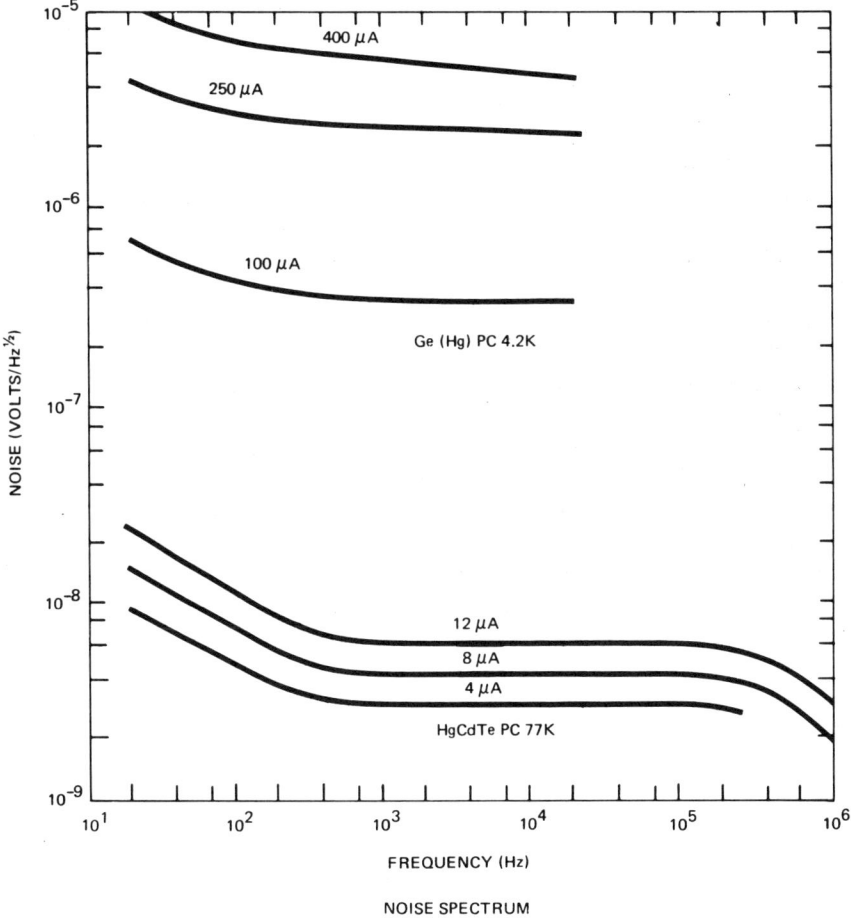

FIG. 14. Frequency spectra of noise voltage at different bias current for two additional photoconductors Ge(Hg) and HgCdTe.

20. THERMAL DETECTORS

Of the thermal detectors, the radiation thermocouple is probably the best known and has the most wide-spread utilization, principally for infrared spectroscopy. Typical D^* range between 10^8 and 10^9 cm Hz$^{1/2}$/W. Generally they are low impedance devices and tend toward fragility. The reader is referred to Stevens' article in Volume 5 for a comprehensive review.[19]

The detectivity of thermistor bolometers has not significantly increased during the past decade. Potter and Eisenman discuss thermistor bolometer characteristics.[16] Figure 7 in that review shows representative data for thermistor bolometers.

Another type of bolometer receiving a large amount of current interest is that which depends upon the pyroelectric effect in certain piezoelectric materials. Putley, in his review in Volume 5[20] and the update, Chapter 7, in this volume, gives a detailed account of this class of bolometer. To a large extent, present day interest is due to the extremely short response time constant (of the order of nanoseconds). Most pyroelectric detectors have very high impedances, thus efficient operation requires coupling with a high impedance, low noise preamplifier. The preamplifier can limit the noise level of the combined circuit, resulting in a detective time constant of the order of milliseconds. At low frequencies, the bolometer appears to be limited by Johnson–Nyquist noise, and any approach toward background limited performance must come by increasing the responsivity at a rate substantially greater than the rate at which device resistance increases.

Although such performance has rarely, if ever, been obtained for thermal detectors, photon noise limits for ideal thermal detectors bring out an interesting point of comparison with photon detectors. Fluctuations of radiant power from the background and the emitted power provide the thermal detector limit. Based on these considerations, ideal detectivities can be calculated for the detector operating temperature and the background temperature. A set of such calculated curves (T. Limperis)[21] is shown in Fig. 15. For comparison the reader is referred to a curve for 300°K background for D_λ^* as shown in Fig. 8. Similar curves for lower temperature backgrounds are given in Fig. 2 of Chapter 1, Volume 5 of this treatise.[15]

Comparing families of curves of these types brings out the fundamental difference between photon detectors and thermal detectors, namely, that

---

[19] N. B. Stevens, in "Semiconductors and Semimetals" (R. K. Willardson and A. C. Beer, eds.), Vol. 5, Chap. 7. Academic Press, New York, 1970.

[20] E. H. Putley, in "Semiconductors and Semimetals" (R. K. Willardson and A. C. Beer, eds.), Vol. 5, Chap. 6. Academic Press, New York, 1970.

[21] T. Limperis, in "Handbook of Military Infrared Technology" (W. Wolfe, ed.), p. 516. Office of Naval Research, Washington, D.C., 1965.

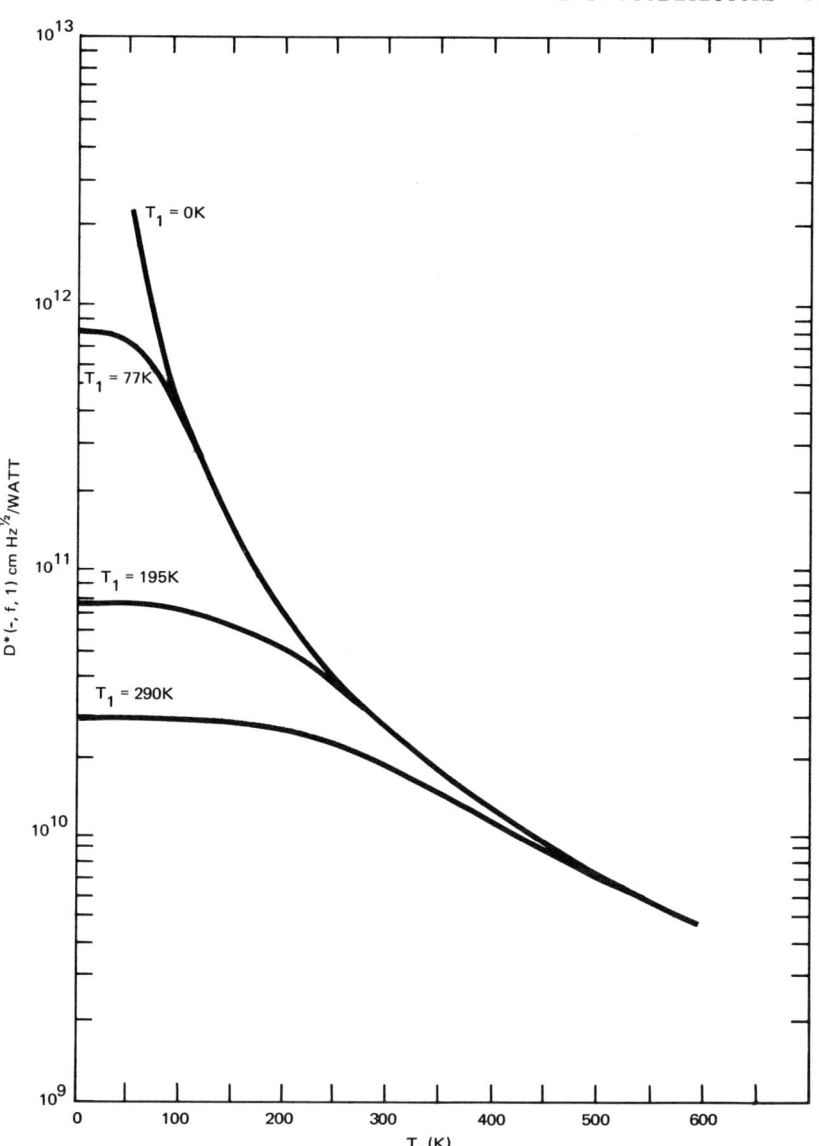

FIG. 15. Photon-noise-limited D^* of thermal detectors as a function of detector temperature T_1 and background temperature T_2. (From Limperis.[21])

the latter accept or "see" radiation from the entire electromagnetic spectrum, while photon detectors cut off the longer wavelengths at their sharp absorption edges at λc. Thus D_λ^* can become very large as λc is lowered for a given background temperature.

VII. Detectors in Advance of the State-of-the-Art

Up to this point the discussion has been about those radiation detectors that are considered conventional because they are relatively readily available from commercial services. In many instances, they can be ordered in various configurations from "off-the-shelf" catalogs. In other cases, special customized packages ranging from simple to very complex can be ordered. In both cases, the expected performance is well-established and documented.

In this part we intend to give a resumé of some less conventional detectors that offer promise for unusual sensor designs of the future.

21. METAL–OXIDE–METAL POINT CONTACT DIODES

The advent of the 10.6 μm CO_2 laser has brought about intensive efforts to produce detectors having high speed of response. The pyroelectric detectors are one such family as already mentioned. Another type is the metal–oxide–metal (MOM) point contact diode. A metal whisker (cross section dimension of the order of 0.1 μm) is brought into contact with a metal plate, a thin metal oxide surface layer provides the insulating barrier through which electrons can tunnel. The wire whisker serves as an antenna for incident radiant power at the desired infrared frequencies. The induced high frequency currents propagate to the junction and are detected as rectified current. The responsivity falls off as the barrier height (work function of the metals) and/or the oxide thickness are lowered. However, the speed of response can be pushed to the 10^{13}–10^{14} Hz range with realizable barrier heights and oxide thicknesses. NEP for direct detection has been reported for a tungsten whisker device as 9×10^{-8} W/Hz$^{1/2}$ for a 10.6 μm beam amplitude modulated at 30 MHz. For heterodyne detection NEP's of 10^{-13} W/Hz$^{1/2}$ have been measured for 30 MHz i.f. with a 10.6 μm local oscillator.[22,23]

22. JOSEPHSON JUNCTION DETECTORS

In Chapter 6 of this volume, Richards gives a detailed review of the Josephson junction detector for far-infrared radiation. We will only indicate the performance characteristics to be expected from this form of a nonlinear junction device. In principle, a dc voltage across the junction causes an ac Josephson current flow. The irradiation of the junction with the high frequency far-infrared signal causes a frequency modulation of the Josephson current that can be sensed in an external circuit. Thus, incident radiation at suitable frequencies will cause the I–V characteristics to change. A simple video broadband detection scheme relies on a shift of the I–V curve. Thus,

---

[22] D. R. Sokoloff, A. Sanchez, R. M. Osgood, and A. Javan, *Appl. Phys. Lett.* **17**, 257 (1970).
[23] R. L. Abrams and W. B. Gardrud, *Appl. Phys. Lett.* **17**, 150 (1970).

for a constant junction bias current, the detector responds with an ac component of the junction voltage corresponding to the input modulation frequency. NEP's down to 10^{-15} W/Hz$^{1/2}$ have been reported. Josephson junction type detectors do offer promise of effective operation at wavelengths of 0.5 mm and longer (to several millimeters). Speed of response can be improved by modification of the junction parameters, but, it must be remembered, at the expense of the sensitivity.

23. SCHOTTKY BARRIER PHOTODIODE

A third type of surface barrier detector is a Schottky barrier on an epitaxial semiconductor. The film metal–semiconductor junction could be an effective competitor to the conventional and well-established p–n junction semiconductor bulk detector. Carriers, excited in the semiconductor film by incident radiation, reach the barrier and give a photovoltaic signal. A PbTe–Pb device has been reported which operates at 77°K.[24] Quantum efficiencies between 50 and 60% have been realized for 4 μm thick PbTe layers on BaF$_2$ substrates. Spectral detectivities for devices with a reduced field of view are shown in Fig. 16. If the same efficiencies can be achieved for PbSnTe films, similar detectors can be expected for operation at 8–14 μm.

24. MIS InSb PHOTOVALTAIC DETECTOR

The final detector to be discussed is also a surface barrier type device. Phelan and Dimmock first reported IR detectors of metal insulator semiconductors (MIS) structures using high quality InSb single crystals.[25] More recently, Lile and co-workers have developed a MIS InSb photovoltaic detector based on thin film technology.[26] Although these devices utilize polycrystalline thin film semiconductors, they exhibit responsivities and detectivities that compare competitively with the bulk crystal diode used by Phelan and Dimmock.[25]

The films themselves are prepared by a recrystallization technique which yields high quality, albeit polycrystalline, materials. An insulating layer, approximately 600 Å thick is deposited by an anodization process. By suitable metallization techniques, comparable to integrated circuit technology, electrodes are deposited and a detector array defined. A schematic representation of an eight element thin film InSb array is shown in Fig. 17.

A peak detectivity of 3×10^{10} cm Hz$^{1/2}$/W has been reported for these thin film devices. This is a lower limit as the measurements were limited

---

[24] E. M. Logothetis, H. Holloway, A. J. Varga, and E. Wilkes, *Appl. Phys. Lett.* **19**, 318 (1971).
[25] R. J. Phelan, Jr., and J. O. Dimmock, *Appl. Phys. Lett.* **10**, 55 (1967).
[26] D. L. Lile and H. H. Wieder, *Thin Solid Films* **13**, 15 (1972); D. L. Lile, *Surface Sci.* **34**, 337 (1973).

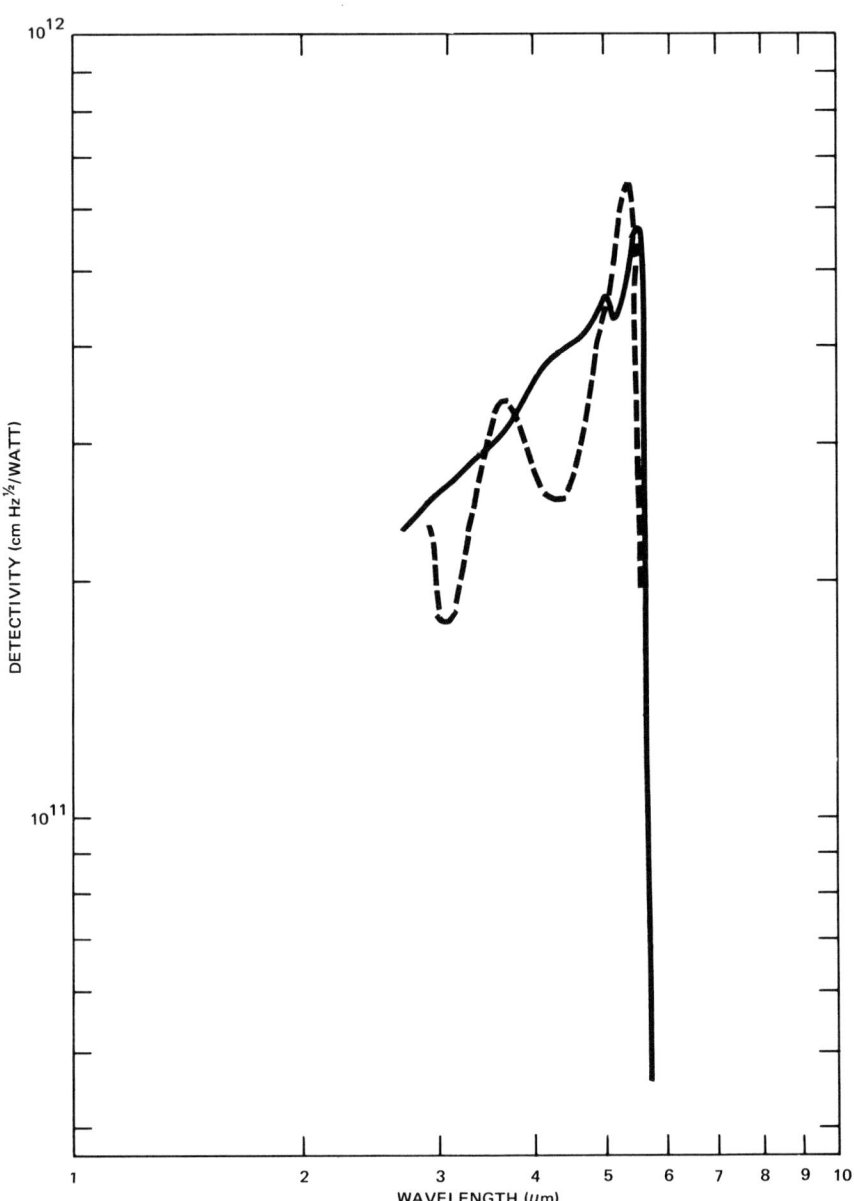

FIG. 16. Spectral detectivities of 0.7 μm-thick (broken line) and 3.8 μm-thick (solid line) PbTe Schottky diodes at 77°K with reduced background. (From Logothetis et al.[24])

by the noise in the electronics (20 nV). The responsive time constant was given as 4 μsec. Lile and Davis also reported that the spread in device response among the individual elements was less than ± 15%.

That detectors made of polycrystalline films have detectivities comparable to single crystal devices is noteworthy from several aspects.

(1) Because carrier transport is normal to the film plane it is essentially parallel to grain boundaries, thus minimizing grain boundary effects.

(2) The polycrystalline thin film nature is likely to minimize ionizing radiation hazards.

(3) The inexpensive nature of preparation from starting material to finished device should lead to low cost, highly reproducible arrays for application in the 3–5 μm spectral region.

(4) Modern detector arrays are often hybridized by having the associated electronics placed in close juxtaposition. Any method forming these arrays that is compatible with integrated circuit technology will lead to significant improvements in monolithic (or integrated) focal plane arrays.

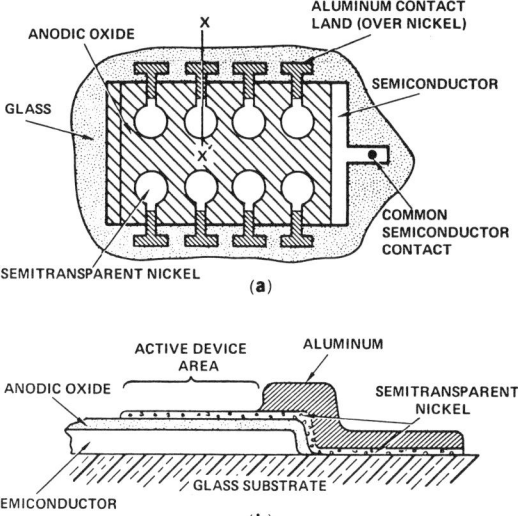

FIG. 17. Schematic representation of thin film MIS detector. (From Lile and Wieder[26]) (a) Configuration of eight devices. (b) Cross section of one device on XX'.

List of Symbols

b	bias value
f	modulation frequency
λ	wavelength of incident radiation
J	incident signal radiation power density
J_{rms}	root-mean-square (rms) value of J
$P(ss)$	signal power incident upon the detector
$P_\lambda(ss)$	spectral radiant power per wavelength; (ss) refers to a spectral source of the incident radiation
H	irradiance of radiant source
V	open circuit signal voltage appearing across detector terminals
$V(b,f)$	modulation frequency and bias dependence of the open circuit voltage appearing across the detector terminals
$V(b, \lambda)$	spectral and bias dependence of the open circuit voltage appearing across the detector terminals
$V(b)$	bias dependence of the open circuit voltage appearing across the detector terminals
$v(f)$	relative modulation frequency dependence of the detector signal, i.e., $v(f) = 1$ at $f = f_{ref}$
$v(\lambda)$	relative spectral dependence of the detector signal, i.e., $v(\lambda) = 1$ at $\lambda = \lambda_{ref}$
$G(f)$	gain of the electronic system at a fixed frequency as a function of the frequency of the calibrating oscillator
G_m	maximum value of the gain of the electronic system at a fixed frequency as a function of the frequency of the calibrating oscillator
E_s	voltage produced at the electronic system output by the detector signal voltage with the detector operated under specified conditions
A	sensitive area of the detector in square centimeters
$R_{bb}(b,f,T)$	responsivity of the detector to a blackbody source at temperature T
E_{ref}	voltage produced at the output of the electronic system by the signal voltage from the reference detector
$\varepsilon(\lambda)$	relative spectral responsivity of the reference detector
E_c	voltage produced at the output of the electronic system by the calibrating signal generator
E_o	root-mean-square noise voltage produced at the output of the electronic system by the detector and the electronic system with the signal radiation source removed
E_a	root-mean-square noise voltage produced at the output of the electronic system by the electronic system only
$g(f)$	voltage gain of the electronic system
$N(b,f)$	root-mean-square open circuit noise voltage of the detector appearing at the terminals of the detector in a 1 Hz bandpass
N_L	root-mean-square open circuit noise voltage of the load resistance appearing at the terminals of the load resistor in a 1 Hz bandpass
Δf	the frequency bandpass of the electronic system
γ_p	spectral power efficiency of the detector
$R(b,f,\lambda)$	responsivity of the detector
$H_N(b,f,\lambda,A)$	noise equivalent irradiance of the detector
$P_N(b,f,\lambda)$	noise equivalent power of the detector
$D^*(b,f,\lambda)$	D-star of the detector
τ	responsive time constant of the detector

CHAPTER 2

Impurity Germanium and Silicon Infrared Detectors

Peter R. Bratt

I.	INTRODUCTION	39
	1. Scope	39
	2. Historical Background	40
II.	IMPURITIES IN GERMANIUM AND SILICON	42
	3. Model for Substitutional Impurities	43
	4. Absorption of IR Radiation by Impurity Atoms	45
	5. Quantum Efficiency	51
III.	IMPURITY PHOTOCONDUCTIVITY	53
	6. Photoconductivity	53
	7. Noise	66
	8. Sensitivity Limits—Detectivity	71
IV.	DEVICE FABRICATION	77
	9. Material Preparation	77
	10. Detector Fabrication	90
V.	OPERATING CHARACTERISTICS	95
	11. Operating Characteristics—Germanium	95
	12. Operating Characteristics—Silicon	108
	13. Germanium–Silicon Alloys	113
	14. Special Features	116
VI.	CONCLUSIONS AND ANTICIPATED FUTURE DEVELOPMENTS	140

I. Introduction

1. SCOPE

This chapter is concerned with infrared (IR) detectors that are produced by the addition of certain impurity atoms to crystals of germanium (Ge) or silicon (Si). Thus the properties of these detectors result more from the characteristics of the impurity atoms rather than the host crystal. For this reason the photosensitivity of these detectors is often called the "extrinsic" sensitivity as opposed to other types of detectors which utilize the "intrinsic" properties of the host crystal.

Impurity Ge and Si IR detectors are useful only when operated in the photoconductive mode. Therefore the discussion will be confined to this mode of operation. Although photovoltaic effects may sometimes be observed in these materials (for example, due to nonuniform impurity distributions or potential barriers at the metal to semiconductor contacts) such effects are of little practical value for making highly sensitive IR detectors. To be sure, photodiodes and phototransistors are also made by introducing impurity atoms into Ge or Si to produce p–n junctions. However, the resulting photosensitivity is of the intrinsic type and extends only to about 1 μm in the near IR. There is voluminous literature on p–n junction devices, but a review of this would be outside the scope of this chapter.

In the following sections we will survey the development of impurity Ge and Si IR detectors since their inception over 20 years ago. The discussion will present the basic physics underlying their operating characteristics, discuss the technology of material preparation and detector fabrication, and give examples of actual device operating characteristics.

For brevity, we will adopt the customary notation for impurity atoms in Ge or Si that places the chemical symbol of the host crystal first, then a colon and the chemical symbol of the impurity atom. For example, gold in germanium becomes Ge:Au. For impurity atoms that may have more than one state of ionization in the host crystal a Roman numeral superscript is sometimes added after the impurity symbol to denote which ionization state is represented.

2. Historical Background

While the initial interest in photoconductivity due to impurities in Ge and Si was mainly for the purpose of discovering the properties of the impurity atoms within the host crystal, it was also realized, at an early date, that the impurity photoconductivity would be extremely useful as a sensitive means of detection of IR radiation. Impurity ionization energies in Ge range from 0.01 to 0.35 eV, and in Si, from 0.04 to 0.55 eV. Therefore, the long wavelength cutoff for photoconductivity in these materials can range from 2 to 120 μm. This remarkable wavelength span of sensitivity gives these materials an unparalleled versatility that has been one of the key factors behind a continuing interest in their development for over 20 years.

The first studies of impurity photoconductivity in Ge and Si were carried out in the early 1950's. The energy levels of group III and V impurities in these materials were already known because of the intensive work that had gone on a decade earlier leading up to the development of the transistor in 1948. For Ge, the energy levels were of the order of 0.01 eV and for Si, they were in the range 0.04–0.06 eV. Si was selected in these early studies because there was little interest at this time in detectors that went to 120 μm. The main goal was to find detectors that were highly sensitive in the 3–5 μm and 8–13 μm atmospheric window regions.

2. IMPURITY GERMANIUM AND SILICON INFRARED DETECTORS

The first results on photoconductivity in impurity Si were reported in 1952 by Burstein and co-workers[1] at the United States Naval Research Laboratory and by Rollin and Simmons[2] in England. About this same time, deep level impurity atoms began to be discovered and the emphasis quickly shifted to impurity Ge containing gold, copper, or zinc. In 1954 Burstein et al.[3] were successful in making IR detectors of Ge:Zn and Ge:Cu and Kaiser and Fan[4] reported on Ge:Au and Ge:Cu detectors. The reason for this change of emphasis was that better sensitivity could be obtained from the Ge than from Si. This was due to the longer photocarrier lifetimes in Ge, which in turn was because at that time, Ge could be made more pure with respect to compensating donor and acceptor atoms than could Si.

Germanium:gold enjoyed a brief popularity as a very sensitive detector for the 3–5 μm region, but subsequently lost favor to other detectors such as PbSe and InSb. Work continued on Ge:Cu and Ge:Zn at other laboratories and both were developed into very sensitive detectors.[5,6] However they had spectral response out to 30 and 40 μm, respectively, which was far beyond that required for the 8–13 μm atmospheric window. They also required cooling with liquid helium which was a definite hindrance to their application in IR systems at that time. Thus the search continued for more suitable impurity energy levels, in particular an energy level of 0.09 eV.

Germanium:cadmium showed some improvement with a long wavelength cutoff at 23 μm and operating temperatures in the liquid hydrogen range. Then in 1957 it was discovered that alloying of Ge with Si caused a shift of the impurity ionization energies to some level between their positions in the pure materials. Germanium–silicon alloys containing Zn or Au were then produced[7] that had the desired 0.09 eV energy level and a spectral cutoff at 13 μm. This approach showed great promise, but the technology of producing these alloys was a difficult one.

Thus there was great excitement when Ge:Hg was discovered[8] in 1962, since this impurity also had a 0.09 eV ionization energy and was somewhat easier to make with reproducible properties. This material has continued

---

[1] E. Burstein, J. J. Oberly, and J. W. Davisson, Silicon as a Far-Infrared Detector, Naval Res. Lab. Rep. 3880, 1952; *Phys. Rev.* **89**, 311 (1953).

[2] B. V. Rollin and E. L. Simmons, *Proc. Phys. Soc. (London)* **B65**, 995 (1952); **B66**, 162 (1953).

[3] E. Burstein, J. W. Davisson, E. E. Bell, W. J. Turner, and H. G. Lipson, *Phys. Rev.* **93**, 65 (1954).

[4] W. Kaiser and H. Y. Fan, *Phys. Rev.* **93**, 977 (1954).

[5] E. Burstein, S. F. Jacobs, and G. Picus, *Proc. 5th Conf. Int. Comm. Opt.*, Stockholm, 1959 (unpublished).

[6] P. Bratt, W. Engeler, H. Levinstein, A. MacRae, and J. Pehek, *Infrared Phys.* **1**, 27 (1961).

[7] G. A. Morton, M. L. Schultz, and W. E. Harty, *RCA Rev.* **20**, 599 (1959).

[8] S. R. Borrello and H. Levinstein, *J. Appl. Phys.* **33**, 2947 (1962).

to serve the need for a sensitive detector in the 8–13 μm atmospheric window for over 10 years.

Development of longer wavelength Ge detectors was also undertaken during the last decade and both Ge:B and Ge:Ga[9] were found to provide good sensitivity out beyond 100 μm. Today, the six impurity elements in Ge, including Au, Hg, Cd, Cu, Zn, and Ga provide a family of highly sensitive IR detectors spanning the wavelength range from 1 to 130 μm.

It is interesting to note in the past few years, a resurgence of efforts to develop impurity Si after several years of dormancy.[10,11] The driving reason is for the purpose of coupling multielement arrays of IR detectors to their respective amplifiers in an integrated circuit. Since the circuits are made of Si, if the IR detectors can also be Si, then detectors and circuit components can all be made together on the same Si chip using standard integrated circuit technology. The ability to fabricate resistors, capacitors, transistors, diodes, and charge coupled devices, along with the detector, would allow a substantial amount of on-chip signal processing. For example, preamplification, time delay and integration of signals, multiplexing, filtering, and post amplification might all be accomplished at the detector focal plane. Such techniques are mandatory if very large multielement arrays (thousands of elements) are contemplated, because the problem of bringing this many leads off the focal plane and out of the detector package would be formidable indeed.

Other advantages offered by the Si host crystal are that higher impurity concentrations are attainable, resulting in higher absorption of radiation per unit path length, as well as lower dielectric constant and better developed device technology, including contacting, surface passivation, etc. This could conceivably lead to better detectors being obtained from Si than are presently made using Ge.

Thus we have seen the historical development of impurity Ge and Si IR detectors coming full circle. Having started in the early 1950's with an emphasis on impurity doped Si and proceeding in the late 1950's and 1960's through a number of impurity doped Ge detectors, once again, some 20 years later, emphasis has switched to impurity doped Si.

II. Impurities in Germanium and Silicon

The subject of impurities in Ge and Si has been reviewed several times[12–16]

[9] W. J. Moore and H. Shenker, *Infrared Phys.* **5**, 99 (1965).
[10] R. A. Soref, *J. Appl. Phys.* **38**, 5201 (1967).
[11] R. A. Soref, *IEEE Trans. Electron Devices* **ED-15**, 209 (1968).
[12] E. Burstein, G. Picus, and N. Sclar, in *Photoconductivity Conf.* (R. G. Breckenridge *et al.*, eds.), p. 353. Wiley, New York, 1956.

and some of the material presented in this part, and also in Part IV, is repetitive of those previous reviews. However, for completeness, it was thought best to include a brief discussion and present the pertinent data in tables and graphs so that the interested reader will not have to refer back to previous literature. The type of data presented will be that which has importance to the preparation of IR detectors from these materials.

3. Model for Substitutional Impurities

For infrared detectors, we are concerned exclusively with substitutional impurities, that is, impurities that replace a Ge or Si atom in the crystal lattice. Such impurity atoms will behave as donors or acceptors depending on whether their number of valence electrons is greater than or less than four.

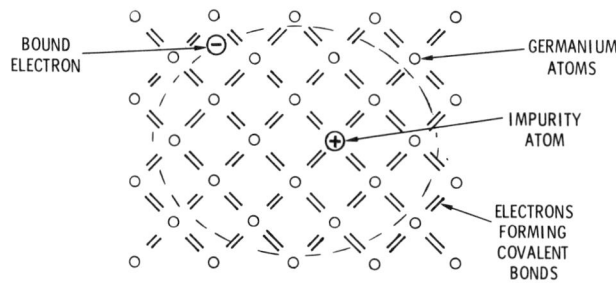

(a) Donor Impurity Atom

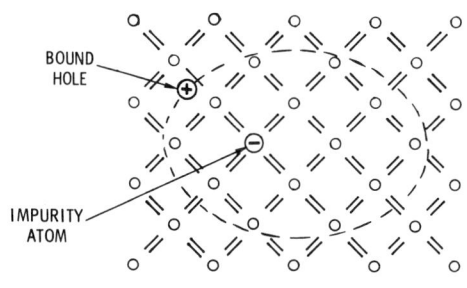

(b) Acceptor Impurity Atom

FIG. 1. Schematic diagram of impurity centers in Ge or Si.

---

[13] W. C. Dunlap, Jr., *Progr. Semiconduct.* **2**, 165 (1957).
[14] T. H. Geballe, in "Semiconductors" (N. B. Hannay, ed.), Chapter 8. Van Nostrand-Reinhold, Princeton, New Jersey, 1959.
[15] M. L. Schultz, *Infrared Phys.* **4**, 93 (1964).
[16] R. Dalven, *Infrared Phys.* **6**, 129 (1966).

Thus, atoms of elements from columns VA of the periodic table produce donor centers while those from the IIIA column produce acceptor centers. A schematic model of the nature of these centers is shown in Fig. 1.

Germanium atoms, with four valence electrons, form covalent bonds by sharing one electron with each of four neighboring atoms. If a group V atom such as As, with a valence of five, is substituted for a Ge atom in the crystal lattice, then four of its valence electrons will be taken up by covalent bonding, but one electron will be left over. This electron will then be loosely bound to the impurity center by the extra positive charge on the impurity atom core. Such a situation is illustrated by Fig. 1a.

In the case of a group III atom, such as B, with a valence of three, there will be one electron lacking from the number necessary to complete the covalent bonding arrangement. However, an electron from a nearby Ge–Ge bond can be accepted to complete the bonding of the impurity atom. Thus, a hole is created which will be loosely bound by the Coulomb attraction of an extra negative charge on the impurity center. This is illustrated by Fig. 1b.

This type of center with a single electron (or hole) bound to a positive (or negative) "nucleus" is in many ways equivalent to a hydrogen atom. Indeed, the theory of the hydrogen atom, with suitable modifications, forms the basis for our understanding of these centers.

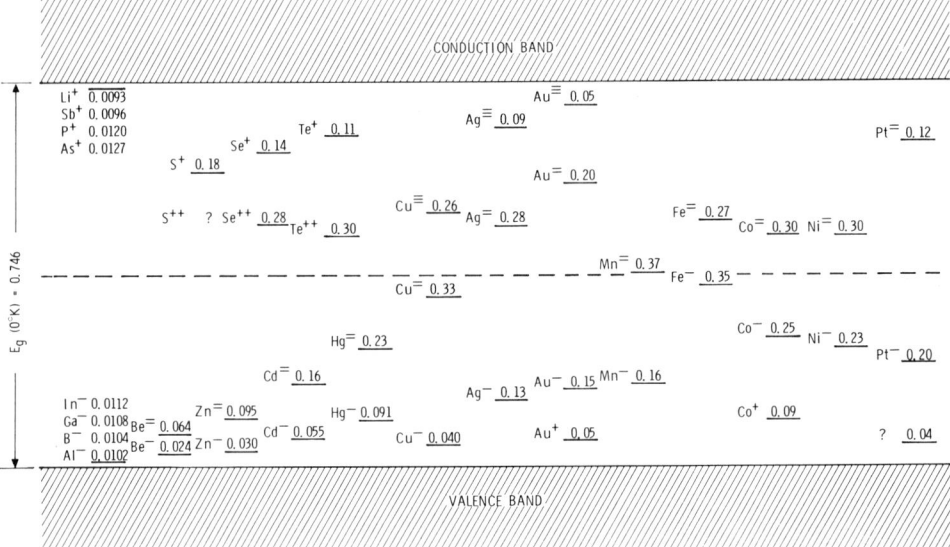

FIG. 2. Energy levels of impurities in Ge. Energy values are measured in electron volts from the nearest band edge.

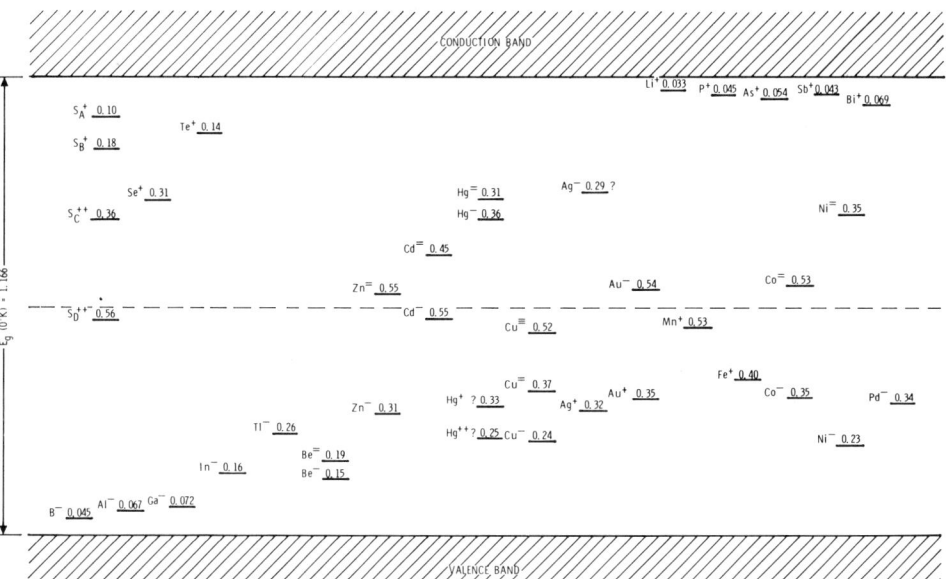

FIG. 3. Energy levels of impurities in Si. Energy values are measured in electron volts from the nearest band edge.

Atoms of elements in column VIA (S, Se, Te) have two electrons more than required for the four covalent bonds with neighboring Ge atoms and therefore, will behave as double donor centers. Similarly, those in column IIB (Zn, Cd, Hg) are double acceptors. A pseudohelium atom model would be appropriate for these types of centers.[16a] In addition, there are some impurities that are triple acceptors (for example, Au in Ge) and some that seem to have a variety of states because of the formation of complexes between two or more impurity atoms or with other types of impurities within the host lattice (for example, S in Si).

Figures 2 and 3 show the location of the various impurity energy levels in the forbidden band for both Ge and Si. The distance of each impurity in electron volts from the nearest bandedge is also given in both of these figures and its charge state after ionization is noted by plus and minus signs.

4. Absorption of IR Radiation by Impurity Atoms

Infrared radiation is absorbed by an impurity atom when an incoming photon has sufficient energy to excite an electron residing in the ground state of a donor-type impurity atom into either an excited state or into the conduction band where it is completely free of the impurity atom's binding

[16a] A. Glodeanu, *Phys. Status Solidi* **19**, 343 (1967).

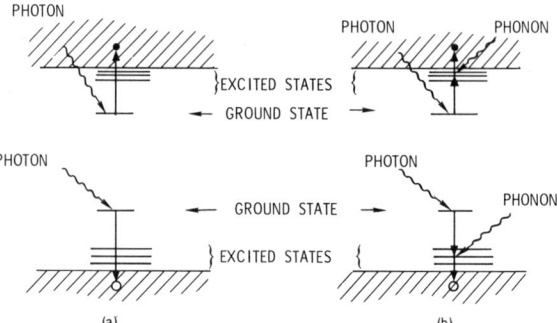

FIG. 4. Diagram showing excitation of an electron or hole from impurity atom ground state to a free state in conduction or valence band. (a) Photon excited transitions, (b) photon excited transitions to an excited state, followed by a phonon stimulated transition to a free state.

forces. In the case of an acceptor-type impurity atom the electron is excited from the valence band into the impurity atom ground state or, conversely, we can think of a hole being excited out of the ground state into an excited state or into the valence band. These photon induced transitions are represented schematically in Fig. 4.

These transitions are normally single photon events, however, cooperative transitions involving absorption of first a photon and then a phonon, resulting in a free electron or hole are also possible. In this case, the photon provides the energy to lift the electron into an excited state of the impurity atom and the phonon provides an additional energy boost to free the electron from the excited state. This type of process leads to what has been called photothermal conductivity and will be discussed further in Section 14.

If the photon energy hv is greater than the impurity atom binding energy E_i, then the electron will be freed to contribute to photoconductivity in the host crystal. Thus, we have the simple criteria $hv \geq E_i$ releases an electron, $hv < E_i$ does not (assuming the absence of photothermal effects). In terms of photon wavelength, this leads to the conclusion that all wavelengths such that $\lambda \leq 1.24/E_i$ may free an electron from the impurity atom, where λ is expressed in micrometers and E_i in electron volts.

Whether an incoming photon gets absorbed by an impurity atom depends on the photoionization cross section $\sigma_i(\lambda)$ and the concentration of impurity atoms N_I. The product of these two quantities is the absorption coefficient α,

$$\alpha(\lambda) = \sigma_i(\lambda)N_I. \tag{1}$$

Various attempts have been made at theoretical calculations of the photoionization cross section and its wavelength dependence. Lax[17] took the

---

[17] M. Lax, in *Photoconductivity Conf.* (R. G. Breckenridge et al., eds.), p. 111. Wiley, New York, 1956.

theory for the cross section of the hydrogen atom and adapted it to suit the case of a "hydrogen-like" center in a semiconductor crystal. This theory agrees with experimental data only for the shallowest impurities such as B in Si or the group III and V impurities in Ge. Lucovsky[18] took a different approach by assuming that the potential energy function of the impurity atom had a delta-function character rather than the Coulombic character of the hydrogen model. He derived an expression for the photoionization cross section that provided a good fit to the very deep impurities such as In in Si.

Finally, Bebb and Chapman[19-21] provided a more general approach by adopting the "quantum defect" method from atomic spectroscopy and combining it with the "effective mass" methods of solid state physics. Their analytical expressions contain both the Lax and Lucovsky results as limiting cases (i.e., the cases of either very shallow or deep impurities, respectively) and also provides solutions for all in-between cases.

The starting point in the quantum defect method is the observed value of impurity binding energy E_i (obs) which is expressed by

$$E_i(\text{obs}) = -R^*/\beta^2, \qquad (2)$$

where R^* is the hydrogenic Rydberg $e^2/2\varepsilon a^*$ and β is an adjustable parameter between 0 and 1. The quantity a^* is the effective Bohr radius in the crystal $a^* = a_0 \varepsilon m/m^*$ where $a_0 = \hbar^2/me^2$. An expression for the photoionization cross section is derived which is

$$\sigma_i(h\nu) = F a(h\nu), \qquad (3)$$

where $a(h\nu)$ is the atomic ionization cross section and the factor F, given by

$$F = \frac{1}{\sqrt{\varepsilon}} \left(\frac{\mathscr{E}_e}{\mathscr{E}_0}\right)^2 \left(\frac{a^*}{a_0}\right)^2, \qquad (4)$$

accounts for the fact that the atom is imbedded in a dielectric medium. The quantity $\mathscr{E}_e/\mathscr{E}_0$ is the ratio of the electric field at the impurity atom to the average field in the medium and ε is the optical dielectric constant. Figure 5 shows comparisons of theory with experiment for B, Al, Ga, and In in Si and for Hg in Ge. These curves are plotted as normalized cross section $\sigma(h\nu)E_i$ that has units of eV-cm^2 versus normalized photon energy which is dimensionless. The value of E_i used in each case is shown on the figures. The F and μ are adjustable parameters in the theory and the values chosen

---

[18] G. Lucovsky, *Solid State Commun.* **3**, 299 (1965).
[19] H. B. Bebb and R. A. Chapman, *J. Phys. Chem. Solids* **28**, 2087 (1967).
[20] H. B. Bebb, *Phys. Rev.* **185**, 1116 (1969).
[21] H. B. Bebb and R. A. Chapman, *Proc. 3rd Int. Conf. Photoconductivity* (E. M. Pell, ed.), p. 245. Pergamon, Oxford, 1971.

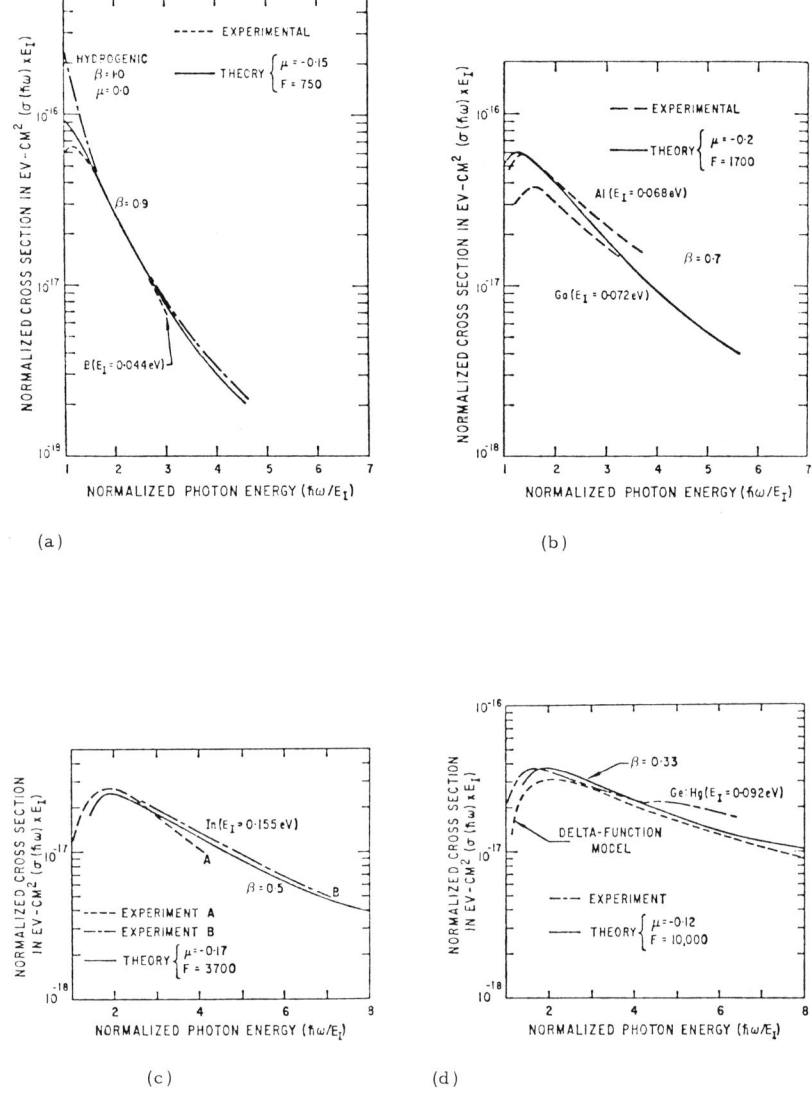

FIG. 5. Comparison of experimental and theoretical photoionization cross sections for impurities in Ge and Si. (a) B in Si; also shown is the theoretical curve for Lax's hydrogenic model. (b) Al and Ga in Si. (c) In in Si. (d) Hg in Ge; also shown is the theoretical curve for Lucovsky's delta function model. (After Bebb and Chapman.[19])

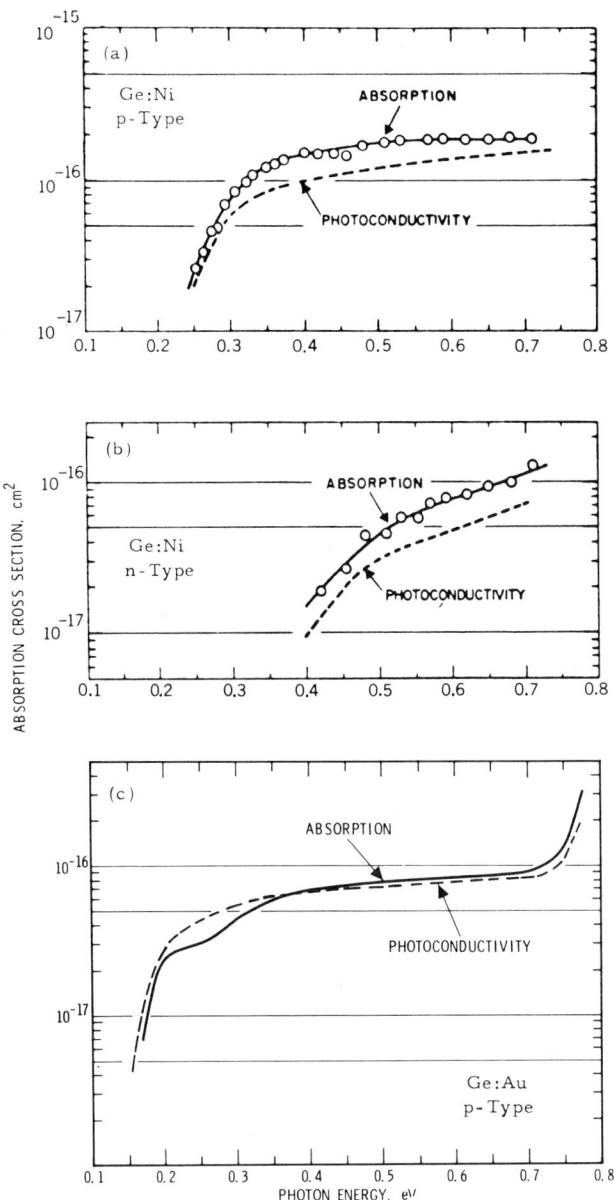

FIG. 6. Photoionization absorption spectra at 77 K of Ni and Au in Ge. (a) p-type Ge:Ni with 6.4×10^{15} neutral Ni centers, (b) n-type Ge:Ni^{II} with 6.1×10^{15} doubly charged Ni centers, (c) p-type Ge:Au with 6.8×10^{15} neutral Au centers. Photoconductivity spectra also shown for comparison. (Ge:Ni data after Newman and Tyler.[22])

[22] R. Newman and W. W. Tyler, *Solid State Phys.* **8**, 49 (1959).

TABLE I

Experimental Values of Photoionization Cross Section for Impurity Atoms in Ge and Si

Impurity atom	Cross section at wavelength peak in cm^2					
	Si	λ_p (μm)	Ref.	Ge	λ_p (μm)	Ref.
B	1.4×10^{-15}	23	a			
Al	8×10^{-16}	15	a			
Ga	5×10^{-16}	15	a			
In	3.3×10^{-17}	4.3	b			
Tl				8.7×10^{-15}	95	g
P	1.7×10^{-15}	27	c	1.5×10^{-14}	98	h
As				1.1×10^{-14}	89	h
Sb				1.6×10^{-14}	124	h
Bi				1.5×10^{-14}	98	h
Zn^I	$\sim 10^{-16}$	2.5	d,e			
Zn^{II}	4.6×10^{-17}	1.4	d			
Hg				3.9×10^{-16}	10	i
Cu				1.0×10^{-15}	23	j
Au				8×10^{-17}	2	k
S(0.61C)	6×10^{-17}	1.8	f			

[a] Burstein et al.[24]
[b] Messinger and Blakemore.[25]
[c] Aggarwal and Ramdas.[26]
[d] Herman and Sah.[27]
[e] Zavadskii and Kornilov.[28]
[f] Sah et al.[29]
[g] Jones and Fisher.[30]
[h] Reuszer and Fisher.[31]
[i] Chapman and Hutchinson.[32]
[j] Kaiser and Fan.[4]
[k] Johnson and Levinstein[23] (corrected for a Au atom concentration of 6.8×10^{15} cm^{-3}).

[23] L. F. Johnson and H. Levinstein, *Phys. Rev.* **117**, 1191 (1960).
[24] E. Burstein, G. Picus, B. Henvis, and R. Wallis, *J. Phys. Chem. Solids* **1**, 65 (1956).
[25] R. A. Messinger and J. S. Blakemore, *Solid State Commun.* **9**, 319 (1971).
[26] R. L. Aggarwal and A. K. Ramdas, *Phys. Rev.* **137**, A602 (1965).
[27] J. M. Herman, III, and C. T. Sah, *J. Appl. Phys.* **44**, 1259 (1973).
[28] Yu. I. Zavadskii and B. V. Kornilov, *Phys. Status Solidi* **42**, 617 (1970).
[29] C. T. Sah, T. H. Ning, L. L. Rosier, and L. Forbes, *Solid State Commun.* **9**, 917 (1971).
[30] R. L. Jones and P. Fisher, *Solid State Commun.* **2**, 369 (1964).
[31] J. H. Reuszer and P. Fisher, *Phys. Rev.* **135**, A1125 (1964).
[32] R. A. Chapman and W. G. Hutchinson, *Solid State Commun.* **3**, 293 (1965).

for best fit are also shown on the figures. The factor F was defined in Eq. (4) and μ is the quantum defect parameter.

The quantum defect theory is able to provide an excellent fit to most experimental data, probably better than the experimental accuracy of the data itself. However, it does not appear to be applicable to the case of transition metal impurities such as Mn, Ni, and Fe that produce very deep levels in both Ge and Si, nor to the case of Au in Ge. These impurity centers have photoionization cross section curves that are much different in shape from those shown previously. Examples are shown in Fig. 6 for Ni and Au in Ge. Note the gradual decrease in cross section as the photon energy approaches the impurity binding energy. This is in marked contrast to the previous examples which exhibit a peak in the cross section as photon energy decreases.

Neither has a detailed theoretical analysis been applied to the case of photoionization of electrons out of the upper levels of multileveled acceptor atoms in which the lower levels have been filled with electrons from compensating donors. This situation again gives rise to a cross section curve that decreases monotonically as photon energy decreases. A further discussion of this situation is given in Section 11.

Table I lists values of absorption cross section obtained experimentally for a number of different impurities in Ge and Si. For the most part these are the peak values at the indicated wavelengths. In cases where there is no peak in the cross section curve, we have listed the value at one-half the long wavelength threshold (or twice the binding energy). It is interesting to note that atoms with similar ionization energies have comparable photoionization cross sections regardless of the host lattice. Compare, for example, Ge:Cu with Si:B and Si:P; Ge:Hg with Si:Al and Si:Ga. Also, as would be expected, atoms with a smaller ionization energy have a larger photoionization cross section. The relation $\sigma_i = 2.5 \times 10^{-18}/E_i^2$ is a rough approximation which may be useful for estimating photoionization cross sections when only the ionization energy is known.

5. Quantum Efficiency

Consider a photon flux density of J photons/cm²/sec incident on a plane parallel slab of doped Ge or Si of thickness d. The absorption coefficient of the slab is α and the reflectivity at front and back surfaces is r. Then, allowing for multiple internal reflections between front and back surfaces, it can be shown that the total photon flux absorbed within the slab J_a is given by

$$J_a = J\left\{\frac{(1-r)[1-\exp(-\alpha d)]}{1-r\exp(-\alpha d)}\right\}. \tag{5}$$

Assuming the internal quantum yield is unity, that is, each photon absorbed by an impurity atom creates only one electron or hole as the case may be, then we define the quantum efficiency η as the fraction of the total incident flux absorbed in the slab, hence

$$\eta = \frac{J_a}{J} = \left\{ \frac{(1-r)[1-\exp(-\alpha d)]}{1 - r\exp(-\alpha d)} \right\}. \tag{6}$$

The quantity η thus defined is also sometimes called the "responsive quantum efficiency." In this review, we will simply use the briefer expression "quantum efficiency." By definition, η is the efficiency of the process whereby a photon stream incident on the detector crystal is converted to free electrons (or holes) inside the crystal. It has the true character of an efficiency since its values may range only between zero, corresponding to no photons absorbed in the crystal, and unity, corresponding to absorption of all photons incident on the crystal. In the event that there are competing absorption processes that do not lead to creation of electrons (or holes) such as, for example, lattice absorption, then the equation for η must be modified. This can easily be done[33]; however, the resulting expression is somewhat cumbersome and will not be needed here because the effect of competing absorption processes is usually quite small excepting for certain spectral bands where the lattice absorption becomes significant.

To optimize detector performance it is desirable to maximize the quantum efficiency. From Eq. (6) it is clear that this can be done by making the absorption coefficient as large as possible and the reflectivity as small as possible. By Eq. (1) a large absorption coefficient would be achieved by adding a large concentration of impurity atoms to the host crystal. However, it turns out that there are limits to the amount of impurity doping that can be achieved in practice. This limit may be set by the maximum solid solubility of the impurity atom in the host crystal or by impurity banding effects which take place when the electron orbits of neighboring impurity atoms begin to overlap. Further discussion of these limits is given in Parts IV and V.

Practical values of α for optimized IR detectors are in the range 1–10 cm^{-1} for Ge and 10–50 cm^{-1} for Si. Thus, to maximize quantum efficiency, the thickness of the detector crystal should be not less than about 0.5 cm for doped Ge and about 0.1 cm for doped Si.

The reflectivity at the front surface can be minimized by application of a suitable antireflection coating. The reflectivity at the back surface should be maximized so as to redirect the photon flux back through the crystal for added absorption. This has sometimes been done in the past by fabricating the detector crystal with the back surface shaped as a "roof top" prism to

---

[33] E. Burstein, G. Picus, and N. Sclar, *in Photoconductivity Conf.* (R. G. Breckenridge *et al.*, eds.), p. 398. Wiley, New York, 1956.

take advantage of total internal reflection. Of course, in this situation Eq. (6) no longer applies since it was derived for equal reflectivities at front and back surfaces. For the case where r is zero at the front surface and unity at the back surface, the quantum efficiency is

$$\eta = 1 - \exp(-2\alpha d). \qquad (7)$$

Thus, in this case, if the αd product can be made equal to one, a quantum efficiency of 86% might be obtained.

Following absorption of photons of IR radiation and the creation of free electrons or holes, the added charge carriers may move through the crystal and subsequently recombine at other ionized impurity atom sites. This produces a change in conductivity of the crystal called the photoconductivity which is the topic of the next part.

III. Impurity Photoconductivity

The theory for impurity photoconductivity was developed many years ago and has been adequately expounded by several authors. Particularly good discussions are given by Burstein et al.,[12] Newman and Tyler,[22] Bube[34] and Putley.[35] The theoretical ideas will be reviewed in this section along with the theories of noise and ultimate sensitivity limits for this type of detector. This will provide a basic foundation for understanding the performance characteristics of actual detectors to be described in Part V.

6. Photoconductivity

Impurity photoconductivity is simplified because it is a majority carrier process; only one carrier type need be considered, either electrons or holes, not both. Additional assumptions usually made in the theoretical analysis are (1) each photon absorbed by an impurity atom excites one and only one charge carrier from a bound state on the impurity atom to a free state in the conduction or valence band, (2) the impurity atoms are distributed uniformly throughout the crystal, (3) the crystal has "Ohmic contacts" attached to it so that majority charge carriers may easily pass in or out at these contacts, and (4) trapping effects are absent. It turns out that all of these assumptions generally conform with the real situation. There have been some cases where contact resistance has presented a problem, but these can be corrected by making better contacts. Also, trapping effects have been observed under certain circumstances. However, this is not a common occurrence and need not be of concern here. For the purpose of this part, the assumptions as stated above are considered to hold.

[34] R. H. Bube, "Photoconductivity of Solids," Chapter 3. Wiley, New York, 1960.
[35] E. H. Putley, *Phys Status Solidi* **6**, 571 (1964).

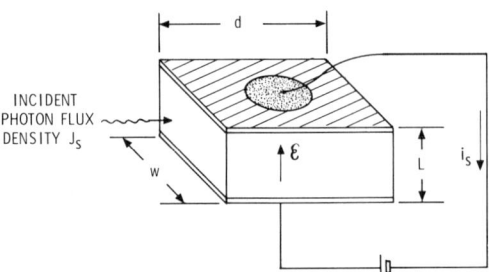

FIG. 7. Simple circuit for photoconductivity.

The equations for photocurrent can be rigorously derived from fundamental charge continuity and current equations and such a derivation can be found in a paper by Rittner.[36] A better physical insight for the process is obtained by adopting a less rigorous, phenomenological approach, which we do here.

Consider the photoconducting crystal of Ge or Si as shown in Fig. 7. The crystal has metal contacts applied on two opposing faces and an electric field $\mathscr{E}$ is impressed across it. Let a single photon enter the front face of area $A = wL$, be absorbed by a neutral impurity atom and create one free charge carrier. This free charge carrier can be said to have a "lifetime" τ during which it may move a distance L_d (called the "drift length") in the direction of the applied field. After this time, the free charge carrier recombines with some other ionized impurity atom and disappears. During this process, an amount of charge will be moved through the external circuit equal to eL_d/L. This must happen in order to preserve charge neutrality in the complete circuit. Now if the charge carrier has a drift velocity $v_d = \mu\mathscr{E}$, where μ is the drift mobility and $\mathscr{E}$ the electric field strength, then $L_d = v_d\tau = \mu\mathscr{E}\tau$ and the charge moved through the external circuit is

$$e(L_d/L) = e(\mu\mathscr{E}\tau/L). \tag{8}$$

Next, consider a steady flux density of photons J incident on the crystal face giving rise to JA photons/sec on the crystal. A fraction η of these will be absorbed and generate free charge carriers. This steady photon stream therefore leads to a steady current in the external circuit, the photocurrent, given by

$$i_s = JA\eta e(\mu\mathscr{E}\tau/L). \tag{9}$$

This is actually the "short circuit" photocurrent. In a practical situation, impurity photoconductors would not be operated this way. However, the

[36] E. H. Rittner, in Photoconductivity Conf. (R. G. Breckenridge et al., eds.), p. 215. Wiley, New York, 1956.

above equation still holds and represents a useful quantity for subsequent theoretical analysis.

The quantity in parentheses in Eq. (8) is a dimensionless ratio called the photoconductive gain G_{pc},

$$G_{pc} = L_d/L = \mu \mathscr{E} \tau / L. \tag{10}$$

The idea of photoconductive gain was put forth by Rose[37] as a simplifying concept for the understanding of photoconductive phenomona and is now widely used in the field. Inspection of Eq. (9) shows that the photoconductive gain is the ratio of the number of free charge carriers per second passing around the circuit to the number of photons absorbed per second. Thus, the detector can be thought of as a transducer which converts a photon stream into an electron current. The product ηG_{pc} gives the efficiency of this transduction process.

The photoconductive gain can be less than or greater than unity depending upon whether the drift length L_d is less than or greater than interelectrode spacing L. Values of L_d greater than L imply that a free charge carrier swept out at one electrode is immediately replaced by injection of an equivalent free charge carrier at the opposite electrode. Thus, a free charge carrier will continue to circulate until recombination at an ionized impurity atom takes place.

The IR detector designer attempts to optimize detector photocurrent by maximizing the ηG_{pc} product. We have previously discussed how the quantum efficiency may be maximized. Equation (10) shows that, to obtain a large photoconductive gain, one desires high carrier mobility and lifetime, electric field strength in the crystal as large as possible, and a minimum interelectrode spacing. The mobility is usually limited by neutral impurity scattering, therefore, since high impurity concentrations are essential to obtain a high quantum efficiency, little can be done to increase the mobility. The lifetime can be maximized by the elimination of compensating impurities of the opposite type from the major doping impurity. This reduces the number of ionized impurity sites that are the recombination centers at which free charge carriers terminate their lifetime. The limitation on electric field strength is impact ionization breakdown. This is produced when free charge carriers accelerated by the electric field attain kinetic energies sufficient to ionize neutral impurities by inelastic collisions. It usually occurs at some critical electric field strength and is accompanied by a precipitous drop in detector resistance and complete loss of photosignal. The interelectrode spacing is determined by the required geometry of the detector's sensitive area and is therefore not a freely variable parameter. However,

[37] A. Rose, *RCA Rev.* **12**, 362 (1951); *Proc. IRE* **43**, 1850 (1955).

if a choice is available, one usually selects the smallest dimension for the interelectrode spacing.

Some representative values of $\mu, \mathscr{E}, \tau$, and L might be as follows: $\mu = 3 \times 10^4$ cm^2/V-sec, $\mathscr{E} = 200$ V/cm, $\tau = 3 \times 10^{-8}$ sec, and $L = 0.1$ cm. Inserting these in Eq. (10) gives a photoconductive gain of 1.8. The range of values observed in doped Ge and Si detectors is normally between 0.01 and 10.

a. Detector as a Circuit Element

In practical use, the photoconductive IR detector circuit is not that of Fig. 7, but is as shown in Fig. 8. The detector is connected in series with a load resistor R_L and a source of direct current such as a battery V_b. A current (called the dark current) flows through the circuit even in the absence of signal radiation. This current results because the detector resistance is finite. Free charge carriers are present due to thermal generation or background radiation. Surface leakage conductance may also contribute to the dark current. The photocurrent produced by incoming signal photons is usually very small compared to the dark current, so one is faced with the problem of measuring very small currents in the presence of a much larger direct current. This is done by modulating the signal photon flux either with a chopper blade or by a scanning mirror. By ac coupling of the detector circuit to the preamplifier, the large direct current is blocked out and only the fluctuating signal current is measured.

Consider now the circuit of Fig. 8. The direct current through it is given by

$$I = V_b/(R + R_L). \tag{11}$$

The voltage drop across the detector is

$$V = IR \tag{12}$$

and a small change in R due to signal radiation will produce a small change in V, which we call the signal voltage, given by

$$\Delta V = I\, \Delta R + \Delta I R. \tag{13}$$

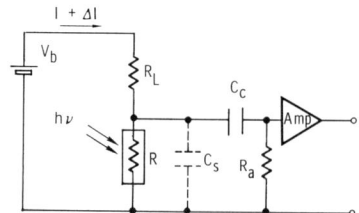

FIG. 8. Practical detector circuit.

By differentiation of Eq. (11), we have

$$\Delta I = -I \, \Delta R/(R + R_L) \tag{14}$$

and when this is substituted into Eq. (13) one obtains

$$\Delta V = IR_L \, \Delta R/(R + R_L), \tag{15}$$

which can also be written as

$$\Delta V = I \frac{\Delta R}{R} \frac{RR_L}{(R + R_L)}. \tag{16}$$

If the signal voltage is taken across the load resistor rather than the detector, as is sometimes the case, then the same result as Eq. (16) is obtained, but with the sign of the signal voltage reversed. Equation (16) is valid provided that the amplifier input resistance R_a is much greater than the detector resistance.

Burstein et al.[5] have pointed out that Eq. (16) holds rigorously only for an "Ohmic" photoconductor, that is, one whose resistance is independent of applied voltage. Most impurity Ge and Si detectors are decidedly "non-Ohmic" and in this case, Eq. (16) should be replaced by

$$\Delta V = I \frac{\Delta R}{R_{dc}} \frac{R_{ac} R_L}{(R_{ac} + R_L)}. \tag{17}$$

R_{ac}, given by dV/dI, is the ac resistance of the detector. R_{dc}, given by V/I, is the dc resistance. A relation between R_{ac} and R_{dc} can be given as

$$R_{ac} = R_{dc} \left[1 - \frac{V}{R_{dc}} \frac{dR_{dc}}{dV} \right]^{-1}. \tag{18}$$

For the derivations of Eqs. (17) and (18), the reader is referred to an article by Putley.[35] Similar equations have been derived by Wallis and Shenker,[38] using conductance in place of resistance.

From Eq. (17) it is apparent that maximum signal voltage will be fed to the amplifier when $R_L \gg R_{ac}$. However, because R_{ac} may typically be 0.1 to 1 MΩ, it could be impractical to satisfy this condition because rather large battery voltages are required to obtain the necessary current through the detector. Thus, one usually settles for $R_L \leqslant R_{ac}$. The case where $R_L = R_{ac}$ is called the "matched load" condition and gives a signal voltage to the amplifier of one-half the maximum.

The frequency response of the circuit of Fig. 8 is sometimes determined by the distributed shunt capacitance C_s and not the detector's frequency

---

[38] R. F. Wallis and H. Shenker, NRL Rep. 5996, U.S. Naval Res. Lab., August 1963.

response. If this is the case, the signal voltage dependence on frequency can be expressed by

$$\Delta V(f) = I \frac{\Delta R}{R_{dc}} \frac{R_{ac}R_L}{[(R_{ac} + R_L)^2 + (2\pi f C_s R_{ac} R_L)^2]^{1/2}}. \quad (19)$$

This gives a signal voltage that decreases for frequencies greater than

$$f = [2\pi C_s R_{ac} R_L/(R_{ac} + R_L)]^{-1}. \quad (20)$$

Another important thing to note about the detector as a circuit element is that the the voltage fed to the amplifier is proportional to $I\,\Delta R/R_{dc}$ and this quantity is identical to the short circuit photocurrent:

$$i_s \equiv -I\,\Delta R/R_{dc}. \quad (21)$$

Furthermore, since

$$R_{dc} = L/ne\mu wd, \quad (22)$$

where n is the concentration of free charge carriers in the detector and the other symbols are as previously defined, we have

$$-\Delta R/R_{dc} = \Delta n/n \quad (23)$$

and

$$i_s = I\,\Delta n/n. \quad (24)$$

Although not immediately obvious, this expression is equivalent to Eq. (9). The following discussion will demonstrate this.

b. Dynamical Properties

Up to this point in the discussion of photoconductivity, we have considered just the steady values of photocurrent after a sample has been illuminated with external radiation. Photoexcitation is actually a dynamic process; free charge carriers are continually being generated by absorbed photons and then disappearing by recombination at charged impurity sites. Other generation processes may also be taking place, such as generation by absorption of phonons (thermal generation). A further study of this dynamical situation will serve to elucidate some important properties of impurity Ge and Si detectors.

The approach taken here is adapted from Burstein et al.[12] and Putley.[35] We start with the rate equation which describes the change with time of free charge carrier concentration,

$$dn/dt = A_T(N_D - N_A - n) + A_I n(N_D - N_A - n) + A_0(N_D - N_A - n) + g \\ - B_T n(N_A + n) - B_I n^2(N_A + n) - B_0 n(N_A + n). \quad (25)$$

2. IMPURITY GERMANIUM AND SILICON INFRARED DETECTORS

The first four terms on the right-hand side represent generation rates and the last three are recombination rates. $A_T(N_D - N_A - n)$ is the rate of generation by phonons from the crystal lattice vibrations, $A_I n(N_D - N_A - n)$ is the rate of generation due to impact ionization by other free charge carriers, $A_0(N_D - N_A - n)$ is the generation rate due to internal photons and g is the generation rate due to external photons. The B terms represent the corresponding recombination rates. Thus, $B_T n(N_A + n)$ is the recombination rate wherein the excess carrier energy is given up as phonons to the crystal lattice, $B_I n^2(N_A + n)$ is the recombination rate involving transfer of energy to other free charges (Auger effect), and $B_0 n(N_A + n)$ is the recombination rate with emission of photons. The assumption is that we are dealing with an n-type material, N_D is the donor concentration, and N_A is the compensating acceptor concentration. Thus $(N_D - N_A - n)$ is the concentration of neutral donors from which free electrons are generated and $(N_A + n)$ is the concentration of ionized donors with which they recombine. Similar relations hold for p-type material and can be obtained by substituting p for n and interchanging N_A with N_D.

In the absence of external radiation and in the steady state such that $dn/dt = 0$, the principle of detailed balance requires that the individual generation and recombination processes be in equilibrium with each other. Thus

$$A_0(N_D - N_A - n) = B_0 n(N_A + n),$$
$$A_T(N_D - N_A - n) = B_T n(N_A + n), \quad (26)$$
$$A_I n(N_D - N_A - n) = B_I n^2(N_A + n),$$

and

$$A_0/B_0 = A_T/B_T = A_I/B_I = n(N_A + n)/(N_D - N_A - n) = K_e, \quad (27)$$

where K_e is the equilibrium constant for the processes. From semiconductor statistical analysis, it can be shown that [39]

$$K_e = (2/\delta)(2\pi m^* kT/h^2)^{3/2} \exp(-E_i/kT), \quad (28)$$

where δ is the ground state degeneracy of the donor impurity, with the provision that $kT \ll E_i$, which is generally a very good approximation at the low temperature of operation of impurity photodetectors. Also, at these low temperature $n \ll N_D, N_A$ so that a combination of Eqs. (27) and (28) yields

$$n = \frac{(N_D - N_A)}{N_A} \frac{2}{\delta} \left(\frac{2\pi m^* kT}{h^2}\right)^{3/2} \exp\left(\frac{-E_i}{kT}\right). \quad (29)$$

This gives the dependence of thermal equilibrium charge carriers on impurity

[39] J. S. Blakemore, "Semiconductor Statistics," Chap. 2. Pergamon, Oxford, 1962.

concentrations and temperature. Additional charge carriers may be added to the crystal by absorption of external radiation or by impact ionization. These can be determined in the following manner.

(1) *Generation by External Radiation.* Consider the case of low temperature and low electric field strengths where the carrier concentration is produced only by external photons, i.e., all the A coefficients are zero. Furthermore, calculations have shown that recombination accompanied by photon emission is extremely unlikely[40] compared with the other recombination mechanisms; thus, the B_0 term can also be neglected. The impact recombination coefficient B_I is also found to be small,[40] so that the recombination rate from this process is negligible excepting for situations where a large carrier concentration is present. The phonon recombination process represented by the B_T term in Eq. (25) has been considered in great detail by Lax[41,42] and by Ascarelli and Rodriguez[43,44] and Brown and Rodriguez.[45] They have treated theoretically a model that consists of the initial trapping of a free charge in one of the excited states of the ionized impurity atom with the emission of a single phonon. By the further emission of phonons, the trapped charge gradually cascades down to the ground state and is permanently captured until freed again by one of the generation processes already discussed. The theory also includes the possibility of the trapped charge absorbing a phonon while still in an excited state and being freed. Theoretical and experimental results indicate that this cascade recombination process is the dominant mechanism for free electron or hole recombination at ionized impurities in Ge and Si.

With these assumptions, Eq. (25) reduces to the simple form

$$dn/dt = g - B_T n(N_A + n). \tag{30}$$

A rate equation such as this can always be written in the form[46]

$$dn/dt = g - (n/\tau), \tag{31}$$

where τ is the mean steady state bulk lifetime of the free carriers generated and is the same lifetime previously mentioned in Section 6. In the steady state, generation and recombination rates are equal so that $dn/dt = 0$, and

$$n = g\tau. \tag{32}$$

[40] N. Sclar and E. Burstein, *Phys. Rev.* **98**, 1757 (1955).
[41] M. Lax, *J. Phys. Chem. Solids* **8**, 66 (1959).
[42] M. Lax, *Phys. Rev.* **119**, 1502 (1960).
[43] G. Ascarelli and S. Rodriguez, *Phys. Rev.* **124**, 1321 (1961).
[44] G. Ascarelli and S. Rodriguez, *Phys. Rev.* **127**, 167 (1962).
[45] R. A. Brown and S. Rodriguez, *Phys. Rev.* **153**, 890 (1967).
[46] J. S. Blakemore, "Semiconductor Statistics," Chap. 4. Pergamon, Oxford, 1962.

It should be noted that τ may also be a function of n so that the specification of n for a given external generation rate g is incomplete until the expression for τ is inserted. Thus, in the case of Eq. (30) we would use

$$\tau = [B_T(N_A + n)]^{-1}. \tag{33}$$

However, in most practical cases, it turns out that $n \ll N_A$ so that Eq. (32) becomes

$$n = g/B_T N_A. \tag{34}$$

The generation rate is proportional to quantum efficiency and external photon flux density and is given by

$$g = \eta J/d. \tag{35}$$

The recombination coefficient is given by

$$B_T = \langle v \rangle \sigma_c, \tag{36}$$

where v is the free carrier velocity, σ_c is the capture cross section of the recombination center and the angle brackets indicate an average over the distribution of free carrier energies in the crystal.[47]

Next we consider what happens when a small change in generation rate is induced in the crystal due to the arrival of signal photons. This will produce an additional concentration of free carriers Δn, whose evolution with time is given by

$$d\,\Delta n(t)/dt = \Delta g - [\Delta n(t)/\tau]. \tag{37}$$

The solution to this equation depends upon the boundary conditions imposed by the nature of Δg. For the situation where signal radiation arrives as a square pulse, the solution is

$$\Delta n(t) = \Delta g\, \tau [1 - \exp(-t/\tau)] \tag{38}$$

for the rising portion of the pulse, and

$$\Delta n(t) = \Delta g\, \tau \exp(-t/\tau) \tag{39}$$

for the decay after the pulse is turned off. A "response time" is usually defined as the time interval taken for the excess carrier concentration to rise from zero to $1 - (1/e)$ of its maximum value, or to decay to $1/e$ from the maximum value. See Fig. 9. The response time defined in this way is then identical to the average free carrier lifetime.

---

[47] More precisely, $B_T = \langle v\sigma_c \rangle$. However, since the free carrier energy distribution function is frequently unknown, it has become conventional to use Eq. (36) and let $\langle v \rangle$ be represented by the average velocity for a Maxwell–Boltzmann distribution, i.e., $\langle v \rangle = (8kT/\pi m^*)^{1/2}$.

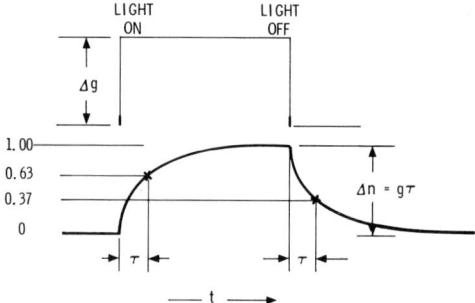

FIG. 9. Response of a detector to a square pulse of radiation.

The changes in $\Delta n(t)$ expressed by Eqs. (38) and (39) produce proportional changes in photocurrent given previously by Eq. (24). Thus, the rise and decay of photocurrent pulses can be used to determine free carrier lifetimes.

Another case of interest is when the signal photon flux on the detector is sinusoidally modulated at some frequency $f = \omega/2\pi$. In this case, the solution to Eq. (37) can be written

$$\Delta n(t) = \frac{\Delta g\, \tau[1 + \sin(\omega t - \phi)]}{2(1 + \omega^2\tau^2)^{1/2}}, \qquad (40)$$

where the phase angle ϕ is equal to $\tan^{-1}(\omega\tau)$.

It was mentioned previously that, at higher free carrier concentrations, recombination by the Auger process may become important. In such a case the rate equation becomes

$$dn/dt = g - B_1 n^2(N_A + n) - B_T n(N_A + n), \qquad (41)$$

which is nonlinear and does not yield a simple solution for n as a function of time. However, if we again consider only small changes in n the equation can be linearized and solved for Δn in the following way.

Consider Eq. (41) to be the equation for n at some time. Then let g be suddenly increased to a new value g_1. The new rate equation becomes

$$dn_1/dt = g_1 - B_1 n_1^2(N_A + n_1) - B_T n_1(N_A + n_1). \qquad (42)$$

Subtraction of Eq. (41) from Eq. (42) gives

$$\frac{d\,\Delta n}{dt} = \Delta g - \Delta n[2B_1 n(N_A + \tfrac{3}{2}n) + B_T(N_A + 2n)], \qquad (43)$$

where $\Delta g = g_1 - g$, $\Delta n = n_1 - n$, and $\Delta n \ll n_1$ and n so that $n_1 \approx n$. This equation is now linear in Δn, so for small signal levels, the rise and decay of photocurrents will again be exponential with a response time

$$\tau = [2B_I n(N_A + n) + B_T(N_A + n)]^{-1}. \tag{44}$$

This response time is dependent on the "dark" carrier concentration n and thus, different sample temperatures or background generation rates could possibly change the low level lifetime τ.

B_T is $\sim 10^{-6}$ cm^3/sec[48-50] at 8°K and Koenig[48] has reported the value for B_I to be $\sim 10^{-17}$ cm^6/sec in the temperature range 4–10°K. However, Ascarelli and Rodriguez[44] calculate a value of $B_I \sim 10^{-21}$ cm^6/sec, several orders of magnitude smaller. Norton and Levinstein[50] have found no evidence of Auger recombination for carrier concentrations up to 2×10^{11}/cm^3 and, on this basis, it would appear that $B_I \ll 10^{-17}$ cm^6/sec. Thus, carrier concentrations $> 10^{12}$/cm^3 (in fact $> 4 \times 10^{15}$/cm^3 by Ascarelli and Rodriguez's calculation[44]) would be required for the Auger recombination rate to be equal to the thermal recombination rate. Such a situation would be unusual, but might be encountered in a detector under intense illumination from a laser.

(2) *Generation by Impact Ionization.* At high electric field strengths, free carriers are accelerated and may acquire enough kinetic energy to ionize neutral impurity atoms by impact, thus creating additional free charge carriers. This effect was first recognized in Ge by Sclar et al.[51] and is manifested by a sharp increase in current through the crystal at some critical electric field strength $\mathcal{E}_c$. The effect is analogous in some respects to a gas discharge. It is a reversible and nondestructive type of ionization breakdown; however, it is detrimental in a photoconductor because the excess carriers generated by impact ionization overwhelm the photon generated carriers making observation of small photocurrents impossible. Impact ionization also produces excessive electrical noise due to the sporadic nature of the breakdown in different regions of the crystal.

Measurements by Sclar and Burstein[52] and others[53,54] established that, for shallow level impurities such as Ge:In with concentrations $\geq 1 \times 10^{15}$ cm^{-3}, the critical breakdown field varies linearly with impurity concentration. Figure 10 shows representative data. A theory was developed[52] by

---

[48] S. H. Koenig, *J. Phys. Chem. Solids* **8**, 227 (1957).
[49] S. H. Koenig, R. D. Brown, III, and W. Shillinger, *Phys. Rev.* **128**, 1668 (1962).
[50] P. Norton and H. Levinstein, *Phys. Rev.* **B6**, 489 (1972).
[51] N. Sclar, E. Burstein, W. J. Turner, and J. W. Davisson, *Phys. Rev.* **91**, 215 (1953).
[52] N. Sclar and E. Burstein, *J. Phys. Chem. Solids* **2**, 1 (1957).
[52a] S. Koenig and G. R. Gunther-Mohr, *J. Phys. Chem. Solids* **2**, 268 (1957).
[53] E. I. Abaulina-Zavaritzkaya, *Zh. Eksp. Teor. Fiz.* **36**, 1342 (1959) [*English Transl.: Sov. Phys.—JEPT* **9**, 953 (1959)].
[54] A. L. McWhorter and R. H. Rediker, *Proc. Int. Conf. Semicond. Phys., Prague, 1960*, p. 134. Czech. Acad. Sci., Prague and Academic Press, New York, 1961.

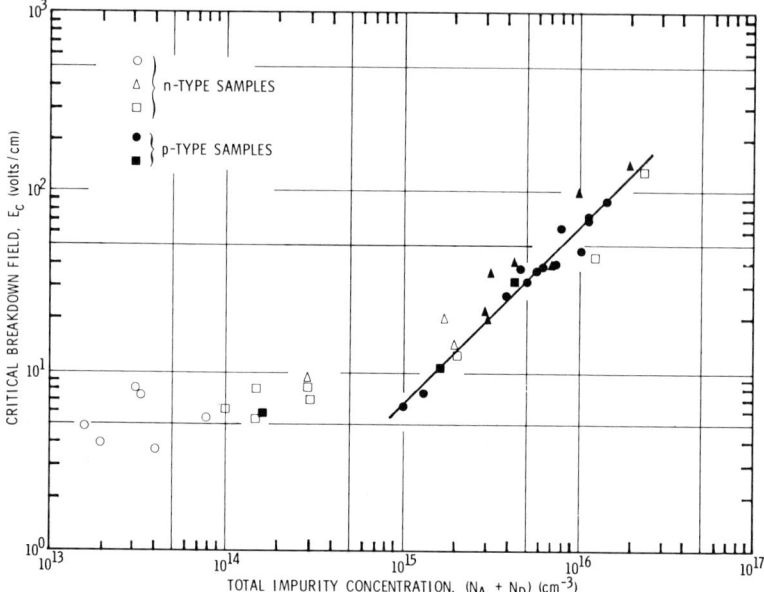

FIG. 10. Critical impact ionization breakdown field for shallow level impurities in Ge at 4 to 5°K. Data on n-type samples from Koenig et al.[49]; Sclar and Burstein[52]; and Koenig and Gunther-Mohr.[52a] Data on p-type samples from Sclar and Burstein[52]; McWhorter and Rediker[54]; and Bratt.[54a]

equating the rate of energy gain by free carriers due to acceleration by the electric field to their rate of energy loss in collisions with the crystal lattice and assuming that, once the average carrier energy reached a certain fraction of the impurity ionization energy, there would be enough free carriers with energies significantly greater than the average to induce impact ionization. The formula for the critical breakdown field obtained in this way is

$$\mathscr{E}_c = \frac{2c_s}{\mu}\left[\frac{\gamma E_i}{2kT} - 1\right]^{1/2}, \tag{45}$$

where c_s is the velocity of sound in the crystal and γ is the fraction of the ionization energy at which breakdown occurs (an experimentally determined quantity). The other symbols are as previously defined.

Although relatively successful at explaining the behavior of more heavily doped uncompensated samples, this theory could not predict the breakdown fields of lightly doped material nor could it account for the effect of differing compensation levels. Koenig[55] pointed out that a better solution can be

[54a] P. Bratt, unpublished report, 1962.
[55] S. H. Koenig, Phys. Rev. 110, 986 (1958).

2. IMPURITY GERMANIUM AND SILICON INFRARED DETECTORS

obtained from consideration of the rate equation, Eq. (25). Under conditions which we have previously considered, i.e., low temperatures such that A_T, A_0, B_0, and $B_1 n$ are negligible, the steady state free carrier concentration is

$$n = g/[B_T N_A - A_1(N_D - N_A)]. \tag{46}$$

At low electric field strengths, the A_1 term is negligibly small and B_T is a slowly decreasing[56] function of $\mathscr{E}$ which leads to a slowly increasing value of n. At sufficiently high electric fields A_1 begins to increase and, as $A_1(N_D - N_A)$ approaches $B_T N_A$, the value for n increases very rapidly. This corresponds to the onset of impact ionization breakdown. However, the increase in n cannot go on indefinitely because at large n values Eq. (46) must be modified to include the Auger recombination term in the denominator. This causes a slowdown in the rate of increase in n.

Equation (46) predicts that an increase in compensation will increase the field required for breakdown as has been observed experimentally.[49,55,57] An increase in majority impurity concentration also increases the breakdown field. This is not shown explicitly in Eq. (46), but is implicit in the impact ionization coefficient A_1. Increases in majority impurity concentration decrease the free carrier mobility through neutral impurity scattering and thereby decrease the rate at which carriers can gain energy from the electric field. Theoretical calculations of A_1 have been made by Zylbersztejn[58] and Cohen and Landsberg,[59] which show good agreement with experimental data.

In making impurity IR detectors, one attempts to keep the compensating impurity concentration to as low a level as possible so as to maximize the free carrier lifetime[60] [see Eq. (33)]. Therefore degree of compensation is not an important issue. Furthermore, doping levels are high enough so that the mobility is limited mainly by neutral impurity scattering rather than by lattice scattering and this, fortunately, permits operation at higher electric field strengths before impact ionization occurs.

For deep level impurities, no detailed theoretical analysis for breakdown is available. The situation is complicated by insufficient knowledge of the energy distribution function of carriers in high electric fields (so called "hot" carriers). When carriers in Ge attain energies in excess of 0.037 eV (the

---

[56] It will be shown in Section 14 that B_T goes roughly as $\mathscr{E}^{-1/2}$.
[57] L. M. Lambert, *J. Phys. Chem. Solids* **23**, 1481 (1962).
[58] A. Zylbersztejn, *Phys. Rev.* **127**, 744 (1962).
[59] M. E. Cohen and P. T. Landsberg, *Phys. Rev.* **154**, 683 (1967).
[60] An exception to this is the special case where additional compensating impurities are purposely added to shorten the lifetime to produce a high speed detector. See Section 14d for a further discussion of this case.

optical phonon energy) inelastic scattering tends to remove these carriers from the high energy side of the distribution.[61,62] Thus, fewer carriers become available to cause impact ionization. Again, this is an advantageous situation and one reason that deep level impurities make better detectors than shallow impurities. Much higher field strengths can be applied, producing higher photoconductive gain. Table II lists the measured breakdown

TABLE II

IMPACT IONIZATION BREAKDOWN FIELD STRENGTHS FOR IMPURITIES IN Ge AND Si

Impurity	Ionization energy (eV)	Breakdown field (V/cm)[a]	Impurity concentration (cm^{-3})	compensating donor concentration (cm^{-3})
Ge:In[b]	0.01	100	1.5×10^{16}	$< 1 \times 10^{14}$
Ge:Zn[c]	0.03	405	1×10^{16}	$< 1 \times 10^{14}$
Ge:Cu[d]	0.042	600	6×10^{15}	$\sim 1 \times 10^{13}$
Ge:Cu[c]		630	1.8×10^{16}	$\sim 1 \times 10^{14}$
Ge:Cd[d]	0.054	900	1.1×10^{16}	$< 1 \times 10^{14}$
Ge:Hg[d]	0.092	1600	6×10^{15}	$< 1 \times 10^{14}$
Si:P[d]	0.045	~200	3×10^{16}	$< 1 \times 10^{14}$
Si:As[d]	0.054	~300	3×10^{16}	$< 1 \times 10^{14}$
Si:Ga[d]	0.073	~2000	3×10^{16}	$< 1 \times 10^{14}$

[a] At a temperature of 4–5°K.
[b] Sclar and Burstein.[52]
[c] Picus.[63]
[d] Data taken at Santa Barbara Research Center.

fields for a number of impurity atoms in Ge and Si. In Fig. 11 the breakdown field is plotted versus ionization energy. Note the approximate linear relationship for energies greater than the optical phonon energy in Ge. In n-type Si, breakdown fields are about a factor of 3 lower than p-type Ge and, in p-type Si, they are higher.

7. NOISE

There is extensive literature on noise in semiconductors and the case of extrinsic photoconductors has been treated by a number of authors. A detailed review is beyond the scope of this chapter; the interested reader is

[61] W. E. Pinson and R. Bray, *Phys. Rev.* **136**, A1449 (1964).
[62] E. G. S. Paige, in *Proc. Int. Conf. Phys. Semicond., Kyoto, 1966*, p. 397. Phys. Soc. of Japan, 1966.
[63] G. Picus, *J. Phys. Chem. Solids* **22**, 159 (1961).

2. IMPURITY GERMANIUM AND SILICON INFRARED DETECTORS

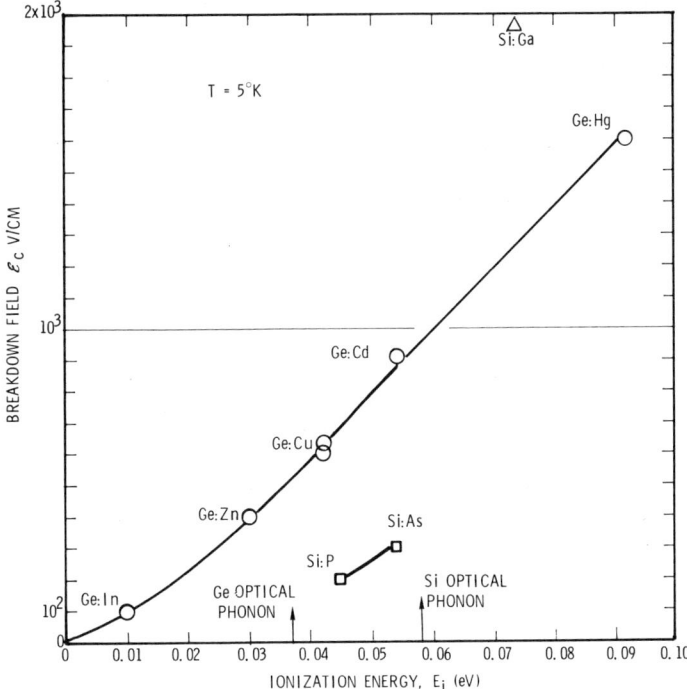

FIG. 11. Critical impact ionization breakdown field for deep level impurities in Ge and Si.

directed toward a selection of key references[64-68] that contain further references to the literature. The approach taken here is to list the four major noise contributions important for impurity Ge and Si IR detectors and give a brief discussion of their physical origins.

The mean squared short circuit noise current in the photoconductor circuit of Fig. 8 can be represented by

$$\langle i_n(f)^2 \rangle = \int_0^\infty S_i(f)\, df, \qquad (47)$$

---

[64] A. van der Ziel, "Fluctuation Phenomena in Semiconductors." Academic Press, New York, 1959.
[65] K. M. van Vliet, *Proc. IRE* **46**, 1005 (1958).
[66] K. M. van Vliet and J. R. Fassett, in "Fluctuation Phenomena in Solids" (R. E. Burgess, ed.), Chapter 7. Academic Press, New York, 1965.
[67] K. M. van Vliet, *Appl. Opt.* **6**, 1145 (1967).
[68] P. W. Kruse, L. D. McGlauchlin, and R. B. McQuistan, "Elements of Infrared Technology," Chapters 7 and 9. Wiley, New York, 1962.

where $S_i(f)$ is the spectral density of the current fluctuations and contains the frequency dependence of the noise current. The frequency dependence, as well as the magnitude, is the distinguishing feature of the different noise currents.

a. Johnson–Nyquist Noise

This type of noise is due to the random thermal motion of charge carriers in the crystal and not due to fluctuations in the total number of these charge carriers. It occurs in any conducting material having resistance R at a temperature T and is essentially due to small changes in the voltage at the terminals of the device due to random arrival of charge at the terminals. The current spectral density is given by

$$S_i(f) = \frac{4kT}{R}\left(\frac{hf}{kT}\right)\left[\exp\left(\frac{hf}{kT}\right) - 1\right]^{-1}. \quad (48)$$

However, at the normal frequencies encountered for IR detectors, $hf \ll kT$ and this expression reduces to

$$S_i(f) = 4kT/R, \quad (49)$$

which is independent of frequency.

b. $1/f$ Noise

Many electrical devices exhibit an excess noise that is large at low frequencies and decreases as the frequency is increased. Carbon resistors, semiconductor devices, evaporated metal films, etc. all show this effect. In vacuum tubes, it is known as "flicker" noise. The spectral intensity for this type of noise can be written

$$S_i(f) = KI^x/f^y, \quad (50)$$

where, in most cases $x \approx 2$, $y \approx 1$, and K is an unknown proportionality factor which must be determined experimentally.

While the exact physical mechanism responsible for this noise is not definitely known, in resistors it is thought to arise at the contacts between carbon granules, while in semiconductor elements, it is believed to be due to conductivity modulating processes taking place at the free surface or at the metal to semiconductor contacts. MacRae and Levinstein[69] have shown that $1/f$ noise in Ge is associated with an inversion layer at the surface of the crystal. In Si devices, it has been attributed to charge transfer between the bulk and surface states in the oxide passivating layer.

[69] A. U. MacRae and H. Levinstein, *Phys. Rev.* **119**, 62 (1960).

c. Generation–Recombination (g–r) Noise

This type of noise, as its name implies, is due to the random generation of free charge carriers by the crystal lattice vibrations and their subsequent random recombination. Because of the randomness of the generation and recombination processes, it is unlikely that there will be exactly the same number of charge carriers in the free state at succeeding instances of time. This leads to conductivity changes that will be reflected as fluctuations in current flow through the crystal.

Van Vliet[65,66] has shown that for the case of an extrinsic semiconductor with both donor and acceptor impurities present and (for the purposes of illustration) with the donors as the major impurity,

$$S_i(f) = 4\left(\frac{I}{N}\right)^2 \frac{N(N_A + N)(N_D - N_A - N)}{(N_D - N_A - N)(N_A + N) + NN_D}\left[\frac{\tau}{1 + (2\pi f \tau)^2}\right]. \quad (51)$$

In this formula, N is the total number of free electrons and N_A and N_D now represent total numbers of acceptor and donor impurity atoms rather than concentrations.[70]

For the usual case at low temperatures, $N \ll N_A, N_D$ and this formula reduces to

$$S_i(f) = 4\frac{I^2}{N}\left[\frac{\tau}{1 + (2\pi f \tau)^2}\right]. \quad (52)$$

This noise is flat at low frequencies where $2\pi f\tau \ll 1$, but decreases with frequency when $2\pi f\tau \gg 1$. At high temperatures where $N \approx N_D - N_A$, Eq. (51) shows that the g–r noise from extrinsic generation approaches zero. However, intrinsic generation of carriers will eventually occur and this will again produce g–r noise.

d. Photon Noise

When free charge carriers are generated by external photons rather than by lattice vibrations, fluctuations in the generation rate will be caused by the random arrival of photons at the detector. The detector will thus reflect the fluctuations in the photon stream. It will be assumed that the photons causing radiation noise are those from the background radiation falling on the detector, not from signal radiation. If the background photons have the energy distribution of a blackbody, then the fluctuation in that portion of the stream in a narrow wavelength interval $\Delta\lambda$ about the wavelength λ

---

[70] Equation (51) is the only one to which this comment applies. Elsewhere in this chapter, N_A and N_D will denote acceptor and donor concentrations.

is given by[71]

$$S_J(f) = 2J_B(\lambda)A[1 + b], \quad (53)$$

where

$$J_B(\lambda) = 2c\Omega b/\lambda^4, \quad (54)$$

$$b = [\exp(hc/\lambda k T_B) - 1]^{-1}. \quad (55)$$

$J_B(\lambda)$ is the photon flux density in the stream, A is the detector's sensitive area, Ω is the effective solid angle of background viewed by the detector and T_B is the background temperature. The factor b arises from the fact that photons obey Bose–Einstein statistics and is sometimes called the "boson factor". For wavelengths such that $hc/\lambda k T_B \gg 1$ the boson factor is much less than unity and the photons behave as if they obeyed Poisson statistics. On the other hand, for $hc/\lambda k T_B \lesssim 1$ the boson factor is of the order of unity or greater and the photon stream is noisier than would be expected for a Poisson distribution. Since only those photons absorbed by the detector can contribute to the noise, Eq. (53) must be multiplied by the quantum efficiency $\eta(\lambda)$ and integrated over wavelength to obtain the total spectral intensity of the photon fluctuations.

Van Vliet[65,67] has shown that the spectral intensity of current fluctuations caused by photon noise is given by

$$S_i(f) = 4\left(\frac{I}{N_B}\right)^2 \frac{\tau^2}{[1 + (2\pi f\tau)^2]} \int_0^\infty \eta(\lambda)J_B(\lambda)A(1 + b)\,d\lambda, \quad (56)$$

where N_B is the equilibrium total number of free charge carriers generated by the background radiation. Note that the integral in Eq. (56) does not actually extend to infinity. It will be terminated at some cutoff wavelength λ_c which is the maximum wavelength to which the detector is sensitive. This is implicit in the quantum efficiency $\eta(\lambda)$ which goes to zero for $\lambda > \lambda_c$. It has been pointed out that this photon generated noise is simply another form of g–r noise. For the case where $b \ll 1$, it can be recognized that the integral in Eq. (56) is just the total generation rate of free carriers by background photons

$$G_B = \int_0^\infty \eta(\lambda)J_B(\lambda)A\,d\lambda \quad (57)$$

and from Eq. (32), $G_B = N_B/\tau$. Thus, Eq. (56) becomes

$$S_i(f) = 4\frac{I^2}{N_B}\left[\frac{\tau}{1 + (2\pi f\tau)^2}\right], \quad (58)$$

which is identical in form to Eq. (52).

[71] W. B. Lewis, *Proc. Phys. Soc.* **59**, 34 (1947).

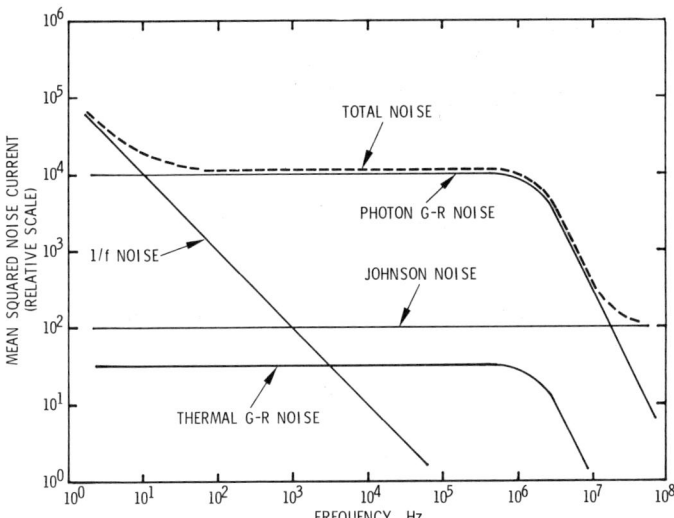

FIG. 12. Illustration of the contribution of various noise sources to the total noise spectrum.

For a background temperature of 300°K, the boson factor is negligible out to 10 μm; it has increased to 0.18 at 25 μm, to 0.60 at 50 μm, and to 1.64 at 100 μm. Thus, it will be important in the noise equation only for the longer wavelength detectors. For lower background temperatures it has even less importance.

The four noise contributions just described will add together to produce the total noise spectrum of the detector. Of these four, three depend on current while one (Johnson–Nyquist noise) does not. In the usual mode of operation, noise current will predominate over Johnson–Nyquist noise except at very high frequencies. Once the rms noise current is known, the rms noise voltage that is fed to the amplifier of Fig. 8 can be obtained from

$$\langle v_n(f)^2 \rangle^{1/2} = \langle i_n(f)^2 \rangle_T^{1/2} R_L R_{ac}/(R_L + R_{ac}). \quad (59)$$

An illustration of the composition of the total noise spectrum is shown in Fig. 12. The relative magnitudes of the various components are approximately what one obtains for impurity Ge and Si detectors. The noise due to background photons is shown as dominant over the thermal g–r noise because this is the most useful mode of operation and detectors are usually cooled to a low enough temperature so that this condition holds.

8. SENSITIVITY LIMITS—DETECTIVITY

In describing the performance of an IR detector, a quantity of interest is the responsivity, defined as the signal response per watt of incident signal power on the detector. Responsivity may be expressed in terms of signal current in which case it has units of amperes per watt or in terms of signal voltage, which has units of volts per watt. The short circuit current responsivity is

$$\mathcal{R}_I(\lambda) = i_s(f)/P(\lambda). \tag{60}$$

In a previous section [see Eq. (24)] it was shown that

$$i_s(f) = [\Delta n(f)/n]I. \tag{61}$$

The photoexcited free carrier density $\Delta n(f)$ can be obtained from Eqs. (40) and (35), thus

$$\Delta n(f) = \eta J_s \tau / d[1 + (2\pi f \tau)^2]^{1/2}, \tag{62}$$

giving

$$i_s(f) = \eta J_s \tau I / nd[1 + (2\pi f \tau)^2]^{1/2}. \tag{63}$$

We also know that

$$P(\lambda) = (hc/\lambda) J_s A \tag{64}$$

and, inserting this along with Eq. (63) into Eq. (60) gives the current responsivity

$$\mathcal{R}_I(\lambda) = \frac{\eta \lambda}{hc} \frac{I}{n(Lwd)} \frac{\tau}{[1 + (2\pi f \tau)^2]^{1/2}} \tag{65}$$

Lwd is the volume of the detector element as shown in Fig. 7. Voltage responsivity may now be easily obtained because

$$\mathcal{R}_V(\lambda) = \mathcal{R}_I(\lambda) R_{ac} R_L / (R_{ac} + R_L). \tag{66}$$

The open circuit voltage responsivity is sometimes required, and this can be obtained from Eq. (66) by letting $R_L \gg R_{ac}$, so that

$$\mathcal{R}_{V_0}(\lambda) = \mathcal{R}_I(\lambda) R_{ac}. \tag{67}$$

While responsivity is an important detector performance parameter, the ultimate sensitivity is determined by the noise as well as the responsivity, i.e., what is of primary interest is the signal-to-noise ratio, as in any radiation receiver. The quantity called detectivity,[72,73] with the symbol D^*, has been adopted as a measure of the sensitivity. It is defined as the signal-to-noise ratio obtained from the detector per watt of input signal power with the detector area normalized to 1 cm² and an electrical frequency bandwidth

---

[72] R. C. Jones, *Advan. Electron.* **5**, 1 (1953).
[73] R. C. Jones, *Proc. IRE* **47**, 1495 (1959).

of 1 Hz. Thus,

$$D^*(\lambda, f) = \frac{i_s(f)}{i_n(f)} \frac{(A\,\Delta f)^{1/2}}{P}. \tag{68}$$

Since both signal and noise may be functions of frequency, then detectivity may be also and this is indicated in the equation. The fundamental limits to the detectivity of impurity Ge and Si IR detectors depend upon the temperature of operation and the background photon flux incident upon the detector. In the remainder of this section we will examine what these limits are.

Maximum performance may be achieved only when the noise in the device is of the g–r type. The rms g–r noise current in a narrow frequency band Δf about some frequency f is obtained from Eq. (47) and either Eq. (52) or (58), and is given by

$$\langle i_n(f)^2 \rangle^{1/2} = 2I \left\{ \frac{\tau\,\Delta f}{nLwd[1 + (2\pi f\tau)^2]} \right\}^{1/2}, \tag{69}$$

where we have neglected the boson factor in the noise. Inserting this noise equation, the signal current Eq. (63), and Eq. (64) into Eq. (68) gives

$$D^*(\lambda) = (\eta\lambda/2hc)(\tau/nd)^{1/2}. \tag{70}$$

When both thermal and photon generation are important, the free carrier density may be written as a sum of two terms, $n = n_T + n_B$, with n_T given by [see Eq. (29)]

$$n_T = \frac{2}{\delta}\left(\frac{N_D - N_A}{N_A}\right)\left(\frac{2\pi m^* kT}{h^2}\right)^{3/2} \exp\left(\frac{-E_i}{kT}\right) \tag{71}$$

and

$$n_B = \eta J_B \tau/d = \eta J_B/dB_T N_A, \tag{72}$$

where Eq. (33) has been used to express the lifetime in terms of B_T, which contains the temperature dependence. These last three equations then allow the detectivity to be written with explicit temperature dependence as

$$D^*(\lambda, T) = \frac{\eta\lambda}{2hc}\left[\eta J_B + \frac{2dB_T}{\delta}(N_D - N_A)\left(\frac{2\pi m^* kT}{h^2}\right)^{3/2}\exp\left(\frac{-E_i}{kT}\right)\right]^{-1/2}. \tag{73}$$

Figure 13 shows a plot of detectivity versus reciprocal temperature calculated for the case of Ge:Hg at a wavelength at 10 μm. Values of the various input parameters used are shown with the figure. At low temperatures, the D^* is background photon limited and independent of temperature. As the temperature is raised, a point is reached where thermal generation becomes important and then D^* decreases rapidly with increasing temperature. In

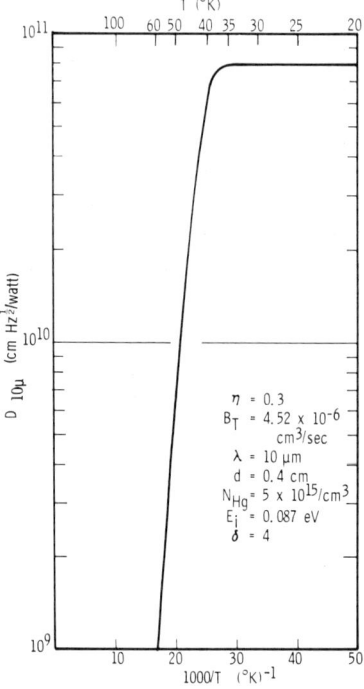

FIG. 13. D^* (10 μm) versus reciprocal temperature for Ge : Hg calculated from Eq. (73).

the background limited range, Eq. (73) shows that D^* is proportional to $J_B^{-1/2}$ and very high values are theoretically attainable if J_B can be made small enough.

Obviously it is desirable to operate this type of detector at a temperature where it is background photon limited, since this gives the maximum detectivity. Under this condition, the D^* can be expressed by

$$D^*(\lambda) = \frac{\eta\lambda}{2hc}\left[\int_0^{\lambda_c} \eta(\lambda)J_B(\lambda)(1+b)\,d\lambda\right]^{-1/2}, \quad (74)$$

where we have now included the boson factor for accuracy at longer wavelengths. Using Eq. (54) for the background photon flux density gives

$$D^*(\lambda) = \frac{\eta\lambda}{2hc}\left[2c\Omega\int_0^{\lambda_c} \frac{\eta(\lambda)}{\lambda^4}b(1+b)\,d\lambda\right]^{-1/2}. \quad (75)$$

Detectors whose detectivity can be described by the above equation are said to be "background limited impurity photoconductors." This phrase was originated by Burstein and Picus[74] and is usually abbreviated as "blip."

---

[74] E. Burstein and G. Picus, presented at *Infrared Informat. Symp. Chicago, Illinois*, February 3, 1958 (unpublished).

2. IMPURITY GERMANIUM AND SILICON INFRARED DETECTORS 75

Although first applied to extrinsic photoconductors, the term is now also used to describe intrinsic photoconductors and photodiodes. However, the formula for the background photon limited detectivity of these devices may be different from Eq. (75).

An "ideal" blip detector is one whose detectivity is given by Eq. (75), but also has a quantum efficiency of unity for wavelengths up to λ_c and zero beyond. This is a theoretically perfect detector that counts every photon incident upon it within its wavelength range of sensitivity. Although unattainable in practice, it is of interest to calculate what the ideal limits of performance are for comparison with real detector performance data. This has been done by numerical integration of Eq. (75) on a digital computer

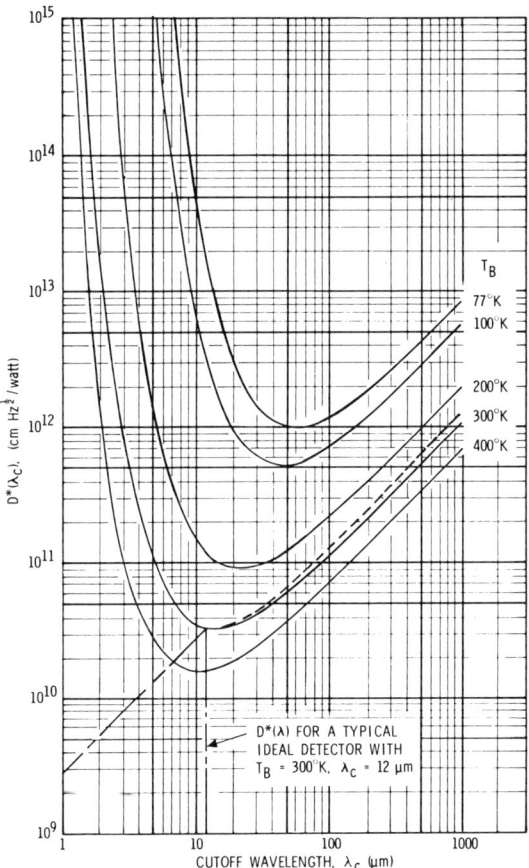

FIG. 14. Detectivity at the wavelength λ_c plotted versus λ_c for ideal detectors operating with different background temperatures and a 180° field of view. The dashed line for 300°K was calculated neglecting the boson factor.

for a number of different background temperatures and the results are shown in Fig. 14. Kruse et al.[68] and Van Vliet[67] have plotted similar curves, but for the case of the type of photon detector which does not exhibit recombination noise (i.e., photodiode, photoelectromagnetic detector, and photoemissive tube). The ideal limit for these devices is $\sqrt{2}$ higher than for an extrinsic photoconductor. The curves of Fig. 14 are correct as drawn[75] for the case of extrinsic photoconductors such as impurity Ge and Si. The consequence of excluding the boson factor has been illustrated for one background temperature.

The field of view for the ideal detector was assumed to be 180°. If a cold shield with a circular aperture is placed over the detector so as to define a conical field of view with cone angle θ. The background photon flux density on the detector varies as

$$\Omega = \pi \sin^2 \tfrac{1}{2}\theta. \tag{76}$$

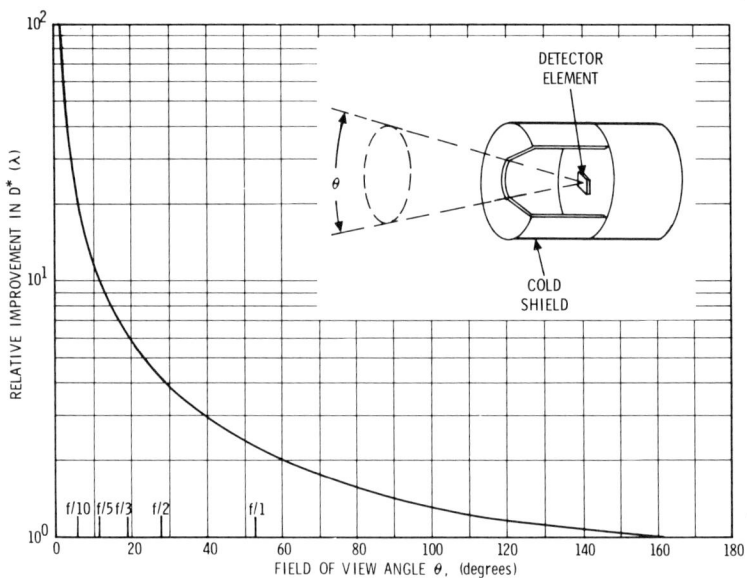

FIG. 15. Relative improvement factor for detectivity with reduction in field of view cone angle for a blip detector.

[75] K. M. van Vliet, *Appl. Opt.* **6** (1967), Fig. 2b, has criticized an earlier calculation of the ideal detector D^* limit [P. Bratt, W. Engeler, H. Levinstein, A. MacRae, and J. Pehek, *Infrared Phys.* **1**, 27 (1961)] as being in error because of neglect of the boson factor. It is true that the boson factor was neglected in this calculation, however, for 300 °K background and $\lambda < 10\ \mu$m, this has negligible effect. Van Vliet was actually comparing his calculation for an ideal detector that does not show recombination noise with a calculation for a photoconductor that does. Therefore, a $\sqrt{2}$ difference between the two curves is to be expected.

Ω is the effective solid angle viewed by the detector (not the true solid angle). Therefore, the D^* varies with field of view as $(\sin \tfrac{1}{2}\theta)^{-1}$. Figure 15 is a curve showing how ideal D^* is improved as the cone angle θ is reduced for any given background temperature.

IV. Device Fabrication

9. Material Preparation

As with most semiconductor devices, the ultimate performance of a doped Ge or Si IR detector depends first of all on the quality of the starting material from which the device is made, and second, on the perfection of the fabrication technology used in making the final device. Without adequate quality starting material, no amount of processing skill can yield a high performance device. On the other hand, given good quality starting material, then proper fabrication technique can yield the desired high performance detector.

Normal transistor grade materials usually are not sufficiently pure to be used in making doped Ge and Si IR detectors. The starting materials must be further zone refined to bring the impurity concentrations down to a very low level, preferably less than 10^{13} impurity atoms per cm^3. Germanium is purified by zone refining ingots in carbon coated quartz boats and Si is purified by the floating zone method.

After refining, doping elements are added. In the case of Ge, zone leveling is the most frequently used method for adding impurities[76]; however slight variations of the method are used depending on the type of impurity to be added. Doped Ge and Si may also be produced by pulling crystals from the melt by the Czochralski technique.

Figure 16 illustrates the zone leveling method. The high purity Ge is placed in a quartz boat along with a seed crystal cut to the proper crystallo-

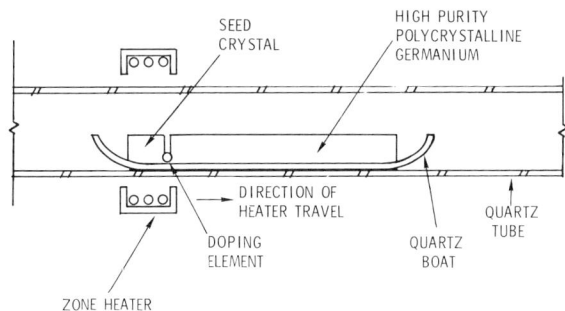

Fig. 16. Diagram of zone leveling method for impurity doping of Ge crystals.

[76] W. G. Pfann, "Zone Melting," Chapter 7. Wiley, New York, 1958.

graphic orientation. The (111) direction is frequently used for the growth axis since the crystal grows well in this direction and slices cut perpendicular to the growth axis are easy to evaluate for dislocations and lineage. This is done by etching the (111) crystal face with suitable chemicals to bring out dislocation etch pits.[77] The boat is placed in a quartz tube through which hydrogen gas is flowed to prevent oxidation of the Ge when it is melted.

The doping element is put at the boundary between seed crystal and high purity Ge where the molten zone will be first formed. Those elements that have a small segregation coefficient such as Au may be added in elemental form while those with a large segregation coefficient must first be alloyed with Ge and then added as a mixture. This allows one to be able to handle easily weighable quantities of doping material.

The molten zone is formed and allowed to remain in position at the seed crystal while the doping element dissolves and the zone reaches a stable size. The zone is then caused to move away from the seed at a slow rate by moving the zone heater. Typical zone speeds are 1 to 2 cm/hr. The doping impurity will then be distributed throughout the recrystallized Ge produced after the molten zone has traversed the length of the polycrystalline Ge charge. The atom concentration C_s of dopant introduced into the recrystallized Ge is given by

$$C_s = k_0 C_m (\rho_s / \rho_m), \tag{77}$$

where C_m is the concentration in the melt, k_0 is the segregation coefficient, and ρ_s and ρ_m are the densities of solid and molten Ge respectively. Values of k_0 for the major dopant impurities in Ge and Si are given in Tables III and IV. The ratio of ρ_s to ρ_m is 0.87 and C_m may be calculated from the formula

$$C_m = W N_a / A V_m, \tag{78}$$

where W is the weight of impurity element added to the melt, A is its atomic weight, N_a is Avogadro's number, and V_m is the volume of the melt.

A uniform concentration of impurity in the doped crystal will be obtained provided that k_0 is small so that the amount of impurity that leaves the melt and enters the solid is neglibible and also that the melt volume is held constant during the crystal growth process.

For impurities with large k_0 values, a nonuniform impurity distribution will be obtained if only one zone pass is made. However, the method of zone leveling,[76] that is, movement of the molten zone back and forth several times may be used to obtain a uniform distribution if desired.

The floating zone method used for Si is similar in principle to the boat

---

[77] See, for example, *Solid State Phys., (Methods Exp. Phys.)* **6A**, 147, 321 (1959).

TABLE III

IMPURITY ATOM CHARACTERISTICS IN Ge

Group	Impurity atom	Tetrahedral radius (Å)[a]	Distribution coefficient[b]	Maximum solid solubility (atoms/cm^3)[b]
	Ge	1.22		
IB	Cu	1.35	1.5×10^{-5}	3.3×10^{16}
	Ag	1.52	4×10^{-7}	9×10^{14}
	Au	1.50	1.3×10^{-5}	$> 2 \times 10^{16\,e}$
IIA	Be	1.06		$\sim 1 \times 10^{19\,f}$
IIB	Zn	1.31	4×10^{-4}	2.5×10^{18}
	Cd	1.48	$3 \times 10^{-6\,d}$	$\sim 1 \times 10^{16\,d}$
	Hg	1.48	$2 \times 10^{-6\,c,d}$	$\sim 1.5 \times 10^{16\,c,d}$
IIIB	B	0.88	17	
	Al	1.26	7.3×10^{-2}	4.1×10^{20}
	Ga	1.26	8.7×10^{-2}	5.0×10^{20}
	In	1.44	1×10^{-3}	3.9×10^{18}
	Tl	1.47	4×10^{-5}	
VB	P	1.10	8×10^{-2}	
	As	1.18	2×10^{-2}	1.8×10^{20}
	Sb	1.36	3×10^{-3}	1.2×10^{19}
	Bi	1.46	4.5×10^{-5}	
VIB	S	1.04		$> 5 \times 10^{15\,e}$
	Se	1.14		$> 5 \times 10^{15\,e}$
	Te	1.32	$\sim 10^{-6}$	$> 2 \times 10^{15\,e}$
Transition metals	Mn		$\sim 10^{-6}$	
	Fe		$\sim 3 \times 10^{-5}$	1.3×10^{15}
	Co		$\sim 10^{-6}$	$\sim 2 \times 10^{15}$
	Ni		3×10^{-6}	8×10^{15}
	Pt		$\sim 5 \times 10^{-6}$	

[a] Pauling.[78]
[b] After Trumbore,[79] except where otherwise noted.
[c] Darviot et al.[80]
[d] Bratt and Cole.[81]
[e] Tyler.[82]
[f] Tyapkina et al.[83]

[78] L. Pauling, "The Nature of the Chemical Bond," p. 246. Cornell Univ. Press, Ithaca, New York, 1960.

[79] F. A. Trumbore, *Bell Syst. Tech. J.* **39**, 205 (1960).

[80] Y. Darviot, A. Sorrentino, and B. Joly, *Infrared Phys.* **7**, 1 (1967).

[81] P. R. Bratt and R. Cole, previously unpublished data taken at the Santa Barbara Research Center, 1965–1966.

[82] W. W. Tyler, *J. Phys. Chem. Solids* **8**, 59 (1959); see also Newman and Tyler.[22]

[83] N. D. Tyapkina, M. M. Krivopolenova, and V. S. Vavilov, *Fiz. Tverd. Tela* **6**, 2192 (1964) [*English Transl.: Sov. Phys.—Solid State* **6**, 1732 (1965)].

TABLE IV

IMPURITY ATOM CHARACTERISTICS IN Si

Group	Impurity atom	Tetrahedral radius (Å)[a]	Distribution coefficient[b]	Maximum solid solubility (atoms/cm^3)[b]
	Si	1.18		
IB	Cu	1.35	4×10^{-4}	1.4×10^{18}
	Au	1.50	2.5×10^{-5}	1.2×10^{17}
IIA	Be	1.06	1.3×10^{-4} [c]	5×10^{17} [c]
	Mg	1.40	8×10^{-6} [c]	1×10^{16} [c]
IIB	Zn	1.31	$\sim 1 \times 10^{-5}$	6.0×10^{16}
	Cd	1.48		
	Hg	1.48		
IIIB	B	0.88	8.0×10^{-1}	6.0×10^{20}
	Al	1.26	2.0×10^{-3}	2.0×10^{19}
	Ga	1.26	8.0×10^{-3}	4.0×10^{19}
	In	1.44	4×10^{-4}	1×10^{18} [d]
	Tl	1.47		$> 10^{17}$
VB	P	1.10	3.5×10^{-1}	1.3×10^{21}
	As	1.18	3×10^{-1}	1.9×10^{21}
	Sb	1.36	2.3×10^{-2}	6.8×10^{19}
	Bi	1.46	7×10^{-4}	8.2×10^{17}
VIB	S	1.04	$\sim 10^{-5}$	3.8×10^{16}
	Se	1.14	$\sim 10^{-8}$ [c]	$\sim 1 \times 10^{17}$ [e]
	Te	1.32	8×10^{-6} [c]	$> 3 \times 10^{16}$
Transition metals	Mn		$\sim 10^{-5}$	3.4×10^{16}
	Fe		8×10^{-6}	3.0×10^{16}
	Co		8×10^{-6}	2.4×10^{16}

[a] Pauling.[78]
[b] After Trumbore,[79] except where otherwise noted.
[c] Aigrain and Balkanski.[84]
[d] Kozlovskaya and Rubenshtein.[85]
[e] Jostad and Moore.[86]

method. However, in this case the crystal is grown in the vertical direction. Figure 17 shows the essential elements of a float zone furnace. Both the seed crystal and polycrystalline feed rod are held tightly by metal chucks

[84] P. Aigrain and M. Balkanski, "Selected Constants Relative to Semiconductors." Pergamon, Oxford, 1961.
[85] V. M. Kozlovskaya and R. N. Rubenshtein, *Fiz. Tverd. Tela* **3**, 3354 (1961) [*English Transl.: Sov. Phys.—Solid State* **3**, 2434 (1962)].
[86] L. L. Jostad and T. G. Moore, unpublished, 1974.

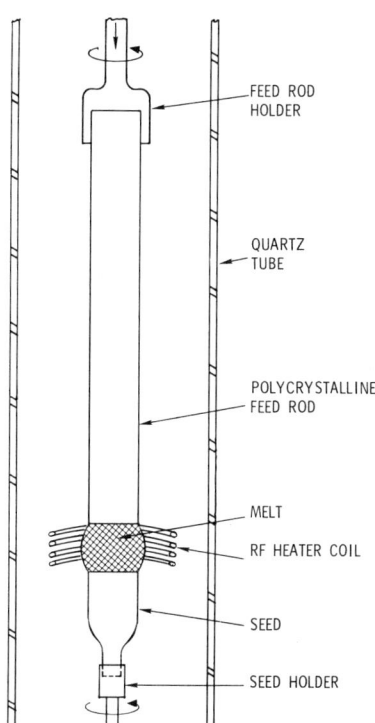

FIG. 17. Diagram of floating zone method for growth of Si crystals with various doping impurities.

which are rotated at a rate of about 30–40 rpm. This assembly is then slowly lowered through a heating coil which produces the molten zone. Surface tension of the molten Si holds the melt in place during crystal growth. The main advantage of this method over the Czochralski method is that the melt is not in contact with the walls of a crucible and contamination from this source is eliminated. Vacuum zone refining is effective for reducing P, B, heavy metal impurities, and oxygen to the lowest levels attainable.

Doping can be accomplished by a variety of methods. Elements with low volatility can be added in pellet form at the seed end at the beginning of the zone leveling run. Elements with high volatility can sometimes be introduced from a gaseous compound. For example, phosphorus doping can be done using phosphine gas (PH_3). Accurate control of dopant concentration can then be achieved by controlling gas pressure and flow rate. Dopants might also be introduced into the melt by sublimation from a nearby source with a separate heater element.

a. Vapor Doping

Doping with elements that have a high vapor pressure such as Zn, Cd, and Hg is best accomplished in a sealed tube with the dopant added from the

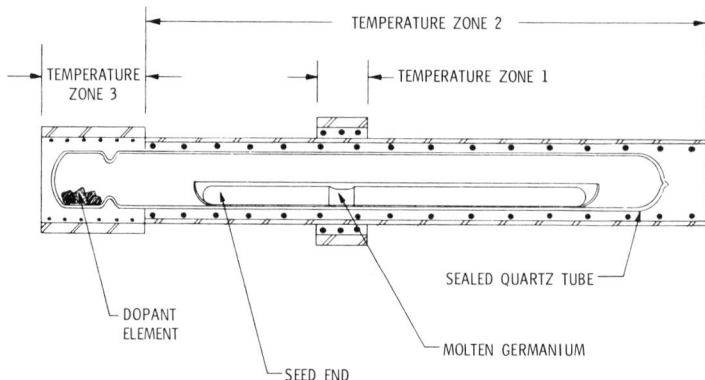

FIG. 18. Vapor doping furnace for introduction of volatile impurities into Ge.

vapor phase. A vapor doping furnace for Ge is shown schematically in Fig 18. The Ge is contained in a carbon coated quartz boat which has been loaded into a quartz tube also containing the pure doping element at one end. The tube is evacuated, sealed off, and placed inside a furnace having three different temperature zones. Zone one is set at a temperature sufficient to produce a molten zone in the Ge ingot (approximately 1000°C). The heater producing this zone is movable so that the molten zone can be swept along the ingot at a controlled rate of speed. Zone two is held at a temperature less than the melting point of Ge, but sufficient to maintain the doping element in the vapor phase. Zone three is at the lowest temperature and therefore controls the dopant vapor pressure in the tube.

A typical sequence of operations for this furnace might be as follows. The temperatures in zones two and three are brought to their desired values. Zone one is moved to a position near the seed end of the ingot and its temperature is raised to produce a molten zone in the Ge. The molten zone is held in this position for a period of time, usually about thirty minutes, to allow dissolution of the vapor dopant in the melt and the establishment of an equilibrium between melt and vapor. After this time, the molten zone heater is set into motion, sweeping the molten zone along the Ge ingot.

The dopant enters the recrystallized Ge by segregation from the melt and the dopant dissolved in the melt is replenished from the vapor phase. Thus, a uniformly doped, single crystal ingot is obtained. On the other hand, if variations in doping concentration are desired, this can easily be achieved with a good degree of precision by changing the temperature in zone three and thereby changing the dopant vapor pressure in the tube.

For the elements Zn, Cd, and Hg, it has been found that the concentration of dopant in the grown crystal increases linearly with dopant vapor pressure

2. IMPURITY GERMANIUM AND SILICON INFRARED DETECTORS 83

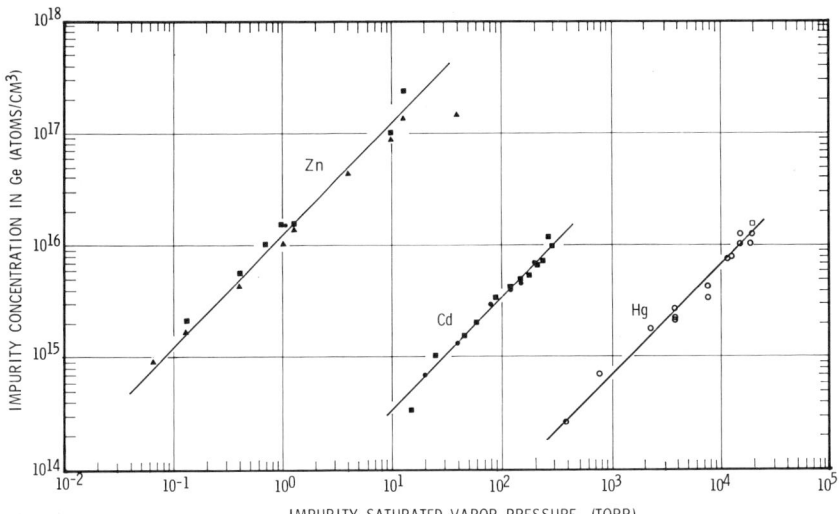

FIG. 19. Impurity concentrations in Ge achieved by vapor doping with saturated vapor pressures of Zn, Cd, and Hg. Solid symbols represent data obtained at Santa Barbara Research Center. Open circles for Hg are after Reynolds et al.[87] The open square for Hg is from the work of Darviot et al.[80]

up to the maximum amount of dopant that can be added before gross dislocations and polycrystalline growth set in. Experimental data are shown in Fig. 19.

The linearity of dopant concentration with vapor pressure is not unexpected and can be explained in a fairly simple manner as follows. The atom fraction F_m of dopant dissolved in the melt at equilibrium is given by Raoult's law for ideal solutions:

$$F_m = P_a / P_a^0(T_m), \tag{79}$$

where P_a is the vapor pressure of dopant over the melt and $P_a^0(T_m)$ is the vapor pressure of the pure dopant at the temperature T_m of the melt.

This is converted to an atom concentration (atoms/cm³ of melt) by multiplication with the factor $N_a(\rho_m/A)_{Ge}$ where ρ_m and A are the melt density and atomic weight for Ge and N_a is Avogadro's number. Therefore

$$C_m = F_m N_a(\rho_m/A)_{Ge}. \tag{80}$$

Combining this equation with Eqs. (77) and (79), we find that the doping

---

[87] R. Reynolds, G. Cronin, W. Hutchinson, and R. Chapman, Texas Instruments, unpublished report, 1966.

TABLE V

Vapor Doping Parameters for Group IIB Impurities in Ge

Impurity element	$P_a^0(T_m)^a$ (Torr)	Slope $(cm^3 Torr)^{-1}$	k_0	k_0 (literature)
Zn	1.27×10^3	1.19×10^{16}	3.4×10^{-4}	$4 \times 10^{-4\,b}$
Cd	4.5×10^3	3.33×10^{13}	3.4×10^{-6}	$>5 \times 10^{-5\,b}$
Hg	1.48×10^5	6.66×10^{11}	2.2×10^{-6}	$2.0 \times 10^{-6\,c}$

$^a T_m$ assumed equal to 960°C; vapor pressures from Nesmeyanov.[88]
^{b} Tyler.[82]
^{c} Darviot et al.[80]

concentration in the solid can be expressed by

$$C_s = k_0 N_a (\rho_s/A)_{Ge} P_a/P_a^0(T_m) \tag{81}$$

This equation states that, with constant melt temperature, the concentration of impurity obtained in the solid is directly proportional to the dopant vapor pressure P_a as has been observed experimentally. From the slope of the experimental curves drawn in Fig. 19, one can obtain a value for the segregation coefficient k_0. Values thus calculated are listed in Table V and, with the exception of Cd, are in good agreement with previously published values obtained by other methods. The lower value found for Cd by the vapor doping method should be the more reliable. It is also closer to the value found for Hg which is to be expected because these two atoms have identical tetrahedral bonding radii.[79]

b. Doping by Diffusion

Some impurities are best introduced by diffusion. Cu and Ni in Ge are two prominent examples. Since the solid solubility of both of these elements in Ge is greater at some temperature below the Ge freezing point, it is possible to introduce higher concentrations of the doping elements by diffusion at this temperature instead of by normal segregation from a doped melt. Diffusion doping is practical only when the diffusion coefficient of the doping element is large enough to achieve saturation of the diffused wafer in a reasonable time.

Figure 20 shows a plot of Cu concentrations obtained in Ge at different diffusion temperatures. These samples were prepared by electroplating Cu onto Ge slices and diffusing in an evacuated quartz ampoule at the temperatures indicated. In this particular case, a diffusion time of 24 hr was used.

---

[88] A. N. Nesmeyanov, "Vapor Pressure of the Elements," p. 193. Academic Press, New York, 1963.

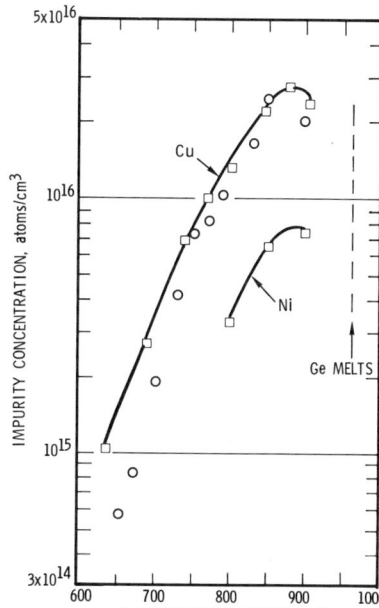

FIG. 20. Concentrations of Cu and Ni obtained by diffusion into Ge at various temperatures. Open circles are results obtained at the Santa Barbara Research Center; open squares are after Newman and Tyler[22] and Tyler.[82]

Also shown in Fig. 20 are data on Cu and Ni as reported by Tyler.[82] In the case of Cu impurities, the data in the higher temperature range is in good agreement. At lower temperatures, Tyler obtained somewhat higher Cu concentrations. This would seem to indicate that the Santa Barbara Research Center samples were not quite saturated in the time allotted for the diffusion.

Most elements that diffuse rapidly into Ge are also fast diffusers in Si. However, up to this point in time there has been little interest in these elements for impurity photoconductivity in Si. Table VI lists representative values of the diffusion coefficients of impurities in both Ge and Si. In order to diffuse an impurity atom through a wafer on the order of 1 mm thick in a reasonable time, say one week, the diffusion coefficient must be greater than $\sim 10^{-9}$ cm^2/sec. Only a few of the impurities listed have this property and for this reason the use of diffusion as a means of impurity doping is limited.

c. Optimizing Impurity Concentrations

It was pointed out in Section 5 that to obtain the highest theoretical quantum efficiency from an impurity IR detector the concentration of

TABLE VI

Diffusion Coefficients for Impurity Atoms in Ge and Si

Impurity atom	Germanium		Silicon	
	Temp. (°K)	Diffusion coefficient (cm²/sec)[a]	Temp. (°K)	Diffusion coefficient (cm²/sec)[a]
B	850	4.0×10^{-12}	1200	2.8×10^{-12}
Al			1200	1.3×10^{-11}
Ga	850	3.5×10^{-13}	1200	4.1×10^{-12}
In	850	5.9×10^{-13}	1200	8.3×10^{-13}
Tl	850	7.0×10^{-14}	1200	8.3×10^{-13}
P	850	2.4×10^{-11}	1200	2.8×10^{-12}
As	850	1.4×10^{-10}	1200	2.7×10^{-13}
Sb	850	6.4×10^{-11}	1200	2.2×10^{-13}
Bi	850	3.0×10^{-11}	1200	2.0×10^{-13}
Cu	700–900	3×10^{-5}	1200	1.5×10^{-4} [e]
Ag	850	1.4×10^{-6}	1200	7×10^{-9} [e]
Au	850	1.5×10^{-9}	1200	10^{-5} [f]
Be	900	1.5×10^{-11} [b]		
Zn	850	4.8×10^{-12}	1200	2×10^{-6} [e]
Cd	850	2×10^{-11} [c]		
Fe	850	2.0×10^{-6}	1200	6×10^{-6} [e]
Ni	850	5×10^{-5}	1200	3×10^{-8} [e]
S	920	10^{-9} [d]	1200	3×10^{-8} [e]
Se	920	10^{-10} [d]	1200	5×10^{-10} [g]
Te	920	10^{-11} [d]		

[a] Where not specifically referenced otherwise, data was taken from the review by Reiss and Fuller.[89] This article should be consulted for the original literature references.

[b] Belyaev and Zhidkov.[90]

[c] V. E. Kosenko[91]

[d] Tyler.[82]

[e] Burger and Donovan.[92]

[f] Gold in Silcon has various diffusion modes depending on the concentration of vacancies in the crystal. See Burger and Donovan.[92]

[g] Jostad and Moore.[86]

[89] H. Reiss and C. S. Fuller, in "Semiconductors" (N. B. Hannay, ed.), p. 244. Van Nostrand-Reinhold, Princeton, New Jersey, 1959.

[90] Yu. I. Belyaev and V. A. Zhidkov, *Fiz. Tverd. Tela* **3**, 182 (1961) [*English Transl.: Sov. Phys.—Solid State* **3**, 133 (1961)].

[91] V. E. Kosenko, *Fiz. Tverd. Tela* **1**, 1622 (1959) [*English Transl.: Sov. Phys.—Solid State* **1**, 1481 (1960)].

[92] R. M. Burger and R. P. Donovan, "Fundamentals of Silicon Integrated Device Technology," Vol. 1. Prentice-Hall, Englewood Cliffs, New Jersey, 1967.

impurity atoms should be made as large as possible. In practice it turns out that there are two things that limit the impurity concentration. One is the maximum solid solubility of the impurity atom in the host crystal and the other is the onset of impurity conduction effects that take place when the electron orbits of neighboring impurity atoms begin to overlap.

The maximum solid solubility of an impurity atom is a function of the atom's size relative to that of the Ge or Si atom of the host crystal. Impurity atoms smaller than the host crystal atom can be introduced in high concentrations while those larger than the host crystal atom are limited to smaller concentrations. When the impurity concentration approaches its maximum solubility limit, crystalline imperfections generally appear consisting of high dislocation densities, lineage, inclusions, and ultimately, polycrystalline growth. Because such imperfections are detrimental to detector performance, the practical limit to impurity concentration is about one-half the actual maximum solubility limit. Tables III and IV gave values for these limits for impurities in Ge and Si. Impurity elements in Ge such as Au, Hg, and Cd are examples of those whose concentration is limited by the atom's solid solubility.

When impurity atoms are added to a semiconductor in concentrations such that the ground state electron wave functions overlap, an "impurity band" is formed[93] in which electronic conduction may take place without recourse to conduction or valence bands. At somewhat lower concentrations, although banding does not occur, impurity conduction can still take place by a "hopping" mechanism,[94-97] whereby electrons jump from occupied to unoccupied neighboring impurity atoms stimulated by interaction with lattice phonons. The degree of hopping conductivity is dependent on temperature as well as donor and acceptor atom concentrations in the material. Figure 21 shows the variation of conductivity with temperature for a series of Ge:In samples. The steeply rising part of these curves at higher temperatures represents valence band conductivity and varies as $\exp(-E_i/kT)$ with $E_i = 0.01$ eV. At lower temperatures, the conductivity goes as $\exp(-E_3/kT)$ with $E_3 \simeq 0.001$ eV. This is the impurity hopping conductivity. Note the rapid increase of impurity conductivity as impurity concentration increases. A ten-fold increase in $(N_A - N_D)$ causes the conductivity to increase by four orders of magnitude.

Impurity conductivity has been found to be "Ohmic" in character and has

[93] N. F. Mott, *Can. J. Phys.* **34**, 1356 (1965).
[94] E. M. Conwell, *Phys. Rev.* **103**, 51 (1956).
[95] H. Fritzsche, *Phys. Rev.* **119**, 1238 (1960).
[96] A. Miller and E. Abrahams, *Phys. Rev.* **120**, 745 (1960).
[97] N. F. Mott and W. D. Twose, *Advan. Phys.* **10**, 107 (1961).

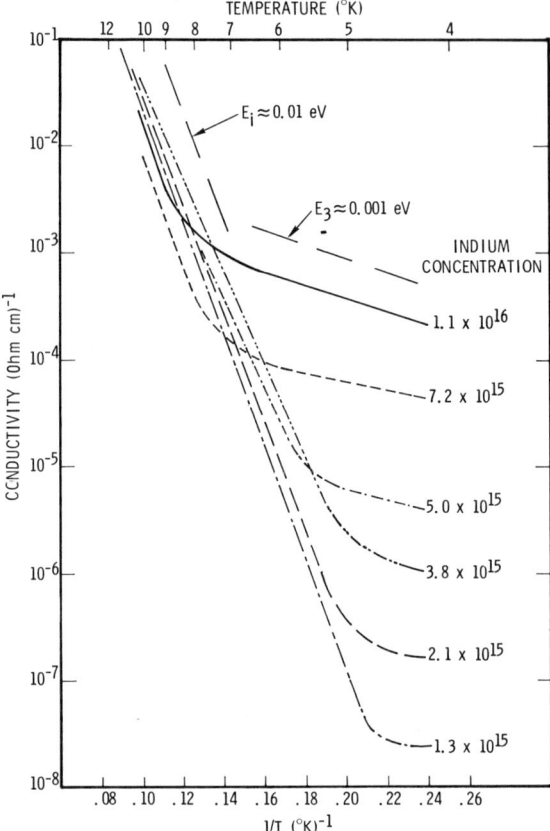

FIG. 21. Conductivity versus reciprocal temperature for a series of Ge:In samples with increasing In concentration. The concentration of compensating donors in these samples is low ($\sim 10^{13}/\text{cm}^3$) and about the same for all.

no excess noise associated with it other than the Johnson noise inherent in any resistive element. Thus, the effect of impurity conductivity is simply to produce an additional conduction path through the crystal in parallel with the normal valence or conduction band. Its effect on the photoconductivity of the material is to reduce the quantity $\Delta\sigma/\sigma$, because σ is increased while $\Delta\sigma$ is unaffected. With impurity conduction present, σ is composed of two terms, $\sigma = \sigma_n + \sigma_i$, where σ_n represents the normal conduction or valence band conductivity and σ_i the impurity hopping conductivity. With increasing impurity concentrations, σ_n is changed only slightly while σ_i rapidly increases.

Figure 22 shows the observed decrease in $\Delta\sigma/\sigma$ for another series of Ge:In samples. Note the rapid decrease in $\Delta\sigma/\sigma$ once the In concentration exceeds

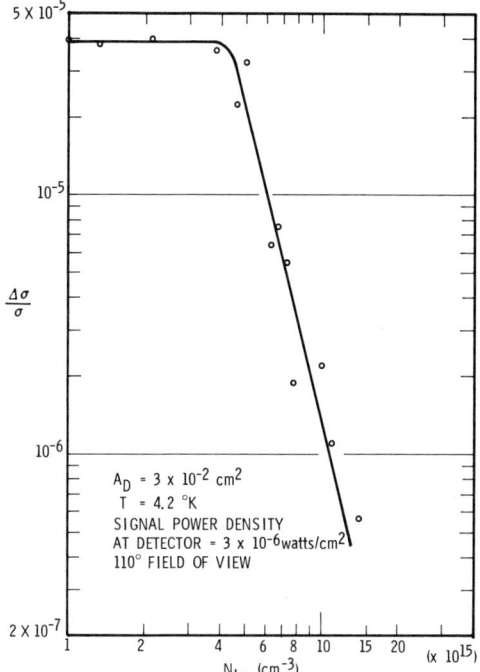

FIG. 22. Experimental data showing the decrease in $\Delta\sigma/\sigma$ when impurity hopping conductivity begins.

about 4×10^{15} atoms/cm³. This results in a proportionate decrease in both photosignal and noise voltages, and ultimately, a loss of detectivity when the noise level approaches system noise.

Data on the impurity concentration limits imposed by impurity conduction effects has been obtained only on a small number of impurity elements. Figure 23 shows these data plotted versus impurity ionization energy. One can see a trend toward a dependence of $N_i(\max)$ on E_i which goes as $E_i^{3/2}$. This would be expected if the radii of the ground state orbits vary with binding energy according to the effective mass formula

$$E_i = \hbar^2/2m^*a^{*2}. \tag{82}$$

The maximum impurity concentration should then be proportional to $(a^*)^{-3}$ or

$$N_i(\max) \propto [(2m^*/\hbar^2)E_i]^{3/2}. \tag{83}$$

However, in view of the known inadequacies of the effective mass model, particularly as it concerns ground state wave functions and also the inexact-

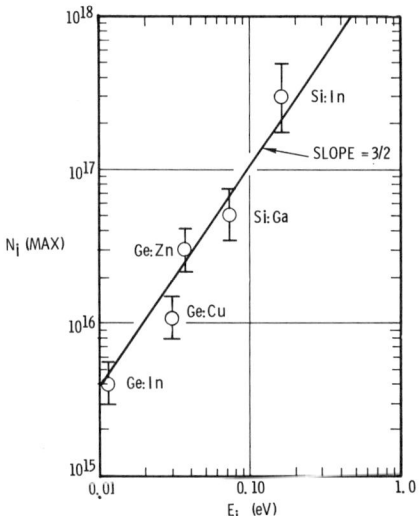

Fig. 23. Approximate values for maximum impurity concentration limits in Ge and Si imposed by impurity hopping conductivity.

ness of the experimental data, one should consider these results only as rough guidelines and seek more accurate data by experimental variation of impurity concentration about the line shown in Fig. 23.

10. Detector Fabrication

Single crystal Ge and Si are fairly strong materials and therefore relatively easy to handle and process into IR detectors. Being single crystal, they are subject to chipping or fracture on a cleavage plane if given a mechanical shock; however, reasonable care in handling obviates any problem of this nature. The crystals can be cut with standard abrasive saws using diamond or carborundum wheels or by string saws. After sawing, the surfaces are usually lapped smooth using a glass lapping plate and a water slurry of 300–600 grit lapping compound. Sometimes polishing with finer abrasives may be done. The final shape of the crystal is usually a rectangular parallelepiped with one face corresponding to the desired sensitive area of the IR detector.

Because the lapping and polishing operations produce a certain amount of damage to the crystal lattice in a thin region near the surface, this damaged layer must be removed by chemical etching. A variety of etch solutions have been developed. It has been found that an etch consisting of three parts concentrated nitric acid, two parts concentrated hydrofluoric acid, two parts concentrated acetic acid, and a small amount of bromine works well on Ge. For Si a mixture of one part concentrated hydrofluoric acid to five parts concentrated nitric acid is suitable. The Ge etch rate at room temperature is highly dependent on the amount of bromine added and just a small amount of vapor from the bottle is usually sufficient to etch off 100 μm in just a few seconds. The Si etch is much slower and requires several minutes to remove 100 μm of material.

Electrical contacts are applied to two opposite faces of the detector crystal by a suitable metallization technique. Surprisingly enough, even after 20 years of development, contacting methods are still not on a firm technological footing and their application is somewhat of an art with different techniques practiced by different detector manufacturers. The general characteristics desired of the contacts can be stated as follows: (1) They should be "Ohmic" for majority carrier flow, that is to say, there should be no potential barriers to impede the flow of majority carriers into or out of the crystal; (2) they should not be a source of excess noise; and (3) they should be mechanically stable.

These requirements are usually fulfilled by applying an acceptor-type metal on p-type crystals or a donor-type metal on n-type crystals. Subsequent heat treatment causes a microalloying which produces a p^+-p or an n^+-n layer as the case may be. Table VII lists a number of contact metals that have been used successfully on impurity Ge and Si detectors. The metals can be either evaporated or electroplated onto the crystal faces. Another method of producing the heavily doped p^+ or n^+ layers is by diffusion or ion beam implantation of suitable dopant atoms at the contact face. When this is done, almost any metal that adheres well to the crystal can then be used to make contact to the diffused surface.

Generally, one of the metallized surfaces is also used to bond the crystal to a metal mount for future assembly into a Dewar package. Figure 24 shows a

TABLE VII

Contact Metals Used for Impurity Ge and Si IR Detectors

	Germanium	Silicon
p-type crystals	In Cu Au Au–In	Al Pd Pt
n-type crystals	Sn Au–Sb Sn–Sb Pb–Sb	Sn–Sb Al on n^+ diffused region

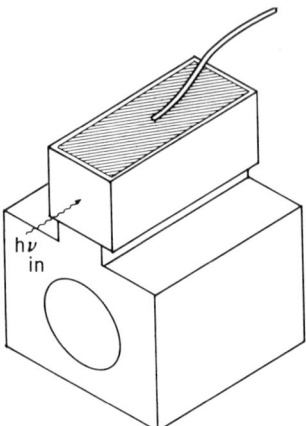

FIG. 24. Diagram of a completed detector crystal made from impurity Ge or Si.

completed detector crystal. Note that electrodes have been applied on the long faces of the crystal giving one of the shorter dimensions for the spacing between electrodes. This gives a lower detector resistance and higher photoconductive gain. The mount used in this case is copper; however, metals such as Kovar, beryllium, and aluminum have also been used. Indium solder is used to bond the crystal to the mount. This material is ductile enough at low temperatures to prevent undue stress on the detector crystal from differences in thermal contraction. Insulating materials such as sapphire or ceramics may also be used as a mount for the detector crystal.

a. Multielement Arrays

Multielement arrays of detectors have been made in a variety of ways. If the element size and spacing between elements is not too small, then an array can be built up by piecing together individual elements. Figure 25 shows a 25-element array fabricated in this manner. To make this array individual elements were pieced together in five-element modules. Five identical modules bolted together then make a 5 × 5 detector matrix. Each detector element in this assembly has a 1 mm² sensitive area.

When the element's size becomes very small, arrays are made from a monolithic slab as shown in Fig. 26. The fabrication procedure is as follows. As shown in Fig. 26a, a slab of Ge or Si that has been metallized on one side is soldered to a metal mount. The slab is then polished down to the appropriate thickness and metallized on the top side. It then has the configuration of Fig. 26b. Next, the individual elements are defined by a slotting operation. This may be done by abrasive sawing or by chemical etching.

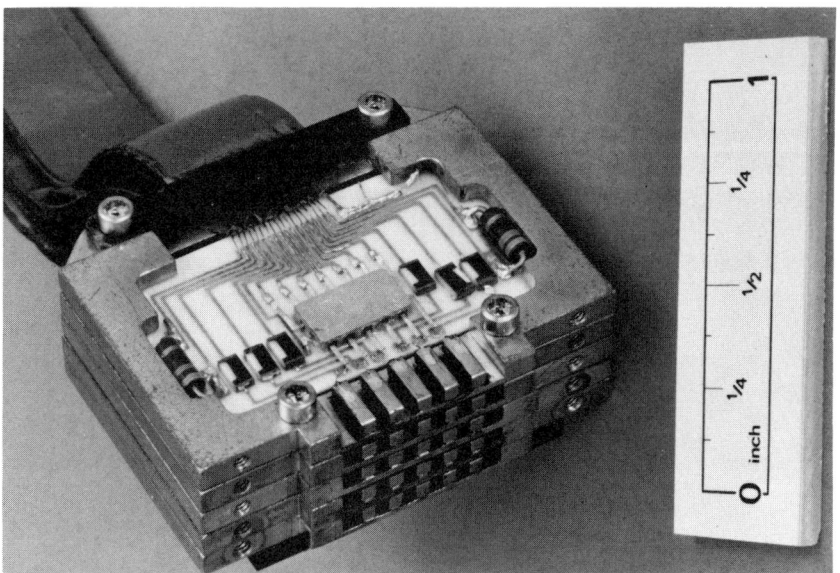

FIG. 25. Photograph of a 25-element Ge:Cd detector array. Note also the cooled electronics assembly with six preamplifier circuits.

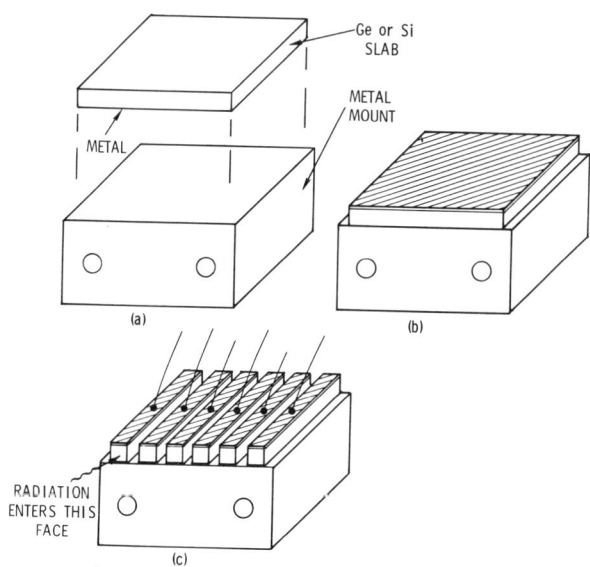

FIG. 26. Steps in multielement array fabrication from impurity Ge and Si.

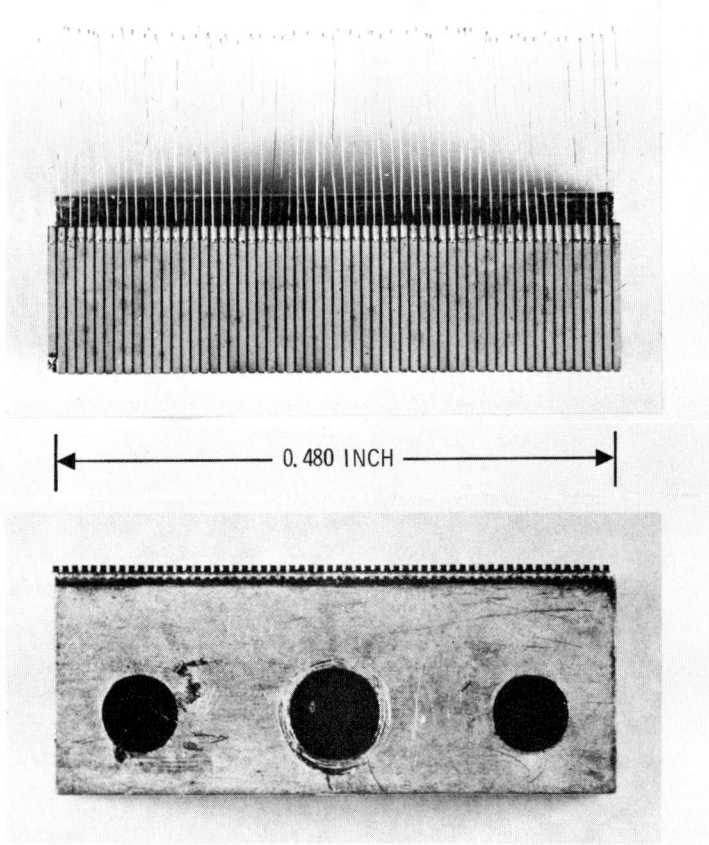

FIG. 27. Photograph of a 60-element linear array made from Ge : Hg. Upper photo : top view, attachment of lead wires. Lower photo : front view showing 0.1 × 0.1 mm sensitive areas.

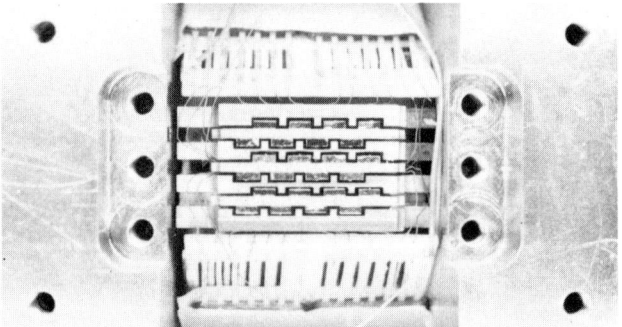

FIG. 28. Photograph of a two-dimensional detector array made from Si : Ga. There are 24 elements in all and each has a sensitive area of 0.25 × 1.0 mm.

The latter approach is preferred for high-density arrays because narrower slots can be achieved. If abrasive sawing is used, the minimum slot width is determined not only by the saw kerf, but by an additional amount of material that must be etched off the sides of the elements to remove the sawing damage. By the use of photolithographic masks and special etch techniques, slots less than 0.1 mm can be made. After the slotting operation, leads are attached to the top of each element and the array is complete as shown in Fig. 26c.

Figure 27 shows a photograph of an actual array made from Ge:Hg. The lower photo shows the sensitive elements in front view (the face through which incoming IR radiation enters) and the upper photo shows the top view with wire leads attached to each element. The sensitive area of each element is 0.1×0.1 mm and the length is 3 mm. As pointed out in previous discussions, such a length is necessary to achieve an αd value of the order of unity so that a reasonable quantum efficiency may be obtained.

A number of array modules such as that shown in the photograph may be pieced together to make a linear array of any desired length. Two-dimensional arrays may also be fabricated; however, a certain spacing between rows must be allowed to bring out the lead wires. Figure 28 shows a two-dimensional array fabricated from Si:Ga.

V. Operating Characteristics

In this part, we discuss the experimentally observed operating characteristics of impurity Ge and Si detectors. While attempting to present as much data as possible so as to give a comprehensive review of what has been done in the past, some selection and elimination had to be done in the interest of conserving space. Apologies are offered to any co-workers in this field who may feel that their work has been slighted.

The part is organized to present data on impurity Ge first, then Si, and then a short discussion of impurity Ge–Si alloys. The part concludes with a discussion of several special features that are common to both groups of impurity detectors.

11. OPERATING CHARACTERISTICS—GERMANIUM

Impurity elements in Ge have ionization energies ranging from 0.01 to 0.35 eV. These have been used to make IR detectors with cutoff wavelengths varying from 4 to 120 μm. No other semiconductor has this remarkable versatility to provide such a wide range of spectral responses and this has been one of the prime reasons for the popularity of impurity Ge over the past 20 years.

In discussing the operating characteristics of these detectors, it is convenient to make a division into three groups according to the wavelength range of sensitivity. The first group covers the shorter infrared region from 1 to 10 μm and is exemplified by the elements Au and Ni; the second group covers the moderately long infrared wavelengths from 10 to 50 μm such as Hg and Zn, and the third group covers the far infrared from 40 to 100 μm and beyond.

a. Short IR Range, 1–10 μm

The doping elements Ag, Au, Mn, Fe, Co, Ni, S, Se, and Te in Ge have been shown to produce impurity photoconductivity to infrared wavelengths shorter than 10 μm. The initial investigations on most of these impurities were performed by the group at General Electric Research Laboratory. A more extensive discussion of the characteristics of these impurities in Ge is given in the excellent review article by Newman and Tyler.[22] As seen in Fig. 2 all of these elements produce a multileveled impurity center in Ge. Silver and gold are triple acceptors and S, Se, and Te are double donors.

By the addition of the appropriate amount of a shallow level impurity

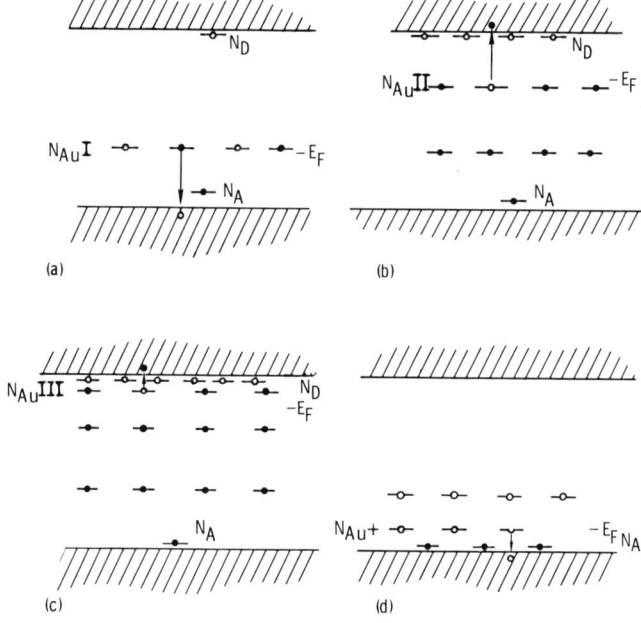

FIG. 29. Photoionization processes for the four different states of the Au impurity in Ge. (a) Photoionization of hole from first Au acceptor level 0.15 eV above valence band. (b) Photoionization of electron from second Au acceptor level 0.20 eV below conduction band. (c) Photoionization of electron from third Au acceptor level 0.05 eV below conduction band. (d) Photoionization of hole from Au A$^+$ state 0.05 eV above valence band.

of the opposite type from the main doping element, the various levels of a multileveled impurity may be activated for photoconductive response. This is illustrated in Fig. 29, where the Au triple acceptor is shown as an example. Even in the highest purity zone refined Ge there will always be some residual shallow acceptor impurities (probably B and Al). These have been shown in the figure by the symbol N_A. If these are not populated with electrons from compensating donor impurities, then they will accept electrons from the valence band creating a number of free holes that will decrease the resistance of the material and thereby decrease its sensitivity. Therefore, a number of compensating donor atoms N_D must be added such that $N_A < N_D \ll N_{Au}$. The situation will then be as shown in Fig. 29a with the Fermi level located near the lower Au level. The neutral Au levels can then be ionized by absorbed photons of energy $h\nu$, producing free holes via the reaction

$$Au^0 + h\nu \rightarrow Au^- + p^+. \tag{84}$$

With the further addition of donor atoms in an amount such that $N_{Au} + N_A < N_D \lesssim 2N_{Au} + N_A$, the second Au level becomes almost completely filled with electrons and the situation is as shown in Fig. 29b. Now the material is n-type and absorbed photons create free electrons via the reaction.

$$Au^{2-} + h\nu \rightarrow Au^- + n^-. \tag{85}$$

Further addition of compensating donors to populate the third Au level leads to the following photoionizing reaction

$$Au^{3-} + h\nu \rightarrow Au^{2-} + n^-. \tag{86}$$

which is shown in Fig. 29c.

The Au atom in Ge has the unusual property of another ionization state which is positive like that of a donor atom. Dunlap[98] has shown that this is a property of the substitutional Au atom in the Ge lattice. More recently, Norton[99] has presented evidence that this state is a positive hole bound to a neutral center to form a positive acceptor ion (A^+) analogous to a negative electron bound to a neutral hydrogen atom to form an H^- ion. (Norton has also demonstrated the existence of negative donor ions in Si:As and Si:P and called them D^- states. A further discussion of these states is presented in Section 14.) The role of the positive Au ion state in Ge is accentuated when shallow acceptor impurities are added to the crystal in amount such that $N_A \lesssim N_{Au}$. The situation for this case is then as shown in Fig. 29d. Holes thermally ionized from the shallow acceptors will be trapped at Au atoms to form A^+ states. These holes can then be

[98] W. C. Dunlap, Jr., *Phys. Rev.* **100**, 1629 (1955).
[99] P. Norton, *J. Appl. Phys.* **47**, 308 (1976).

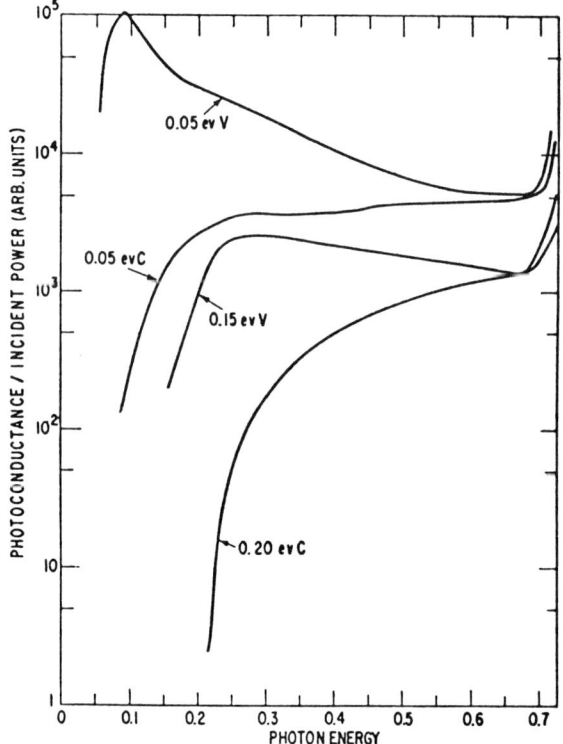

FIG. 30. Photoconductive spectral response of the four different ionization states of Au in Ge. The temperature was 77°K for the 0.15 eV and 0.20 eV levels and 20°K for the others. (After Newman and Tyler.[22])

photoionized out of the A^+ state by radiation of appropriate wavelengths via the reaction

$$Au^+ + h\nu \rightarrow Au^0 + p^+. \tag{87}$$

These four different photoionization processes of the Au atom give four different spectral response curves as shown in Fig. 30. Note in this figure the difference in shape of the curves near the low energy ionization threshold. Those curves involving hole excitation to the valence band have a peaked response near threshold while those involving electron transitions to the conduction band fall off in a gradual manner. Burstein et al.[100] proposed that this is due to differences in the magnitude of the photoionization cross section near threshold due to differences in the Coulomb interaction

[100] E. Burstein, G. Picus, and N. Sclar, in Photoconductivity Conf. (R. G. Breckenridge et al., eds.), p. 383. Wiley, New York, 1956.

between the ionized center and the freed electron or hole. For electrons, the potential energy is repulsive [see Eqs. (85) and (86)] while for holes, it is either attractive or neutral [Eqs. (84) and (87)]. This will lead to differences in the wave functions of the two types of charge carriers that then makes the photoionization cross section near threshold larger for Au^0 or Au^+ centers than for Au^{2-} or Au^{3-} centers. At higher photon energies, the cross sections will be more alike because the wave functions of the more energetic free carriers thus created will be less sensitive to the Coulomb potential of the Au center.

The different Coulomb interaction between free carriers and the various ionization states of a multileveled impurity atom also produces large differences in the capture cross sections and therefore the free carrier lifetimes may be markedly different in these materials. Again using the Au impurity atom as an example, the recombination processes will be the opposite of the excitation processes shown in Fig. 29 and described by Eqs. (84)–(87). A Coulomb attraction between free carrier and recombination site will lead to a large capture cross section, while a repulsion will lead to a smaller capture cross section.

Johnson and Levinstein[23] have obtained a value for the $Au^- + p^+$ recombination process of 1×10^{-13} cm^2 and for the $Au^- + n^-$ process, $\approx 10^{-17}$ cm^2, both values being measured at 80°K. The $Au^{2-} + n^-$ process would have an even smaller cross section, but no data has been reported for this ionization state. Neuringer and Bernard[101] have reported a value of 3.8×10^{-18} cm^2 for the $Au^0 + p^+$ process at 97°K. Capture cross sections for other impurity atoms are presented in Table VIII.[102-109]

Free hole lifetimes in p-type Ge:Au utilizing the Au^I acceptor level typically range from 10^{-9} to 4×10^{-7} sec.[23,110] Free electron lifetimes in n-type material utilizing the Au^{II} level range 10^{-5}–10^{-3} sec.[23] These values are for operating temperatures in the vicinity of 80°K.

Figure 31 shows photoconductive spectral response curves for Mn, Ni, Co, and Fe as reported by Newman and Tyler.[22] Again note the difference in

---

[101] L. J. Neuringer and W. Bernard, *J. Phys. Chem. Solids* **22**, 385 (1961).
[102] P. Norton and H. Levinstein, *Phys. Rev. B* **6**, 489 (1972).
[103] L. Johnson and H. Levinstein, *Phys. Rev.* **117**, 1191 (1960).
[104] J. M. Brown and A. G. Jordan, *J. Appl. Phys.* **37**, 337 (1966)
[105] G. K. Wertheim, *Phys. Rev.* **109**, 1086 (1958).
[106] J. S. Blakemore, *Can. J. Phys.* **34**, 938 (1956).
[107] P. Norton, T. Braggins, and H. Levinstein, *Phys. Rev. Lett.* **30**, 488 (1973).
[108] W. D. Davis, *Phys. Rev.* **114**, 1006 (1959).
[109] K. D. Glinchuk, A. D. Denisova, and N. M. Litovchenko, *Fiz. Tverd. Tela* **5**, 1933 (1963) [*English Transl.: Sov. Phys.—Solid State* **5**, 1412 (1964)].
[110] T. P. Vogl, J. R. Hansen, and M. Garbuny, *J. Opt. Soc. Amer.* **51**, 70 (1961).

shape of the curves near the ionization threshold when the material is converted from p- to n-type by addition of compensating donor atoms.

Of all the impurity atoms available for making doped Ge detectors for the range 1–10 μm, only Au has been exploited for commercial use. All the other atoms have a lesser solid solubility in the Ge lattice and thus produce detectors with lower quantum efficiency. Detectors utilizing the second Au level 0.20 eV below the conduction band enjoyed a brief popularity in the 1950's[111] and were quite sensitive to wavelengths out to about 5 μm. However, they are no longer available commercially. Nickel doped germanium detectors utilizing the 0.23 eV acceptor level have been made at Santa Barbara Research Center which showed quite good sensitivity, but have

TABLE VIII

CAPTURE CROSS SECTION OF IMPURITY ATOMS IN GE AND SI

Impurity atom	Type of capture[a]	Germanium		Silicon	
		Temp. (°K)	Capture cross section (cm²)	Temp. (°K)	Capture cross section (cm²)
B	σ_p^-			4.2	$8 \times 10^{-12\,f}$
Al	σ_p^-	10	$2 \times 10^{-12\,b}$		
Ga	σ_p^-	[(4–15)]	$\sigma_p^- \propto T^{-2.1}$		
In	σ_p^-			77	$10^{-13\,g,h}$
P	σ_n^+				
As	σ_n^+	10	$10^{-12\,c}$	10	$10^{-11\,i}$
Sb	σ_n^+	[4–10]	$\sigma_n^+ \propto T^{-2.5}$	[(2–10)]	$\sigma_n^+ \propto T^{-1.5}$
Cu	σ_p^-	10	$5 \times 10^{-12\,b}$		
		[(3–20)]	$\sigma_p^- \propto T^{-1.6}$		
Au	σ_p^-	80	$1 \times 10^{-13\,d}$	77	$1 \times 10^{-13\,j}$
	σ_p^{2-}	90	$10^{-17\,d}$		
Zn	σ_p^-			80–200	$10^{-13\,k}$
Cd	σ_p^-	8	$1 \times 10^{-11\,e}$		
Hg	σ_p^-	20	$3.6 \times 10^{-12\,e}$		

[a] The superscript + or − indicates the charge on the capturing center and the subscript n or p indicates whether the center is capturing an electron or hole.
[b] Norton and Levinstein.[102]
[c] Koenig et al.[49]
[d] Johnson and Levinstein.[103]
[e] Calculated from the data in Fig. 43.
[f] Brown and Jordon.[104]
[g] Wertheim.[105]   [h] Blakemore.[106]
[i] Norton et al.[107]   [j] Davis.[108]
[k] Glinchuk et al.[109]

[111] M. E. Lasser, P. Cholet, and E. C. Wurst, Jr., *J. Opt. Soc. Amer.* **48**, 468 (1958).

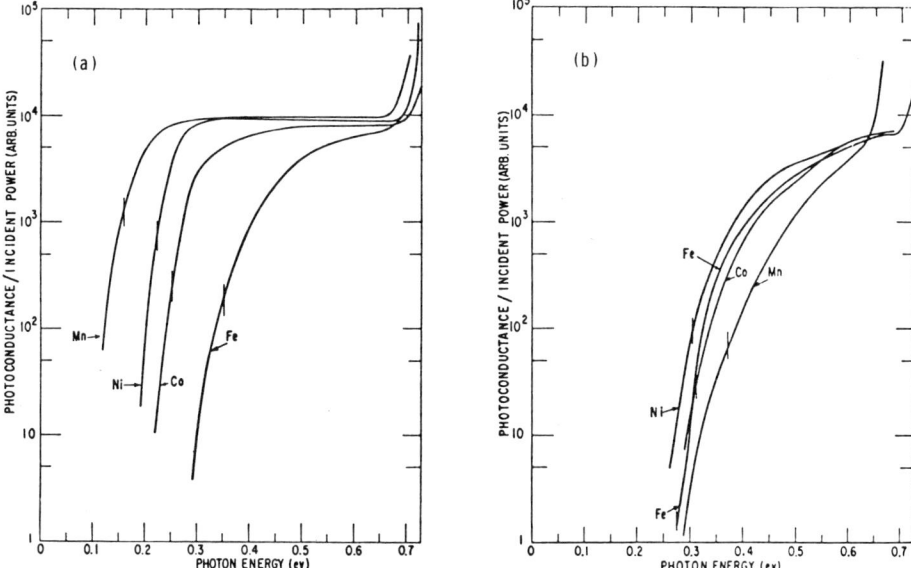

FIG. 31. Photoconductive spectral response for Mn, Ni, Co, and Fe in Ge. (a) *p*-type levels, (b) *n*-type levels. The vertical bars indicate thermal activiation energies. All data taken at 77°K. (After Newman and Tyler.[22])

found no particular application and are not offered commercially. Detectors made from Ge:Au using the 0.15 eV acceptor level have been manufactured since about 1956.[112] This type is still useful as a general purpose detector for the 1–10 μm range. It is popular for use in laser research because of its fast response and resistance to damage by high intensity illumination. Figure 32 shows detectivity curves obtained for these three types of detectors.

For applications requiring the very highest sensitivity in the 1–10 μm range, intrinsic detector materials such a PbS, PbSe, InAs, InSb, and the ternary alloys such as HgCdTe offer several advantages over impurity Ge or Si. For this reason, intrinsic detector materials are much more widely used for this wavelength range.

b. *Medium IR Range, 10–50 μm*

The impurity elements, Hg, Cd, Cu, Zn, and Be exhibit photoconductivity in the range 10–50 μm when introduced into the Ge crystal lattice. Of these five, Hg, Cu, and Zn have found the widest practical use with Hg of most interest because its spectral response nicely matches the 8–13 μm

[112] W. Beyen, P. Bratt, H. Davis, L. Johnson, H. Levinstein, and A. MacRae, *J. Opt. Soc. Amer.* **49**, 686 (1959).

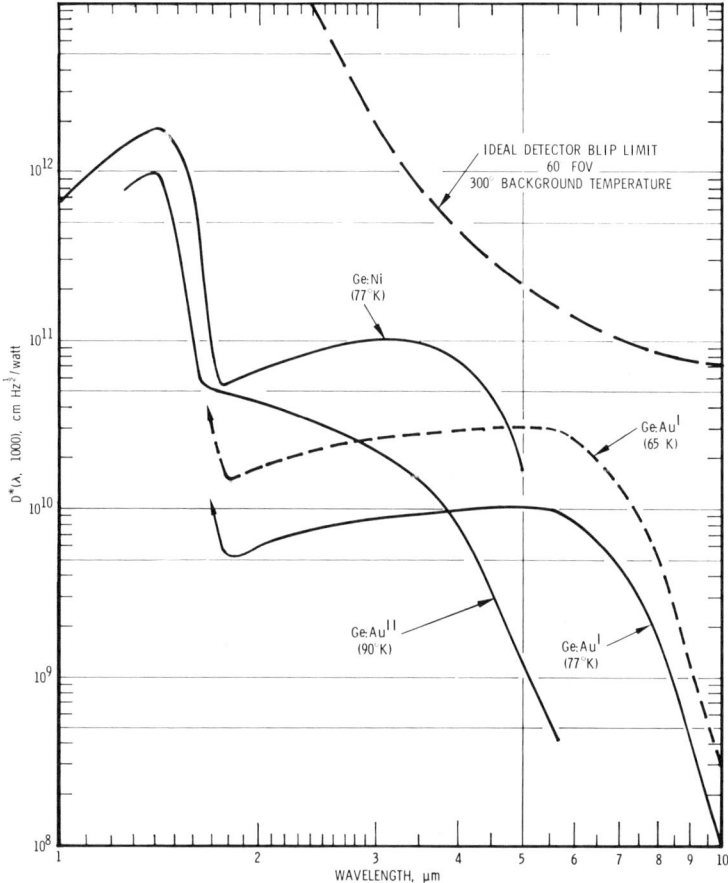

FIG. 32. Spectral variation of D^* for selected short wavelength impurity Ge detectors.

atmospheric window. The second Zn level at 0.09 eV above the valence band was also of great interest at one time for the same reason. However, the need for complete compensation of the lower Zn level with a donor impurity such as Sb leads to practical difficulties in the growth of these crystals. Furthermore, in such a system the free carrier recombination proceeds by the process

$$p^+ + Zn^{2-} \rightarrow Zn^-. \tag{88}$$

The doubly negative charge on the Zn recombination center gives it a large cross section for capture of a positive hole and thus leads to very short free hole lifetimes in this material. Experience has shown that better detectors

2. IMPURITY GERMANIUM AND SILICON INFRARED DETECTORS

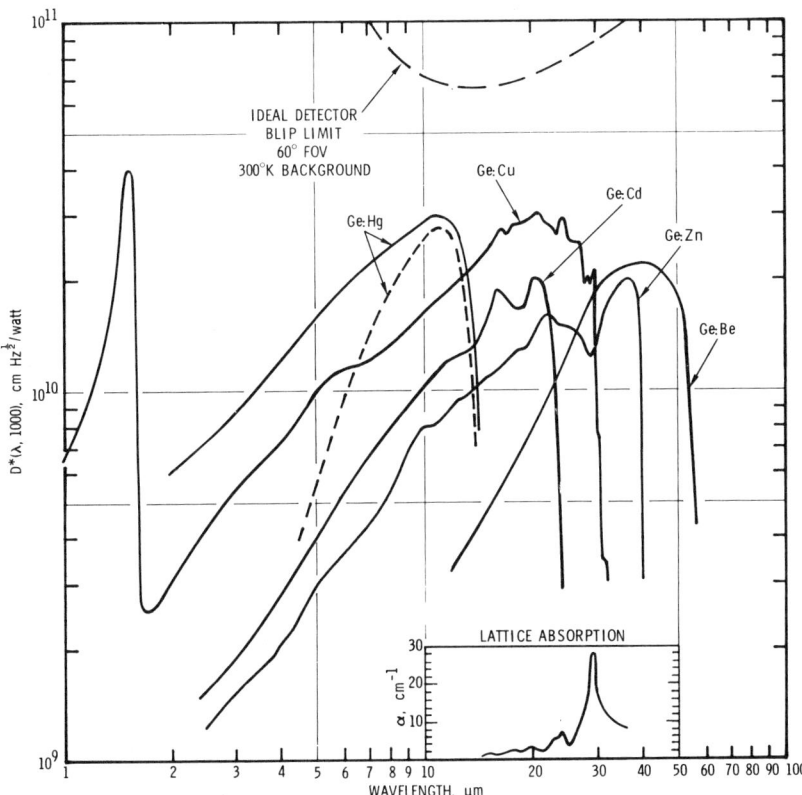

FIG. 33. Spectral variation of D^* for medium wavelength impurity Ge detectors. (Ge : Be after Shenker et al.[113]; Ge : Hg, Ge : Cu, Ge : Cd, and Ge : Zn after Bratt et al.[114]; Ge : Zn after Burstein et al.[5]; the dashed curve represents early Ge : Hg after Levinstein.[118])

are usually obtained when one uses only the lowest acceptor level of a given impurity. Thus, Hg has surpassed Zn^{II} as a dopant element for commercial use.

Figure 33 shows detectivity versus wavelength curves obtained for the five midrange impurity elements. Data on Hg, Cd, and Cu was taken at Santa Barbara Research Center, while that on Zn was taken from the work of Burstein et al.[5] and that on Be from the work of Shenker et al.[113]

In the early work of Borrello and Levinstein, Hg was introduced into the Ge melt at atmospheric pressure and therefore Hg concentrations were limited to less than 5×10^{14} atoms/cm³. With the development of pressure controlled doping techniques (see Section 9) greater concentrations of Hg

---

[113] H. Shenker, E. M. Swiggard, and W. J. Moore, *Trans. Met. Soc. AIME* **239**, 347 (1967).

could be introduced and thus the quantum efficiency was improved, particularly toward the short wavelength direction. Figure 33 compares the early spectral response curve[8] of lightly doped material with that obtained in present day material. Germanium:mercury detectors must be cooled to temperatures below 35°K for background limited operation under normal 300°K background.

Copper and zinc were among the earliest deep level impurities in Ge to be discovered and, because both of these may be introduced into the Ge lattice in relatively large concentrations, they also can be used to make detectors with high quantum efficiency. Burstein et al.[3,5] and Kaiser and Fan[4] were the first to demonstrate the photoconductive response of these materials. Subsequent work has brought them to a high level of performance.[114,116-118] The 0.04 eV Cu acceptor level provides sensitivity to 30 μm and requires cooling to temperatures below 18°K for background limited operation. The 0.03 eV Zn level provides sensitivity to 40 μm and must be cooled below 10°K for background limited operation.

Cadmium in germanium was investigated as a replacement for Cu that would have good response in the 8–13 μm region, but operate at higher temperatures. Early workers had difficulty getting high concentrations of Cd into the Ge because the doping was done at atmospheric pressure. As in the case of Hg, high pressure doping overcame this problem and Ge:Cd detectors with high quantum efficiency were ultimately made.[115] After the discovery of the properties of Hg in Ge, interest in Ge:Cd lagged and it is now used only in specialized applications where a spectral response intermediate between that of Ge:Hg and Ge:Cu is necessary. Germanium: cadmium must be cooled to temperatures below 25°K for background limited operation.

The photoconductive properties of Be in Ge were reported by Shenker et al.[113,119] This acceptor has an ionization energy of 0.024 eV and can be introduced into Ge in concentrations greater than 3×10^{16} atoms/cm^3. As shown in Fig. 33, its spectral response extends to 52 μm. Although background limited performance for this type of detector was not confirmed, there is no reason to believe it cannot be made so.

c. Long IR Range, 50–120 μm

Infrared detectors for the 50–120 μm range have been made using Ge

[114] P. Bratt, W. Engeler, H. Levinstein, A. MacRae, and J. Pehek, *Infrared Phys.* **1**, 27 (1961).
[115] P. R. Bratt and P. P. Debye, *Proc. IRIS* **6**, No. 1, 103 (1961) (unclassified paper).
[116] H. D. Adams, W. J. Beyen, and R. L. Petritz, *J. Phys. Chem. Solids* **22**, 167 (1961).
[117] D. E. Bode and H. A. Graham, *Infrared Phys.* **3**, 129 (1963).
[118] H. Levinstein, *Appl. Opt.* **4**, 639 (1965).
[119] R. F. Wallis and H. Shenker, NRL Memorandum Rep. 1630. U. S. Naval Res. Lab. 1965; NRL Memorandum Rep. 1712. U.S. Naval Res. Lab., 1966.

doped with impurity elements from groups III and V of the periodic table. Some early work using Ge:In was reported by Burstein et al.,[3] but their measurements only extended to 38 μm because of spectrometer limitations. Fray and Oliver[120] made a far IR detector from Ge:Sb and measured long wavelength sensitivity. However, their material was apparently highly compensated with p-type impurities and as a result, exhibited high resistance and low sensitivity. The first good detectors for this long wavelength range were made by Shenker et al.[9,121] at the U.S. Naval Research Laboratories using B and Ga as dopant elements. Lifshitz et al.[122] and Nagasaka et al.[123] subsequently reported results on both group III and group V dopants.

The basic difficulty in the fabrication of long wavelength detectors is caused by impact ionization. One cannot apply high field strengths to these crystals to get a large photoconductive gain because of impact ionization breakdown. (See Section 6b.) Thus, detector sensitivity must be obtained by making material with a minimum of compensating impurities so as to obtain long free carrier lifetimes.

Swiggard and Shenker[124] developed methods for doing this using the horizontal zone refining technique and both B and Ga impurities. One method takes advantage of the large difference in segregation coefficient between the desired dopant impurity and the unwanted compensating impurities. Boron and gallium have segregation coefficients of 17 and 0.087, respectively, while the major donor impurities, As and Sb, have the respective values 0.02 and 0.003. Thus, multiple zone refining passes will reduce the donor concentrations in the crystal to a greater extent than the intentionally added acceptor impurity. The second method utilizes multiple zone passes to reduce all impurities to a very low level and then on the final pass a pellet of the desired p-type dopant is dropped into the melt from a quartz spoon extending into the furnace tube. This minimizes the introduction of any unwanted impurities during the doping operation. High purity seed crystals must also be available for this method to be successful. These techniques allow the preparation of long wavelength material with group III dopants in the $10^{14}-10^{16}$ atoms/cm^3 range and compensating donor impurities as low as $1-5 \times 10^{12}$ atoms/cm^3.

Low noise contacts to long wavelength detectors are difficult to achieve. This is thought to be due to sporadic impact ionization breakdown occurring in localized regions near the metal to semiconductor interface as a result

---

[120] S. J. Fray and J. F C. Oliver, *J. Sci. Instrum.* **36**, 195 (1959).
[121] H. Shenker, W. J. Moore, and E. M. Swiggard, *J. Appl. Phys.* **35**, 2965 (1964).
[122] T. M. Lifshitz, F. Ya. Nad', and V. I. Sidorov, *Fiz. Tverd. Tela* **8**, 3208 (1966) [*English Transl.: Sov. Phys.—Solid State* **8**, 2567 (1967)].
[123] K. Nagasaka, Y. Oka, and S. Narita, *Solid State Commun.* **5**, 333 (1967).
[124] E. M. Swiggard and H. Shenker, *J. Electrochem. Soc.* **113**, 92 (1966).

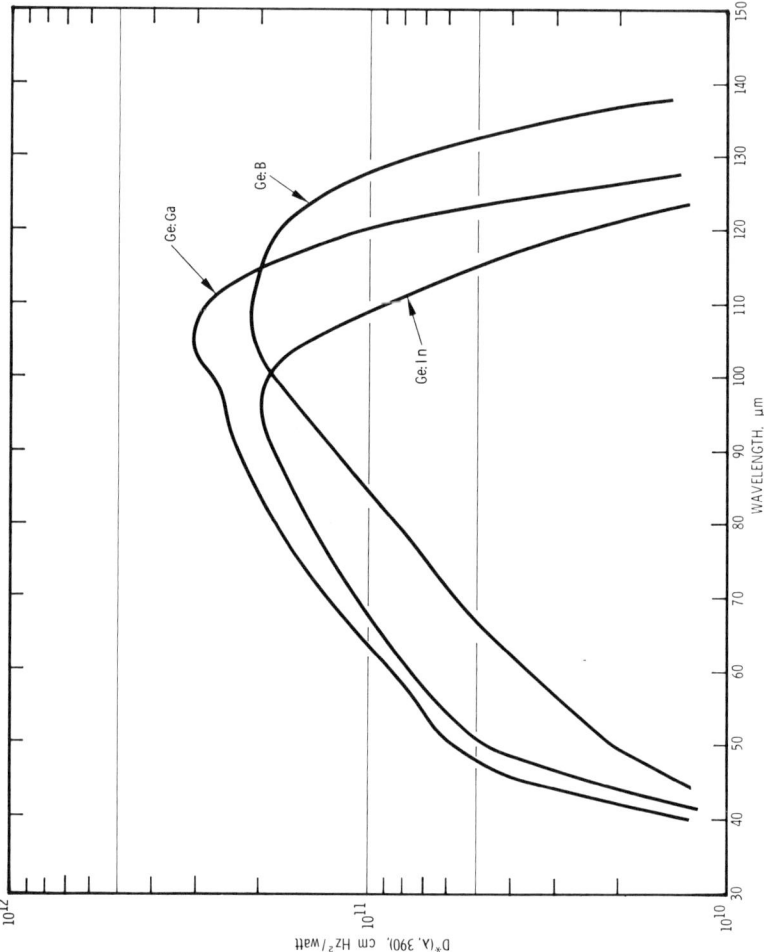

FIG. 34. Spectral variation of D^* for long wavelength impurity Ge detectors. Measurements were made using cold crystal quartz and black polyethylene filters to block out background radiation below 30 μm. (Data on Ge: B and Ge: Ga after Moore and Shenker[9] and Shenker et al.[121]; data on Ge: In after Wallis and Shenker.[124a]

of nonuniform electric fields. Special care must be exercised to provide a uniform metal to semiconductor transition. Some workers have tried diffusion of a p-type impurity to produce a p^+-p contacts while other have actually grown p^+ regions on the two ends of a section of crystal. Moore and Shenker[9] used B doped Ge to grow p^+ contacts on their Ge:Ga detectors.

Figure 34 shows the spectral dependence of D^* for long wavelength impurity Ge detectors. The shape of the curves near the long wavelength cutoff was found to vary somewhat with impurity concentration and electric field strength.[122] Little data has been reported on the detectivity of these devices. However, we have shown in Fig. 34 what is believed to be the state of the art in performance. Moore and Shenker[9] reported a value of D^* (104 μm, 390 Hz) = 3.1×10^{11} cm-Hz$^{1/2}$/W for their best Ge:Ga detector and 2.1×10^{11} cm-Hz$^{1/2}$/W for a Ge:B detector. Both were measured under background filtered conditions (i.e., cold crystal quartz plus black polyethylene filter and 0.15 sr f.o.v. cold shielding). Operating field strengths were 1 to 2 V/cm. They estimated the Ge:Ga D^* to be a factor 2.4 below the photon noise limit. Lifshitz et al.[122] indicated they had made detectors with sensitivities comparable to those made by Moore and Shenker[9], but no data were reported.

Because of the very large photoionization cross section for shallow impurities in Ge, high doping concentrations are not required for good quantum efficiency. Best performance was obtained[9,121] in Ge:B and Ge:Ga detectors with impurity concentrations of about 1×10^{14}/cm^3. This gives good sensitivity at long wavelengths, near the peak response, but results in a low quantum efficiency at short wavelengths. However, such a situation may be desirable in some applications.

Photoresponse beyond the expected long wavelength cutoff was first observed by Lifshitz and Nad'[125] in Ge:Sb. This photoconductivity showed a line spectrum extending to 160 μm similar to the absorption spectrum caused by photon induced transitions of electrons from the ground state to excited states of the impurity atom and was correctly attributed to a photothermal effect. The same effect was subsequently observed in p-type materials, in particular, Ge:B and Ge:In.[126] A further discussion of the photothermal effect is given in Section 14c. Nagasaka and Narita[127] confirmed the photothermal effect in Ge:Sb and also observed an additional photoconductivity beyond the long wavelength cutoff that was structureless and had little

---

[124a] R. F. Wallis and H. Shenker, NRL Memo. Rep. 1385, U.S. Naval Res. Lab., 1963.
[125] T. M. Lifshitz and F. Ya. Nad', *Dokl. Akad. Nauk. SSSR* **162**, 801 (1965) [*English Transl.: Sov. Phys.—Dokl.* **10**, 532 (1965)].
[126] V. I. Sidorov and T. M. Lifshitz, *Fiz. Tverd. Tela* **8**, 2498 (1966) [*English Transl.: Sov. Phys.—Solid State* **8**, 2000 (1967)].
[127] K. Nagasaka and S. Narita, *Solid State Commun.* **7**, 467 (1969).

temperature dependence below 4.2°K. They hypothesized that this might be due to photoionization of a very shallow state caused by clustering of a number of impurity atoms. This photoconductivity is relatively weak and probably not of use in a practical IR detector.

12. Operating Characteristics—Silicon

As in the previous section, it will be convenient to divide this discussion of impurity silicon operating characterisitics into two parts according to wavelength range. The first part will deal with detectors for 1–10 μm and the second with those beyond 10 μm.

a. Short IR Range, 1–10 μm

A number of impurity elements produce energy levels in Si that could be used for short IR detectors. However, as may be seen from Fig. 3, most of these energy levels are greater than 0.25 eV from either band edge which means the detector's spectral response will not extend beyond 5 μm. Many of these deep levels have multiple states of ionization and follow the same pattern as in Ge with the exception that, in Si, they have larger ionization energies. For example, the first acceptor state of Zn in Si is 0.31 eV from the valence band while in Ge, it is only 0.03 eV from the valence band. The second Zn acceptor level in Si lies above midband while in Ge it is still in the lower half of the band at 0.09 eV. Similar behavior is seen in Cd. For Hg, both acceptor states lie in the upper half of the band. Gold presents an interesting example in that its lowest acceptor state is above midband at a point 0.54 eV below the conduction band. The two higher acceptor levels which were identified in Ge have not been found in Si and are presumed to lie in the conduction band and thus be unobservable by a photoexcitation process. The positive acceptor ion state (A^+) for Au in Si has a much larger binding energy (0.35 eV) than in the case of Ge (0.05 eV).

The reason for this behavior is, in part, due to the smaller dielectric constant of Si. This allows a stronger Coulomb attraction between the impurity atom and its bound electron (or hole) that results in larger binding energies. This applies not just to the deep states, but to the shallower impurity states as well, to be discussed later.

Examples of photoconductive response curves for a number of these deep impurity levels are shown in Fig. 35. Note the differences in shape of the spectral response between n- and p-type states of the same impurity level just as was observed in the case of Ge.

These deep level impurities have not been exploited for use in IR detectors, the reason being the same as that given for deep levels in Ge. There are well-developed intrinsic detector materials available that have several advantages over the extrinsic material.

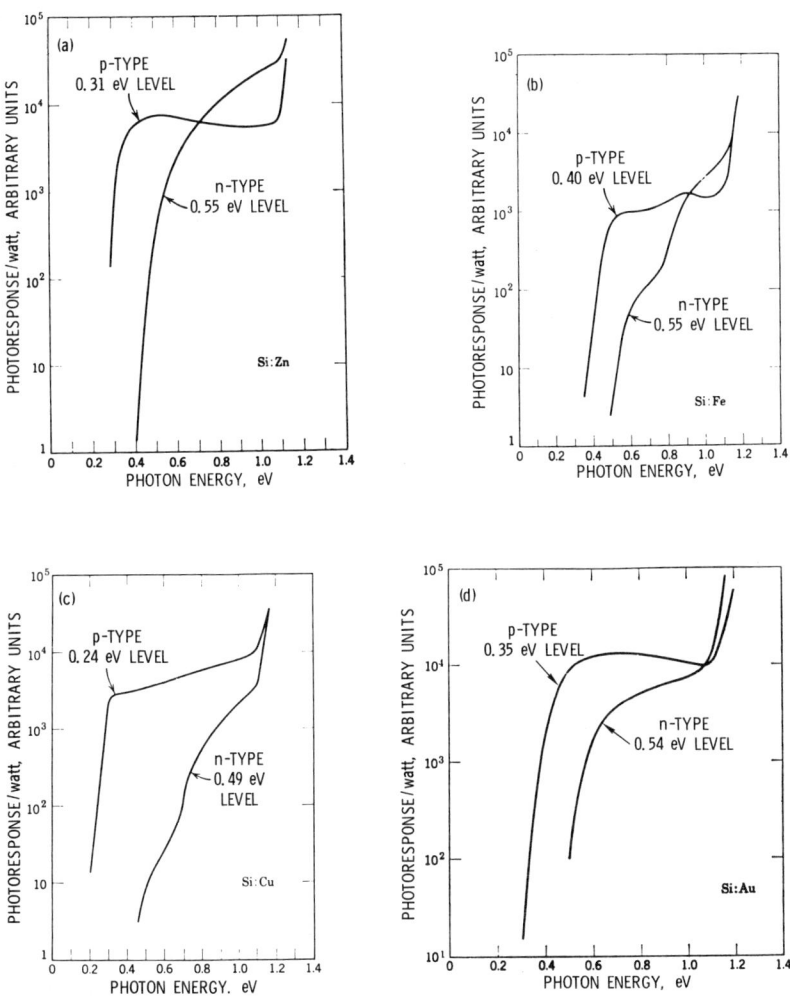

FIG. 35. Examples of photoconductive response of deep level impurities in Si. (a) Si : Zn (after Carlson.[128]); (b) Si : Fe and (c) Si : Cu (after Collins and Carlson[129]); (d) Si : Au (after Collins et al.[130]).

One possible exception to the above remark is the impurity In. This has received more attention over the years by a number of investigators, and recently, detector performance data has been published. Blakemore and

[128] R. O. Carlson, Phys. Rev. **108**, 1390 (1957).
[129] C. B. Collins and R. O. Carlson, Phys. Rev. **108**, 1409 (1957).
[130] C. B. Collins, R. O. Carlson, and C. J. Galagher, Phys. Rev. **105**, 1168 (1957).

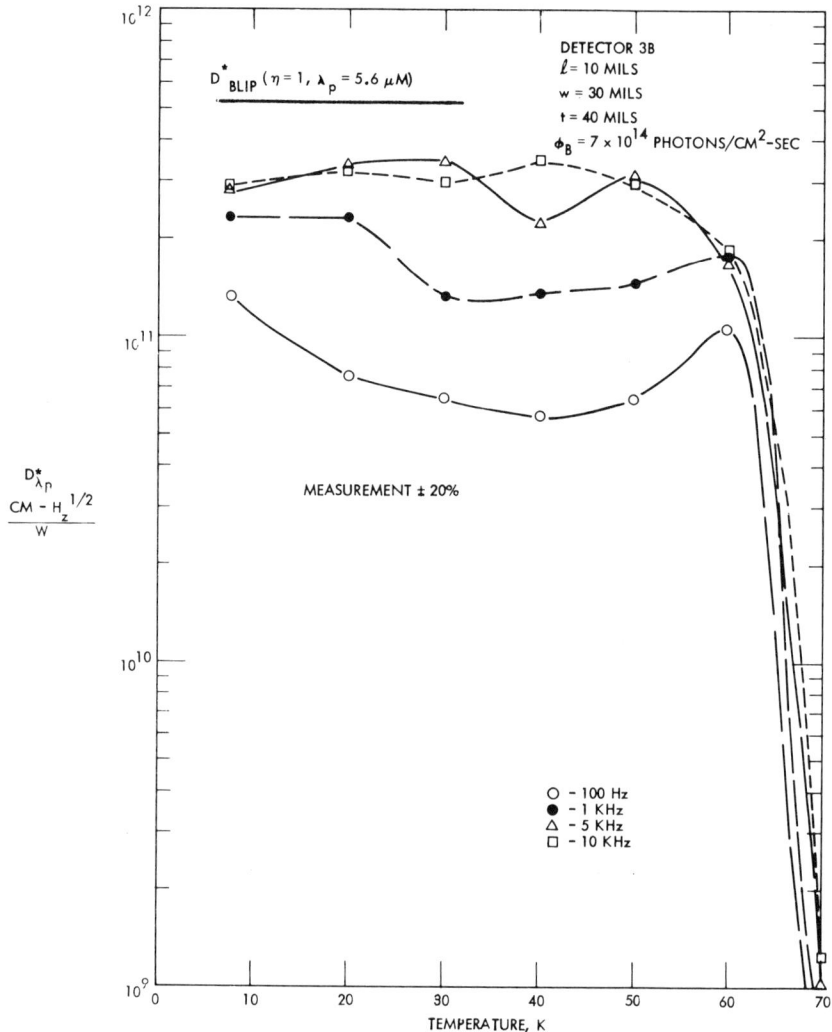

FIG. 36. $D^*(\lambda_p)$ as a function of temperature and frequency for a Si:In detector; $\lambda_p = 5.6$ μm. (After Pines and Baron.[136])

co-workers[25,131,132] have thoroughly investigated the optical absorption cross section and hole recombination cross section. The resulting values were shown in Tables I and VIII. The g–r noise of Si:In has been

[131] J. S. Blakemore and C. E. Sarver, *Phys. Rev.* **173**, 767 (1968).
[132] H. J. Mason, Jr., and J. S. Blakemore, *J. Appl. Phys.* **43**, 2810 (1972).

measured[133-135] and found to fit with the theory outlined in Section 7. Figure 36 shows detectivity versus temperature data for a Si:In detector containing 2.7×10^{17} In atoms/cm^3. Background limited performance is achieved at temperatures below about 60°K and a D^* (5.6μ, 10kHz) of 3×10^{11} cm-Hz$^{1/2}$/W was measured in this range with an effective background photon flux on the detector of 7×10^{14} photons/sec/cm^2. An ideal detector with unity quantum efficiency would have a D^* of 5.3×10^{11} cm-Hz$^{1/2}$/W under these conditions. The relative spectral response of this detector at 7°K is shown in Fig. 37.

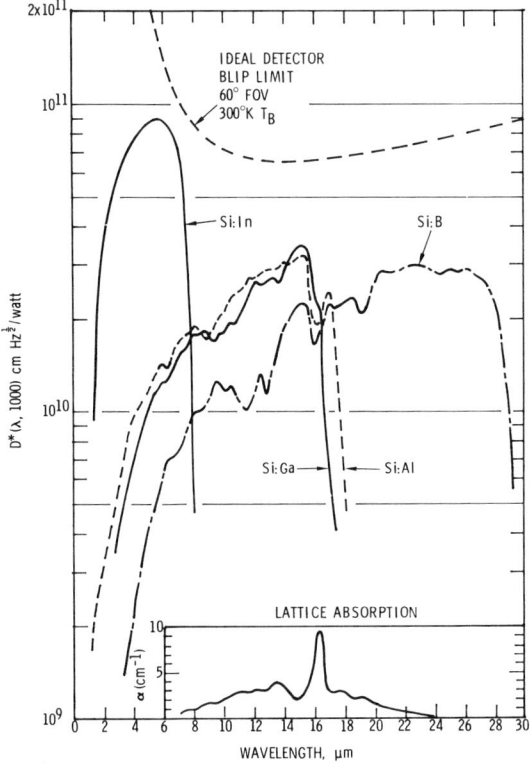

FIG. 37. Spectral variation of D^* for p-type impurity Si detectors.

[133] E. E. Godik and Ya. E. Pokrovskii, *Fiz. Tverd. Tela* **6**, 2358 (1964) [*English Transl.: Sov. Phys.—Solid State* **6**, 1870 (1965)].
[134] V. V. Proklav, E. E. Godik, and Ya. E. Pokrovskii, *Fiz. Tverd. Tela* **7**, 407 (1965) [*English Transl.: Sov. Phys.—Solid State* **7**, 326 (1965)].
[135] H. Preier, *J. Appl. Phys.* **39**, 194 (1968).

b. Medium IR Range, 10–30 μm

The shallow level impurities in Si can be used to produce IR detectors having long wavelength cutoffs ranging from 16 μm–30 μm. These are B, Al, Ga, Sb, P, As, and Bi. After the original discovery work of the 1950's little interest was developed for these impurity Si IR detectors. Then in 1967 Soref[10] reexamined the situation and found that detectors with sensitivity rivaling that of impurity Ge could be made. In fact, the higher solubility of these impurities in Si meant that doped crystals with larger absorption coefficients could be obtained which would lead to detectors with even higher quantum efficiencies than impurity Ge. Subsequent work has established that this is indeed true and has led to the consideration of impurity Si for some applications that were once filled by impurity Ge.

Figures 37 and 38 show detectivity versus wavelength for p- and n-type impurity Si detectors. Comparison of these data with that for the mid-IR

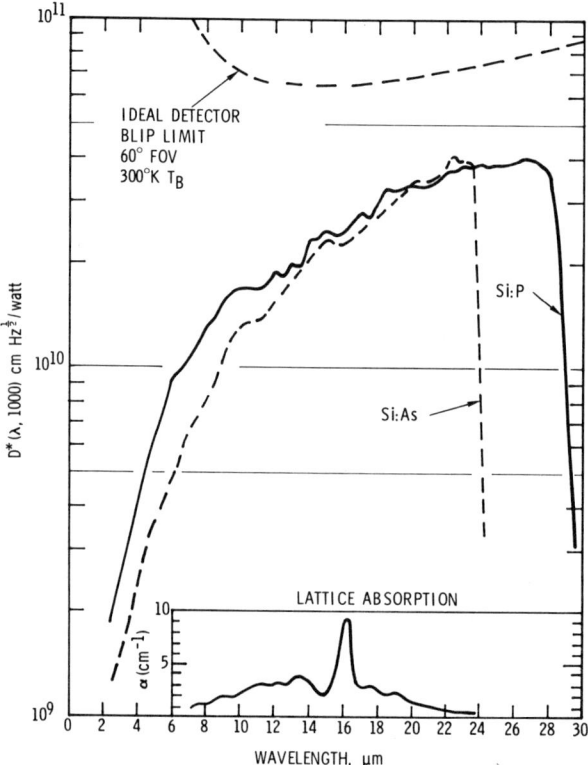

FIG. 38. Spectral variation of D^* for n-type impurity Si detectors.

range impurity Ge detectors shows that Si:P is similar to Ge:Cu, Si:As is similar to Ge:Cd, and Si:Ga is close to Ge:Hg except that its cutoff wavelength is somewhat longer. In a 300°K environment, these doped Si detectors must be cooled below 30°K for background limited performance. Sclar[135a] has recently presented more extensive data on a number of doped Si detectors including temperature dependence of D^*.

13. GERMANIUM–SILICON ALLOYS

When Ge is mixed with Si, a single phase alloy is obtained for all proportions of the two elements. The energy bandgap of the resulting semiconductor varies continuously from 0.7 eV for pure Ge to 1.2 eV for pure Si.[137,138] It has also been found that the ionization energy for impurity atoms in the alloy generally increases as the Si content is increased. Thus, if the energy level of a given impurity in Ge is too low for a particular application, it could be increased by the addition of Si to the Ge host crystal.

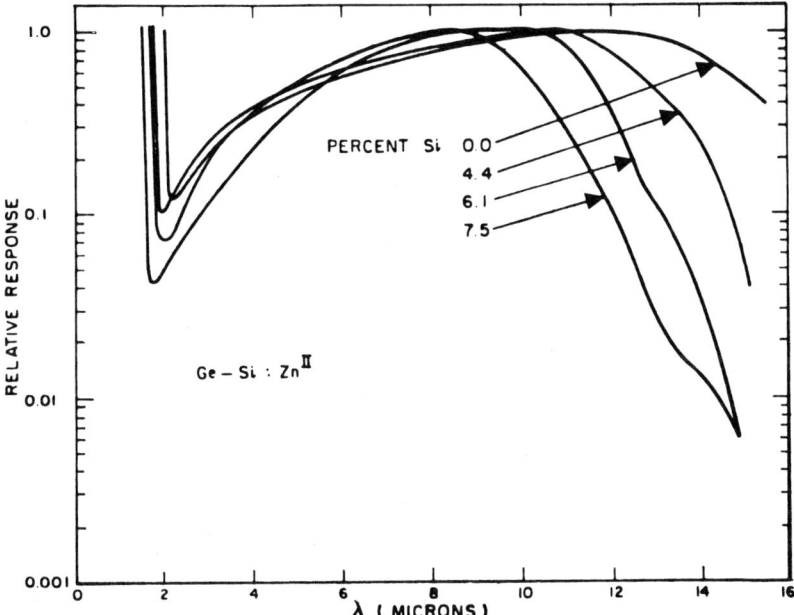

FIG. 39. Relative spectral response curves for Ge, Si : Zn^{II} with various Si contents. (After Morton et al.[7])

[135a] N. Sclar, *Infrared Phys.* **16**, 435 (1976).
[136] M. Y. Pines and R. Baron, *IEEE Int. Electron Devices Meeting, Washington, D.C.*, 1974.
[137] E. R. Johnson and S. M. Christian, *Phys. Rev.* **95**, 560 (1954).
[138] F. Herman, *Phys. Rev.* **95**, 847 (1954).

Morton et al.[7] utilized this result to prepare impurity Ge–Si alloy IR detectors with spectral response optimized for the 8–13 μm atmospheric window. They studied a number of impurities, but found the most useful ones to be the second Zn acceptor level which has an ionization energy of 0.09 eV in pure Ge and the low lying Au donor level which has an ionization energy of 0.05 eV. When alloyed with the appropriate amount of Si (i.e., 5% Si for Zn^{II} impurity and 11% Si for the Au impurity) detectors were obtained having a $D^*(\lambda_p)$ at 10 μm of 1×10^{10} cm-$Hz^{1/2}$/W for Zn^{II} and 6×10^9 cm-$Hz^{1/2}$/W for Au when operated at temperatures of 48–50°K. Figure 39 shows how the relative spectral response of Ge–Si:Zn^{II} changes with alloy composition.

The characteristics of Hg impurity in Ge–Si alloys has also been studied.[139] Detectors were obtained with $D^*(\lambda_p)$ of 2×10^{10} cm-$Hz^{1/2}$/W when operated at 40°K with an $f/2$ field of view. The relative spectral response for the Ge–Si:Hg detector is shown in Fig. 40 where it is compared with that of a Ge:Hg detector with no Si added. The striking thing about this comparison is the gradual falloff toward long wavelengths of the alloy detector as compared to the detector with no Si. Similar behavior is to be noted in the Ge–Si:Zn^{II} curves of Fig. 39. This unusual behavior must be due to some feature of the alloy crystals which is different from either pure Ge or Si.

A possible explanation is afforded by the following hypothetical model. Consider that the Si atoms are distributed randomly throughout the Ge

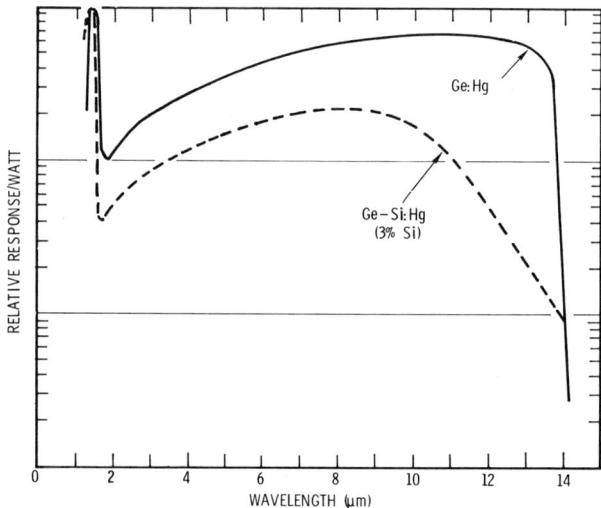

FIG. 40. Comparison of the relative spectral response of Ge, Si:Hg ($N_{Hg} = 4.4 \times 10^{15}$ atoms/cm³) with Ge:Hg ($N_{Hg} = 5.8 \times 10^{15}$ atoms/cm³) at 5°K.

[139] P. R. Bratt, unpublished report, 1968.

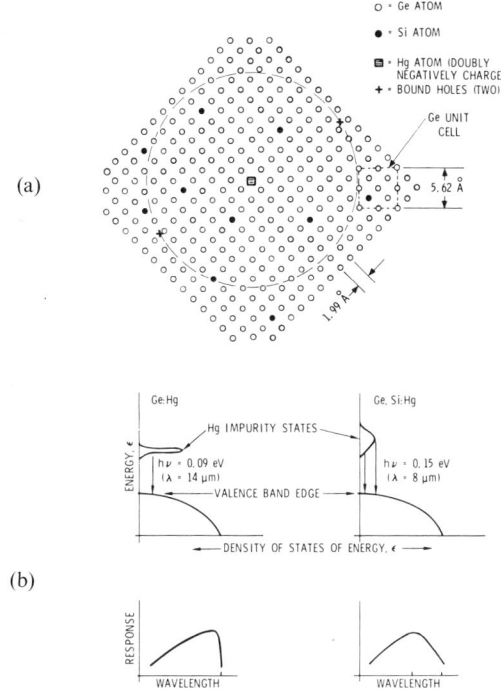

FIG. 41. Model of randomly distributed Si atoms in Ge. (a) Projection of Ge lattice on (100) plane showing 3% Si atoms randomly distributed around one Hg atoms. (b) Density of states diagrams for Ge : Hg and Ge, Si : Hg showing hole transitions from Hg ground state to the valence band and resultant spectral response shapes.

lattice as shown in the diagram of Fig. 41a which serves to illustrate the region about one Hg atom. The Hg atom, being a double acceptor, is assumed to produce a pseudo-helium atom-type of center, that is, a doubly negative charged "nucleus" surrounded by two "positive electrons" or holes. The radius of the ground state orbit of the holes is approximately 15 Å. It is apparent that these bound holes will encounter many Si atoms during their peregrinations around the Hg nucleus. Encounters with Si atoms could be expected to produce energy perturbations, causing the ground state energy level of the system to be broadened as well as raised in energy. The density of states function for the Hg energy levels might then look somewhat as shown in Fig. 41b. Since transitions would take place preferentially where the density of states function is largest, such a model would account for the shift of the peak spectral response to about 8 μm while still retaining a "tail" response out to 14 μm. Further support for this model is provided by the fact that no evidence of excited state transitions is observed in Ge–Si impurity detectors. The ground state broadening apparently causes a "smearing out" of all excited state transitions so they are unobservable.

Because of complications such as this, plus the difficulty of growing uniform Ge–Si alloy crystals of high crystalline perfection, what once appeared to be a very promising approach for making impurity IR detectors has now been essentially abandoned.

14. Special Features

In this section, a number of special features that are common to both impurity Ge and Si detectors are discussed. These are (a) non-Ohmic behavior, (b) characteristics under low background applications, (c) effects of excited states on the spectral response, and (d) performance as a high speed detector.

a. Non-Ohmic Behavior

Impurity Ge and Si detectors generally display a non-Ohmic behavior, that is to say, the current through the device is not always proportional to applied voltage. Although non-Ohmic effects can sometimes arise at the metal to semiconductor contacts, the bulk conductivity is also found to vary with the electric field strength in the crystal. The reasons for this will be discussed in the following paragraphs.

The current density through a detector (assumed to be p-type) is given by

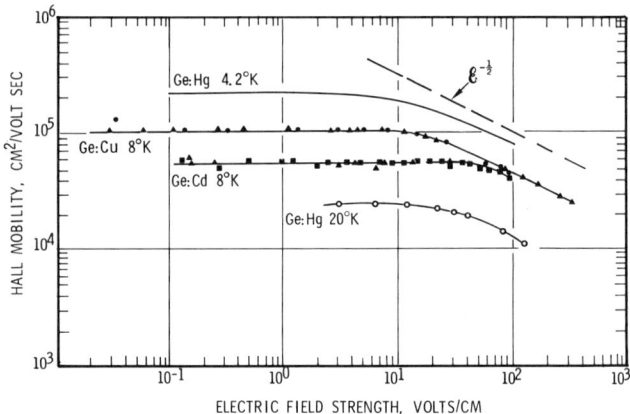

Fig. 42. Variation of Hall mobility in p-type Ge with electric field strength. (Ge : Hg, 4.2°K data after Reynolds et al.[140]; Ge : Hg, 20°K data after Yariv et al.[141])

[140] R. A. Reynolds et al., Texas Instruments, unpublished report, 1966.
[141] A. Yariv, C. Buczek, and G. S. Picus, in Proc. 9th Int. Conf. Phys. Semicond., Moscow, p. 500. Nauka, Leningrad, 1968.

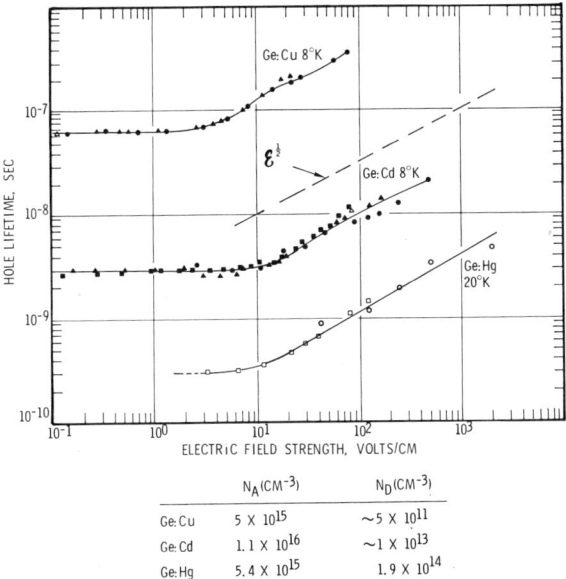

FIG. 43. Variation of hole lifetime in p-type Ge with electric field strength. (Ge : Hg, 20°K data after Yariv et al.[141])

$$j = pe\mu\mathscr{E} \qquad (89)$$

and, for Ohm's law to hold, the conductivity $pe\mu$ would be invariant with electric field. This is not the case with impurity Ge and Si because the quantities p and μ may both be field dependent. When p is determined by background radiation, then it is proportional to free carrier lifetime according to

$$p = \eta J_B \tau / d. \qquad (90)$$

The lifetime is field dependent and, since all other quantities in this equation are constants, this field dependence is reflected in p.

Figures 42 and 43 show measured variations of μ and τ with electric field strength for typical impurities in Ge. Most of this data was obtained by photo–Hall measurements which give values for μ and p directly. If the photon flux on the sample and its quantum efficiency are known, then τ values can be obtained by the use of Eq. (90). Alternatively, τ values may be obtained from the rate of change of photoconductivity, using Eq. (39) or (40). The τ data points in the high field region of the Ge:Hg curve of Yariv et al.[141] were obtained in this way from high frequency heterodyne signal measurements using CO_2 lasers.[142] The wide variation in lifetime values

[142] C. J. Buczek and G. S. Picus, *Appl. Phys. Lett.* **11**, 125 (1967).

for the three curves in Fig. 43 is due to different levels of compensating donors in these crystals. For field strengths less than about 10 V/cm, μ is constant and, for less than about 5 V/cm, τ is constant. For field strengths greater than 10 V/cm, μ decreases as $\mathscr{E}^{-1/2}$ while τ increases as $\mathscr{E}^{1/2}$.

The change in μ and τ with electric field strength is due to carrier "heating." Under acceleration by the field, carriers may gain kinetic energy faster than they are able to give it up to the lattice by collisions. Thus, their average kinetic energy increases and they are said to be "warm" or "hot" depending upon how much the average energy is increased above the thermal equilibrium (lattice temperature) value. Average energy may be related to temperature by the equation

$$\langle E \rangle = \langle m^* v^2/2 \rangle = 3kT/2, \tag{91}$$

where the brackets indicate an average over all carrier energies. If the average energy is known, the "temperature" of the free carrier distribution can be obtained. Impurity Ge and Si detectors are usually operated at field strengths where the free carrier temperature is considerably above that of the lattice.

At low electric fields, the mobility is determined mainly by neutral impurity scattering. The collision rate with neutral impurities is independent of carrier velocity.[143] At higher electric fields, carrier velocities are increased and scattering by acoustic[144] vibrational modes of the crystal lattice predominates. In this regime the mobility is represented by

$$\mu = 4el/3\sqrt{\pi}\, m^* v_T, \tag{92}$$

where l is the mean free path between collisions and $v_T = (2kT/m^*)^{1/2}$ is the thermal velocity (actually the rms velocity along one crystal axis). At high electric fields, Shockley[145] has shown that v_T increases as $\mathscr{E}^{1/2}$. Since the mean free path is constant, this means that $\mu \propto \mathscr{E}^{-1/2}$.

A mobility decreasing as $\mathscr{E}^{-1/2}$ was first observed by Ryder[146] in both Ge and Si. At much higher field strengths, Ryder found the mobility goes over to an $\mathscr{E}^{-1}$ dependence. The data in Fig. 42 for three different impurities in Ge also shows a well-defined $\mathscr{E}^{-1/2}$ slope. The $\mathscr{E}^{-1}$ range is not seen here because this requires field strengths above the normal operating range for IR detectors which would produce impact ionization breakdown. Conwell[147] has shown that optical mode lattice scattering is also important, particularly

---

[143] M. S. Sodha and P. C. Eastman, *Phys. Rev.* **108**, 1373 (1957).

[144] W. Shockley, "Electrons and Holes in Semiconductors," p. 277. Van Nostrand-Reinhold, Princeton, New Jersey, 1950.

[145] W. Shockley, *Bell Syst. Tech. J.* **30**, 990 (1951).

[146] E. J. Ryder, *Phys. Rev.* **90**, 766 (1953).

[147] E. Conwell, *J. Phys. Chem. Solids* **8**, 234 (1959).

in the temperature range above 100°K. However, in the low temperature range of interest here, very few optical phonons are present. On the other hand, when free carriers are heated to the point at which their energy becomes comparable to the optical phonon energy (0.037 eV for Ge), then energy loss to optical modes begins.

The observed variation of lifetime with electric field strength is also in accord with theoretical expectations. In the cascade capture theory,[41–45] the capture cross section is given by

$$\sigma_c = \sum_j \sigma_c(j) S(j), \qquad (93)$$

where $\sigma_c(j)$ is the cross section for capture into the jth excited state and $S(j)$ is the sticking probability for that state, i.e., the probability that the captured charge will cascade down to the ground state and remain captured until freed again by some excitation mechanism. The initial capture takes place when a free carrier suffers an energy losing collision with the lattice in the vicinity of the capturing center and then becomes bound by the Coulomb attractive force. On this basis it would be expected that

$$\sigma_c(j) = \text{const.}/\langle E \rangle = \text{const.}/v_T^2. \qquad (94)$$

Since the lifetime is $\tau = [\sigma_c \langle v \rangle N_A^-]^{-1}$, it follows that $\tau \propto v_T$. As we have already seen, $v_T \propto \mathscr{E}^{1/2}$, therefore $\tau \propto \mathscr{E}^{1/2}$, which is the dependence observed experimentally and shown in Fig. 43.

Yariv et al.[141] have used this type of analysis along with experimental data on Ge:Hg to deduce a relation between effective carrier temperature and electric field strength, which is

$$T_e = 2.4\mathscr{E}. \qquad (95)$$

Thus at $\mathscr{E} = 100$ V/cm, $T_e = 240$°K, which is much greater than the lattice temperature of 20°K. The average energy corresponding to this temperature is $3kT_e/2 = 0.031$ eV, still smaller than the optical phonon energy of 0.037 eV. At higher field strengths, carrier energies will exceed the optical phonon energy and the simple analysis given above no longer holds.

In Si the optical phonon energy is 0.058 eV, so one might think that free carriers could be made considerably "hotter" by an electric field than in Ge. This will be true for the deeper level dopants such as Ga, Al, In, and Bi which have ionization energies > 0.06 eV. However, for the shallower dopants, such as B, P, As, and Sb, impact ionization will occur before the optical phonon collisions begin to play a dominant role. This is probably the reason for the large difference in breakdown field exhibited by the impurity Si detectors in Fig. 11.

It should be noted that not all samples show the $\tau \propto \mathscr{E}^{1/2}$ dependence described above. Some show $\tau \propto \mathscr{E}^x$ with x between 1 and 2. Norton[148] has obtained data on Si:P and Si:As with this behavior and the author[149] has seen similar effects in Ge:Cu. Thus additional complexities enter under some circumstances. The full picture has not yet emerged; however, it may be the result of trapping at shallow levels between the major impurity level and the band edge.

As the electric field strength is further increased, impact ionization breakdown begins. The free carrier density increases rapidly with small increments of field strength and, therefore, current density rises sharply. This effect has already been discussed in Section 6.

The causes of non-Ohmic behavior just described will be present in greater or lesser degree depending upon the host material, impurity element, compensation level, and operating temperature. Therefore it is not possible to give a universal specification for the current–voltage curve. Some examples are shown in Fig. 44 to illustrate what has been observed.

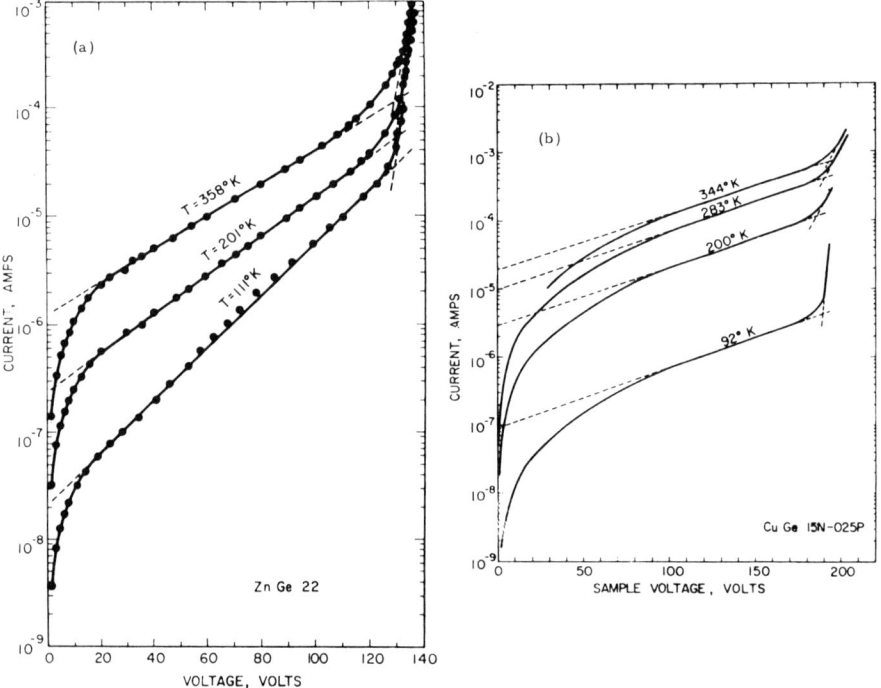

FIG. 44. Current–Voltage curves for (a) Ge:Zn and (b) Ge:Cu at 5°K with different background temperatures. (After Picus.[63])

[148] P. Norton, unpublished data, 1973.
[149] P. R. Bratt, unpublished data, 1969.

b. Low Background Performance

We have seen in Section 8 how the detectivity of impurity Ge or Si detectors is limited by the background photon flux which falls on the detector. Because these photons arrive randomly in time, they cause a fluctuating generation rate of free charge carriers which produce "noise" in the current flowing through the detector. Figure 14 showed how a reduction in background photon flux would allow increases in detectivity because of the reduced noise.

To get a feeling for the limitations placed on an IR detector by 300°K background photons, it is interesting to compare what the detector "sees" in the IR with what the human eye sees in the visible. In a typical case, an IR detector operating in the 8–12 μm spectral band with an $f/5$ field of view will have about 10^{16} background photons/sec-cm^2 falling on its sensitive area. The human eye looking at the clear sky on a bright day will have a comparable number of photons/sec-cm^2 falling on it. Thus, the detector's situation is somewhat analogous to the human eye looking at the daylit sky and attempting to see objects such as the moon or stars. Our experience is that a large object such as the moon can be dimly seen, but small objects such as stars cannot be seen. Upon removal of the background photons (nighttime) these objects become clearly visible. Similarly, reduction of background radiation on the IR detector can greatly increase its ability to distinguish objects. In the remainder of this section, we discuss the operating characterisitics of impurity Ge and Si detectors under low background conditions.

An expression was derived in Section 8 for detectivity when the limiting noise source was background photon g–r noise [Eq. (74)]. For wavelengths in the vicinity of 10 μm, this reduces to

$$D^*(\lambda) = \frac{\eta \lambda}{2hc} \left[\int_0^{\lambda_c} \eta(\lambda) J_B(\lambda) \, d\lambda \right]^{-1/2}. \tag{96}$$

Thus, $D^*(\lambda)$ should increase inversely as the 1/2 power of the total background photon flux on the detector. Quist and co-workers[150,151] have reported measurements on Ge:Cu detectors cooled to liquid helium temperature and operated at reduced background photon flux levels. The experimental arrangement used is shown in Fig. 45. Various background flux levels were obtained down to 5×10^9 photons/sec/cm^2 by suitable combinations of neutral density filters and field of view limiting apertures. The detector's

---

[150] T. M. Quist, *Proc. IEEE* **56**, 1212 (1968).
[151] R. J. Keyes and T. M. Quist, *in* "Semiconductors and Semimetals" (R. K. Willardson and A. C. Beer, eds.), Vol. 5, p. 321. Academic Press, New York, 1970.

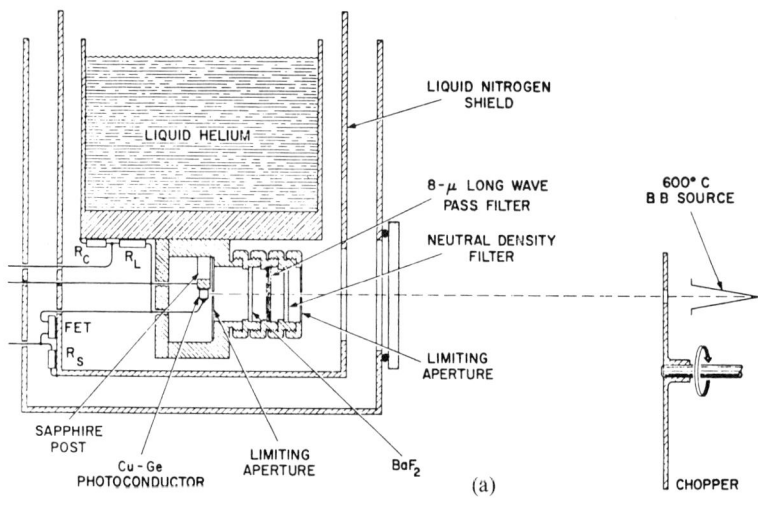

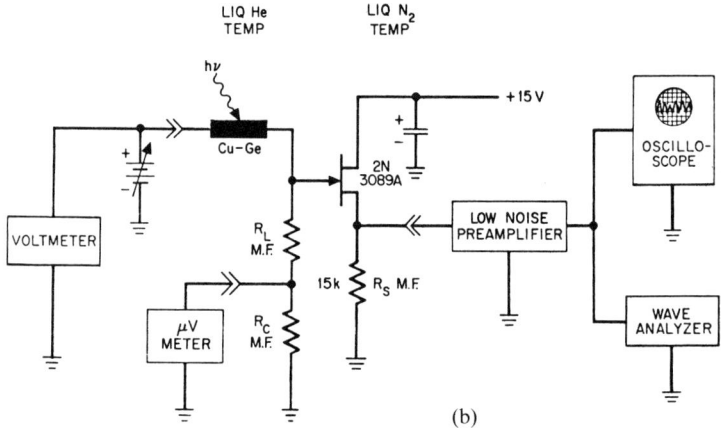

FIG. 45. Experimental arrangement for measurement of D^* under low background conditions. (a) Physical arrangement for the low background measurement of Cu–Ge photoconductors at liquid helium temperatures and for the 8- to 12-μm region. (b) The electrical circuit for the measurement of responsivity and low noise. (After Quist.[150])

sensitive wavelength was restricted to the 8–12 μm range by an 8 μm turn-on filter and a BaF_2 turn-off filter.

At low backgrounds, the detector resistance becomes very large, greater than 10^{10} Ω. Therefore, special electronic circuitry is required. Quist's circuit is shown in Fig. 45b. A high impedance load resistor was used (in some cases 40 MΩ) and a low noise field-effect transistor (FET) operated in the source

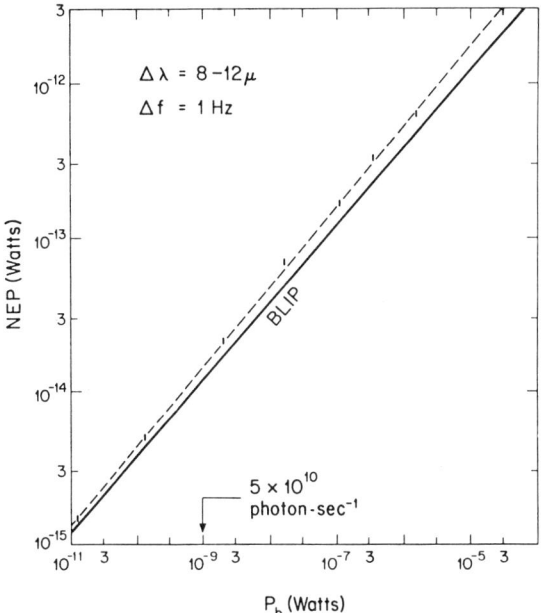

FIG. 46. Measured NEP (NEP $=\sqrt{A}/D^*$) for a Ge:Cu detector versus background flux. (After Quist.[150])

follower mode (voltage gain of 0.9) transformed the high impedance to a value in the kilohm range to be fed out of the Dewar to a low noise preamplifier. The FET and its source resistor were located on the liquid nitrogen shield of the Dewar so as to be in close proximity to the detector and to minimize thermal noise from these components.

Measured data obtained by Quist are shown in Fig. 46. He obtained a noise equivalent power (NEP) of about 1×10^{-15} W/Hz$^{1/2}$ at a background flux of 5×10^8 photons/sec. This corresponds to a D^* of about 1×10^{14} cm-Hz$^{1/2}$/W, assuming a 1 mm^2 sensitive area for the detector. Since NEP is the inverse of D^*, the dependence of NEP on background photon flux follows the expected behavior given by Eq. (96). Calculated values of NEP, using a quantum efficiency of 0.5 (obtained from the measured absorption coefficient of the Ge:Cu crystal), are shown by the solid line in the figure. This is within 30% of the experimentally measured data. The discrepancy can be attributed to a small amount of extraneous noise from the circuitry.

When the background photon g–r noise current is decreased, other sources of noise, such as the load resistor and FET noise, become important and it is necessary to minimize these contributions if high D^* is to be achieved.

The Johnson–Nyquist noise current from the load is given by

$$\langle i_{n,J}^2 \rangle^{1/2} = (4kT_L \Delta f/R_L)^{1/2} \tag{97}$$

and the g–r noise current, obtained from Eqs. (22), (69), and (72) can be written as

$$\langle i_{n,g-r}^2 \rangle^{1/2} = 2e(\tau\mu\mathscr{E}/L)(\eta J_B wL \Delta f)^{1/2}. \tag{98}$$

Note that the quantity $(\tau\mu\mathscr{E}/L)$ is the photoconductive gain, J_B is the total integrated photon flux density on the detector, and wL is its sensitive area. As an example, consider a case where $\eta = 0.5$, $J_B = 10^{10}$ photons/sec-cm^2, $wL = 10^{-1}$ cm^2, $\tau\mu\mathscr{E}/L = 0.5$, and $\Delta f = 1$ Hz. Equation (98) then gives a g–r noise current of 3.6×10^{-15} A/Hz$^{1/2}$. For the load resistor noise current to be less than this, we require $(4kT_L/R_L)^{1/2} < 3.6 \times 10^{-15}$, or $R_L > 21$ MΩ (assuming that T_L is at 5°K). Thus, the choice of 40 MΩ satisfies this requirement.

The frequency response of this circuit will be $R_L C$ limited with R_L being the load impedance and C the sum of detector capacitance, FET input capacitance and distributed capacitance of the wires. A typical total capacitance is 5 pf. Thus, the upper frequency -3 dB point is $(2\pi R_L C)^{-1} = 800$ Hz if R_L is chosen to be 40 MΩ.

Another property that limits the performance of impurity Ge and Si IR detectors at low backgrounds is gain saturation due to carrier sweep-out and dielectric relaxation. The importance of the effects of dielectric relaxation on the photoresponse of Ge:Hg was pointed out by Williams.[152,153] He observed that, when a detector was illuminated by a square-wave pulse of illumination, its photoresponse curve included a slow component that was small at low bias fields, but increased to dominate the whole curve at high bias fields. The effect is illustrated schematically in Fig. 47. A somewhat simplified theoretical explanation was offered by Williams[153] which goes as follows.

Consider a detector element cooled to a low temperature and exposed to a low level of background radiation. Assume that the detector impurity is an acceptor so this radiation will produce an equilibrium hole density p_0. Let a pulse of signal photons fall on the element for a time short compared to all other time periods of interest. This pulse will produce Δp excess holes/cm^3 and a corresponding density of negatively ionized acceptors ΔN_A^-. Now if the hole drift length before recombination is greater than the length between electrodes, all excess holes will be swept out of the element, leaving behind a uniform distribution of excess ionized acceptors. This space charge gives rise to a nonuniform electric field component $\Delta\mathscr{E}$

[152] R. L. Williams, *J. Appl. Phys.* **38**, 4802 (1967).
[153] R. L. Williams, *J. Appl. Phys.* **40**, 184 (1969).

2. IMPURITY GERMANIUM AND SILICON INFRARED DETECTORS

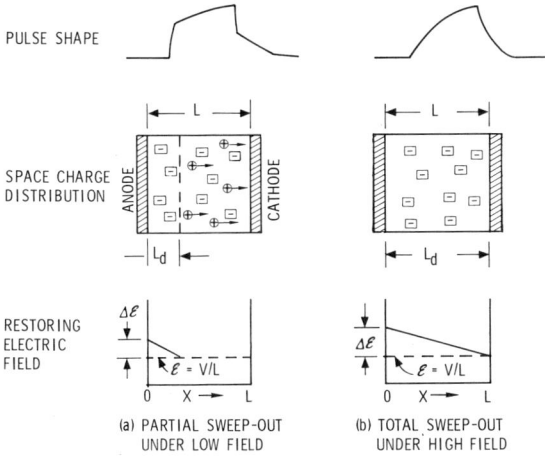

FIG. 47. Schematic illustration of carrier sweep-out. (Adapted from Williams.[153])

which adds to the external bias field $\mathscr{E}$ already present. The situation is shown in Fig. 47.

The magnitude of $\Delta\mathscr{E}$ can be obtained by solving Poisson's equation

$$d\mathscr{E}/dx = -e\,\Delta N_A^{-}/\varepsilon\varepsilon_0. \tag{99}$$

Including the unperturbed bias field $\mathscr{E} = V/L$, the solution is

$$\mathscr{E}(x) = \frac{-e\,\Delta N_A^{-}}{\varepsilon\varepsilon_0}(L-x) - \frac{V}{L}. \tag{100}$$

The excess field strength

$$\Delta\mathscr{E}(x) = \frac{-e\,\Delta N_A^{-}}{\varepsilon\varepsilon_0}(L-x) \tag{101}$$

provides the restoring force that returns the crystal to the neutral charge condition.

The restoring current can be obtained from the continuity equation

$$\frac{d}{dt}\Delta N_A^{-} = \frac{1}{e}\frac{dj_x}{dx}, \tag{102}$$

where j_x is the current density in the x direction and generation and recombination terms normally appearing in this equation are neglected. The current density is given by

$$j_x = p_0 e\mu\mathscr{E}. \tag{103}$$

If $\mathscr{E}$ is the only parameter in this equation assumed to vary with x, substitution of Eq. (103) into Eq. (102) gives

$$\frac{d}{dt}\Delta N_A^- = p_0 \mu \frac{d\mathscr{E}}{dx}, \tag{104}$$

which when combined with Eq. (99) yields

$$\frac{d}{dt}\Delta N_A^- = \frac{-p_0 e\mu}{\varepsilon\varepsilon_0}\Delta N_A^-. \tag{105}$$

The solution to this equation is

$$\Delta N_A^-(t) = \Delta N_A^-(0)\exp(-t/\tau_\rho), \tag{106}$$

where $\Delta N_A^-(0)$ is the initial excess charge density and τ_ρ is the dielectric relaxation time

$$\tau_\rho = \varepsilon\varepsilon_0/p_0 e\mu = \varepsilon\varepsilon_0 \rho. \tag{107}$$

Thus, the rate of return to charge neutrality is determined by the dielectric relaxation time which, in turn, depends on the material resistivity ρ and dielectric constant ε.

The key elements to the theory are the sweep-out of mobile excess charge carriers, creation of a space charge of ionized acceptor atoms, and subsequent neutralization of this charge in a time of the order of the dielectric relaxation time. A major assumption which is implied, but has not yet been stated, is that $\Delta p = 0$ at the anode. Thus, there are no excess carriers immediately available to replace those swept out. The condition of charge neutrality previously invoked in Section 6 only holds for times longer than the dielectric relaxation time. For times shorter than this, unneutralized space charge can exist.

During the sweep-out time period, the current in the right-hand side of the crystal is conduction current while that on the left-hand side in the space charge region is displacement current (see Fig. 47). Displacement current depends on the rate of change of electric field strength in the crystal and, expressed as current density, is given by

$$j_D = \varepsilon\varepsilon_0 \, d\mathscr{E}/dt. \tag{108}$$

For the linear sweep-out example of Fig. 47, we can represent $d\mathscr{E}/dt$ by

$$d\mathscr{E}/dt = \Delta\mathscr{E}(x)/\Delta t, \tag{109}$$

where $\Delta\mathscr{E}(x)$ is given by Eq. (101) and Δt is the time interval taken for sweep-out to be completed, which is $\Delta t = L/\mu\mathscr{E}$. Thus, the displacement current density flowing in at the anode ($x = 0$) is

$$j_D = -e\mu\mathscr{E}\,\Delta N_A^-. \tag{110}$$

2. IMPURITY GERMANIUM AND SILICON INFRARED DETECTORS

The conduction current density flowing out at the cathode is

$$j_c = \Delta p \, e \mu \mathscr{E} \tag{111}$$

and since $\Delta p = -\Delta N_A^-$, these two currents are equal and in the same direction.

It is important to remember that, under low background conditions, impurity Ge and Si detectors are more like insulators than semiconductors. For example, the equilibrium free hole density in the crystal is given by $p_0 = \eta J_B \tau / d$. Consider a typical detector having $\eta = 0.3$, $\tau = 5 \times 10^{-8}$ sec, $d = 0.3$ cm, and let $J_B = 10^{10}$ photons/sec/cm^2 be incident on it. This produces only 500 free holes/cm^3. A typical detector volume might be $0.05 \times 0.05 \times 0.3 = 7.5 \times 10^{-3}$ cm^3. Thus, the total number of free holes in this crystal is 4. The dielectric relaxation time, obtained from Eq. (107) is 0.35 sec (where we have used $\varepsilon = 16$ and $\mu = 5 \times 10^4$ cm^2/V-sec as typical values for Ge).

The extent of the space charge region is determined by the drift length $L_d = \tau \mu \mathscr{E}$ of the carriers. If the electric field is low, then only a portion of the carriers will be swept out. This situation is also shown in Fig. 47. Those carriers within a distance L_d of the cathode will be swept out. The ionized acceptors left behind will be neutralized by other holes moving in from the bulk of the crystal. Thus, a space charge region of length L_d is created which will subsequently be neutralized in a time $\sim \tau_\rho$.

Sweep-out of holes without replenishment from the anode causes the photoconductive gain to be limited to values less than unity. The expression for gain under these conditions has been derived by Blouke et al.[154] and is

$$G_{pc} = \frac{L_d}{L} \left\{ 1 - \frac{L_d}{L} [1 - \exp(-L/L_d)] \right\}. \tag{112}$$

For the sweep-out case where $L_d \gg L$, a series expansion of the exponential term leads to $G_{pc} \approx 1/2$. Photoconductive gains greater than this can only be obtained in time intervals longer than the dielectric relaxation time. This implies that the photoconductive gain is frequency dependent.

A more exact microscopic theory of sweep-out and dielectric relaxation limited performance has been worked out by Milton.[155,156] The detailed analysis is quite complex and too lengthy to describe here. However, plots of gain versus normalized frequency taken from Milton's paper are shown in Fig. 48. The zero frequency gain G_0 is the usual quantity $\tau \mu \mathscr{E}/L$, the frequency

---

[154] M. M. Blouke, E. E. Harp, C. R. Jeffus, and R. L. Williams, *J. Appl. Phys.* **43**, 188 (1972).
[155] A. F. Milton, *Appl. Phys. Lett.* **16**, 285 (1970).
[156] A. F. Milton and M. M. Blouke, *Phys. Rev. B* **3**, 4312 (1971).

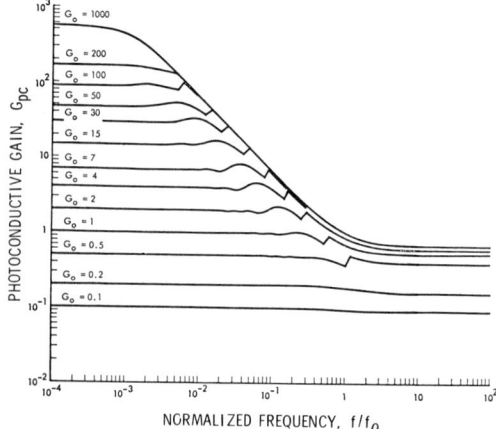

FIG. 48. Plot of photoconductive gain versus signal modulation frequency (normalized) for different values of zero frequency gain, $G_0(\tau_\rho \gg \tau)$. (After Milton and Blouke.[156])

f_ρ is $1/2\pi\tau_\rho$. Note that when signal modulation frequencies exceed f_ρ, the gain is limited to values less than unity. For $G_0 > 1$ the limiting value $G(f) = \frac{1}{2}$ is only reached when $f \gg f_\rho$. Also, it is important to note that for $G_0 \gg 1$, the frequency at which $G(f)$ starts to decrease from its G_0 value is substantially less than f_ρ.

There has been some difficulty in getting good agreement between Milton's theory and experimental data.[156] This problem has been attributed to nonuniform samples and/or a partial failure of the boundary condition, $\Delta p = 0$ at the anode. Also, his expression for gain versus frequency is very cumbersome. Blouke et al.[154] have proposed a simpler semiempirical method for describing the gain versus frequency curve. Their approach also includes the effect of the RC roll-off of the measuring circuit. Their formula for gain is

$$G(f) = \frac{G_0[1 + (2\pi f \tau_\rho')]^{1/2}}{[1 + (2\pi f \tau_\rho)^2]^{1/2}[1 + (2\pi f \tau_{RC})^2]^{1/2}}, \quad (113)$$

where $\tau_\rho' = \tau_\rho/2G_0$, $\tau_\rho = R_D C_D$ (C_D of detector only), and $\tau_{RC} = R_L C$ (C of detector plus circuit).

Determination of three parameters, G_0, τ_ρ, and τ_{RC} then allows a complete specification of the gain versus frequency curve. Figure 49 shows experimental data on a Ge:Hg detector operated at 27°K. The agreement between these data and Eq. (113) is very good for two different background levels. Values of τ_ρ computed from detector resistance and capacitance define a frequency $f_\rho = 1/2\pi\tau_\rho$ at which the gain is 3 dB down from its zero frequency value. This is different from Milton's result where f_ρ was much closer to

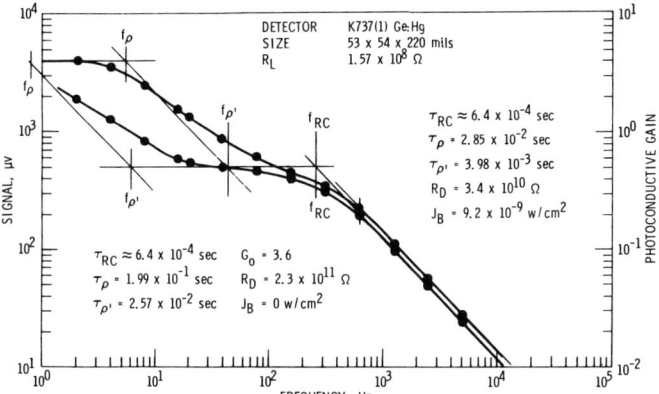

FIG. 49. Variation of signal with frequency for a Ge:Hg detector at 27°K. Lower curve, low background; upper curve, background flux increased so as to reduce detector resistance by about one order or magnitude. (After Blouke et al.[154])

frequencies where the gain has saturated to one-half. The reason for this is not presently known.

Gain saturation effects have been observed in Ge:Hg, Ge:Cu, Ge:Cd,[154] Ge:Au^{II},[157] and Si:B[158] detectors and is believed to be a general property of all detectors of this class. Under low background conditions, it sets an upper limit on the D^*-bandwidth product that can be obtained from these detectors. A calculation of this limit has been presented by Williams.[159] However, he has treated an idealistic situation which is not achievable in practice. He neglected the effects of circuit capacitance and also allowed the bulk recombination lifetime to vary over an unrealistic range of values. The D^*-bandwidth product obtained by Williams was

$$D^*_{max}/\tau_{min} = 1.7 \times 10^{18} \quad \text{cm-Hz}^{1/2}/\text{W-sec},$$

where D^*_{max} and τ_{min} are the maximum detectivity and minimum response time obtainable under a given background photon flux density. This value is more than an order of magnitude higher than can be achieved with real detectors.

c. Effects of Excited States

The excited states of impurity atoms in Si were first observed by absorption

---

[157] S. A. Kaufman, N. Sh. Khaikin, and G. T. Yokavleva, *Fiz. Tekh. Poluprov.* **3**, 571 (1969) [*English Transl.: Sov. Phys.—Semicond.* **3**, 485 (1969)].

[158] M. M. Blouke and R. L. Williams, *Appl. Phys. Lett.* **20**, 25 (1972).

[159] R. L. Williams, *Infrared Phys.* **9**, 37 (1969).

spectroscopy during the pioneering work of Burstein and co-workers[24,160,161] in the 1950's. They appeared as a spectrum of absorption lines at wavelengths just beyond the long wavelength fundamental absorption edge. Subsequently, excited state spectra were also observed for a number of impurities in Ge[162,163] A good review of the pre-1959 work on Si has been given by Hrostowski.[164] Considerable theoretical work has gone into the explanation of these spectra. A review of early work is given by Kohn[165] and more recent work has been done by Faulkner.[166]

Excited state spectra are easily observed for the shallower impurities. They are more difficult to resolve for deep impurities and have been observed only for a very few of these. Specifically, all group IIIB and VB impurities and Li show an excited state spectrum in both Ge[31,167,168] and Si.[26,168,169] The spectra of Cu, Zn^I, Zn^{II}, Cd, and Hg in Ge also have been observed[170-173]; however, that of Au and deeper lying acceptors has not. Excited state absorption lines have been observed for S^I and S^{II} donor levels in both Ge[174] and Si.[175] The S donor and In acceptor in Si are the only deep level impurities for which excited state spectra have been observed.

The effect of excited states in the cascade capture theory for recombination of free charge carriers has already been discussed. Excited states may also play a role in the generation of free charges during photoexcitation. Two ways in which this may occur are (1) during photothermal excitation, and (2) during field assisted excitation. In the following we briefly describe these two processes.

(1) *Photothermoconductivity.* Photothermoconductivity was first observed by Lifshitz and Nad[125] in Ge:Sb. Their spectral response measurements at wavelengths beyond the normal long wavelength cutoff showed an unexpected photoconductivity which had considerable structure in it. Figure

---

[160] E. Burstein, E. E. Bell, J. W. Davisson, and M. Lax, *J. Phys. Chem.* **57**, 849 (1953).

[161] G. Picus, E. Burstein, and B. Henvis, *J. Phys. Chem. Solids* **1**, 75 (1956).

[162] H. Y. Fan and P. Fisher, *J. Phys. Chem. Solids* **8**, 270 (1959).

[163] W. S. Boyle, *J. Phys. Chem. Solids* **8**, 321 (1959).

[164] H. J. Hrostowski, *in* "Semiconductors" (N. B. Hannay, ed.), p. 437. Van Nostrand-Reinhold, Princeton, New Jersey, 1959.

[165] W. Kohn, *Solid State Phys.* **5**, 257 (1958).

[166] R. A. Faulkner, *Phys. Rev.* **184**, 713 (1969).

[167] P. Fisher and H. Y. Fan. *Phys. Rev. Lett.* **2**, 456 (1959).

[168] R. L. Aggarwal, P. Fisher, V. Mourzine, and A. K. Ramdas, *Phys. Rev.* **138**, A882 (1965).

[169] H. J. Hrostowski and R. H. Kaiser, *J. Phys. Chem. Solids* **4**, 148, 315 (1958).

[170] P. Fisher and H. Y. Fan, *Phys. Rev. Lett.* **5**, 195 (1960).

[171] W. J. Moore, *Solid State Commun.* **3**, 385 (1965).

[172] R. A. Chapman and W. G. Hutchinson, *Solid State Commun.* **3**, 293 (1965).

[173] R. A. Chapman, W. G. Hutchinson, and T. L. Estle, *Phys. Rev. Lett.* **17**, 132 (1966).

[174] W. E. Krag and H. J. Zeiger, *Phys. Rev. Lett.* **8**, 485 (1962).

[175] W. H. Kleiner and W. E. Krag, *Phys. Rev. Lett.* **25**, 1490 (1970).

2. IMPURITY GERMANIUM AND SILICON INFRARED DETECTORS

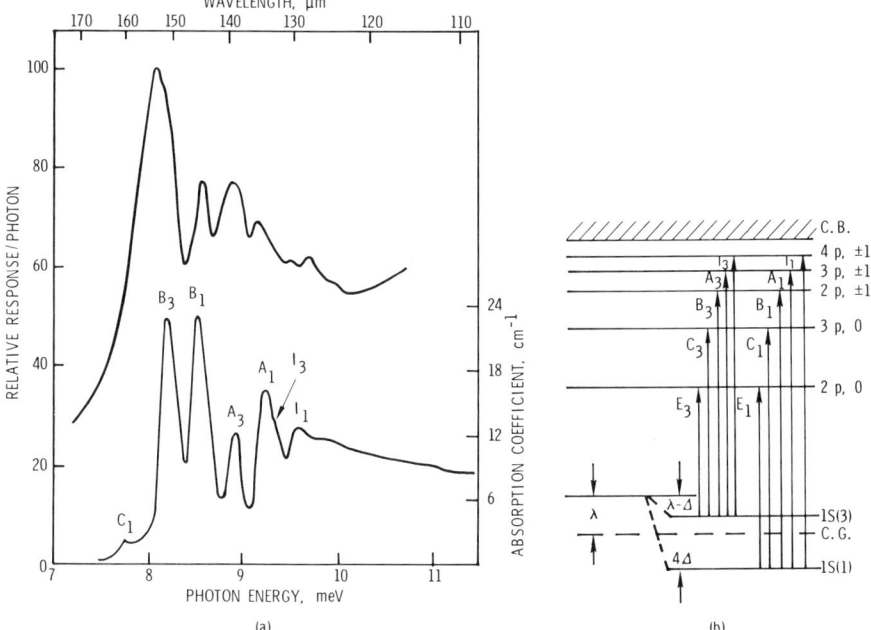

FIG. 50. (a) Upper curve shows photothermoionization spectrum of Ge:Sb with $N_{Sb} \approx 3 \times 10^{15}$ cm^{-3}, $T = 10°$K. (After Lifshitz and Nad'.[125]) Lower curve shows the optical absorption spectrum of Ge:Sb with $N_{Sb} \approx 7 \times 10^{14}$ cm^{-3}, $T = 9°$K. (After Reuszer and Fisher.[31]) (b) Excited state energy level diagram for donor atoms in Ge or Si.

50 shows the observed data. Also shown in the figure is the optical absorption spectrum. The correspondence of the peaks in the two curves is quite evident. The peaks are labeled according to the excited state energy diagram shown beside the curves. The letters assigned to each peak are those used by early workers as a means for labeling the absorption lines and have no other special significance. Figure 51 shows similar data on Ge:In and also the corresponding excited state energy level diagram for p-type Ge. In these curves, the photothermoconductivity peaks are greatest around 10°K and decrease as temperature is lowered to become nearly unobservable at 4°K. However, for very small impurity concentrations (10^{11}–10^{12}/cm^3) the lines become narrower and higher and good spectra are obtained at 4°K.

A theory for photothermoexcitation from impurity atoms has been presented by Kogan and Sedunov.[176] They argue that an incoherent process is most likely; that is, a photon is absorbed first and then a phonon, rather than a simultaneous absorption of both photon and phonon. Therefore, the model for the process is as shown previously in Fig. 4b. After photoexcitation to

---

[176] Sh. M. Kogan and B. I. Sedunov, *Fiz. Tverd. Tela* **8**, 2382 (1966) [*English Transl.: Sov. Phys.—Solid State* **8**, 1898 (1967)].

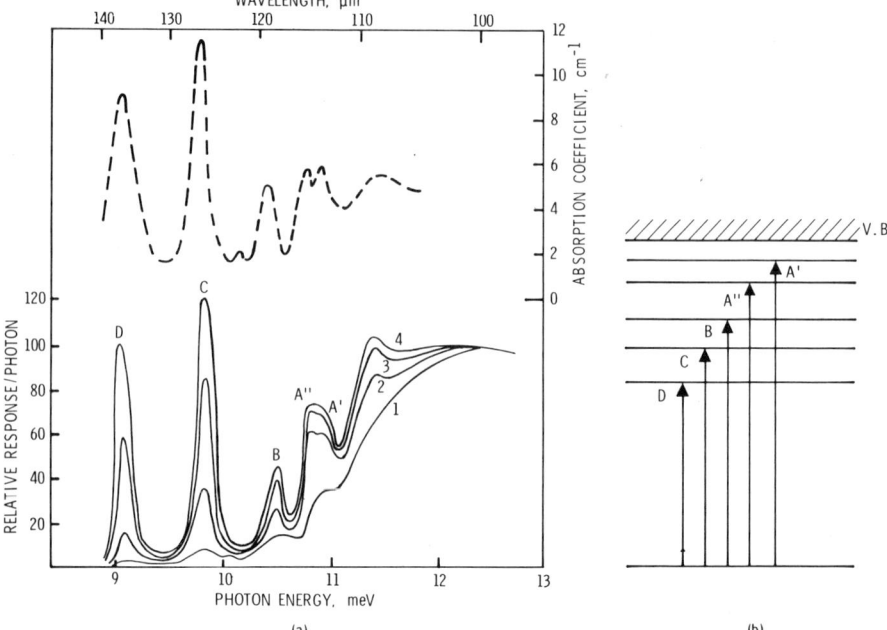

FIG. 51. (a) Lower curves, photothermoionization spectrum of Ge:In at four different temperatures; (1) 4.2°K, (2) 6.2°K, (3) 8.4°K, and (4) 10.7°K. (After Lifshitz et al.[177]) Upper curve, optical absorption spectrum of Ge:In. (After Jones and Fisher.[178]) (b) Excited state energy level diagram for acceptors in Ge or Si.

the jth excited state, the electron (or hole) has a probability $I(j)$ of absorbing a phonon(s) and becoming freed, or, a probability $S(j)$ (the "sticking" probability) of giving up its energy to a phonon(s) and returning to the ground state. This sticking probability is the same quantity used by Lax in his theory for cascade recombination. Since either one or the other event must happen

$$I(j) + S(j) = 1. \tag{114}$$

An essential result of the theory is that

$$\sigma_i(v_j) = \sigma(v_j)I(j), \tag{115}$$

where $\sigma_i(v_j)$ is the photothermoionization cross section for a line at frequency v_j and $\sigma(v_j)$ is the optical absorption cross section for the same line. Lifshitz et al.[177,179] have used this result to obtain values of $I(j)$ from experimental

[177] T. M. Lifshitz, N. P. Likhtman, and V. I. Sidorov, in Proc. 9th Int. Conf. Phys. Semicond. Moscow, p. 1081. Nauka, Leningrad, 1968.
[178] R. L. Jones and P. Fisher, J. Phys. Chem. Solids 26, 1125 (1965).
[179] T. M. Lifshitz, N. P. Likhtman, and V. I. Sidorov, Fiz. Tekh. Poluprov. 2, 782 (1968) [English Transl.: Sov. Phys.—Semicond. 2, 652 (1968)].

2. IMPURITY GERMANIUM AND SILICON INFRARED DETECTORS 133

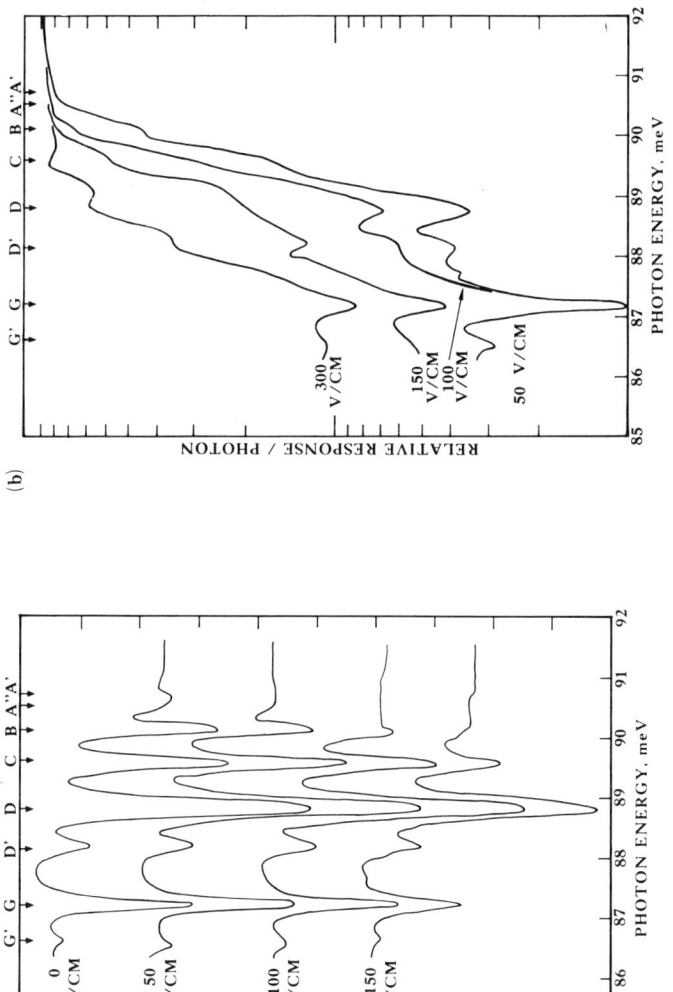

FIG. 52. Electric field dependence of (a) optical transmittance, and (b) photoconductivity for Ge:Hg in the spectral region beyond the long wavelength cutoff $N_{Hg} = 7.3 \times 10^{15}$ cm^{-3}, $T = 6.8°$K. Transmittance curves in (a) have been displaced vertically for clarity. Photoconductivity curves in (b) have been normalized at 92 meV. The D' and G' lines are satellites to the stronger D and G lines caused by a splitting of the Hg ground state energy level.[173]

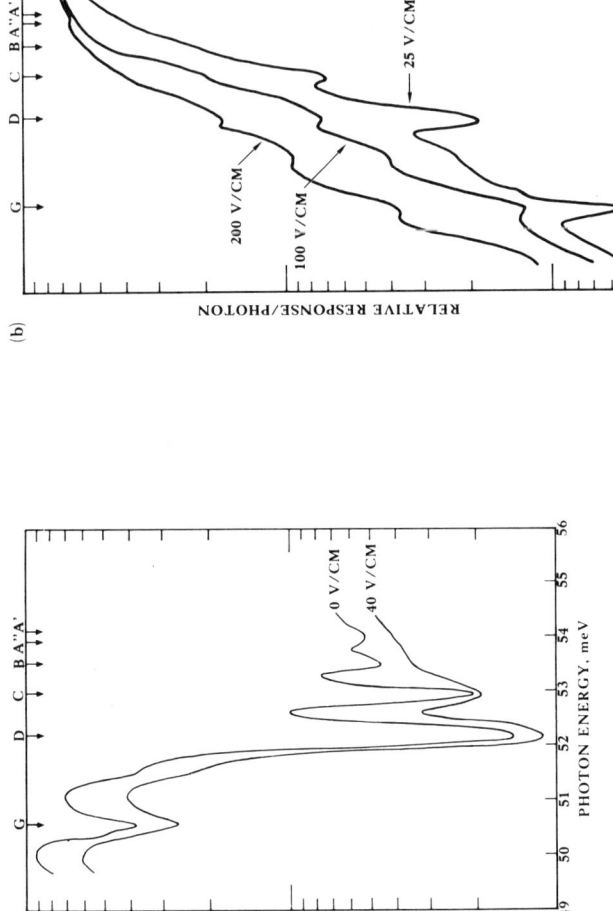

FIG. 53. Electric field dependence of (a) optical transmittance, and (b) photoconductivity for Ge:Cd in the spectral region beyond the long wavelength cutoff. $N_{Cd} = 6 \times 10^{15}$ cm^{-3}, $T = 5.7$°K. Transmittance curves in (a) have been displaced vertically for clarity. Photoconductivity curves in (b) have been normalized at 56 meV.

data. The quantity $\sigma_i(v_j)$ can be obtained from photothermoconductivity data and $\sigma(v_j)$ from optical absorption data. Thus, $I(j)$ follows directly from Eq. (115) and $S(j)$ can then be obtained from Eq. (114). Experimentally determined values of $S(j)$ for Ge:In ranged between 1.0 and 0.4 over the temperature interval from 4 to 10°K and the temperature dependence was in reasonably good agreement with values calculated from Lax's theory.

Lifshitz et al.[179] have made use of photothermoconductivity for the spectroscopic quantitative analysis of very small impurity concentrations in Ge. Positive identifications can be made for concentrations as low as 1×10^{11} atoms/cm^3, less than 1 atom in 10^{10}! Photothermoionization through excited states of S in Si has also been identified.[29] However, these transitions are weak and require a very sensitive method for detection. The method used was transient photocapacitance of a p–n junction.

(2) *Field Assisted Photoconductivity.* The second mechanism for photoresponse beyond the long wavelength cutoff is field assisted photoconductivity. Such an effect was first identified by Loh and Picus[180] in Ge:Cu. The effect has also been observed[181] in Ge:Hg and Ge:Cd. Figures 52 and 53 show these data. Both optical transmission and photoconductivity were measured as a function of electric field strength on the same sample by using the technique shown in Fig. 54. Two detectors are mounted in a liquid He Dewar, one in front of the other. The front detector is the one under investigation and the photoconductivity spectrum is measured on it. The back detector is made from a material having a longer cutoff wavelength than the front one and is used to measure the transmittance through the front detector. Germanium:copper was used as the back detector in these experiments.

Examination of the transmittance curves in Figs. 52 and 53 shows clear

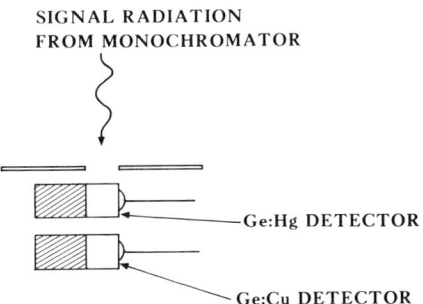

FIG. 54. Experimental arrangement for study of field assisted photoconductivity and optical transmittance of impurities in Ge and Si.

[180] E. Loh and G. Picus, *Bull. Amer. Phys. Soc.* **7**, No. 3, 173 (1962).
[181] P. R. Bratt and P. J. Schreiber (to be published).

evidence of the effect on higher excited state levels as electric field strength is increased. The A and B lines are seen to disappear completely and some diminuation of the other lines is also observed. The photoconductivity spectra are a little more complex. Note that, at low electric field strengths, the excited states appear as dips in the photoconductive response curve, while at higher field strengths, they become peaks. Apparently, two different mechanisms are operative. There is a weak photoconductivity beyond the edge and, at low field strengths, photon induced transitions to excited states produce optical absorption lines which compete with this photoconductivity at certain wavelengths. Thus, we get dips in the photoresponse curves. At higher electric field strengths, true field assisted photoconductivity comes into play and the dips then become peaks as expected.

The question then remains as to the cause of the low level photoconductivity beyond the cutoff edge. A possible explanation is afforded by the presence of other shallower impurity atoms within the crystal which act as hole traps. This situation is diagrammed in Fig. 55. Even though enough compensating donor atoms may be present to fill all the shallow acceptors, room temperature background radiation can generate free holes, some of which are then trapped on the shallow level impurities. These trapped holes can then be freed by signal photons from the monochromator beam and give rise to the observed low level photoconductivity beyond the cutoff edge. Copper impurity in Ge:Hg and Ge:Cd detectors has been identified as one cause of this behavior. Group III acceptor impurities may also play a role at low temperatures.

Because of field assisted photoconductivity, the cutoff wavelength of a detector, defined by the usual convention to be that wavelength where the photoresponse is down to one-half to its peak value, is a variable dependent on the field strength used when measuring the spectral response. The shift in cutoff wavelength between a low and a high field measurement is only a fraction of a micron for Ge:Hg, but may be 1 μm or more for Ge:Cd or

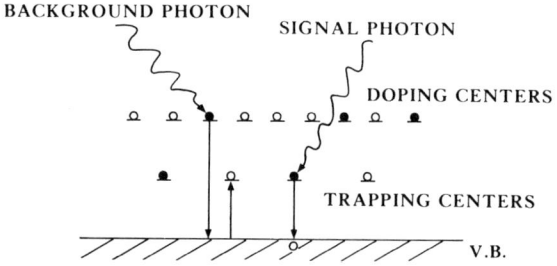

FIG. 55. Diagram showing how background radiation populates shallow trapping centers with holes which then produce a low level photoconductivity when freed by monochromatic signal photons.

Ge:Cu. Similar effects have been seen for impurities in Si. It was also mentioned earlier that an electric field dependence of cutoff wavelength has been observed in the very long wavelength detectors made using group III impurities in Ge.[122]

Thus, if one intends to use an impurity Ge or Si detector in a precision radiometer that is not provided with an optical bandpass filter, the detector must be operated with the same bias field strength at which the spectral response calibration was made. Otherwise a shift of the long wavelength edge could reduce the accuracy of the calibration data.

d. Donor Ion States

Norton[99] has used photoconductivity and photo-Hall measurements to identify and confirm the existence of negative donor ion (D^-) and positive acceptor ion (A^+) states in Si and Ge. It has already been discussed in Section 11 how, in the case of Ge:Au, the A^+ state, which has a binding energy of 0.05 eV, gives rise to photoconductivity for wavelengths out to 25 μm. D^- states in Si have much smaller binding energies and produce photoconductivity for wavelengths in the 100–600 μm range.

In lightly doped Si:P and Si:As the isolated D^- ion is found to have a binding energy of 0.0017 eV and a long wavelength cut-off of about 600 μm.[181a] When the donor concentration is increased above 3×10^{15} atoms/cm^3, random clusters of donor atoms are formed that have overlapping D^- wave functions. Electrons are bound more strongly to these donor groups than to isolated donors and the long wavelength cutoff is shifted toward shorter wavelengths. For example, with 1×10^{16} donor atoms/cm^3 the long wavelength cutoff is shifted to 200 μm.

Norton[99] has demonstrated the feasibility of these states for use as a far IR detector with fast response time, on the order of 1 nsec. He has estimated the NEP of this type of detector to be on the order of 10^{-11} W/Hz$^{1/2}$ for operation at temperatures $\leqslant 2°$K and exposed to unfiltered 300°K background radiation.

e. High Speed Performance

It was shown in Section 6 how the excess carrier lifetime depends inversely on the number of ionized impurity atoms since these are the recombination centers with which the excess carriers must recombine. We have also mentioned that best performance at low frequencies is obtained by minimizing the number of recombination centers so that the excess carrier lifetimes are maximized. However, if one needs to have a detector with a very fast response, then this can be achieved by the intentional addition of compensating impurity atoms so as to increase the number of recombination centers. This

---

[181a] P. Norton, *Phys. Rev. Lett.* **37**, 164 (1976).

has been done by adding Sb donor atoms to Au, Cu, and Hg doped Ge to produce high speed detectors that have been used for both pulse detection and optical heterodyne receiving of laser radiation.

Wood,[182] at Bell Telephone Laboratories, used p-type Ge:Au to detect He–Ne laser beat frequencies out to 1.5 GHz. Yardley and Moore[183] at the University of California (Berkeley) performed similar measurements on Ge:Cu and deduced detector response times less than two nanoseconds. Buczek and Picus[142] at Hughes Research Laboratories measured response times of Ge:Cu using CO_2 laser radiation sine wave modulated by a GaAs electrooptic modulator.[184] They found detector response times ranging from 3.3 nsec down to less than 0.4 nsec depending on the amount of compensating Sb atoms added to the Ge:Hg crystal. Similar measurements were made by Bridges et al.[185] at Bell Telephone Laboratories on all three detector types, but using square wave pulse modulation on the GaAs electrooptic modulator instead of sine wave modulation. This group observed response times less than 0.5 nsec for their more heavily compensated detectors, but detectors with less compensation had response times of several nanoseconds.

The development of high speed impurity Ge detectors made possible the first demonstrations of sensitive optical heterodyne receivers with CO_2 laser radiation at 10.6 μm. Teich et al.[186] demonstrated a noise equivalent power of 1.3×10^{-19} W in a 1-Hz bandwidth using Ge:Cu and Buczek and Picus[142] obtained 1.7×10^{-19} W using Ge:Hg. Both of these values are within a factor of 10 of the theoretical limit for a perfect quantum counter. Similar results were achieved by Arams et al.[187]

(1) *Practical Considerations.* The high speed response of impurity Ge detectors is obtained only at the expense of decreased responsivity. The reason for this can be seen by examination of Fig. 8 and Eq. (19). Because these detectors usually have a fairly high impedance, their response time when using a load resistance value $R_L \approx R$ will be determined by the circuit response time $\tau_c = R_\| C$, where $R_\|$ is the parallel combination of detector and load resistance. In a typical case, $R_\|$ might be 5×10^4 Ω and C about 20 pF. Then the circuit response time will be 1 μsec. Thus, the inherent high speed response of the detector can only be realized by reducing the value of the load resistor so that τ_c will be much less than τ. By Eq. (19) this reduc-

---

[182] R. A. Wood, *J. Appl. Phys.* **36**, 1490 (1965).
[183] J. T. Yardley and C. B. Moore, *Appl. Phys. Lett.* **7**, 311 (1965).
[184] A. Yariv, C. A. Mead, and J. V. Parker, *IEEE J. Quantum Electron.* **QE2**, 243 (1966).
[185] T. J. Bridges, T. Y. Chang, and P. K. Cheo, *Appl. Phys. Lett.* **12**, 297 (1968).
[186] M. C. Teich, R. J. Keyes, and R. H. Kingston, *Appl. Phys. Lett.* **9**, 357 (1966).
[187] F. R. Arams, E. W. Sard, B. J. Peyton, and F. P. Pace, *IEEE J. Quantum Electron.* **QE3**, 484 (1967); also in "Semiconductors and Semimetals" (R. K. Willardson and A. C. Beer, eds.), Vol. 5, p. 409. Academic Press, New York, 1970.

tion in R_L will also result in a lower value of responsivity for the detection circuit. Continuing with the example above, if we now reduce the load resistor to 50 Ω, the circuit response time will be reduced to 1 nsec, but the responsivity will also be reduced by a factor $R_L/R_{\|} = 50/(5 \times 10^4) = 10^{-3}$. This reduction in responsivity is an unavoidable consequence when high speed operation is required.

Further reduction in the capacitive loading of a high speed detector circuit can be obtained by using coaxial transmission lines going inside the Dewar directly to the detector. The paper by Wood[182] shows a scheme used for Ge:Au in a liquid nitrogen Dewar housing. Buczek and Picus[142] used a specially constructed liquid helium Dewar which is shown in Fig. 56.

Ultimately, the detector's response time is limited by its dielectric relaxation time. With no bias voltage applied to the detector, this is given by $\tau_\rho = \rho\varepsilon\varepsilon_0 = \varepsilon\varepsilon_0/ne\mu$. Since the dielectric constant is fixed, the relaxation time is determined by the free carrier density and mobility. Typical values for impurity Ge under normal ambient background conditions would be $n \approx 10^{10}/\text{cm}^3$ and $\mu \approx 10^5$ cm^2/V-sec. Using a value for ε of 16, the dielectric relaxation time is found to be 8.8×10^{-9} sec.

How can this value be reconciled with the subnanosecond response times actually measured? Two reasons can be given.

(1) As pointed out in the discussion of Section 14b, response times less than the dielectric relaxation time can be observed with a photoconductive

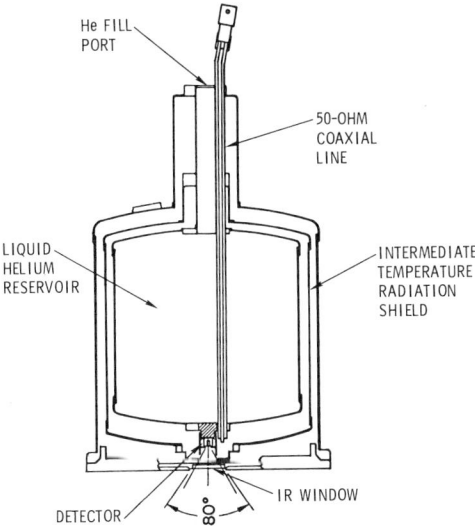

FIG. 56. Liquid helium Dewar with integral 50 Ω coaxial line for use with high speed impurity Ge and Si IR detectors.

gain less than one-half. Because of the reduced free carrier lifetime in high speed material, the photoconductive gain is usually less than one-half and, in fact, may commonly be about 0.1.

(2) Frequently these detectors are illuminated with fairly high photon flux densities from the laser which are sufficient to raise the free carrier densities significantly above the unilluminated value. This will then result in a lower dielectric relaxation time due to the lower detector resistivity. However, this is only the case while the detector is under illumination. When the radiation is turned off, the dielectric relaxation time will return to its original value as the excess carriers die away. This can possibly lead to unequal rise and decay times for these types of detectors under conditions of high pulse illumination. Furthermore, when used in the heterodyne mode, the average radiant power from the local oscillator may also cause a decrease in detector resistivity, thereby decreasing the dielectric relaxation time.

VI. Conclusions and Anticipated Future Developments

Extrinsic photoconductivity due to impurities in Ge and Si has been observed and studied for more than 20 years. Most of the details of the physical processes involved, photon excitation, free carrier transport, and recombination are now fairly well understood. These studies have provided a fertile field for both solid state physicists seeking to understand the properties of impurity atoms in semiconductors and device engineers attempting to develop more sensitive detectors. The two areas of endeavor have enjoyed a complementary relationship and this is a good example of how a practical requirement has stimulated fundamental research which in turn has benefited the device development effort.

Our discussion of impurity Ge and Si detectors has been necessarily brief and somewhat incomplete; a whole book could have been devoted to the subject. The approach taken was to give a "broad brush" treatment with sufficient scope to give the reader a good overview of the field. We purposely avoided a simple cataloging of detector properties in the hope of providing a better appreciation for the physics and technology of these devices. The literature references cited are not all inclusive, but represent a selection of key articles from a historic and technical point of view and give a guide to further literature.

Using selected impurity elements, a variety of IR detectors have been developed having cutoff wavelengths ranging from about 4 to 120 μm. The sensitivity limits of these detectors, determined by the signal-to-noise ratio per unit signal power, have been theoretically established and practical detectors have been found to conform well with the established theory. The best detectors are found to have detectivity values about one-half the theoretical limit which implies a quantum efficiency of 0.25. This value is a little puzzling since, based on measured values of absorption coefficient, one would expect a number closer to 0.5. It would appear that some photons are being lost to other absorption mechanisms within the crystal, but the mechanism for this is not presently understood.

This type of detector invariably requires cooling to cryogenic temperatures in order to operate effectively. The optimum operating temperature may vary anywhere from 4 to 100°K or higher, depending upon the cutoff wavelength (or ionization energy) of the impurity atom involved. A detailed discussion of cooling methods was felt to be outside the scope of this chapter. However, it should be pointed out that advances in cooling techniques have kept pace with detector developments. Efficient, light-weight mechanical displacement-type refrigerators have been developed that can achieve any design temperature down to about 15°K. For applications where 4–10°K operation is required, somewhat larger cryoengines are available, but where weight and power restrictions are important, liquid helium is still used. The availability of this cryogen has greatly increased and its cost has decreased steadily over the past few years. Liquid nitrogen is readily available and other liquid cryogens such as neon (27°K) and hydrogen (20°K) are more frequently used, although neon is relatively expensive and hydrogen is potentially explosive. For further information on detector cooling techniques, the reader is referred to the books by Hudson[188] and Wolfe[189] and a conference proceedings on cryogenics and IR detectors held in 1969.[190]

The applications of impurity Ge and Si IR detectors are fairly well known and again, a detailed review was not intended for this chapter. They have been covered in excellent fashion by Hudson[188] and also reviewed in other publications.[191] A brief listing of applications where this type of detector has been of use is given in Table IX. Of all the different impurities, Ge:Hg has been used most extensively in IR imaging systems for night viewing and airborne reconnaissance.

The technology of impurity Ge IR detectors is somewhat more mature than that of impurity Si. However, as we mentioned in the introduction, there has been a renewed interest in Si IR detectors over the past few years because Si offers some important potential advantages over Ge. These are: higher practical doping concentrations, resulting in higher absorption

[188] R. D. Hudson, Jr., "Infrared Systems Engineering," Chapter 11. Wiley, New York, 1969.
[189] W. L. Wolfe (ed.), "Handbook of Military Infrared Technology," Chapter 12. Office of Naval Res., Dept. of the Navy, Washington, D.C. (1965) (available from Superintendent of Documents, U.S. Govt. Printing Office, Washington, D.C.).
[190] W. H. Hogan and T. S. Moss (eds.), "Cryogenics and Infrared Detection." Boston Tech. Publ., Cambridge, Massachusetts, 1970.
[191] P. J. Ovrebo, *Proc. IRE* **47**, 1610 (1969).

TABLE IX

Partial Listing of Applications for Impurity Ge and Si IR Detectors

Military	Reconnaissance and surveillance
	Night viewing
	ICBM launch detection and tracking
	Target/decoy discrimination
	Submarine detection
Scientific	IR astronomy
	Laser research
	Spectroscopy
	Meteorological research
Medical	Measurement of skin temperature
	Detection of cancer or infection
	Location of blood circulation blockage
Industrial	Remote temperature sensing for process control
	Nondestructive testing
	Heat flow problem analysis
	Air pollution monitoring

coefficients; lower dielectric constant which gives better performance under dielectric relaxation limited operating conditions; better developed device technology, including contacting, surface passivation, and photoetch techniques; and compatibility with large scale integrated circuits. The integration of high density multielement detector arrays with associated signal processing circuitry on the same Si chip would offer a substantial reduction in complexity and cost of future high performance imaging systems. This goal is currently being pursued in a number of research laboratories and it will be interesting to observe the impending developments during the next few years.

Acknowledgments

The author has benefited over the years by discussions and exchange of technical data with many colleagues in the infrared detector field. To attempt to list all their names would be inappropriate here. Nevertheless, we are grateful for these associations, for without them this chapter would not have been possible.

We would also like to acknowledge the assistance of a number of associates on the staff of the Santa Barbara Research Center; in particular, Peter J. Schreiber for furnishing detector spectral response data and computer calculations of ideal detector D^*; Richard L. Nielsen for data on operating characteristics of both impurity Ge and Si detectors; and Donald E. Bode for a critical reading of the manuscript. The Publications Department staff provided invaluable assistance in the preparation of illustrative material.

We are grateful to Michael Y. Pines of the Hughes Aircraft Company for several stimulating discussions and for furnishing data on Si:In detectors. Also, Paul Norton of Bell Telephone Laboratories provided a number of helpful suggestions for improving the manuscript and furnished a preprint of his work on negative donor ions in Si.

CHAPTER 3

InSb Submillimeter Photoconductive Detectors

E. H. Putley

I.	PHYSICAL PRINCIPLES	143
II.	DESCRIPTION OF DETECTOR	147
III.	PERFORMANCE OF DETECTOR	150
	1. *Responsivity and Speed of Response*	150
	2. *Noise Equivalent Power or Detectivity*	154
	3. *Behavior in Strong Magnetic Fields*	162
	4. *Heterodyne Detection*	163
IV.	APPLICATIONS AND FUTURE DEVELOPMENTS	163
	APPENDIX. AMPLIFIERS FOR USE WITH THE INSB SUBMILLIMETER DETECTOR	164
	ADDENDUM	166

I. Physical Principles

An account of the physical mechanism of hot electron photoconductivity has been given in Volume 1 of this treatise[1] so that only a short description will be given here.

The absorption of far-infrared radiation by free carriers in a semiconductor, with the characteristic λ^2 variation of the absorption coefficient is well-known (λ in the wavelength). It is a consequence of the high frequency behavior of the electrical conductivity σ. The low frequency conductivity σ_L can be written in the form

$$\sigma_L = ne^2\tau_e/m^* \qquad (1)$$

where n is the free carrier concentration, e the electronic charge, m^* the effective mass, and τ_e the conductivity relaxation time is related to the electron mobility μ,

$$\mu = e\tau_e/m^*. \qquad (2)$$

Then the conductivity $\sigma(\omega)$ at an angular frequency ω is

$$\sigma(\omega) = \sigma_L(1 + \omega^2\tau_e^2)^{-1}. \qquad (3)$$

---

[1] E. H. Putley, "Semiconductors and Semimetals" (R. K. Willardson and A. C. Beer, eds.), Vol. 1, pp. 289–313. Academic Press, New York, 1966.

Hence if $\omega\tau_e \ll 1$,
$$\sigma(\omega) = \sigma_L, \tag{4}$$
but if $\omega\tau_e \gg 1$,
$$\sigma(\omega) = ne^2/\omega^2\tau_e m^*. \tag{5}$$

Provided σ is sufficiently small so that the conductive component of current flow is small compared with the capacitative component, the optical absorption coefficient can be written as
$$\alpha = 4\pi\sigma(\omega)/c(K)^{1/2}, \tag{6}$$
where c is the velocity of light and K the dielectric constant. Hence
$$\alpha = ne^2\lambda^2/\pi c^3\tau_e m^* K^{1/2} \tag{7}$$
where λ is the wavelength. This expression shows that α varies as λ^2. Although free carrier absorption can be observed in many semiconductors over wide ranges of temperature, it is not usually associated with any form of photoconductivity. In most circumstances the energy absorbed by the electrons is rapidly dissipated to the crystal lattice so that the most one might expect is to be able to use the absorption to improve the far-infrared performance of some form of bolometer. This was suggested by Novak[2] as a mechanism for a 10 μm InSb detector operating at 77°K and it probably occurs in the carbon and germanium liquid helium cooled bolometers.[3] However, an exception occurs in semiconductors with very high mobility carriers at low temperature, a good example being n-type InSb. In these materials the coupling between the electron distribution and the lattice is weak so that it is possible to create a steady state in which the mean energy of the electrons is appreciably greater than their thermal energy. Since the mobility will normally be energy-dependent, the mobility and, hence, the conductivity will depend on the intensity of absorbed radiation. Thus, if the energy is supplied via far-infrared free carrier absorption, a significant change of conductivity can occur which may be used to detect the radiation. This process has close analogies to that occurring in a bolometer. It has, however, one important difference. The response time of a bolometer or other thermal detector depends on thermal capacity of the detecting element. Because in practice this can never be reduced below a certain limit, the response time will always be relatively long, normally not less than 1 msec. In the case of the free carrier absorption detector, the relevant thermal capacity is that of the electron gas which is very small. Hence, the response time associated with this process is very much shorter, usually less than 1 μsec. It is perhaps a somewhat arbitrary matter whether one regards this process as a form of photoconductivity or as a thermal process. Since the change of conductivity

[2] R. Novak, *Meeting Int. Comm. Opt.*, 5th, Stockholm, August 1959.
[3] P. L. Richards, Chapter 6, this volume.

3. InSb SUBMILLIMETER PHOTOCONDUCTIVE DETECTORS

occurs directly as the result of electronic transition and since the fast response is comparable with that of other types of photoconductor, this process can be regarded as a photoconductive one, although I sometimes describe it as an "electronic bolometer."

Since the absorption coefficient Eq. (7) increases as λ^2, the performance of a device utilizing this effect should improve as the wavelength increases, leveling off when $\omega\tau_e \gtrsim 1$. Consider a numerical example: For a pure sample of n-type InSb at $4°K$, $n \sim 5 \times 10^{13}$ cm^{-3}, $\mu \sim 10^5$ cm^2 V^{-1} sec^{-1}, $m^* = 1.4 \times 10^{-2}$ m (m is the free space electron mass) and $K \sim 15$. Then $\omega\tau_e = 1$ for $\lambda = 1.6$ mm and when $\lambda = 1$ mm (1000 μm) $\alpha = 22$ cm^{-1}, but for $\lambda = 100$ μm, $\alpha = 0.30$ cm^{-1}. The value for α at 1 mm wavelength is comparable with that found in extrinsic germanium photoconductive detectors, but the value found at 100 μm is so small that it would be difficult to make an efficient detector at this wavelength. From these values it would appear that this process should be useful at wavelengths of 1 mm or somewhat less, i.e., in the region of the spectrum beyond the range of the extrinsic photoionization detector, but at wavelengths so short that simple microwave techniques are inadequate.

Expressions for the voltage responsivity and the response time have been derived by a number of workers.[4,5] The non-Ohmic behavior of the conductivity can be expressed by writing

$$\sigma = \sigma_0(1 + \beta E^2), \tag{8}$$

where E is the applied electric field. If E is sufficiently small, β is a constant. For InSb at $4°K$ this holds for fields less than 0.1 V cm^{-1}. At larger fields Eq. (8) can still be used to describe the behavior of σ for small changes in E and β is then simply defined as

$$\beta = (1/\sigma)[d\sigma/d(E^2)]. \tag{9}$$

Kogan[4] then derived expressions for the responsivity R:

$$R = \beta V/v\sigma \tag{10}$$

and the response time τ

$$\tau = \frac{3}{2}\left(\frac{k}{e}\right)\beta\bigg/\frac{d\mu}{dT}. \tag{11}$$

Here V is the applied voltage, v the volume of the detector, and k is Boltzmann's constant.

---

[4] Sh. M. Kogan, *Fiz. Tverd. Tela* **4**, 1891–1896 (1962) [*English Transl.: Sov. Phys.—Solid State* **4**, 1386–1389 (1963)].
[5] B. V. Rollin, *Proc. Phys. Soc. (London)* **77**, 1102–1103 (1961); M. A. Kinch and B. V. Rollin, *Brit. J. Appl. Phys.* **14**, 672–676 (1963).

When this hot electron photoconductive effect was first observed,[6] it was found that a relatively small magnetic field markedly enhanced the effect, although if the magnetic field was increased beyond a certain value the effect was reduced. It was later found[7] that in large magnetic fields a resonant effect occurred near the cyclotron resonance frequency ω_c

$$\omega_c = eB/m^* \tag{12}$$

where B is the magnetic induction.

Measurement of the direct current–voltage characteristics enables the parameters appearing in Eqs. (10) and (11) to be calculated. The results obtained are in good agreement with the measured performance of submillimeter detectors and confirm that the principal effect of a small magnetic field is to reduce σ, hence causing R to increase [Eq. (10)]. Equation (11) predicts a value of τ less than 1 μsec. and this is relatively insensitive to the effect of a magnetic field. The effect of a magnetic field on the performance of this device is a consequence of the magnetic freeze-out effect. On applying a field the separation between the impurity states and the conduction band increases so that the free carrier concentration is reduced. This reduces the free carrier absorption, but at first the effect of this is masked by the rapid increase in resistivity which causes the responsivity [Eq. (10)] to increase. On further increasing the magnetic field, the reduction in absorption becomes more important and the responsivity falls off. When this stage is reached, study of the spectral response in the cyclotron resonance region reveals the existance of a resonant photoconductive effect. This was first observed by Kimmitt and has proved of practical value for certain purposes.[8,9]

The advantage of operating with an applied magnetic field is that the responsivity is increased without impairing the response time. Since the amplifier is the main source of noise it is essential to obtain as large a responsivity as possible to achieve the best possible signal-to-noise ratio. In applications where the short response time is not required (as in most spectroscopy) an alternative method of obtaining a high effective responsivity is to step up the impedance of the detector by means of a transformer cooled in the liquid helium. It is cooled to eliminate noise associated with its loss components. This technique was first used with superconductive bolometers, but its application to the InSb submillimeter detector was proposed by Rollin.[5] The transformer is usually tuned to about 1 kHz so that the effective response

---

[6] E. H. Putley, *Proc. Phys. Soc. (London)* **76**, 802–805 (1960); *J. Phys. Chem. Solids* **22**, 241–247 (1961).

[7] M. A. C. S. Brown and M. F. Kimmitt, *Brit. Commun. Electron.* **10**, 608–612 (1963).

[8] M. F. Kimmitt and G. B. F. Niblett, *Proc. Phys. Soc. (London)* **82**, 938–946 (1963).

[9] H. Yoshinaga, *Meeting Int. Comm. Opt., Reading, 1969*.

3. InSb SUBMILLIMETER PHOTOCONDUCTIVE DETECTORS

time is about 1 msec. However since one normally uses a much longer integration time than this in most spectroscopic applications, this is no disadvantage.

II. Description of Detector

Figure 1 shows the design of detector and cryostat originally developed at the Royal Radar Establishment (RRE) and extensively used in plasma diagnostic studies and other applications. The detector element is fabricated from high purity n-type InSb (carrier concentration $\sim 5 \times 10^{13}$ cm^{-3} electron mobility $\sim 5 \times 10^5$ cm^2 V^{-1} sec^{-1} as measured at 77°K). Optimum results were obtained with elements 1–2 mm thick and with areas to match the optics, usually 3–4 mm^2. Leads were attached using indium solder. The element was mounted at the center of a superconducting solenoid placed at the bottom of a glass Dewar. Radiation was condensed onto the element by means of a tapered light-pipe. Filters when required could be placed at the end of the light-pipe to restrict the transmission of unwanted parts of the spectrum. The transmission efficiency was about 50% which compares favorably with what can be achieved using far-infrared transmitting windows. The light-pipe construction is simpler to manufacture than a cryostat with windows. Also, bearing in mind that the operating temperature is below 2°K, the light-pipe design is more efficient thermally, since it is easier to control the admission of stray radiation. In the original design a niobium superconducting solenoid was used. This enabled a magnetic induction of up to 8 dT to be used. This was one of the first practical applications of a superconducting solenoid. Better materials than niobium are now available so that this material is not now used to construct the solenoid. The output from the detector was taken via low capacity leads to a low noise head amplifier. The design of suitable amplifiers is discussed in the appendix to this chapter.

When it is more convenient to use a cooled step-up transformer this is inserted in the cryostat in place of the solenoid. The size of the transformer is similar to that of the solenoid so that overall size is not altered. There is no advantage in using both a solenoid and a step-up transformer. This is because the efficiency of the transformer circuit falls off if the load resistance becomes too high so that there is no advantage in starting at a resistance much higher than that of the cooled element in zero magnetic field (typically $< 100\ \Omega$).

Commercial versions of this apparatus are now available from a number of sources. Mullard, Southampton, England manufactures a developed version of the RRE design (Fig. 2). Raytheon and Advanced Kinetics, Costa Mesa, California, both manufacture detectors of this type in the U.S.A. while

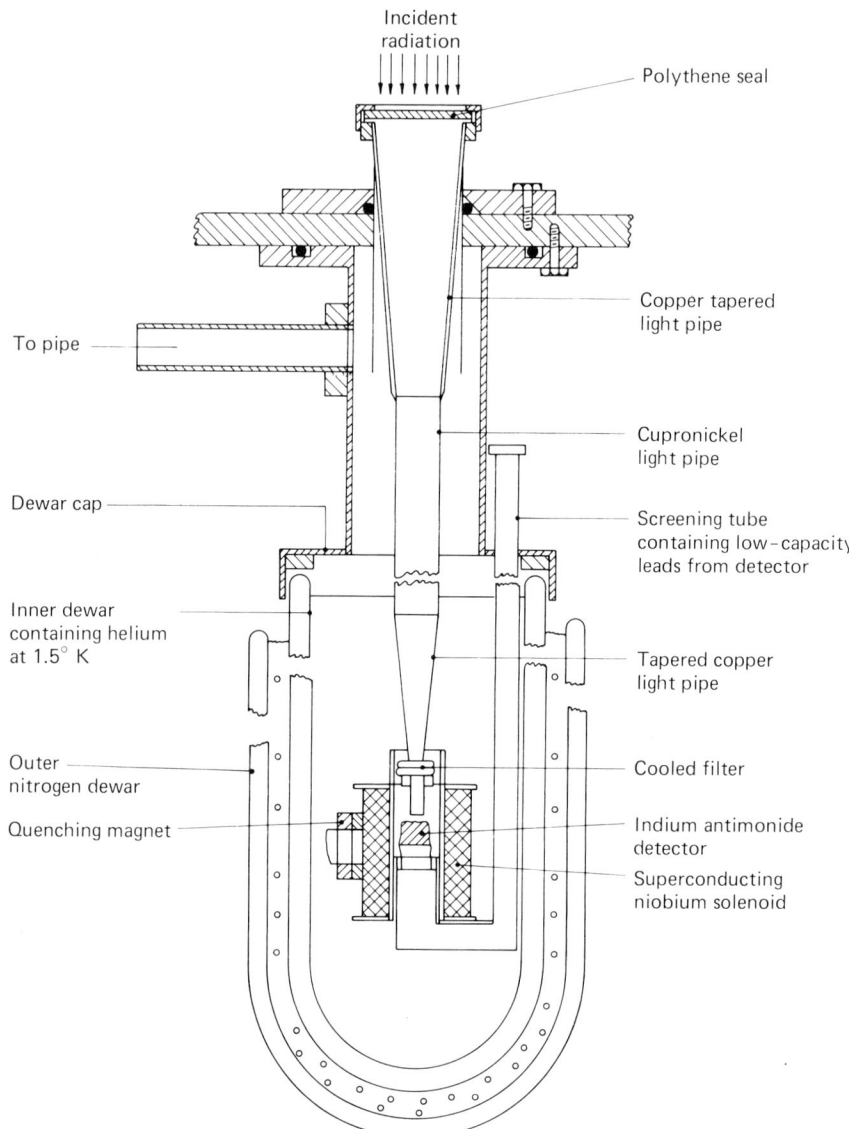

Fig. 1. Diagram of InSb submillimeter detector cryostat as used at RRE.

3. InSb SUBMILLIMETER PHOTOCONDUCTIVE DETECTORS

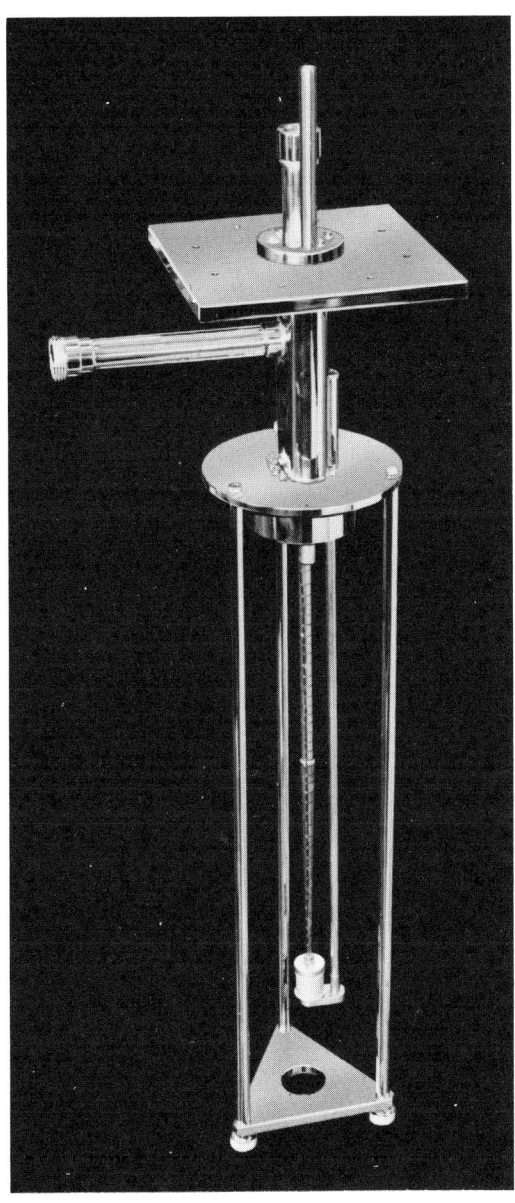

FIG. 2. Photograph of Mullard RPY 23 InSb submillimeter detector.

S.A.T.[10] in France and the Institute of Radio Engineering and Electronics[11] in Moscow have also manufactured versions of it. In addition, several laboratories have constructed models for their own use. These include the Physics Department of Queen Mary College, London; Texas Instruments, Dallas[12]; and the N.R.L.,[13] Washington, who in addition to designing a version for laboratory use have also succeeded in operating a rocket-borne version in space.[14]

III. Performance of Detector

1. Responsivity and Speed of Response

The responsivity of the broad-band detector at a fixed wavelength depends on both the applied electric and magnetic fields and it is found that the

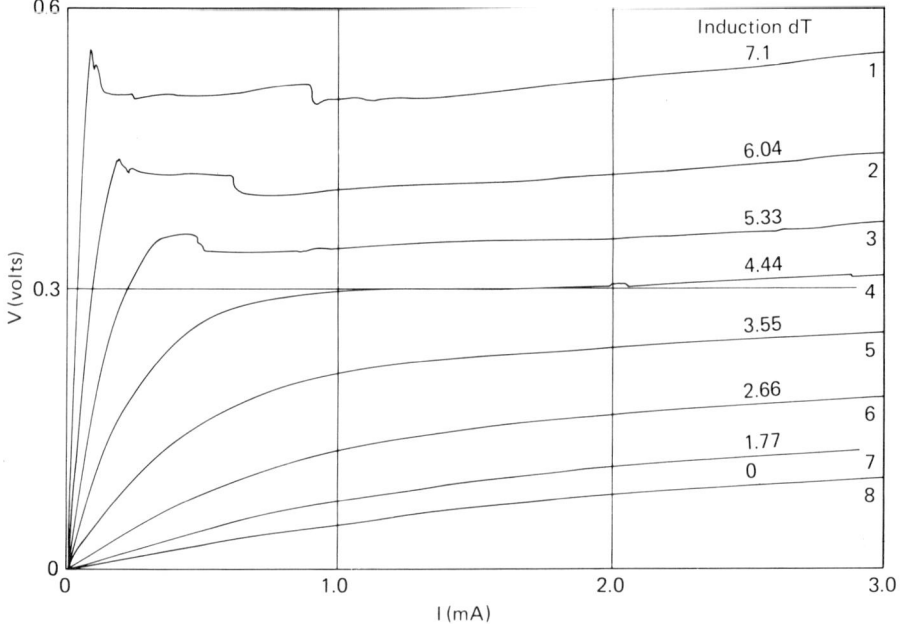

FIG. 3. Plot of electric field against current density for InSb submillimeter detector element at 1.82 K temperature.

[10] J. Besson, R. Cano, M. Matteoli, R. Papoular, and B. Philippeau, *L'Onde Elect.* **45**, 107–115 (1965).
[11] *Prib. Tekh. Eksp.* **4**, 228 (1966).
[12] M. A. Kinch, *Appl. Phys. Lett.* **12**, 78–80 (1968).
[13] R. Kaplan, *Appl. Opt.* **6**, 685–690 (1967).
[14] K. Shivanandan, J. R. Houck, and M. O. Harwit, *Phys. Rev. Lett.* **21**, 1460–1462 (1968).

3. InSb SUBMILLIMETER PHOTOCONDUCTIVE DETECTORS

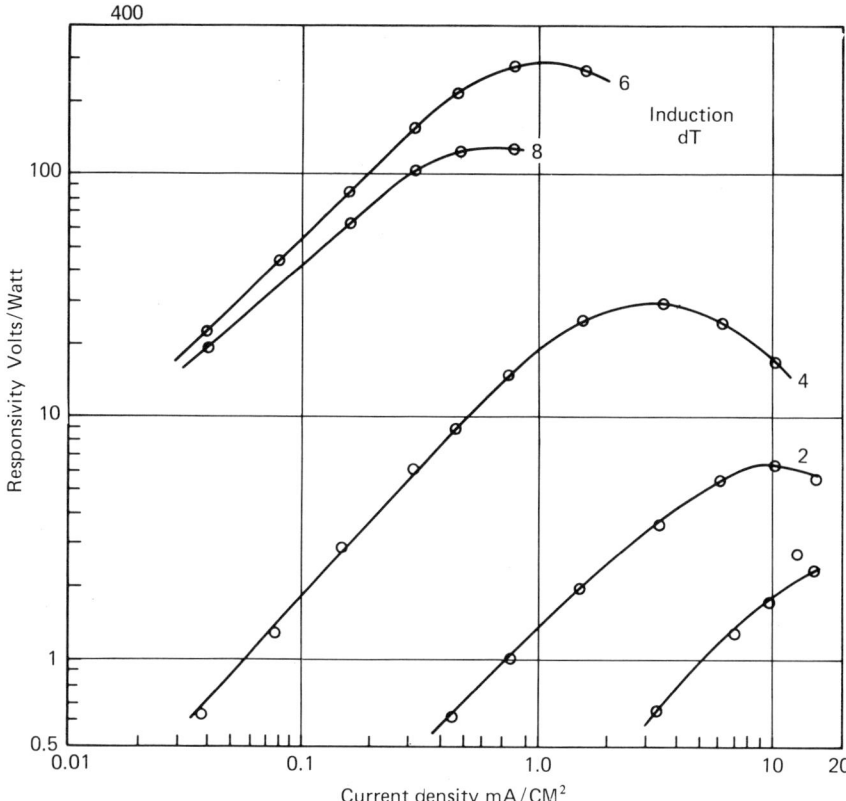

FIG. 4. Dependence of responsivity upon current density and magnetic induction at 0.2 mm wavelength and 1.65°K temperature.

optimum values are related to each other, but are relatively independent of the wavelength. When using the resonant mode the magnetic field is set for the wavelength required and the electric field adjusted accordingly.

Figure 3 gives a set of current–voltage characteristics for an InSb sample. The optimum photoconductive effect occurs just before the abrupt discontinuity in the curves. If larger electric fields are applied, noisy and unstable behavior results and oscillation is sometimes seen.[15] Figure 4 shows how the responsivity at a fixed wavelength varies with electric and magnetic fields while Fig. 5 shows the spectral variation of responsivity.

These results show that there are well-defined operating conditions for optimizing the responsivity and that at wavelengths less than about 1 mm

[15] E. H. Putley, *Proc. Int. Conf. Phys. Semicond. 7th*, pp. 443–450. Dunod, Paris, 1964.

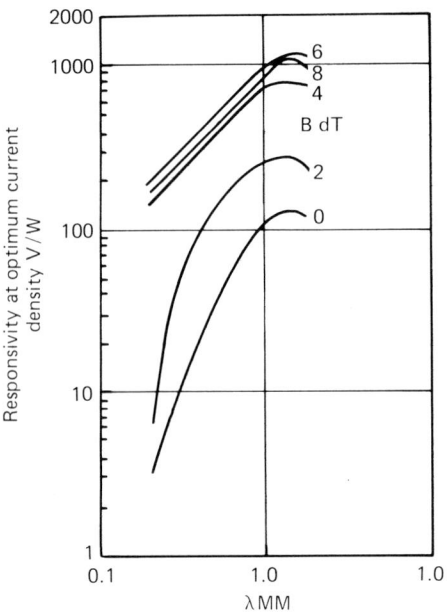

Fig. 5. Variation of responsivity with wavelength for various magnetic inductions (current density optimum in each case).

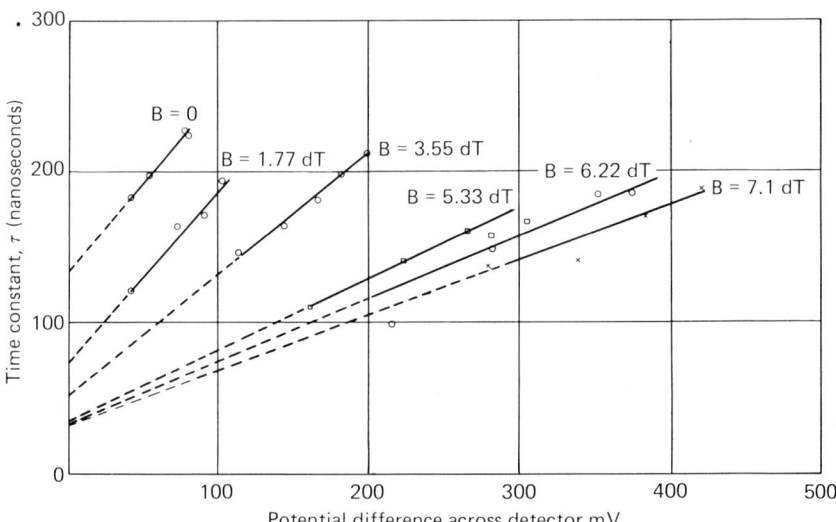

Fig. 6. Dependence of response time upon magnetic induction at wavelength 0.9 mm (carcinotron) and 1.24°K temperature.

3. InSb SUBMILLIMETER PHOTOCONDUCTIVE DETECTORS

the responsivity falls off roughly as λ^2, as might be expected from Eq. (7). At longer wavelengths the responsivity appears constant. The practical long-wave limit is set by the smallest diameter of the light-pipe which at long enough wavelengths behaves as a waveguide beyond cutoff. With practical dimensions this maximum wavelength usually falls within the range 5 mm–1 cm. Although one does not normally think of this apparatus as a millimeter wave detector, it does have the advantage over conventional millimeter wave detectors of being extremely broad-band and this is useful for some types of spectroscopy.

The absolute value of the responsivity was determined using a grating spectrometer calibrated using a glass blackbody radiator similar to the procedures described by Lichtenberg and Sesnic[16] and Platt.[17] The values obtained were in reasonable agreement with those calculated from Eq. (10).

Equation (11) predicts that the response time should be less than 1 μsec. Detailed measurements using a modulated 1 mm carcinotron source have confirmed this. The results obtained are shown in Fig. 6. These show that there is some dependence on the applied electric and magnetic field, but when operated under optimum conditions the response time is about 0.2 μsec. It does not vary very significantly with temperature below 4°K.

Above 4°K the response time falls with increasing temperature. It has been possible to determine it indirectly from measurements of the electrical properties[18,19] and from pulsed measurements of the current–voltage characteristic.[20] Thus at 20°K the response time is about 3×10^{-9} sec. This might be very useful, but unfortunately the responsivity falls at approximately the same rate. Thus Eqs. (10) and (11) can be combined by eliminating β to give

$$R = \frac{2}{3}\left(\frac{\tau}{nk}\right)\left(\frac{1}{\mu}\frac{d\mu}{dT}\right)(V/v). \tag{13}$$

Now n the carrier concentration is independent of temperature below about 100°K and $[(1/\mu)(d\mu/dT)]$ varies only slowly with T. Hence R is practically proportional to τ. Since [as discussed in Section 2] the noise equivalent power (NEP) varies as $1/R$ reducing τ will produce a corresponding reduction in detectivity. Measurement of the responsivity at 20°K has confirmed that it is about two orders smaller than the value of 4°K.[19]

---

[16] A. J. Lichtenberg and S. Sesnic, *J. Opt. Soc. Amer.* **56**, 75–79 (1966).
[17] C. M. R. Platt, *Infrared Phys.* **9**, 1–10 (1969).
[18] M. A. Kinch, *Brit. J. Appl. Phys.* **17**, 1257–1267 (1966).
[19] E. H. Putley and N. Shaw, 1965, Private Communication; A. N. Vystavkin and V. N. Gubankov, *Fiz. Tekh. Pol.* **2**, 1158–1163, 1968, [*English Transl.: Sov. Phys.—Semicond.* **2**, 968–972 (1969)].
[20] G. D. Peskett and B. V. Rollin, *Proc. Phys. Soc. (London)* **82**, 467–469 (1963).

Non-Ohmic behavior has also been observed at 77°K.[21,22] It is possible to estimate from the current–voltage characteristics that the response time should be of the order of 10^{-10} sec, but measurements of frequency conversion at 2 and 4 mm have shown that it is less than 5×10^{-11} sec.[22a]

Figure 5 shows the spectral variation of the responsivity obtained with a sample of high purity InSb. The responsivity falls as λ is reduced below 1 mm so that for applications at wavelengths below 500 μm the performance is significantly degraded. One way of improving the performance at the shorter wavelengths is by operating in a larger magnetic field when a tunable resonant effect is obtained (see Section 3), but it is possible to extend the broad-band response by using more heavily compensated material.[23] In this material the mobility at 4°K may be as small as 10^4 cm^2 V^{-1} sec^{-1}. The shoulder in the spectral response is determined by the conductivity relaxation time since it occurs where $\omega\tau_e \sim 1$. In good quality material with mobility $\sim 10^5$, $\tau_e \sim 7 \times 10^{-13}$ so that $\omega\tau_e \sim 1$ for $\lambda = 1.4$ mm. If a sample with $\mu \sim 2 \times 10^4$ were used, τ_e would be $\sim 1.5 \times 10^{-13}$ making $\omega\tau_e \sim 1$ for λ 280 μm. This improvement in spectral response would be obtained at the price of reduced responsivity since β [Eq. (8)] is given approximately by[21]

$$\beta \sim \mu_L \mu / V_s^2 \qquad (14)$$

where μ_L is the lattice scattering mobility, while μ is the measured mobility (including impurity scattering effects), and V_s is the acoustic phonon speed. Thus β will be reduced by a factor of 5 and hence the responsivity [Eq. (10)] will be reduced. The time constant [Eq. (11)] will also be reduced by the same factor, so apart from improving the spectral response, a response time of about 40 nsec should be obtained. Thus the use of heavily compensated material operating at 4°K or below provides another method for reducing the response time which may be of use providing some sacrifice in responsivity is acceptable.

2. Noise Equivalent Power or Detectivity

The noise equivalent power (NEP) attainable by an ideal submillimeter detector has been discussed by Putley.[24] This can be calculated in a similar way to that used for shorter wavelength infrared detectors, but the effects of the restricted field of view of the light-pipes normally used and of restrictions

---

[21] R. J. Sladek, *Phys. Rev.* **120**, 1589–1599 (1960); J. B. Gunn, *Progr. Semiconduct.* **2**, 213–247 (1957).

[22] Y. Kanai, *J. Phys. Soc. Japan* **15**, 830–835 (1960).

[22a] A. M. Belyantsev, V. A. Valov, V. N. Genkin, A. M. Leonov, and B. A. Trifonov, *Sov. Phys.—JETP* **34**, 471–473 (1972).

[23] H. Shenker, N. R. L. Memo. Rep., No. 1743, 1967.

[24] E. H. Putley, *Infrared Phys.* **4**, 1–8 (1964).

3. InSb SUBMILLIMETER PHOTOCONDUCTIVE DETECTORS

on the spectral response set by the use of cooled filters must be taken into account. If the spectral response is limited to wavelengths λ satisfying the condition

$$\lambda_1 < \lambda < \lambda_2,$$

then the NEP P_N is given by

$$P_N = [2(kT)^{5/2}/ch^{3/2}][BA\alpha \cos \theta]^{1/2}[J_4(x_1) - J_4(x_2)]^{1/2}, \quad (15)$$

where k is Boltzmann's constant, h is Planck's constant, and c is the velocity of light; T is the absolute temperature of the background to which the detector is exposed (not the temperature of the detector itself); A is the effective receiving area of the detector; and B is the bandwidth of its output circuit. It is assumed that the light-pipes or other optical components restrict the field of view to a small solid angle α in a direction making an angle θ with the normal to A. The function $J_4(x)$ is

$$J_4(x) = \int_0^x \frac{x^4 e^x}{(e^x - 1)^2} dx,$$

where $x = hc/kT\lambda$. This function has been tabulated by Rogers and Powell,[25] but for long wavelengths x is small and an approximate form can be used. In the limit $x \to 0$, $J_4(x) \to \frac{1}{3}x^3$. If $\lambda = 100$ μm and $T = 300°$K,

$$x(=hc/kT\lambda = 1.44/T\lambda) = 0.48.$$

Hence the approximation $x \to 0$ is adequate for the submillimeter region and Eq. (15) simplifies to

$$P_N = [2(kT)^{5/2}/ch^{3/2}][BA\alpha \cos \theta]^{1/2}[\tfrac{1}{3}(x_1 - x_2)(x_1^2 + x_1 x_2 + x_2^2)]^{1/2}$$
$$= 1.60 \times 10^{-18}(T)^{5/2}[(AB\alpha \cos \theta)(x_1 - x_2)(x_1^2 + x_1 x_2 + x_2^2)]^{1/2}. \quad (16)$$

This expression is valid when the Rayleigh–Jeans approximation is applicable. The presence of the factor $A\alpha \cos \theta$ implies that the detector can be treated as a diffuse reflector obeying Lambert's law. While this may be applicable to short and medium wavelength infrared detectors, it ceases to be true when the microwave region is approached. In considering a submillimeter detector similar to that shown in Fig. 1, it is not clear whether A should refer to the area of the detecting element itself or to the area of aperture of the light-pipe. Radiation entering the light-pipe may be absorbed by the detecting element after multiple reflections in the cavity enclosing the detector element so that the optical characteristics of this system are more closely analogous to those of a microwave horn antenna then to those of a shorter wavelength infrared detector. If therefore the design of the optical

---
[25] W. H. Rogers and R. L. Powell, N.B.S. Circ. No. 595, 1958.

system for the submillimeter detector succeeds in matching the element into the radiation field, microwave antenna theory[26] can be applied, which shows that

$$A\alpha \cos \theta \sim \lambda^2. \quad (17)$$

If this result is combined with Eq. (16), and it is assumed that the detector is matched into the background over the frequency band v_1 to v_2, then the expression for the NEP reduces to

$$P_N = 2^{1/2}(kT)[B(v_1 - v_2)]^{1/2}, \quad (18)$$

which is similar to that for a square law video receiver.[27] When Eq. (18) applies, the receiver may respond only to one plane of polarization when the value of P_N is reduced by $2^{1/2}$. This result is independent of the area of the detecting element so that the concept of D^* does not apply. Since InSb elements respond to all frequencies less than some limit set by the falling off of free carrier absorption, $v_2 \ll v_1$ and Eq. (18) becomes

$$P_N = 2^{1/2}(kT)(Bv_1)^{1/2} = 5.86 \times 10^{-21}(Bv_1)^{1/2}. \quad (19)$$

v_1 may be set by the characteristics of the InSb element or it may be set by the use of a cooled filter inserted at the base of the light-pipe. Equation (19) thus gives a value for the NEP attainable from a perfect submillimeter detector. Although it has been derived with the InSb detector in mind, similar considerations will apply to other types of detecting systems in this part of the spectrum so that the result is of more general application. Choosing suitable values for v_1 (with $B = 1$ Hz) gives the values for P_N shown in the accompanying tabulation.

λ (μm)	v_1 (Hz)	P_N (W)
100	3×10^{12}	1.0×10^{-14}
500	6×10^{11}	4.5×10^{-15}
1000	3×10^{11}	3.2×10^{-15}

The performance of a real detector could only approach this limit if all other sources of noise were small compared to that of the radiation noise. Regarding the hot electron detector as a form of thermal detector, this means that the noise terms associated with the radiation component of the thermal conductance of the electron gas to its surroundings must be greater than any others. The main noise sources to be considered are the noise of the thermal conductance coupling the electrons to the lattice, the

[26] See, for example, J. C. Slater, "Microwave Transmission." McGraw-Hill, New York, 1942.
[27] R. A. Amith, *Proc. IEE* **98**, 43–54 (1951).

3. InSb SUBMILLIMETER PHOTOCONDUCTIVE DETECTORS

Johnson noise, and amplifier noise. The radiative conduction noise leads to the NEP given by Eq. (19) so that by calculating the contributions the other noise sources make to the NEP we can see how closely the performance of an actual detector approaches the ideal.

Consider the equivalent circuit of Fig. 7. A biasing current I is passing through the detector element D and the output signal is applied to the amplifier A. Suppose the signal power incident upon the detector is P. If only background radiation fluctuations need be considered, then the NEP P_{NR} would be that given by Eq. (19). However, suppose there are noise sources associated with the Johnson noise of the resistance of the detector, shot noise in the bias current, temperature noise associated with the thermal coupling between the electron gas and the lattice, and amplifier noise and that these can be represented by the rms noise voltage generators ΔV_J, ΔV_I, ΔV_T, and ΔV_A, respectively. If the responsivity of the detector is R, the equivalent signal power required to equal the noise output from one of these sources is

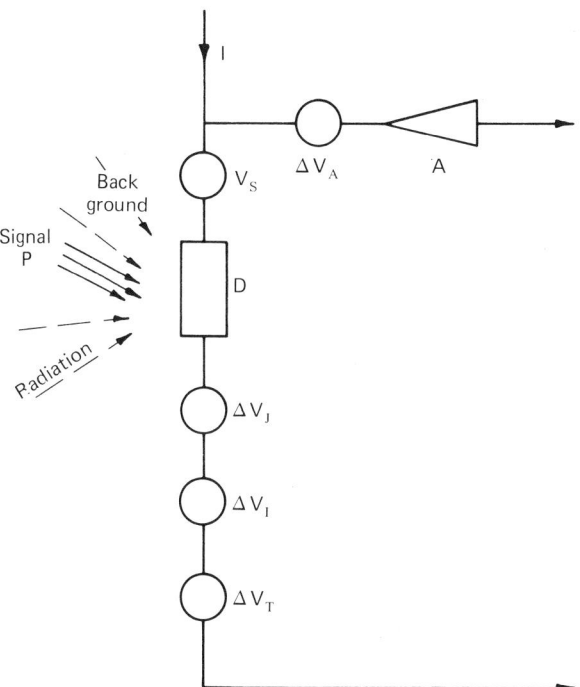

FIG. 7. Equivalent circuit showing noise generators associated with the InSb submillimeter detector.

$$P_{Ni} = \Delta V_i/R. \tag{20}$$

Hence calculating P_{Ni} for each source in turn, and comparing it with P_{NR} derived from the background radiation noise, the relative importance of these noise sources can be established.

Consider a numerical example. Suppose in an applied field of 6 dT a detector has a resistance of 5 kΩ at 2°K and that its responsivity is about 1000 V/W with a biasing current of 10^{-5} A. To a first approximation the responsivity is proportional to the current I. Hence we can write $R = 10^8 \, I$. The Johnson noise is given by

$$\Delta V_J = \{4kTBr\}^{1/2} = 7.43 \times 10^{-12}(TBr)^{1/2} \tag{21}$$

so that with $B = 1$ Hz,

$$V_J = 7.0 \times 10^{-10} \quad V,$$

giving from Eq. (20), $P_{NJ} = 7.0 \times 10^{-18} \, I^{-1}$.

For an amplifier having a equivalent noise input resistance of 100 Ω, $\Delta V_A \sim 1.3 \times 10^{-9}$ V so that from Eq. (20), $P_{NA} = 1.3 \times 10^{-17} \, I^{-1}$. The shot noise of the biasing current is

$$\Delta i = \{2eIB\}^{1/2} = 5.58 \times 10^{-10}(IB)^{1/2} \tag{22}$$

so that if the resistance of the element is r, and $B = 1$ Hz,

$$\Delta V_I = r \, \Delta i = 5.58 \times 10^{-10}(I)^{1/2} r$$

and hence

$$P_{NI} = 2.8 \times 10^{-14} I^{-1/2}.$$

The temperature fluctuation noise is obtained by estimating the thermal conductance coupling the electron gas to the lattice. The NEP P_{NT} is given by

$$P_{NT} = 2T(kGB)^{1/2} = 7.43 \times 10^{-12} T(GB)^{1/2}. \tag{23}$$

Estimates for G^{24} give values of about 10^{-5} W/°K hence $P_{NT} \sim 3.5 \times 10^{-14}$ W. The background radiation limit P_{NR} is given by Eq. (19). Both P_{NT} and P_{NR} are independent of the biasing current. The values of P_{Ni} are plotted in Fig. 8. This shows that the bias current should be made as large as possible. If it could be increased indefinitely the NEP would be determined by the temperature fluctuation noise. However the maximum current is set by the onset of avalanche ionization to a value between 10^{-5} and 10^{-4} A. Hence when operating with the maximum permissible current, the NEP would be about 10^{-11}, and determined principally by the bias current shot noise. This is several orders above the value set by temperature fluctuation. When the

3. InSb SUBMILLIMETER PHOTOCONDUCTIVE DETECTORS

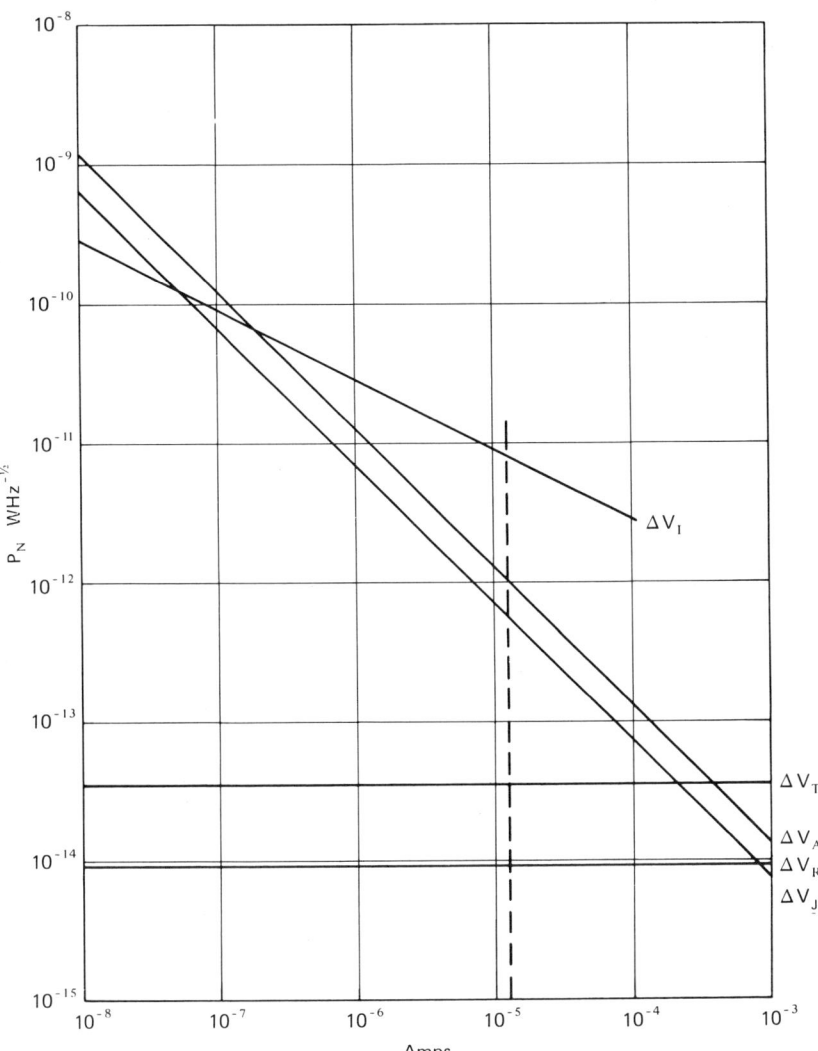

FIG. 8. Variations of NEP with bias current for 6 dT magnetic induction. The lines show the contributions to the NEP associated with the various noise sources: Johnson ΔV, amplifier ΔV_A, current shot noise ΔV_I, temperature fluctuation noise ΔV_T, and radiation noise ΔV_R. The value of the optimum bias current is also shown.

detector is operated without a magnetic field, the responsivity will be less (~ 100 V/W at 100 μA bias current) and the resistance will be smaller ($\sim 10 \Omega$). If a cooled stepup transformer with a stepup ratio of 100 is used, the effective amplifier noise voltage referred to the detector terminals will be reduced

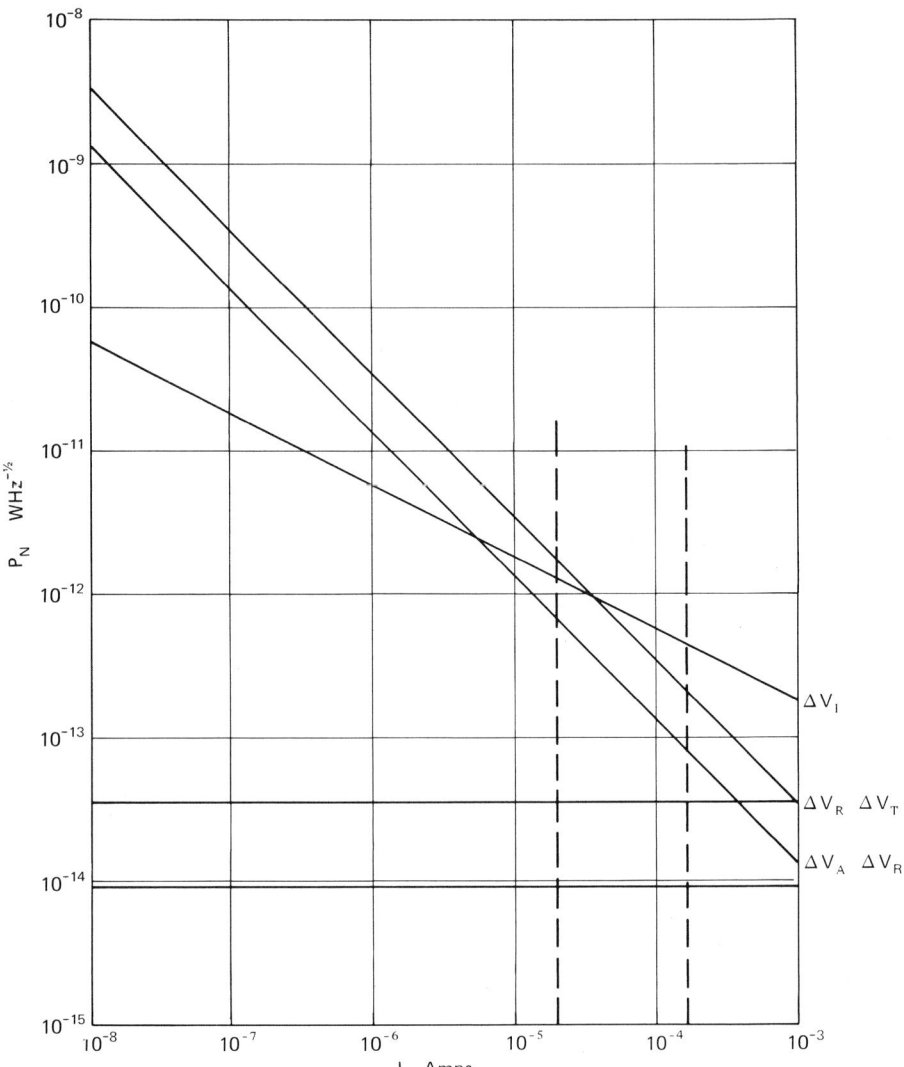

FIG. 9. Variations of NEP with bias current for zero magnetic induction and using a cooled step-up transformer. The lines show the contributions to the NEP associated with the various noise sources: Johnson ΔV_J, amplifier ΔV_A, current shot noise ΔV_I, temperature fluctuation noise ΔV_T, and radiation noise ΔV_R. The value of the optimum bias current is also shown.

by a factor of 100. Calculating the values of P_{Ni} produces the results shown in Fig. 9. The maximum bias current is now greater than 10^{-4} A. Hence at the maximum current P_N is about 5×10^{-13} which is over an order better than the value obtained in a magnetic field. This result stems from the facts that a larger biasing current can be used, but because the resistance is lower the shot noise of this current is less important. Also to achieve this result the effective amplifier noise contribution has been reduced by two orders by means of the cooled transformer.

Examination of Figs. 8 and 9 shows that even if the biasing current could be made arbitrarily large the NEP would be limited by the temperature fluctuation noise which is almost five times greater than the radiation background noise.

As Eq. (23) shows, the temperature fluctuation noise is determined by the thermal conductance G so that to reduce the temperature fluctuation noise G must be reduced to less than the radiative conductance. However, G is determined by the interaction between the conduction electrons and the lattice since this determines the rate at which energy is exchanged between them. An approximate expression derived for G which assumes that ionized impurity scattering is the dominant scattering mechanism is[24]

$$G = \tfrac{3}{2}(\sigma_0/\beta T), \tag{24}$$

where the electrical conductivity shows non-Ohmic behavior described by

$$\sigma = \sigma_0(1 + \beta E^2). \tag{25}$$

Thus to reduce G, σ_0 must be small and β large. This shows that material of the highest purity is required. Since in fact the best available material is used we cannot obtain an improvement in G unless methods of improving the purity can be developed. Hence while it may be possible to improve the performance of the InSb submillimeter detector a few times by improving the amplifier or possibly by improvement in the optics, larger improvements are unlikely without better material.

The performance estimates given in Figs. 8 and 9, although based on measured parameters, are derived by a rather simple and idealistic treatment. However, the values predicted are comparable with measured values and in fact recent work by Kinch[12] shows that Fig. 9 is somewhat pessimistic. Kinch reports an NEP at 4 mm wavelength of 1.3×10^{-13} W for a detector operating at 1.2°K in zero magnetic field and with a cooled input transformer. Platt[17] has measured the performance of an InSb radiometer at 1.2 mm operating in a magnetic induction of 6 dT and obtained a NEP[27a] of 7×10^{-12}, which agrees well with the estimate of Fig. 8. Thus the estimates obtained from Figs. 8 and 9 agree well with measured values, although the precise values for the various parameters will differ somewhat for different detectors.

[27a] Platt and other astronomers appear to use a somewhat different definition of NEP than that used here in that they do not include factors such as amplifier noise or transmission losses in their estimate and so obtain an apparently much better value for the NEP. The quantity Platt calls then "minimum detectable power" corresponds to our definition of NEP.

3. Behavior in Strong Magnetic Fields

As Fig. 4 shows, the broad-band response falls when the magnetic field is increased above a certain point. It was first observed by Kimmitt (private communication) and later studied in more detail by Brown and Kimmitt[28] and others[29,30] that although the broad-band response falls with increasing field there is a sharp increase in response near the cyclotron resonance wavelength. This is a consequence of the large absorption at cyclotron resonance by the electrons magnetically frozen out onto the donor impurity centers. Careful study of the absorption peak shows that although it contains fine structure[31] produced by transitions involving the excited states of the donor impurity centers no resonance occur at any harmonics of the fundamental absorption frequency.[28] This means that the effect is of practical use in far-infrared spectroscopy as a means of eliminating errors caused by overlapping grating orders which can be very difficult to eliminate by conventional filtering techniques. It was first used for this purpose by Kimmitt and Niblett[8] to estimate the position of a sharp plasma cutoff edge in a theta pinch experiment.

Figure 10 shows the relative performance in a large magnetic field. The

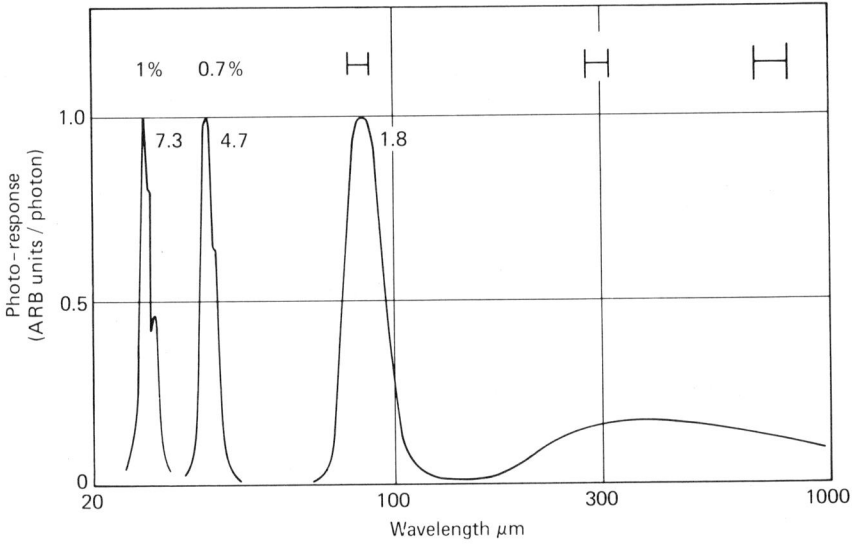

FIG. 10. Resonant photoconductive effect.

[28] M. A. C. S. Brown and M. F. Kimmitt, *Infrared Phys.* **5**, 93–97 (1965).
[29] R. Kaplan, *Proc. Int. Conf. Phys. Semicond., Kyoto, 1966*.
[30] J. Yamamoto and H. Yoshinaga, *Jap. J. Appl. Phys.* **7**, 498–505 (1968); J. Yamamoto, H. Yoshinaga, and S. Kon, *Jap. J. Appl. Phys.* **8**, 242–249 (1969).
[31] P. L. Richards, private communication, 1964.

dip in performance near 50 μm is caused by the InSb restrahlen reflection band. The shortest attainable resonance wavelength is determined by the size of the available magnetic field 7.6 T, producing a resonance at 26 μm.

4. Heterodyne Detection

The use of the InSb submillimeter detector in heterodyne systems has been described by Arams et al.,[32] by Putley,[33] and by Kinch[12] and its use in a heterodyne system with the 337 μm CN maser has been described by Gebbie et al.[34]

When used to detect a coherent signal power P_s the voltage output from a photoconductive detector used in the heterodyne mode is

$$V = R(P_s P_L)^{1/2}, \qquad (26)$$

where P_L is the local oscillator power and R is the responsivity given by eq. (10).

Thus if the NEP, when the device is operated as a video detector, is P_N, the signal power that can now be detected is given by

$$(P_s P_L)^{1/2} = P_N, \qquad (27)$$

so that if P_L can be made large, P_s will be correspondingly smaller. This simple relation applies because in this case the main noise source in that of the amplifier external to the detector itself. Optimum results are obtained when the local oscillator power is of the same order as the dc biasing power in the detector, i.e., of the order of a few microwatts. It is then possible to detect signal powers of 10^{-18} W or less. Although the performance may be as much as two orders worse than that of an ideal superheterodyne receiver, it nevertheless compares very favorably with that of any other receiver for the 1 mm region. Use of the InSb detector in the heterodyne mode should form the basis for a very useful 1 mm radiometer, but so far as is known this application has not yet been developed.

IV. Applications and Future Developments

The InSb submillimeter detector was first developed to meet the requirement for a fast detector in submillimeter plasma diagnostics[35] and has been used in a number of experiments of this type. It is also useful for studying

---

[32] F. Arams, C. Allan, B. Peyton, and E. Sard, *Proc. IEEE* **54**, 612–622 (1966).
[33] E. H. Putley, *Proc. IEEE* **54**, 1096–1098 (1966).
[34] H. A. Gebbie, N. W. B. Stone, E. H. Putley, and N. Shaw, *Nature (London)* **214**, 165–166 (1967).
[35] M. F. Kimmitt, A. C. Prior, and V. Roberts, "Plasma Diagnostic Techniques," Chapter 9. Academic Press, New York, 1965.

submillimeter lasers and masers. Because of the high sensitivity that can be obtained near 1 mm wavelength, it has been used in some spectroscopic applications where speed is not important. However, for many applications of this type one of the cooled bolometers is often more convenient since the bolometers have a flatter spectral response. This is not the case in 1 mm radio astronomy where the usable spectral bandwidth is restricted by atmospheric transmission. The InSb detector appears at present to be the most sensistive detector available for this application, as is illustrated by the recent detection of radiation from α-Cygnus.[36] It has also been used in rocket-borne astronomy[14] to investigate the $3°K$ thermal background radiation. This is an application where its short response time is important.

Although at wavelengths longer than 2 mm a conventional microwave receiver will normally be preferred, the InSb detector has a much wider bandwidth than the usual microwave detectors so that for applications, such as millimeter wave interferometry, requiring uniform response over the whole millimeter region the InSb detector is more convenient.

In principle, further developments of this type of detector are possible, by improving the InSb, developing more sophisticated structures for the detector elements, improving the electronics, or by developing other materials with low effective mass carriers. At present none of these approaches seem practicable without considerable effort, which is not likely to be forthcoming unless there is a considerable increase in interest in the submillimeter region.

Appendix. Amplifiers for Use with the InSb Submillimeter Detector

As Section 2 shows, the best available low noise amplifier in the frequency range up to a few megahertz is required. When this device was first developed, low noise triode amplifiers, such as the 3A/167M or the E81OF, were the best available, but recent progress with field effect transistors has reached the point where the best FET's are superior to the valve amplifiers. They are also more convenient to use and may, in some cases, be operated at cryogenic temperatures[37–39] so that they can be placed close to the detector. While this may not reduce the amplifier noise significantly, it will simplify the practical problems of reducing pickup and interference, which is often very serious when operating the detector near a large plasma generator or other pulsed source.

The noise characteristics of an amplifier can be represented in general by the combination of a voltage and a current noise source at the input.

---

[36] J. A. Bastin, private communication, 1965.
[37] C. G. Rogers, *Solid State Electron.* **11**, 1079–1091 (1968).
[38] E. Elad and M. Nakamura, *Nucl. Instrum. Methods* **54**, 308–310 (1967).
[39] F. E. Kingston and K. Lee, *Rev. Sci. Instrum.* **39**, 599–601 (1968).

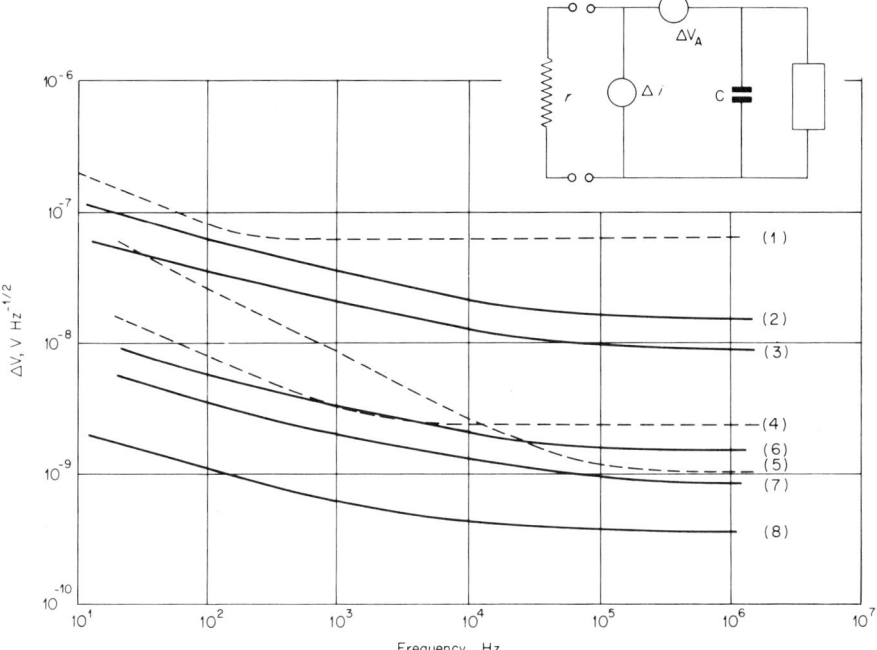

FIG. 11. Noise characteristics of some amplifying devices: The equivalent input circuit is shown in the inset. ΔV_A represents the series voltage noise generator. Δi represents the shunt current noise generator. r is the output resistance of the detector. The input impedance of the amplifier is represented by the capacitance C. Δi can be replaced by another voltage generator $\Delta V_1 = r \Delta i$ in series with ΔV_A. If r is a large resistance $r \Delta i > \Delta V$. This becomes important when r is at low temperature (1–2°K) when Johnson noise in r ($\Delta V_j = (4kTr^{1/2}$ in unit bandwidth) will be small, but the value of ΔV_1 is independent of the temperature of r. When r is large, the frequency response of the circuit may be determined by the time constant rC which can in some cases become longer than the response time of the detector.

The main figure shows the frequency dependence of the voltage noise generator ΔV; the corresponding values for the current noise generator Δi (which is not frequency dependent) and the input capacitance C are given in the table. The lowest value of Δi is that of the miniature electrometer pentode XE 5886 [curve (1)]. Curves (4) and (5) refer to the Phillips low noise pentode E810F, which was typical of the amplifiers used for the early work with the InSb submillimeter detector. All the other curves are taken from the Texas Instruments (Bedford, UK) range of FETs which have been specially developed for use with infrared detectors and nuclear particle counters. In situations where ΔV_A makes the principal contribution to the amplifier noise, the modern FETs are very superior to the thermionic valves. Only if the smallest possible valve for Δi were needed, would a valve still prove superior and even then the inconvenience of using it may offset any performance advantage.

Curve no.	Device	Δi (A Hz$^{-1/2}$)	C (pF)
1	Hivac XE5886 pentode	3×10^{-17}	~2–3
2	Texas VX11013	1×10^{-16}	0.4
3	Texas BF800 (E 1600)	3×10^{-16}	3
4	Phillips E810F (I_k, 2 mA)	5×10^{-14}	24
5	Phillips E810F (I_k, 32 mA)	3×10^{-14}	24
6	Texas BF810	2.5×10^{-15}	5.0
7	Texas BF817	2.5×10^{-15}	11
8	Texas VX11010B	4×10^{-15}	65

These two sources may not be entirely independent of each other since both may depend to some extent on the operating conditions. The importance of the current generator will depend on the value of the detector resistance r [Fig. (11)]. The current flowing through r produces a voltage $\Delta V_I = r \Delta i$ so that for the value of r making $\Delta V_I = \Delta V_\lambda$ the two sources are of equal importance. In most amplifiers the value of r required to satisfy these conditions is usually so large that where r is at room temperature the Johnson noise from r masks the presence of ΔV_I. However, where r is at a temperature near 1°K the Johnson noise is much reduced and hence the current generator is of greater relative importance. The noise characteristics of some suitable amplifiers are shown in the body of Fig. 11.

Addendum

Since this chapter was prepared, the InSb submillimeter detector has continued in use by radio astronomers[40-45] and others[46,47] requiring a sensitive, reliable, and comparatively fast submillimeter detector. The basic characteristics of the device have not changed since they depend critically upon the purity of the available InSb, which has not been improved upon. More attention has been given to heterodyne techniques where it has been shown that the insertion loss of the InSb mixer can be as low as 7 dB,[48] which is substantially less than that of other readily available millimeter wave mixers. It has been shown[49] that use of microwave biasing is convenient compared with either direct current biasing or heterodyning and that the use of heavily compensated material gives a high responsivity as well as a short response time in the absence of a magnetic field.[42] The performance

---

[40] T. G. Phillips and K. B. Jefferts, *Rev. Sci. Instrum.* **44**, 1009–1014 (1973).

[41] J. E. Beckman and J. A. Shaw, *Infrared Phys.* **12**, 219–234 (1972).

[42] A. N. Vystavkin, V. N. Gubankov, V. N. Listvin, and V. V. Migulin, *PIB Symp.* **20**, 321–329 (1971).

[43] N. G. Afonchenkov, A. N. Vystavkin, V. N. Listvin, and A. D. Morenkov, *Cryogenics* **12**, 415–416 (1971).

[44] P. E. Clegg and J. S. Huizinga, *Conf. Infra-Red Techniques, IERE Conf. Proc.* No. 22, 21–30 (1971).

[44a] P. E. Clegg and J. S. Huizinga, *Infrared Detector Techniques Space Res.* (V. Manno and R. Ring, eds.), pp. 132–140. Reidel, Dordrecht, Holland, 1972.

[45] V. F. Zabolotnyi et al., *Sov. Astron.—AJ* **16**, 795–802 (1973).

[46] A. N. Vystavkin, V. N. Gubankov, V. N. Listvin, and V. V. Migulin, *Electron Technol. (Warsaw)* **2**, 247–248 (1969).

[47] J. R. Stockton and C. C. Bradley, *J. Phys. E* **6**, 1049–1052 (1973)

[48] V. P. Voronenko, A. N. Vystavkin, B. G. Zyabrev, and V. I. Navrotskiy, *Radio Eng. Electron Phys.* **17**, 1287–1292 (1972).

[49] I. I. Eldumiati and G. I. Haddad, *IEEE Trans. Electron Devices* **ED-19**, 257–267, 1061–1063 (1972).

has also been improved by careful design of the element mounting to optimize the absorption of the radiation.[41,42,50]

Recent work has included further development and study of the InSb hot carrier detector, both with and without a biasing magnetic field.[51-64] The existence of an effect at 77°K has been demonstrated.[65] Forms of hot electron photoconductivity have also now been observed in GaAs[66] and in Ge[67-69].

Interest in the heterodyne mode of operation has continued[70-73] leading to a very interesting application to sensitive mm wave radiometers for use in radio astronomy.[74,75] These radiometers working at wavelengths of 4 mm or

[50] A. S. Vardanyan, A. N. Vystavkin, I. A. Iskhakov, V. N. Listvin, N. A. Savich, and A. V. Sokolov, *Sov. Astron.—AJ* **16**, 806–808 (1973).
[51] V. M. Afinogenov and V. I. Trifonov, *Sov. Phys.—Semicond.* **6**, 1099–1105 (1973).
[52] Y. Tsunawaki, T. Gamo, A. Tanimoto, and H. Yoshinaga, *Jap. J. Appl. Phys.* **11**, 1746 (1972).
[53] S. Kobayashi, *PIB Symp.* **20**, 331–344 (1971).
[54] A. N. Vystavkin, V. N. Gubankov, and N. M. Margolin, *Sov. Phys.—Semicond.* **5**, 915–919 (1971).
[55] A. N. Vystavkin, Yu. S. Galipern, and V. N. Gubankov, *Sov. Phys.—Semicond.* **2**, 1373–1378 (1969).
[56] T. Murotani and Y. Nisida, *J. Phys. Soc. Japan* **32**, 986–998 (1972).
[57] L. C. Robinson and L. B. Whitbourn, *J. Phys. A: Gen. Phys.* **5**, 263–271 (1972).
[58] Y. Tsunawaki, T. Gamo, A. Tanimoto, and H. Yoshinaga, *Jap. J. App. Phys.* **11**, 1746 (1972).
[59] I. I. Chusov and Yu. V. Gulyaev, *Sov. Phys.—Semicond.* **8**, 435–438 (1974).
[60] I. I. Chusov, *Sov. Phys.—Semicond.* **8**, 506–507 (1974).
[61] S. D. Lazarev and G. D. Efremova, *Sov. Phys.—Semicond.* **8**, 474–477 (1974).
[62] K. Yamada and T. Sekiguchi, *J. Phys. Soc. Japan* **37**, 95–103 (1974).
[63] E. Gornik, W. Müller, and F. Kohl, *IEEE Trans. Microwave Theory Tech.* **MTT-22**, 991–995 (1974).
[64] D. F. Walls, *J. Phys. A: Math. Nucl. Gen.* **8**, 751–758 (1975).
[65] A. M. Belyantsev, V. N. Genkin, A. M. Leonov, and B. A. Trifonov, *Zh. Eksp. Teor. Fiz. Pis'ma Red.* **18**, 616–620 (1973).
[66] J. M. Ballantyne, J. P. Baukus, and J. M. Lavin, *Appl. Opt.* **12**, 2486–2493 (1973).
[67] Ye. M. Gershenzon, G. N. Gol'tsman, Yu. A. Gurvich, S. L. Orlova, and N. G. Ptitsina, *Radio Eng. Electron. Phys.* **16**, 1346–1353 (1971).
[68] V. M. Afinogenov, *Sov. Phys.—Semicond.* **6**, 1916–1917 (1973).
[69] L. V. Berman and A. G. Zhukov, *Sov. Phys.—Semicond.* **8**, 1287–1288 (1975).
[70] J. J. Whalen and C. R. Westgate, *Polytechnic Inst., Brooklyn Symp. Sub-mm Waves, PIB Symp.* **20**, 305–320 (1971).
[71] J. J. Whalen and C. R. Westgate, *IEEE Trans. Electron Devices* **ED-17**, 310–319 (1970).
[72] A. M. Belyantsev, V. A. Valov, V. N. Genkin, A. M. Leonov, and B. A. Trifonov, *Sov. Phys.—JETP* **34**, 471–473 (1972).
[73] G. Biskupski, H. Dubois, and D. Ferre, *C. R. Acad. Sci. Ser.* **276**, 749–751 (1973).
[74] T. G. Phillips, and K. B. Jefferts, *Rev. Sci. Instrum.* **44**, 1009–1014 (1973).
[75] V. P. Voronenko, A. N. Vystavkin, B. G. Zyabrev, and V. I. Navrotskiy, *Radio Eng. Electron Phys.* **17**, 1287–1292 (1972).

shorter have achieved equivalent noise temperature of about 0.1°K with an integration time of one second and must be amongst the most sensitive radiometers available for this part of the spectrum. The direct mode of operation is still also of importance in astronomy[76,77] and consequently effort is still being applied to the difficult problem of absolute calibration in the sub-mm region.[78]

---

[76] K. Shivanandan, D. P. McNutt, R. J. Bell, *Infrared Phys.* **15**, 27–32 (1975).
[77] A. S. Vardanyan, A. N. Vystavkin, I. A. Iskhakov, V. N. Listvin, N. A. Savich, and A. V. Sokolov, *Sov. Astron.—AJ* **16**, 806–808 (1973).
[78] M. Daehler, Calorimetric Method of Sensitivity Calibration for Far-Infrared Detectors, NRL Rep. 7976, April 16, 1976.

CHAPTER 4

Far-Infrared Photoconductivity in High Purity GaAs*

G. E. Stillman, C. M. Wolfe, and J. O. Dimmock

I.	GENERAL INTRODUCTION	169
	1. *Photoconductive Mechanisms*	170
	2. *Hydrogenic Model for Impurities*	172
	3. *Impurity Excited States in Photoconductivity* . . .	175
II.	MATERIAL	176
	4. *Crystal Properties*	177
	5. *Preparation*	191
	6. *Characterization*	196
III.	PHOTOCONDUCTIVITY	208
	7. *Introduction*	208
	8. *Excited States and Photoconductivity*	213
	9. *Dependence of Spectral Response on Doping* . . .	227
	10. *Magnetic Field Effects*	241
IV.	DETECTOR PERFORMANCE	262
	11. *Introduction*	262
	12. *Analysis of Detector Performance*	263
	13. *Noise Mechanisms*	267
	14. *GaAs Extrinsic Photodetectors*	270
	15. *Performance in Reduced Background Conditions* . .	277
	16. *Response at Millimeter and Centimeter Wavelengths* . .	284
	17. *Response Time*	288
V.	SUMMARY	288

I. General Introduction

Far-infrared photoconductivity in high purity GaAs has been used to study the shallow donor states in this material and to extend the long wavelength limit for extrinsic photoconductive detection. In this chapter we will review the information concerning the properties of shallow donor levels in GaAs obtained from these measurements as well as the performance that has been achieved with GaAs extrinsic photoconductive detectors.

The far-infrared part of the spectrum is generally considered to fall in the wavelength range 40–1000 μm, and the GaAs detectors are useful in the region between about 50 and 350 μm. The former range seems to be a large region of the spectrum when considered in terms of wavelength, but it is a relatively narrow region in terms of wavenumbers (250 to 10 cm^{-1}) or

*This work was sponsored by the Department of the Air Force.

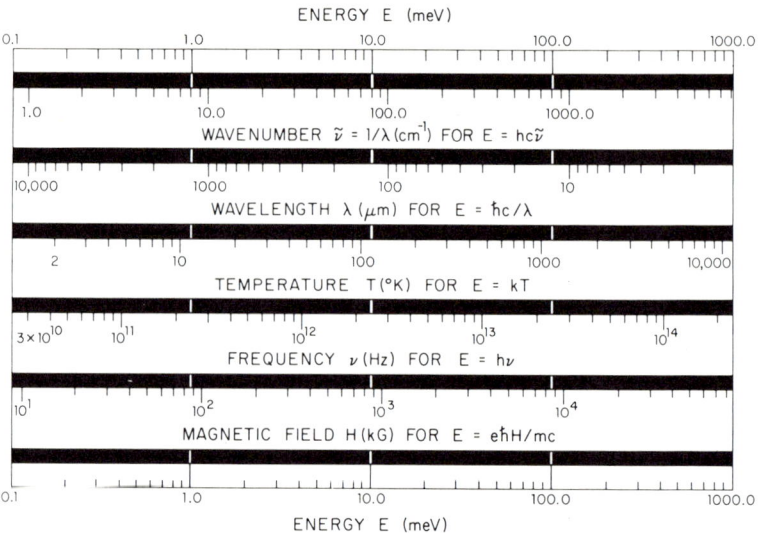

FIG. 1. Energy conversion chart for quantities used in describing far-infrared radiation.

energy (31 to ~1 meV). The relationships of various units and quantities that will be used in discussing far-infrared radiation and photoconductivity are shown in the energy conversion chart of Fig. 1.

1. Photoconductive Mechanisms

There are three main types of photoconductive processes, and these are illustrated schematically in Fig. 2.[1] Intrinsic photoconductivity involves the excitation of electrons from the valence band of a semiconductor to the conduction band. The conductivity of the sample is increased because both the free electrons and free holes created by this excitation are free to move through the crystal. Since intrinsic photoconductivity can only occur when the energy of the incident photons is greater than the energy gap between the valence and the conduction bands, this process is characterized by a a long wavelength cutoff corresponding to the semiconductor band gap. The second photoconductive mechanism, extrinsic photoconductivity, occurs in doped semiconductors which are at a temperature that is low enough so that most of the uncompensated impurity levels are neutral. Photoconductivity, in this case, results from the excitation of an electron bound to a neutral donor center into the conduction band for an n-type semiconductor or from the corresponding excitation of a hole bound to a neutral acceptor into the

[1] For a more complete discussion of far-infrared photoconductivity in general, see the review article by E. H. Putley, *Phys. Status Solidi* **6**, 571 (1964).

4. FAR-INFRARED PHOTOCONDUCTIVITY IN HIGH PURITY GaAs

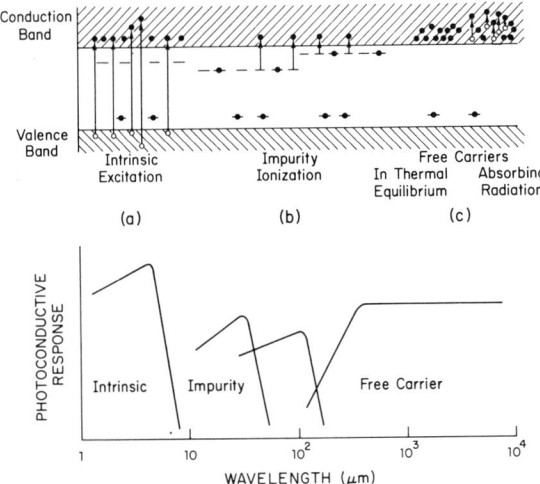

FIG. 2. Schematic diagram of different photoconductive processes in semiconductors and corresponding spectral responses. (After Putley.[1])

valence band for a p-type semiconductor. In either case, only one type of free carrier is produced by this mechanism in contrast to intrinsic photoconductivity where both free electrons and holes are generated. Since the impurity levels occur within the band gap of the semiconductor, the long wavelength threshold for this impurity photoionization occurs at a longer wavelength than that for intrinsic photoconductivity in the same material, and this threshold corresponds to the ionization energy of the impurity center. The third type of photoconductive process involves the absorption of radiation by free carriers in a semiconductor. At low temperatures, the coupling between the lattice and electrons can be very weak in semiconductors characterized by a small effective mass and high mobility, and the free electrons can attain an energy considerably higher than their equilibrium energy by interaction with the incident radiation and/or the applied electric field. If the electron mobility is limited by an energy dependent scattering process such as ionized impurity scattering, an increase in the electron energy will change the electron mobility and therefore the conductivity of the sample. Since this photoconductive mechanism depends on the absorption of radiation by free carriers, it will not be characterized by a long wavelength threshold, as are the other two processes, but instead should show an increase with λ^2 at sufficiently short wavelengths up to some constant value at very long wavelengths. Because this process is due to "heating" of the free electrons, detectors using this type of photoconductivity are referred to as free electron bolometers.

The long wavelength threshold for intrinsic photoconductivity occurs at about 0.9 μm in GaAs, so it will not be discussed further. Extrinsic photoconductivity is very important from the standpoint of far-infrared detectors, and the most important material for this application is Ge. The long wavelength cutoff for the shallowest impurities in Ge occurs at about 120 μm. Stringent purity requirements must be met by any material with shallower impurity levels that might be used to extend the long wavelength threshold for this mechanism. Such an extension is desirable, however, because the InSb free electron bolometer, which utilizes the third process, has its optimum response only at wavelengths longer than 1000 μm.[1,2]

2. Hydrogenic Model for Impurities

The simple hydrogenic model can be used to estimate the donor binding energy and thus the long wavelength threshold for extrinsic photoconductivity. In this model a donor atom, relative to the host atom it replaces, has one extra valence electron and one extra positive charge on its nucleus. Since an extra electron and the extra nuclear charge in the host material is analogous to a hydrogen atom in a vacuum, the binding energy can be approximated by the Bohr model, with the dielectric constant and effective mass appropriate to the semiconductor. In this approximation, the donor binding energy is given by

$$E_D = m^* e^4 / 2\varepsilon_0^2 h^2 = 13.6(m^*/m)/\varepsilon_0^2 \quad \text{eV} \tag{1}$$

and the corresponding Bohr radius for the donor electron is

$$a_B = 5.29 \times 10^{-9} \varepsilon_0/(m^*/m) \quad \text{cm}, \tag{2}$$

where m^*/m is the effective mass ratio for electrons and ε_0 is the static dielectric constant for the semiconductor. This simple picture of a hydrogenic impurity has been justified in the effective mass approximation by a quantum mechanical derivation starting from the expansion of the wave function of a bound electron as a linear combination of Wannier functions,[3] and equivalently by Kohn and Luttinger[4] who used a slightly different set of functions for the expansion. For a cubic crystal with the conduction band minimum at the center of the Brillouin zone, the effective mass m^* is isotropic. In this case, the energy eigenvalues are given by the solutions to the equation for the bound donor electron in the Coulomb potential $V(r) = -e^2/\varepsilon_0 r$,

$$\left[-\frac{h^2}{2m^*} \nabla^2 - \frac{e^2}{\varepsilon_0 r} \right] \Psi = E\Psi, \tag{3}$$

[2] E. H. Putley, this volume, Chapter 3.
[3] J. C. Slater, *Phys. Rev.* **76**, 1592 (1949).
[4] W. Kohn and J. M. Luttinger, *Phys. Rev.* **97**, 869 (1955).

TABLE I

IONIZATION ENERGIES AND BOHR RADIUS CALCULATED FROM THE SIMPLE HYDROGENIC MODEL

Semiconductor	m^*/m	ε_0	E_D (meV)	a_B (Å)	a_B/a_0	a_0 (Å)
Si	0.4[a]	11.40	41.7	15	2.8	5.43072
Ge	0.2[a]	15.36	11.5	41	7.2	5.65754
GaAs	0.0665	12.5	5.77	99	17.6	5.65315
InSb	0.013	16	0.689	650	100.3	6.47877

[a] The effective masses used for Si and Ge are average scalar effective masses. For more accurate values of E_D and a_B the nonisotropic effective masses and revised effective mass theory must be used.

where Ψ represents the hydrogen-like envelope functions.[4] This equation is identical to that for the hydrogen atom, except for the substitution of m^* for m and e^2/ε_0 for e^2, so that the solution for the ground state energy is the same as that given in Eq. (1). The ionization energy and Bohr radius calculated from these relations for donors in several different semiconductors are given in Table I.

It should be noted that the hydrogenic model predicts that all donors in a given material will have the same ionization energy. Although it is clear that the potential seen by an electron close to the donor atom will not be the simple Coulomb potential, for shallow donor levels (large Bohr radii), very little contribution to the binding energy comes from the potential at values of r less than a lattice spacing. The requirement for the validity of the solutions of Eq. (3) in the effective mass approximation is just that the solutions vary slowly relative to the lattice spacing a_0 of the semiconductor, and this is true if $a_B \gg a_0$. For donors with large binding energies, this will not be satisfied and the actual potential close to the donor atom will be important in determining the true value of E_D. The shift in the binding energy of the donor from the energy calculated using the hydrogenic model is commonly referred to as a central cell shift or central cell correction, and is generally different for different impurities.

From the comparison of the ratio of the Bohr radius to the lattice constant in Table I, it would be expected that the calculated values would be more accurate for donors in Ge than in Si and indeed this is the case. For group V donors, there is very little species dependence of the ionization energy in Ge (~ 9.6 meV for Sb to ~ 12 meV for P) while in Si there is a considerably larger variation (from ~ 40 meV for Sb to ~ 70 meV for Bi). The calculated ionization energies obtained from Eq. (1) for Si and Ge are not really appropriate, however, since an average scalar effective mass was used. The actual effective masses in Si and Ge are not isotropic, and for better values of the ionization energy the appropriate form of the effective mass equation must be used to calculate the hydrogenic ionization energies. The resulting equation can only be solved numerically and this has been done by several groups of workers.[5-7] However, even such a numerical calculation does not give accurate values for the donor ground state energy. In Ge, for example, there are four equivalent conduction band minima and therefore *in the effective mass approximation* the ground state is fourfold degenerate. The wave functions of the four ground states are constructed by forming linear combinations of the wave functions for each of the four equivalent valleys. Group theoretical arguments can be applied to show that under a perturbation the fourfold degenerate ground state must split into a singlet state A_1 and triplet states T_1 of the T_D group. The singlet state is experimentally found to be the lowest energy state and this can be understood at least qualitatively because it is the state which would be affected most by the breakdown of the effective mass approximation, since its wave function does not vanish at the donor nucleus. The energy between the singlet and triplet states is called the valley–orbit splitting, while the shift of the center of gravity of the singlet and triplet states is the chemical shift or central cell correction, but these terms are often used interchangeably for the splitting of the singlet and triplet states. The perturbation that causes the valley–orbit splitting is not included in the effective mass theory, while the chemical shift is due to a breakdown of the effective mass theory in the central cell and occurs even for a simple parabolic conduction band.

Since GaAs and InSb both have nearly isotropic effective masses and nondegenerate lowest conduction band minima, the complication of the valley–orbit splitting is not present and Eq. (3) is applicable. The solutions to Eq. (3) for both of these materials should be more nearly correct than for Ge, since a_B is larger.

There is another possible complication in GaAs which might have an effect on the validity of Eq. (3) and its solutions. The next higher set of conduction band minima in GaAs occur in the $\langle 100 \rangle$ direction and are only about 0.33 eV above the lowest conduction band minimum at the center of the Brillouin zone. If the effective mass equation is centered on this set of conduction band minima, the same problems mentioned above for Ge must be considered. The fact that the shallow donor levels in GaAs, as discussed in Part III, are so well described by the effective mass approximation centered on the lowest conduction band minimum indicates that, at least at atmospheric pressure, the effect of the higher minima is negligible.

[5] C. Kittel and A. H. Mitchell, *Phys. Rev.* **96**, 1488 (1954).
[6] M. A. Lampert, *Phys. Rev.* **97**, 352 (1955).
[7] W. Kohn, *Solid State Phys.* **5**, 258 (1957).

The donors in InSb should be accurately described by the hydrogenic effective mass approximation. However, in this case another problem arises. Because of the large Bohr radius of donors in InSb, even in the purest crystals available ($N_D \approx 5 \times 10^{12}$ cm^{-3}) the donor concentration is so high that the donors are impurity banded. Thus, carrier freeze-out effects as well as impurity photoionization photoconductivity cannot be observed (in the absence of a magnetic field). For isolated donors to be observed the separation between donor atoms must be somewhat larger than the Bohr radius and this imposes an upper limit on the number of impurities in the material. It is for this reason that until recently simple extrinsic photoionization photoconductivity had not been observed beyond the long wavelength threshold for shallow impurities in Ge. However, the development of new epitaxial growth techniques described in Part II has now made it possible to prepare GaAs of sufficient purity to observe carrier freeze-out and extrinsic photoconductivity. In addition to extending the long wavelength threshold for this photoconductivity mechanism, because of its simple band structure GaAs is an ideal material in which to study the properties of shallow donor levels.

3. IMPURITY EXCITED STATES IN PHOTOCONDUCTIVITY

The hydrogenic model described above predicts that in addition to the ground state there is a series of bound excited states with energies given by

$$E_n = \frac{E_D}{n^2} = 13.56 \left(\frac{m^*}{m}\right) \frac{1}{\varepsilon_0^2 n^2} \quad \text{eV}, \tag{4}$$

where $E_1 = E_D$ and E_n is the energy of the nth state. Since the "Bohr radii" of the excited states are given by

$$a_n = a_B n^2 = 5.29 \times 10^{-9} (m/m^*)\varepsilon_0 n^2 \quad \text{cm}, \tag{5}$$

the effective mass hydrogenic model should be more accurate for the excited states than for the ground state. This is found to be the case in Si and Ge where, although the effective mass theory fails quite badly in predicting the energy of the ground state of the shallow donor levels, the agreement between the excited state energies predicted theoretically and those determined from optical absorption measurements is within experimental error.[8]

Although the excited states are responsible for significant absorption bands at wavelengths longer than the photoionization wavelength, it was not surprising that photoconductivity was not observed at wavelengths corresponding to ground state transitions to these bands in the early work on Si.[9] At the liquid He temperatures used for these measurements there was

[8] R. A. Faulkner, *Phys. Rev.* **184**, 713 (1969).
[9] E. Burstein, E. E. Bell, J. W. Davisson, and M. Lax, *J. Phys. Chem.* **57**, 849 (1953).

not sufficient thermal energy available to ionize the charge carriers bound in these excited states before they would return to the ground state (with the possible exception of those carriers in excited states very close to the conduction band). Also in the early far-infrared photoconductivity measurements in Ge, no additional photoconductivity was observed corresponding to the absorption bands due to excited states.[10] This was a little more surprising, since for Ge the excited states are much closer to the band than for Si. In addition, the bound carriers in Ge are easily impact ionized from the ground state at applied voltages only a little larger than that used for the photoconductivity measurements. Photoconductivity at wavelengths corresponding to excited state absorption bands was observed in later work on Ge at slightly higher temperatures.[11] Excited state photoconductivity has now been observed in several other materials, and these measurements and the mechanisms involved are discussed in some detail in Part III.

Excited state photoconductivity is a very powerful tool for studying the properties of shallow impurity states, especially in samples which are too pure or too thin for precise absorption measurements. In most cases, this type of photoconductivity can only be observed in a specific temperature range. For GaAs, this temperature range is $\lesssim 4.2°K$ and the impurity photoconductivity spectrum at liquid He temperatures is *dominated* by excited state photoconductivity. This results in an extension of the effective long wavelength limit for extrinsic photoconductivity for this material from the photoionization wavelength threshold of about 214 μm to beyond 300 μm, since most such detectors are operated at 4.2°K, and makes GaAs an ideal material in which to study the behavior of shallow donor states.

In Part II we summarize the properties of GaAs which are relevant to shallow donors. Preparation of high purity GaAs is then reviewed, and the characterization of GaAs samples is then discussed. Characterization is very important, since without it the conclusions drawn from experimental measurements are at best uncertain and at worst completely unreliable. In Part III we discuss the far-infrared photoconductivity in GaAs and closely related topics such as optical absorption and recombination. In Part IV we discuss the performance and application of GaAs far-infrared detectors, and in Part V we give a comparison of the GaAs detectors with other far-infrared detectors.

II. Material

Because GaAs has been used for a large number of different applications, a significant amount of theoretical and experimental work has been done in

---

[10] H. Shenker, W. J. Moore, and E. M. Swiggard, *J. Appl. Phys.* **35**, 2965 (1964).
[11] T. M. Lifshits and F. Ya. Nad', *Dokl. Akad. Nauk SSSR* **162**, 801 (1965) [*English Transl.: Sov. Phys.—Dokl.* **10**, 532 (1965)].

4. FAR-INFRARED PHOTOCONDUCTIVITY IN HIGH PURITY GaAs

understanding the properties of this material. Some of the microwave device applications of GaAs have placed stringent demands on the purity of the material so that new epitaxial growth techniques were developed. At the present time it is only by these epitaxial techniques that GaAs of sufficient purity to observe far-infrared extrinsic photoconductivity has been prepared. In this part we summarize the properties of GaAs which are important to the understanding of shallow donor levels and far-infrared photoconductivity, as well as the methods of preparation and characterization of high purity GaAs.

4. Crystal Properties

a. Crystal Structure

GaAs crystallizes in the zinc-blende structure which is similar to the diamond structure of Si and Ge. Ga and As atoms occupy adjacent positions in the lattice, and the bonding between the atoms is tetrahedral, as for column IV elements. Each Ga atom is surrounded by four As atoms and each As atom is surrounded by four Ga atoms in the perfect lattice. This crystal structure is illustrated in Fig. 3, in which it can be seen that the As and Ga atoms lie on two interpenetrating face centered cubic sublattices displaced from each other by the tetrahedral spacing. While the bonding in Si and

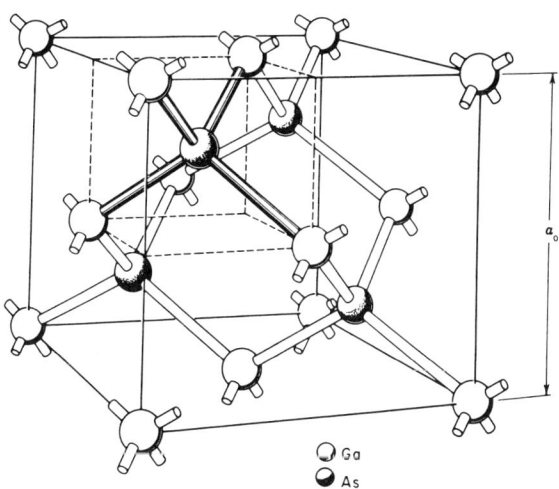

FIG. 3. Zinc-blende crystalline structure of GaAs. (After W. Shockley.[11a])

[11a] W. Shockley, "Electrons and Holes in Semiconductors." Van Nostrand-Reinhold, Princeton, New Jersey, 1950.

Ge is strictly covalent, the binding of GaAs has a significant ionic component. There is another distinction between the structure of GaAs and Si and Ge. In the ⟨111⟩ directions the atoms are alternately singly and triply bonded. Since it is energetically favorable for the surfaces to consist of the triply bonded atoms, the {111} surfaces are not equivalent. Thus, one set of surfaces consists entirely of Ga atoms, the {111} A or {111} Ga, while the opposite set consists entirely of As atoms, the {111} B or {111} As. The same distinction can be made for some of the higher order surfaces such as {211}, {311}, etc., and the growth, impurity incorporation, and etching properties of these Ga and As surfaces can be very different.

b. Band Structure

In 1960 the experimental data available on GaAs were reviewed and analyzed by Ehrenreich[12] to yield the band structure in the vicinity of the

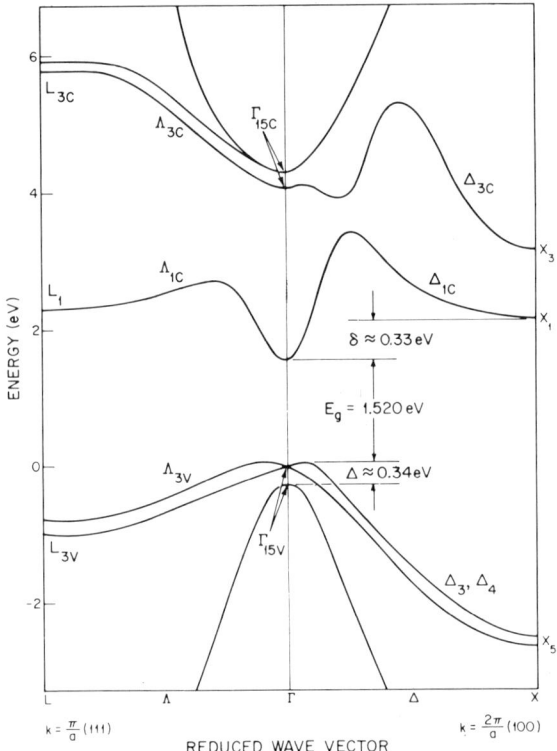

FIG. 4. Energy band structure of GaAs in two principal directions in the Brillouin zone.

[12] H. Ehrenreich, *Phys. Rev.* **120**, 1951 (1960).

band edges. Although some of the parameters characterizing the energy bands have since been determined more accurately and more information is available concerning higher conduction band minima, the band structure presented in that paper is still generally valid. Later reviews of the band structure have been given by Hilsum[13,14] and Madelung.[15] The energy band structure is shown in Fig. 4. The lowest conduction band minimum Γ_1 occurs at the center of the Brillouin zone, $k = 0$, with higher minima occurring in the $\langle 100 \rangle$ or Δ and $\langle 111 \rangle$ or Λ directions. At low temperatures E_g, the energy separation between the Γ_1 minimum and the highest valence band, is about 1.5214 eV.[16] The three higher X_1 minima at the edges of the Brillouin zone in the $\langle 100 \rangle$ directions are about 0.33 eV above the Γ_1 minimum.[14,17] There is probably also a set of three X_3 conduction band minima at the edges of the zone in the $\langle 100 \rangle$ or Δ directions, as well as a set of four L_1 equivalent minima at the edges of the zone in the $\langle 111 \rangle$ or Λ directions, several tenths of an electron volt higher in energy. These conduction band minima do not appear to have much of an effect on the *shallow* donor levels, although deeper donor levels due to S which may be associated with the X_1 conduction band minima have been observed in GaAs and $GaAs_{1-x}P_x$.[18,19] The valence band of GaAs consists of heavy hole and light hole bands, degenerate at $k = 0$, and a spin–orbit split band about 0.34 eV lower in energy.[14,20]

The energy surfaces of the conduction band minimum at the center of the Brillouin zone are spherical, and to a first approximation the dependence of the conduction band energy $E(k)$ on the magnitude of the wave vector **k** is parabolic. Thus, $E(k) \approx \hbar^2 k^2/2m^*$ and m^*, the conduction band effective mass, is independent of energy or **k**.

However, a more accurate approximation to the wave vector dependence of $E(k)$ for small k is given by Kane's $\mathbf{k} \cdot \mathbf{p}$ calculation[21] which was originally applied to InSb. Because of the large energy gap, the nonparabolic corrections to the conduction band of GaAs are much smaller than those for InSb, and it is usual to approximate them in GaAs by only including terms

---

[13] C. Hilsum and A. C. Rose-Innes, "Semiconducting III–V Compounds," p. 57. Pergamon, Oxford, 1961.
[14] C. Hilsum, in "Semiconductors and Semimetals" (R. K. Willardson and A. C. Beer, eds.), Vol. 1, p. 4. Academic Press, New York, 1966.
[15] O. Madelung, "Physics of III–V Compounds." Wiley, New York, 1964.
[16] J. A. Rossi, private communication, 1970.
[17] G. D. Pitt and J. Lees, *Solid State Commun.* **8**, 491 (1970).
[18] A. R. Hutson, A. Jayaraman, and A. S. Coriell, *Phys. Rev.* **155**, 786 (1967).
[19] M. G. Craford, G. E. Stillman, J. A. Rossi, and N. Holonyak, Jr., *Phys. Rev.* **168**, 867 (1968).
[20] M. Reine, R. L. Aggarwal, B. Lax, and C. M. Wolfe, *Phys. Rev. B* **2**, 458 (1970).
[21] E. O. Kane, *J. Phys. Chem. Solids* **1**, 249 (1957).

to k^4 in $E(k)$. For GaAs in this approximation the conduction band energy dependence is given by[22]

$$E(k) = \frac{\hbar^2 k^2}{2m_0^*} - \left(1 - \frac{m_0^*}{m}\right)^2 \left(\frac{\hbar^2 k^2}{2m_0^*}\right)^2 \left\{\frac{3E_g + 4\Delta + 2\Delta^2/E_g}{(E_g + \Delta)(3E_g + 2\Delta)}\right\}, \quad (6)$$

where Δ is the spin–orbit splitting of the valence band, E_g is the energy gap, and m is the free electron mass. In this expression the zero of energy is taken at the bottom of the Γ_1 conduction band. Therefore, m_0^* is the electron effective mass at $k = 0$ (i.e., at the bottom of the conduction band), given by

$$\frac{1}{m_0^*} = \frac{1}{m} + \frac{2P^2}{3\hbar^2}\left(\frac{2}{E_g} + \frac{1}{E_g + \Delta}\right), \quad (7)$$

where P is the matrix element connecting the conduction band and the light hole and spin–orbit split valence bands.

In a magnetic field the conduction band is split into a series of magnetic subbands, each with a continuous energy variation in the direction of the magnetic field. The $E(k)$ variation for nonparabolic conduction bands described by Eq. (6), with the magnetic field H_z in the z direction, is given by

$$E(k) = \frac{\hbar^2 k_z^2}{2m_0^*} + \left(N + \frac{1}{2}\right)\hbar\omega_0 - \left(1 - \frac{m_0^*}{m}\right)^2$$
$$\times \left(\frac{3E_g + 4\Delta + 2\Delta^2/E_g}{(E_g + \Delta)(3E_g + 2\Delta)}\right)\left(N + \frac{1}{2}\right)^2 (\hbar\omega_0)^2 \pm \frac{1}{2}g^*\beta H_z. \quad (8)$$

In this expression, N is the quantum number of the magnetic subbands or Landau levels, $\omega_0 = e\hbar H_z/m_0^* c$ is the cyclotron frequency, g^* is the effective g-factor for conduction band electrons, and β is the Bohr magneton. The last term in the right member of Eq. (8) represents the splitting of each magnetic subband due to the spin–orbit coupling of the electron spin momenta ($s = \pm\frac{1}{2}$). Since in GaAs the effective g-factor for conduction electrons is small,[23] this splitting is small and thus difficult to observe. The variation of $E(k)$ with magnetic field is important in the interpretation of cyclotron resonance data and other effective mass measurements as discussed below.

c. Effective Mass

The most commonly used value of the effective mass for the Γ conduction band minimum in GaAs until recently was the value $m^* = 0.072\,m$. This was the best value deduced by Ehrenreich[12] from a careful reexamination of

---

[22] Q. H. F. Vrehen, *J. Phys. Chem. Solids* **29**, 129 (1968).
[23] W. Duncan and E. E. Schneider, *Phys. Lett.* **7**, 23 (1963).

earlier infrared reflectivity data of Spitzer and Whelan,[24] and was in good agreement with the average value of the effective mass determined by Moss and Walton[25] from Faraday rotation measurements $[m^*/m = 0.072 (+0.008, -0.005)]$ and the cyclotron resonance value of $0.071 \pm .005$ determined by Palik et al.[26] It was pointed out by Ehrenreich that the discrepancy between the original values determined by Spitzer and Whelan and the results of Moss and Walton was probably due to the nonparabolicity of the conduction band. The concept of a simple effective mass to describe the motion of an electron fails in a nonparabolic conduction band, even for the case of GaAs where the conduction band is still spherical. It has been shown by Kopec,[27] however, that the motion of an electron in a nonparabolic band can be described by three different mass coefficients termed differential effective masses. The three differential effective masses are energy dependent, and the usual density-of-states effective mass is given by an appropriate energy integral of one of the differential effective masses. However, at the bottom of the conduction band all of the differential effective masses are equal so it is the bottom of the band mass which is of interest. In most of the experimental determinations some attempt has been made to reduce the measured effective mass to a bottom-of-the-band mass, m_0^*.

Some of the values of the electron effective mass for GaAs determined from many different magnetooptical measurements as well as a thermoelectric power measurement are summarized in Table II. The mass determination by Emelyanenko et al.[28] used the thermoelectric power and the Hall coefficient measured in the high magnetic field limit. The value obtained, although in good agreement with earlier measurements, may have been influenced somewhat by the presence of impurity bands. Infrared reflectivity was used by Spitzer and Whelan[24] and Rashevskaya et al.[29] and the values for m_0^* obtained by analysis of this data agreed well with the early determinations by Cardona[30] and Moss and Walton[25] using Faraday rotation measurements and by Palik et al.[26] using cyclotron resonance. Some of the experimental problems with previous Faraday rotation measurements were pointed out by DeMeis[31] who obtained a smaller value for the effective

---

[24] W. G. Spitzer and J. M. Whelan, *Phys. Rev.* **114**, 59 (1959).
[25] T. S. Moss and A. K. Walton, *Proc. Phys. Soc.* **74**, 131 (1959).
[26] E. D. Palik, J. R. Stevenson, and R. F. Wallis, *Phys. Rev.* **124**, 701 (1961).
[27] Z. Kopec, *Acta Phys. Pol.* **19**, 295 (1960).
[28] O. V. Emelyanenko, D. N. Nasledov, V. I. Sidorov, V. A. Skirpkin, and G. N. Talalakin, *Phys. Status Solidi* **12**, K93 (1965).
[29] E. P. Rashevskaya, V. I. Fistul', and M. G. Mil'vidskii, *Fiz. Tverd. Tela* **8**, 3135 (1966) [*English Transl.: Sov. Phys.—Solid State* **8**, 2515 (1967)].
[30] M. Cardona, *Phys. Rev.* **121**, 752 (1961).
[31] W. M. DeMeis, Thesis, Harvard Univ., and Tech. Rep. No. HP-15, Gordon McKay Lab., Div. of Eng. and Appl. Phys., Harvard Univ., Cambridge, Massachusetts, 1965.

TABLE II

Conduction Band Edge Effective Mass of GaAs Determined by Various Experimental Methods

Method	Temperature (°K)	Effective mass value (m_0^*/m)	Reference
Thermoelectric power and Hall coefficient	110–310	0.070 ± 0.002	Emelyanenko et al.[28]
Infrared reflectivity and Hall coefficient	RT	0.07	Rashevskaya et al.[29]
	RT	0.072	Spitzer and Whelan[24] and Ehrenreich[12]
Intraband magneto-optical absorption	T	0.067 ± 0.002	Vrehen[22]
Magnetophonon measurements	70	0.067_5	Stradling and Wood[34]
	280	0.065_3	
Faraday rotation and Hall coefficient	RT	$0.072(+0.008, -0.005)$	Moss and Walton[25]
	293	$0.068 (+0.005, -0.002)$	Ukhanov[33]
	296	$0.06_0 \pm 0.002$	Piller[32]
	RT	0.064 ± 0.002	DeMeis[31]
Cyclotron resonance (frequency or wavelength of measurement in parentheses)	77	0.071 ± 0.005 (65, 70 and 85 cm^{-1})	Palik et al.[26]
	4–115	0.0648 ± 0.0015 (2mm and 1mm)	Chamberlain and Stradling[35]
	28	0.0667 ± 0.0003 (337 μm)	Chamberlain et al.[37]
	4.2	0.06649 ± 0.00003 (337 μm)	Fetterman et al.[36]
	4.2	0.068 (220 μm)	Poehler[40]
Zeeman effect on shallow donor levels	1.5–4.2	0.0665 ± 0.0005	Stillman et al.[38]
	4.2	0.067	Narita and Miyao[167 a]

[a] See p. 243 for the reference.

mass for GaAs of $m_0^*/m = 0.064 \pm 0.002$ at 300°K. The experimental values from other Faraday rotation determinations by Piller[32] and Ukhanov[33] as

[32] H. Piller, *Proc. Int. Conf. Phys. Semicond., Kyoto, 1966 (J. Phys. Soc. Japan. Suppl.)* **21**, p. 206. Phys. Soc. Japan, Tokyo, 1966.
[33] Yu. I. Ukhanov, *Fiz. Tverd. Tela* **5**, 108 (1963) [English Transl.: *Sov. Phys.—Solid State* **5**, 79 (1963)].

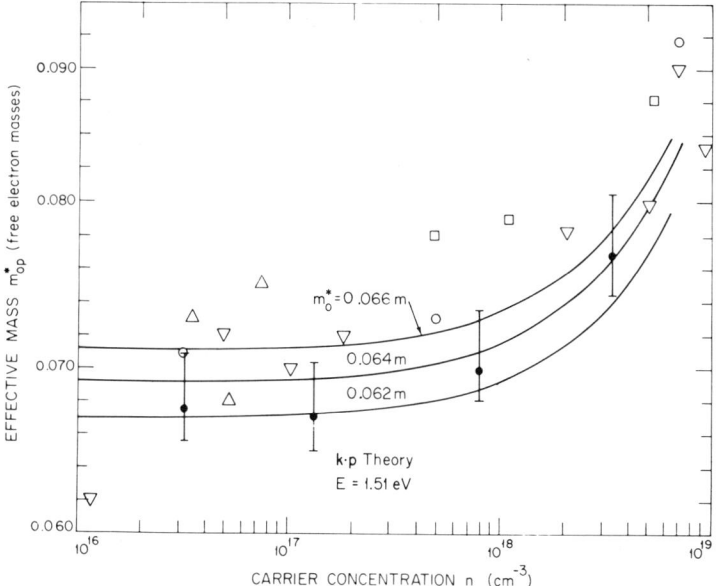

FIG. 5. Experimental and theoretical variation with carrier concentration of the optical effective mass of GaAs at room temperature m_{op}^*. The experimental values were determined from Faraday rotation and infrared reflectivity measurements, and the curves were calculated from $\mathbf{k}\cdot\mathbf{p}$ theory for GaAs using $E = 1.51$ eV for the appropriate energy gap and the value for m_0^* shown on the curves. (After DeMeis.[31]) Data: ●, DeMeis[31]; △, Moss and Walton[25]; □, Spitzer and Whelan[24]; ○, Piller[32]; and ▽, Ukhanov.[33]

well as the experimental and calculated values determined by DeMeis[31] are shown in Fig. 5. It can be seen from this figure that the effective mass determined from these measurements, i.e., the optical effective mass m_{op}^*/m, increases with increasing carrier concentration as expected from the nonparabolicity. Although there is considerable scatter in the data of Fig. 5, which may be due in part to uncertainties in the carrier concentration, it is clear from the calculated curves that m_0^*/m is considerably smaller than the value of 0.072 commonly used in the past. The interband magnetooptical absorption measurements of Vrehen[22] and magnetophonon measurements of Stradling and Wood[34] have also given smaller values for m_0^*/m. Higher purity material has been used for several recent determinations of m_0^* by means of cyclotron resonance measurements. Chamberlain and Stradling[35] have measured the cyclotron resonance at wavelengths of 2 and 1 mm of a sample with a net donor concentration $(N_D - N_A)$ of 1.3×10^{13} cm^{-3}.

[34] R. A. Stradling and R. A. Wood, *J. Phys. C (Proc. Phys. Soc.)* **1**, 1711 (1968).
[35] J. M. Chamberlain and R. A. Stradling, *Solid State Commun.* **7**, 1275 (1969).

After allowing for the shift of the resonant peak due to the closeness of the plasma frequency to the measuring frequency and correcting for the nonparabolicity of the conduction band, they determined that $m_0^*/m = 0.0648 \pm 0.0015$. A narrow cyclotron resonance absorption line ($\omega \langle \tau \rangle \approx 40$) was observed at 337 and 311 μm using an HCN far-infrared laser by Fetterman et al.[36] in a high purity GaAs sample. By using NMR techniques to determine the magnetic field very accurately and making appropriate corrections for nonparabolicity, they determined a value of $m_0^*/m = 0.06649 \pm 0.00003$. This value of the effective mass is the most precise yet determined and is in good agreement with the value of $m_0^*/m = 0.0667 \pm 0.0003$ determined later by Chamberlain et al.[37] using cyclotron resonance at 337 μm, and with the earlier value of $m_0^*/m = 0.0665 \pm 0.0005$ determined by Stillman et al.[38] from Zeeman measurements on shallow donors in GaAs. Cyclotron resonance has also been observed at 195, 118, and 79 μm by Waldman[39] using far-infrared lasers. To observe cyclotron resonance at these wavelengths with the sample at 4.2°K, the number of free carriers had to be increased. This was achieved by biasing the sample into avalanche breakdown. Under these measurement conditions, the lines displayed a marked asymmetry. A splitting of the line at 118 and 79 μm was also observed, and it was proposed that this splitting was due to the presence of transitions between excited spin-split Landau levels which are shifted because of nonparabolicity and populated because of the applied electric field. However, the magnitude of the splitting cannot be accounted for by the nonparabolicity alone. Because of these effects, it is difficult to determine an accurate value of m_0^* from cyclotron resonance at these shorter wavelengths at liquid He temperatures. A broader cyclotron resonance peak has been observed at 220 μm for somewhat less pure GaAs by Poehler[40] using an H_2O laser. The effective mass determined from this measurement was $m^* = 0.068m$, but no attempt was made to determine a value for m_0^*, the mass at the bottom of the band.

The temperature variation of the optical effective mass in GaAs is due to two opposing effects.[30] The nonparabolicity of the band tends to increase the effective mass measured by optical methods m_{op}^* with increasing temperature as the conduction band electrons are excited to higher energies. However, the corresponding decrease in the band gap with increasing temperature

[36] H. Fetterman, J. Waldman, D. M. Larsen, P. E. Tannenwald, and G. E. Stillman, Solid State Res. Rep., Lincoln Lab., M.I.T., p. 37, 1970 :3.
[37] J. M. Chamberlain, P. E. Simmonds, R. A. Stradling, and C. C. Bradley, J. Phys. C (Proc. Phys. Soc.) **4**, L38 (1971).
[38] G. E. Stillman, C. M. Wolfe, and J. O. Dimmock, Solid State Commun. **7**, 921 (1969).
[39] J. Waldman, Thesis, M.I.T., unpublished, 1970; private communication, 1970.
[40] T. O. Poehler, Appl. Phys. Lett. **20**, 69 (1972).

4. FAR-INFRARED PHOTOCONDUCTIVITY IN HIGH PURITY GaAs   185

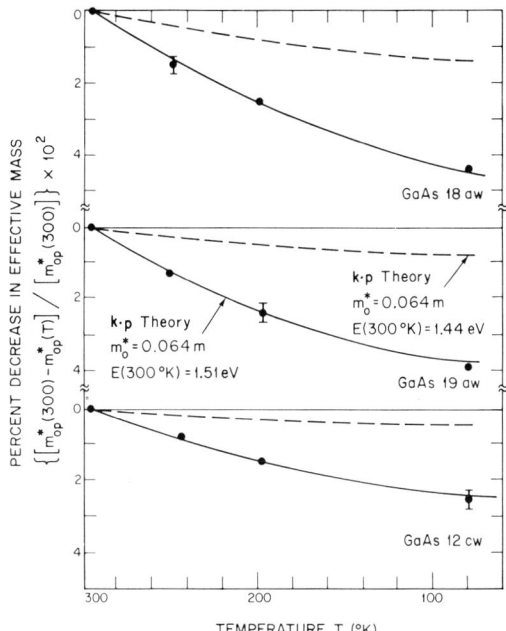

FIG. 6. Percent decrease of m_{op}^* with temperature for three n-type GaAs samples. The carrier concentrations of samples 18aw, 19aw, and 12cw are 1.3×10^{17}, 7.3×10^{17}, and 3.4×10^{18} cm^{-3}, respectively. The points are experimental and the dashed and solid curves were calculated from $\mathbf{k} \cdot \mathbf{p}$ theory using two different values for the effective room temperature bandgap. (After DeMeis.[31])

tends to decrease m_0^* which in turn decreases m_{op}^*. Therefore, the net temperature variation of m_{op}^* is expected to be very small. The percentage decrease in the optical effective mass determined by DeMeis[31] over the temperature range from about 300 to 80°K is shown in Fig. 6 for three different GaAs samples. Two different sets of theoretical curves are also shown in this figure, which were calculated using the optical band gap at 300°K (~1.44 eV) and the band gap at 300°K obtained by correcting the band gap at 0°K only for lattice expansion (~1.51 eV). The better agreement of the calculated variation with $E_g(300°K) = 1.51$ eV indicates that, in the extension of the zero temperature Kane $\mathbf{k} \cdot \mathbf{p}$ theory to finite temperature, the appropriate energy gap to use is the energy separation between quantized states. The calculated curves assume that Λ and P are constant with temperature and that m_0^* at 295°K is 0.064, although the results are only weakly affected by this choice for m_0^*. Included in these calculations is the temperature variation of m_0^*, but this variation was not determined explicitly, although it could be obtained from the data available. Using Eq. (7) and the zero temperature energy gap corrected only for dilation of the crystal lattice as above, the $\sim 2\%$ theoretical decrease in m_0^* for a temperature increase from 60 to 300°K is in good agreement with the $\sim 2.7\%$ experimental decrease determined from magnetophonon measurements by Stradling and Wood.[34] Chamberlain and Stradling[35] have been able to measure the cyclotron resonance over the temperature range 4–115°K, and they determined that there was no indication of any change in m_0^* over this lower temperature range. This is in agreement with the variation predicted by Eq. (7), since the appropriate energy gap varies negligibly over this temperature range. In measurements of the cyclotron resonance linewidth as a function of temperature a narrowing was observed, indicating that the electron scattering time increased as the temperature was lowered, although the drift mobility decreased continuously for temperatures below 50°K. This indicates that ionized impurity scattering, which is the dominant scattering mechanism in this temperature range, is not determining the observed linewidth. This is further substantiated by the observation that the linewidth broadens with increased microwave power while the drift mobility increases with increasing electric fields. Poehler[40] has reached a similar conclusion by comparing the $\langle \tau \rangle$ determined from the cyclotron resonance linewidth with the dc scattering time. However, the measured mobility which he used to determine the dc scattering time at 4.2°K was only 2×10^2 cm^2/V-sec and this value is certainly not the electron mobility in the conduction band. In GaAs, a measured 4.2°K mobility this low has to result from a two-band conduction process involving an impurity band. The true electron drift mobility in the conduction band is probably more than two orders of magnitude higher, so the observed linewidth is not as anomalously narrow as Poehler's calculations indicate.

From the above discussion, we can conclude that the electron effective mass at the bottom of the GaAs conduction band is now extremely well-known. From an examination of the available data we conclude that the best value for the effective mass in GaAs at low temperatures is $m_0^*/m = 0.06650(\pm 0.00005)$. The temperature dependence of m_0^* over the temperature range 4–115°K is negligible, with only about a 2 to 3% decrease expected over the temperature range 60–300°K.

d. Refractive Index and Dielectric Constant

The refractive index of GaAs has been measured in several different ways. The infrared refraction/prism method was used by Hambleton et al.,[41] and Marple.[42] The measurements of Hambleton et al. were made on high resistivity GaAs at four different wavelengths. The increase in the refractive index at the higher energies or shorter wavelengths is due to the band gap

[41] K. G. Hambleton, C. Hilsum, and B. R. Holeman, *Proc. Phys. Soc. London* **77**, 1147 (1961).
[42] D. T. F. Marple, *J. Appl. Phys.* **35**, 1241 (1964).

4. FAR-INFRARED PHOTOCONDUCTIVITY IN HIGH PURITY GaAs

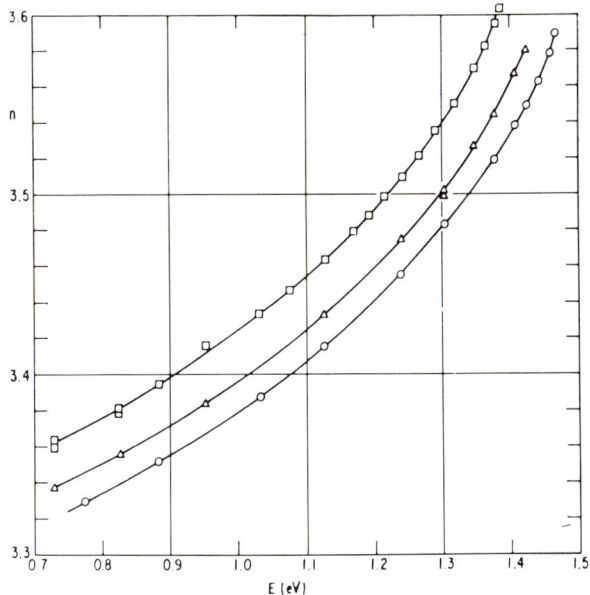

FIG. 7. Variation of refractive index of GaAs with photon energy at three different temperatures. □, 300°K; Δ, 187°K; ○, 103°K. (After Marple.[42])

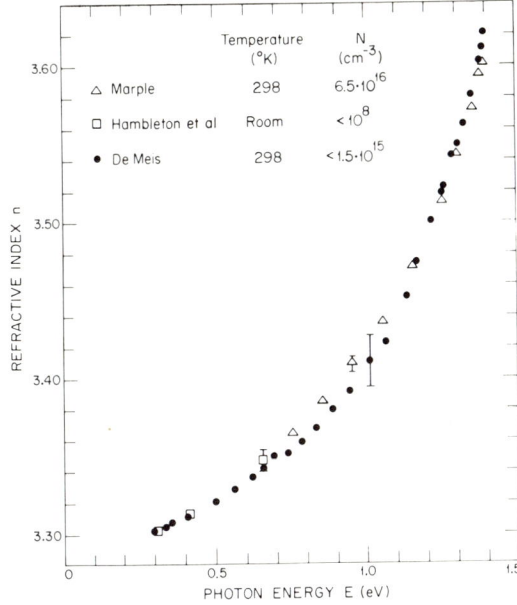

FIG. 8. Comparison of the variations of the refractive index of GaAs with photon energy at room temperature determined by several different workers: □, Hambleton et al.[41]; Δ, Marple[42]; ●, DeMeis. (After DeMeis.[31])

absorption edge, and Hambleton et al. reasoned that their measurements at wavelengths of 4 and 5 μm were far enough from both the band gap absorption at about 0.9 μm and the lattice absorption at ~33 μm to give an accurate value for the high frequency dielectric constant as $\varepsilon_\infty = 10.90 \pm 0.04$. Figure 7 shows the refractive index results of Marple on an undoped GaAs sample at three different temperatures. The results of Hambleton and Marple are compared with those obtained by DeMeis[31] using infrared transmission measurements in Fig. 8. There is generally agreement within experimental error among the different sets of data except in the range 0.7–1.15 eV where the deviation may be due to strains or defects in the sample used by DeMeis.

The data of Hambleton et al. have been analyzed by Hilsum,[43] using the dispersion predicted by the classical theory for a single oscillator. This theory predicts that the quantity $1/(n^2 - 1)$ should vary linearly with $1/\lambda^2$. The straight line fit to the refractive index data of Hambleton et al. obtained by Hilsum[43] is shown in Fig. 9 along with the experimental data of Marple and DeMeis. It can be seen that the straight line also agrees well with the

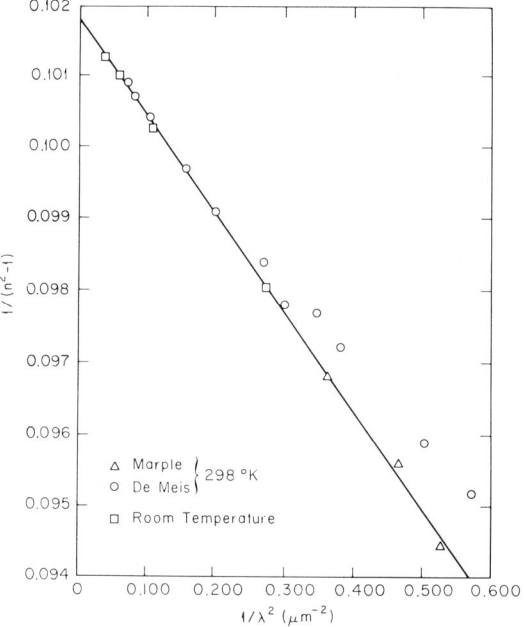

FIG. 9. Variation of $1/(n^2 - 1)$ with $1/\lambda^2$ determined from the refractive index data: □, Hambleton et al.[41]; △, Marple[42]; and ○, DeMeis.[31] The straight line is the fit to the data of Hambleton et al. obtained by Hilsum.[43] The intercept of this straight line gives the limit of the high frequency dielectric constant as $\varepsilon_\infty = 10.82$.

[43] C. Hilsum, *Prog. Semicond.* **7**, 137 (1965).

data of Marple and with that of DeMeis for $1/\lambda^2 \lesssim 0.2$ ($E \lesssim 0.56$ eV). The intercept of the extrapolation of the straight line fit obtained for the plot of $1/(n^2 - 1)$ versus $1/\lambda^2$ in Fig. 9 gives the limit of the square of the refractive index at high frequencies, which is just the high frequency dielectric constant ε_∞, as 10.82 (corresponding to $n = 3.290$).

While the above determinations of the refractive index agree quite well, there is considerable disagreement among different determinations of the static dielectric constant ε_0. A direct measurement of ε_0 was made by Hambleton et al.[41] on high resistivity GaAs using capacitance measurements at 20 kKz and 3 MHz, and a value of $\varepsilon_0 = 12.5(3) \pm 0.10$ at room temperature was determined. More recent measurements using the same method have been made by Jones and Mao[44] who determined a value of 13.2 ± 0.2. The difference between these two results was attributed to different metallization methods used in forming the capacitor. Such a direct method of measurement should be more accurate than the method of analyzing infrared reflection curves in the restrahlen region,[12] which gives values for both ε_0 and ε_∞. This is because in the reflectivity analyses the theoretical fit is not usually good over the whole restrahlen region.[45] The dielectric constant at microwave frequencies, which should be the same as that at low frequencies, is of interest since there is considerable use of GaAs in microwave devices. Because of this device interest, a report of resonance behavior at about 9.5 GHz[46] prompted a series of measurements at microwave frequencies. Although there was considerable scatter in the values obtained, those measurements failed to confirm the existence of such a resonance in other samples of GaAs.[47-51] The high and low frequency dielectric constants have both been determined by Johnson et al.[52] from far-infrared transmission measurements and they obtained $\varepsilon_\infty = 10.9 \pm 0.4$ and $\varepsilon_0 = 12.8 \pm 0.5$ at 300°K.

The temperature variation of ε_0 has been determined over the range 100–600°K.[47,53] This variation, as shown in Fig. 10, can be represented by

$$\varepsilon_0(T) = \varepsilon_0(0)\{1 + \alpha T\} \qquad (9)$$

[44] S. Jones and S. Mao, *J. Appl. Phys.* **39**, 4038 (1968).
[45] S. Iwasa, I. Balslev, and E. Burstein in "Physics of Semiconductors" (*Proc. 7th Intern. Conf.*), p. 1077. Dunod, Paris and Academic Press, New York, 1964.
[46] R. D. Larrabee and W. A. Hicinbothem, Jr., *Appl. Phys. Lett.* **10**, 334 (1967).
[47] K. S. Champlin, R. J. Erlandson, G. H. Glover, P. S. Hange, and T. Lu, *Appl. Phys. Lett.* **11**, 348 (1967); K. S. Champlin and G. H. Glover, *Appl. Phys. Lett.* **12**, 231 (1968).
[48] N. Braslau, *Appl. Phys. Lett.* **11**, 351 (1967).
[49] S. Jones and S. Mao, *Appl. Phys. Lett.* **11**, 351 (1967).
[50] C. B. Rogers, G. H. B. Thomson, and G. R. Antell, *Appl. Phys. Lett.* **11**, 353 (1967).
[51] M. R. E. Bechara, *Appl. Phys. Lett.* **12**, 142 (1968).
[52] C. J. Johnson, G. H. Sherman, and R. Weil, *Appl. Opt.* **8**, 1667 (1969).
[53] T. Lu, G. H. Glover, and K. S. Champlin, *Appl. Phys. Lett.* **13**, 404 (1968).

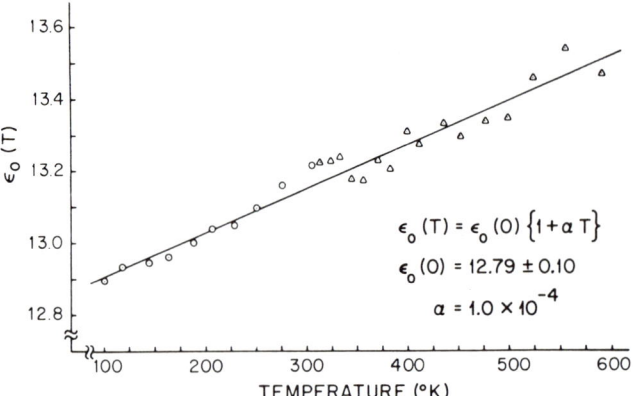

FIG. 10. Temperature variation of the static dielectric constant of GaAs. The data points ⊙ and △ indicate two sets of data obtained on the same sample. (After Lu et al.[53])

where $\varepsilon_0(0) = 12.79 \pm 0.10$ and $\alpha = 1.0 \times 10^{-4}$. A check on the consistency of values of ε_0 and ε_∞ can be made by use of the Lyddane–Sachs–Teller relation,[54]

$$\varepsilon_0/\varepsilon_\infty = (\omega_L/\omega_T)^2 \qquad (10)$$

where ω_L and ω_T are the fundamental longitudinal and transverse phonon frequencies which have been accurately measured by Raman scattering.[55] Also, ω_L and ω_T have the same temperature dependence.[56] Using Eq. (10) and the room temperature values for ω_L and ω_T from Raman measurements,[55] and $\varepsilon_\infty (300°K) = 10.82$, we can calculate that $\varepsilon_0 (300°K) = 12.78$. From an analysis of the Zeeman splitting of the $(2p_{-1}, 2p_0)$ donor levels, Stillman et al.[57] determined that $\varepsilon_0 (4°K) = 12.56 \pm 0.04$. Using the temperature variation for ε_0 shown in Fig. 10 we can calculate a value for ε_0 at room temperature of 12.94. The various values of the static dielectric constant are summarized in Table III. Although there is considerable uncertainty in the experimental values of ε_0, the more recent measurements indicate that the actual room temperature value is somewhat higher than the earlier measurement of Hambleton et al.[41]

---

[54] R. H. Lyddane, R. G. Sachs, and E. Teller, *Phys. Rev.* **59**, 673 (1941).
[55] A. Mooradian and G. Wright, *Solid State Commun.* **4**, 431 (1966).
[56] R. K. Chang, J. M. Ralston, and D. E. Keating, *Proc. Int. Conf. Light Scattering Spectra Solids*, p. 369. Springer, New York, 1969.
[57] G. E. Stillman, D. M. Larsen, C. M. Wolfe, and R. C. Brandt, *Solid State Commun.* **9**, 2245 (1971).

TABLE III

Static Dielectric Constant of GaAs at Room Temperature Determined by Various Experimental Methods[a]

Method	Value	Frequency	Reference
Capacitance	12.53 ± 0.10	1 MHz	Hambleton et al.[41]
Capacitance	13.2 ± 0.2	1 MHz	Jones and Mao[44]
Capacitance	11.6 – 13.3	1.9–13 GHz	Braslau[48]
Microstrip line	9.8	28–40 GHz	Jones and Mao[49]
	10.6	5–18 GHz	
Resonant cavity	12.35 ± 0.07	9.4 GHz	Rogers et al.[50]
Resonant cavity	13.3 ± 0.4	10 GHz	Jones and Mao[44]
	13.25 ± 0.4	33 GHz	
TE_0^0–mode reflection	12.95 ± 0.10	70.12 GHz	Champlin et al.[47]
FIR transmission	12.8 ± 0.5	—	Johnson et al.[52]
IR transmission	13.05	—	Iwasa et al.[45]
IR reflection	13.13	—	
Effective mass analysis	12.94 ± ∼0.04[b]	—	Stillman et al.[57]
From ε_∞	12.78 ± 0.1[c]	—	

[a] After Champlin et al.[47]

[b] The value determined from the analysis was 12.56 ± 0.04 at ∼4°K and the temperature variation of Lu et al.[53] was used to determine a room temperature value.

[c] This value was determined from $\varepsilon_\infty = 10.82$ with the Lyddane–Sachs–Teller relation[54] $\varepsilon_0/\varepsilon_\infty = (\omega_l/\omega_t)^2$ using the values of ω_l and ω_t of Mooradian and Wright.[55]

5. Preparation

It was first demonstrated by Knight et al.[58] and Effer[59] that epitaxial techniques could be used to grow GaAs purer than that available from previous bulk growth methods using an $AsCl_3$–Ga–H_2 vapor phase system. Since then much higher purity material has been prepared by many workers using this technique. It was subsequently shown that high purity material could be prepared with a GaAs–Ga liquid phase epitaxial system by Kang and Greene,[60] and the best material prepared by these two methods is now of comparable quality, with total ionized impurity concentrations in the 10^{13} cm^{-3} range. The preparation and properties of high purity GaAs have recently been reviewed by Wolfe and Stillman[61] and only the general methods and characteristics of these two epitaxial techniques will be summarized here.

[58] J. R. Knight, D. Effer, and P. R. Evans, *Solid State Electron.* **8**, 178 (1965).

[59] D. Effer, *J. Electrochem. Soc.* **112**, 1020 (1965).

[60] C. S. Kang and P. E. Greene, in "Gallium Arsenide" *(Proc. 2nd Int. Symp., Dallas, Texas, 1968)*, p. 18. Inst. Phys. and Phys. Soc., London, 1969.

[61] C. M. Wolfe and G. E. Stillman, in "Gallium Arsenide and Related Compounds" *(Proc. 3rd Int. Symp., Aachen, 1970)*, p. 3. Inst. Phys. and Phys. Soc., London, 1971.

a. Vapor Phase Epitaxy

An $AsCl_3$–Ga–H_2 vapor phase epitaxial reactor system is shown schematically in Fig. 11. The reactor for such a system is usually made of high purity quartz and is designed to minimize the number of joints, valves, etc. to reduce the possibility of air leaks. A two-zone furnace is used to maintain the Ga melt and the seed crystals at the appropriate temperatures. In operation, the seeds are kept in the dead space between the two oil bubblers (which prevent the back diffusion of air into the system) while the furnace is heated up, to keep them free of the initial reaction products. The purified H_2 bubbles through the $AsCl_3$ which is usually maintained below room temperature by a constant temperature bath. When the saturated gas enters the hot zone of the furnace (at ~ 800–$850°C$) the hydrogen reduces the $AsCl_3$ according to the reaction

$$4AsCl_3 + 6H_2 \rightarrow 12HCl + As_4. \tag{11}$$

As the resulting gases flow over the Ga melt, the As_4 is absorbed until the melt is saturated. The volatile gallium chlorides, $GaCl$ and $GaCl_3$, are also formed during this time and are carried downstream together with the As_4 by the gas flow and deposited in the cool end of the furnace. When the seeds are inserted into the cooler zone at the proper temperature (~ 700–$750°C$) and with the proper temperature gradient (~ 5–$15°C/cm$), epitaxial growth can take place with the reversible reaction

$$6GaCl + As_4 \rightleftarrows 4GaAs + 2GaCl_3. \tag{12}$$

The seeds can also be vapor etched according to this reaction by heating the growth zone of the furnace to a higher temperature when the seeds

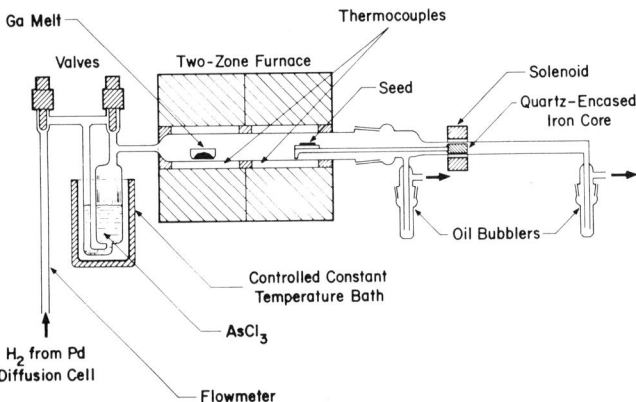

FIG. 11. $AsCl_3$–Ga–H_2 vapor phase epitaxial system. (After Wolfe and Stillman.[61])

are inserted. This makes the reaction in Eq. (12) proceed from right to left, and the temperature of this zone can then be lowered to shift the reaction in the other direction for growth. The solenoid and quartz-encased iron core attached to the seed holder allow the seeds to be inserted or removed from the growth zone without introducing air leaks.

Several factors are important in determining the purity of epitaxial layers grown in a system such as this. The purity of the starting materials is important, and $AsCl_3$ has often been thought to be the source of high residual impurity concentrations. However, it is now possible to prepare high purity GaAs with most commercially available $AsCl_3$.[61] The most frequent cause of unreproducible results and/or high resistivity compensated layers is the presence of small air leaks in the system, and this is the reason for the emphasis on minimizing the possibility of these leaks. In an airtight system with commercially available starting materials, the epitaxial layers which are not intentionally doped are almost always *n*-type, with the residual impurity concentration dependent more on the growth conditions (i.e., temperature, substrate crystal orientation, growth rate, $AsCl_3/H_2$ ratio, etc.) than on the *source* of the starting materials. However, with a small, nearly undetectable air leak the total residual impurity concentration in the epitaxial layers is usually much higher and the layers are more heavily compensated. In extreme cases, the layers may even be *p*-type. The growth rate and the impurity incorporation vary significantly with the crystalline orientation of the seed and some of the higher index orientations such as the {211} Ga incorporate fewer residual impurities than the commonly used low index orientations.[61] In fact, comparisons of the residual impurity content of epitaxial layers grown at the same time in the same reactor, but with different seed orientations, indicate that the GaAs epitaxial systems are relatively high residual impurity systems and that at the present time the purity of the epitaxial layer is determined primarily by the number of the residual impurities that are *incorporated* in an electrically active form rather than by the elimination of the residual impurities in the system. These comments, of course, do not apply to systems with air leaks or other inordinately high sources of impurities. An important factor in the incorporation of these impurities is the composition of the $AsCl_3/H_2$ gas mixture in the furnace, since the residual impurity concentration of the epitaxial layers has been varied several orders of magnitude by changing the relative amounts of $AsCl_3$ and H_2.[62,63] Some of these effects will be discussed further below.

b. Liquid Phase Epitaxy

There are several different types of reactors commonly used for liquid

---

[62] B. Cairns and R. Fairman, *J. Electrochem. Soc.* **115**, 327C (1968); **117**, 197C (1970).
[63] J. V. DiLorenzo, G. E. Moore, Jr., and A. E. Machala, *J. Electrochem. Soc.* **117**, 102C (1970).

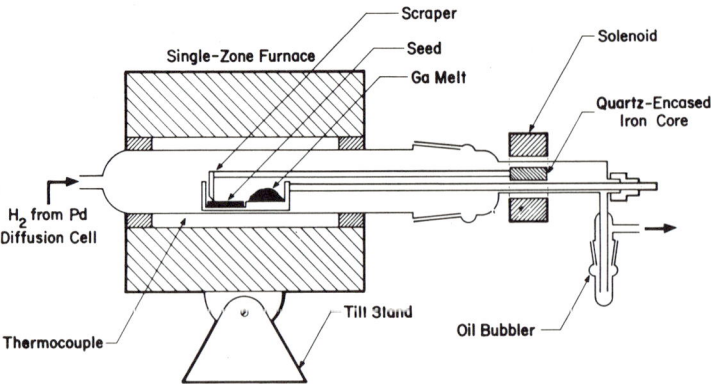

FIG. 12. GaAs–Ga liquid phase epitaxial system. (After Wolfe and Stillman.[61])

phase epitaxial growth, but most of the high purity material has been prepared in horizontal tilt-type reactors rather than the vertical dip reactor or other types of horizontal reactor. One such GaAs–Ga liquid phase epitaxial system is shown schematically in Fig. 12. The reactor tube is usually made of high purity quartz, and the container for the Ga melt and seed crystal is made of high purity graphite or quartz. For this type of reactor, only a single zone furnace is required. In operation, the Ga melt and seed crystals are placed in the furnace and the temperature is raised to a temperature slightly higher than the saturation temperature for the concentrations of Ga and GaAs in the melt. The furnace is then tipped so that the saturated Ga melt flows over the GaAs seed and the temperature is then lowered. The decreasing temperature causes a decrease in the solubility of GaAs in the Ga melt so that epitaxial growth on the seed results. A temperature gradient normal to the seed–melt interface is usually necessary to avoid constitutional supercooling. The growth starting temperatures are usually in the range 700–900°C and the cooling rates are 0.1–7°C/min. Growth is stopped at a final temperature between 400 and 750°C, and the furnace is then tipped back and the Ga melt is removed from the epitaxial layer on the seed with the magnetically coupled scraper. Under these growth conditions the resulting epitaxial layer is usually n-type.

Some of the same factors important in determining the purity of vapor phase epitaxial crystals are also important in liquid phase epitaxial systems. The presence of air leaks in this type of system decreases the purity of the crystals, although oxygen has been reported to increase the carrier concentration in this type of system instead of increasing the compensation as in vapor phase systems.[64] The purity of the starting materials may not be too

[64] R. Soloman, in "Gallium Arsenide" (*Proc. 2nd Int. Symp., Dallas, Texas, 1968*), p. 11. Inst. of Phys. and Phys. Soc., London, 1969.

important because of the low segregation coefficients between the melt and the epitaxial layers for many impurities, but considerable effort is spent in cleaning and baking the carbon or quartz boats and in baking the Ga melts at high temperatures in a hydrogen atmosphere in an effort to reduce the volatile residual impurities. The substrate or seed orientation has a considerable effect on the incorporation of impurities in this system also, with some of the higher index orientations incorporating fewer residual impurities than the more commonly used low index orientations.

c. Types of Defects and Control of Purity

There are many possible defects in a compound semiconductor such as GaAs as can be seen by considering the crystal structure shown in Fig. 3, and in this section only a general description of the problem will be given. The stoichiometric defects of Ga or As vacancies and associated complexes of these defects with other donors or acceptors or interstitial atoms have been observed in photoluminescence measurements on relatively impure material[65] but there is no evidence that any of these defects are important as residual impurities in high purity GaAs. It is well known that a large number of elements can form electrically active impurities in GaAs. The simple centers, which consist of a substitutional atom and contribute only one electron or hole to the conduction process, come from three different groups of the periodic table. When group II elements substitute for a Ga atom, there is a deficiency of one electron to complete the bonds with the neighboring As atoms so these elements can contribute one hole to the conduction process. The simple hydrogenic model predicts an acceptor binding energy of about 34 meV. When group VI elements substitute for an As atom, there is one electron left over after the bonds are completed so these elements act as donor atoms. The group IV atoms have the possibility of being incorporated either on As sites, in which case they will be acceptors or on Ga sites, in which case they will be donors. The optical activation energies for the various acceptor impurities have been summarized by Williams and Bebb[65] and are in reasonable agreement with the value determined from the hydrogenic model. The electrical activation energies are less than the optical values, and this is probably due to banding in the acceptor excited states which essentially decreases the separation between the acceptor ground state and the continuum, even though the acceptors have relatively large binding energy. The problem of identifying and determining the activation energies of specific donor impurities in GaAs is much more difficult. Because all of the shallow donor impurities in GaAs have essentially the same binding energy and because it is very small (~ 5 to 6 meV), the

[65] E. W. Williams and H. B. Bebb, *in* "Semiconductors and Semimetals" (R. K. Willardson and A. C. Beer, eds.), Vol. 8, p. 321. Academic Press, New York, 1972.

concentration of impurities must be very low to observe the binding energies and differences between the binding energies for different donor species. The binding of the excited states reduces the donor activation energy even for donor concentrations as low as 5×10^{13} cm^3.[66] The often quoted impurity energy levels for Te, Si, Ge, Sn, and Se shallow donors in GaAs, compiled by Neuberger[67] and published by Sze and Irvin,[68] have no relationship to the actual energy levels of these elements in GaAs, but only represent the thermal activation energy for a particular sample with particular donor and acceptor concentrations. Some more recent attempts at the determination of the energy levels of specific donor impurities in GaAs will be discussed in Section 9.

6. CHARACTERIZATION

The purity of the GaAs material prepared by the methods described above can vary over wide limits. Because of this variation it is important that the material used for experiments such as photoluminescence, photoconductivity, photoreflectance, absorption, and cyclotron resonance, etc. be characterized before an interpretation of these experiments is attempted. Characterization can help prevent errors of identification or interpretation as well as make it possible for other workers to duplicate the experimental results on similar samples. This is especially important when the experiment and/or interpretation involves shallow donor levels. The determination of just the room temperature carrier concentration and mobility is not sufficient, since these two parameters give little indication of the quality or purity of the material.

a. Hall Coefficient Measurements and Analyses

One of the best means of determining the impurity content of high purity GaAs is an analysis of Hall effect and resistivity measurements over a suitable temperature range. By fitting a theoretical carrier concentration equation to the experimental variation with temperature as determined from Hall constant measurements, it is possible to determine the electrically active donor and acceptor concentrations N_D and N_A, and the donor thermal ionization or activation energy E_{D_t}. For the usual model of an n-type semiconductor with N_D discrete donors near the conduction band and N_A discrete acceptors near the valence band edge, the acceptors are fully ionized, and the concentration of electrons in the conduction band, n, is given by[69]

[66] G. E. Stillman, C. M. Wolfe, and J. O. Dimmock, *Proc. 3rd Int. Conf. Photocond., Stanford, 1969*, p. 265. Pergamon, Oxford, 1971.
[67] M. Neuberger, Gallium Arsenide Data Sheets, and Suppl. DS-144. Electron. Properties Inform. Center, Hughes Aircraft, 1967.
[68] S. M. Sze and J. C. Irvin, *Solid State Electron.* **11**, 599 (1968).
[69] J. S. Blakemore, "Semiconductor Statistics," p. 138. Pergamon, Oxford, 1962.

$$\frac{n(n+N_A)}{(N_D - N_A - n)} = \frac{N_c}{g_1} \exp(-E_{D_t}/kT) \tag{13}$$

where $N_c = 2(2\pi m_D^* kT/h^2)^{3/2}$. In this equation m_D^* is the conduction band density-of-states effective mass, g_1 is the degeneracy of the ground state of of the impurity center, and all excited states have been neglected. This equation is only valid for temperatures such that the hole concentration p is much less than the electron concentration n. The electron concentration is determined from the experimental Hall coefficient with the relation

$$n = r_H/eR_H. \tag{14}$$

The Hall coefficient factor r_H in Eq. (14) is a numerical coefficient that depends on the energy dependence of the free carrier scattering. For an energy independent scattering mechanism $r_H = 1$. In general, however, r_H is a function of the magnetic field, the temperature, and the degeneracy. In the limit of high magnetic fields $r_H \to 1$, and the carrier concentration can be determined directly from the measured Hall constant R_H. For scattering mechanisms in which a relaxation time can be defined, the condition $\mu H > 1$ is generally a valid condition for the approximation $r_H = 1$. The temperature dependence of r_H for polar mode scattering, in which the relaxation time approximation is not valid, has been studied by Stillman et al.[70] These results indicate that

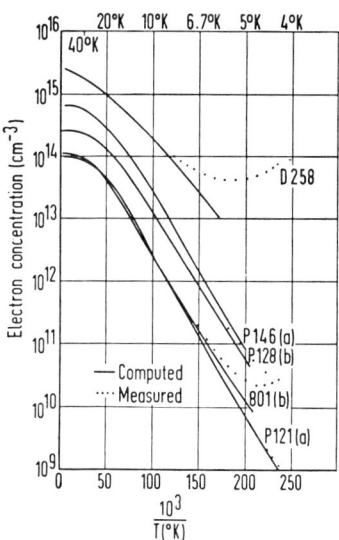

FIG. 13. Electron concentration as a function of temperature for vapor phase epitaxial GaAs. (After Bolger et al.[71,72])

[70] G. E. Stillman, C. M. Wolfe, and J. O. Dimmock, J. Phys. Chem. Solids 31, 1199 (1970).

TABLE IV

Computed Parameters and Measured Mobilities for n-Type GaAs Vapor Phase Samples of Fig. 13[a]

Sample	N_D-N_A (cm^{-3})	N_D (cm^{-3})	N_A (cm^{-3})	E_D (meV)	$\mu_{295°K}$ (cm^2/V-sec)	$\mu_{78°K}$ (cm^2/V-sec)
P146(a)	6.6×10^{14}	8.6×10^{14}	2.0×10^{14}	4.3	8600	106,000
P121(a)	9.7×10^{13}	4.9×10^{14}	3.9×10^{14}	4.2	8500	101,000
P128(b)	2.6×10^{14}	6.1×10^{14}	3.5×10^{14}	3.8	8200	98,000
801(b)	1.1×10^{14}	9.9×10^{14}	8.8×10^{14}	3.7	8300	63,000
D258	2.5×10^{15}	3.0×10^{15}	5.0×10^{14}	2.8	7650	42,000

[a] After Bolger et al.[71]

the variation of r_H with temperature is significant in high purity GaAs at magnetic fields normally used for Hall measurements. Nevertheless, the assumption is usually made that the Hall coefficient factor r_H is unity.

Equations (13) and (14) have been used successfully by several workers for the analysis of high purity GaAs Hall coefficient data. The experimental data and calculated curves obtained by Bolger et al.[71,72] for n-type vapor epitaxial GaAs using these relations are shown in Fig. 13. The value of m_D^* used for the calculated curves was 0.072 m, and the degeneracy factor g_1 was 2. The agreement between the experimental data and the calculated curves is excellent, except at the lowest temperatures where impurity band conduction begins to occur. The computed parameters and the measured mobilities are given in Table IV.

Good results have also been obtained by Stillman et al.[66] using the same values for r_H, m_D^*, and g_1, and these results are shown in Fig. 14. The calculated values of N_D, N_A, and E_{D_t} for these samples as well as the experimental values of carrier concentrations and mobility for several temperatures are given in Table V. These same values of r_H, m_D^*, g_1, and Eqs. (13) and (14) were also used by Maruyama et al.[73] to analyze Hall constant data on their high purity vapor phase epitaxial GaAs, and they also obtained good agreement between the calculated and experimental values.

In these three cases it was felt that the fit was sufficiently good, in view of the approximation $r_H = 1$, to justify the neglect of excited states in the calculation. However, other workers have sometimes found that the experimental variation of the carrier concentration, determined from Hall coeffi-

---

[71] D. E. Bolger, J. Franks, J. Gordon, and J. Whitaker, in "Gallium Arsenide" *(Proc. Int. Symp., Reading, 1966)*, p. 16. Inst. of Phys. Soc., London, 1967.
[72] See also J. Whitaker and D. E. Bolger, *Solid State Commun.* **4**, 181 (1966).
[73] M. Maruyama, S. Kikuchi, and O. Mizuno, *J. Electrochem. Soc.* **116**, 413 (1969).

4. FAR-INFRARED PHOTOCONDUCTIVITY IN HIGH PURITY GaAs

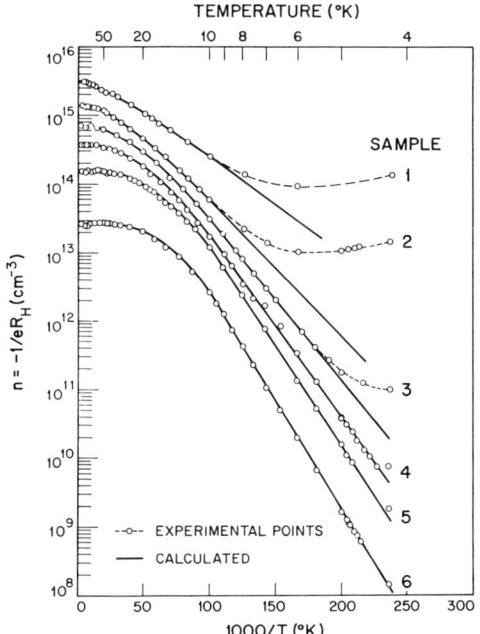

FIG. 14. Carrier concentration variation with temperature for n-type vapor phase epitaxial samples. (After Stillman et al.[66])

cient measurements on high purity GaAs using Eq. (14), cannot be fit over the entire temperature range 300–4°K using Eq. (13). Eddolls[74,75] found that this equation could only give a good fit to the experimental data on high purity vapor phase epitaxial GaAs for temperatures below 20°K. Although an accurate value could be determined for E_{D_t} by analyzing the data over only this temperature range, the values for N_D and N_A were not well defined. To obtain a better fit in the higher temperature range he included the effect of excited states in the calculation.

The problem of the inclusion of excited states in the donor statistics was first considered by Shifrin[76] and later by Landsberg.[77] When the excited states are included, Eq. (13) is modified to

$$\frac{n(n + N_A)}{(N_D - N_A - n)} = \frac{N_c \exp(-E_{D_t}/kT)}{g_1(1 + F)}, \tag{15}$$

[74] D. V. Eddolls, *Phys. Status Solidi* **17**, 67 (1966).
[75] See also D. V. Eddolls, J. R. Knight, and B. L. H. Wilson, in "Gallium Arsenide" (*Proc. Int. Symp., Reading, 1966*), p. 3. Inst. Phys. and Phys. Soc., London, 1967.
[76] K. S. Shifrin, *Zh. Tekh. Fiz.* **14**, 43 (1944).
[77] P. T. Landsberg, *Proc. Phys. Soc. London* **B69**, 1056 (1956).

TABLE V
PROPERTIES OF GaAs SAMPLES OBTAINED FROM HALL COEFFICIENT MEASUREMENTS SHOWN IN FIG. 14[a]

Sample	N_D (cm^{-3})	N_A (cm^{-3})	$n_{300°}$ (cm^{-3})	$\mu_{300°}$ (cm^2/V-sec)	$\mu_{77°}$ (cm^2/V-sec)	$n_{4.2°}$ (cm^{-3})	$\mu_{4.2°}$ (cm^2/V-sec)	E_{D_1} (10^{-3} eV)
1	4.72×10^{15}	1.61×10^{15}	3.14×10^{15}	6220	33,800	1.4×10^{14b}	415[b]	1.89
2	2.06×10^{15}	6.78×10^{14}	1.42×10^{15}	6240	46,300	1.3×10^{13b}	495[b]	3.29
3	1.06×10^{15}	3.27×10^{14}	7.1×10^{14}	7290	72,000	1.0×10^{11b}	2000[b]	3.88
4	5.02×10^{14}	1.36×10^{14}	3.68×10^{14}	7740	107,000	7.6×10^9	52,000	4.51
5	2.04×10^{14}	4.07×10^{13}	1.62×10^{14}	8160	153,000	2.0×10^9	70,000	5.09
6	4.80×10^{13}	2.13×10^{13}	2.67×10^{13}	8620	210,000	1.2×10^8	85,000	5.52

[a] After Stillman et al.[66]
[b] These samples exhibited considerable impurity-band conduction at 4.2°K.

where

$$F = \sum_{r=2} (g_r/g_1)e^{-E_r/kT}. \tag{16}$$

In these equations, g_r is the degeneracy factor for the rth state of the impurity center and E_r is the energy of the rth state measured *above* the ground state energy. The sum in Eq. (16) is taken over all discrete excited states (omitting those which are banded) of the impurity center. Eddolls[74] included the excited states in his calculation using Eq. (15) by assuming that the excited states were related to E_{D_t} by the usual Bohr equation, so that $E_r = E_{D_t}(1 - r^{-2})$ and, with the degeneracy factors from the hydrogenic model,

$$(1 + F) = \sum_{r=1} r^2 \exp[-E_{D_t}(1 - r^{-2})/kT]. \tag{17}$$

By including up to two excited states in the calculation he was able to obtain a good fit over the entire temperature range. It sould be noted that, in addition to the other parameters N_D, N_A, and E_{D_t}, N_c/g_1 was treated as an adjustable parameter in these calculations. Eddolls observed a minimum in the R_H versus $1/T$ curve for a temperature of about 80°K in measurements which were made at a magnetic field of 3 kG, somewhat like the behavior observed by Stillman et al.[70] The minimum was attributed to the magnetic field dependence of R_H, but in this case the minimum disappeared when a small magnetic field (465 G) was used[74] in contrast to the observation of Stillman et al. and the predictions of theory. Therefore, the analyses of Eddolls were made on data taken at low magnetic fields, where the error due to the assumption $r_H = 1$ will be largest.

Carballes et al.[78] also found it necessary to include excited states in order to fit their experimental data. These workers used the same model as Eddolls including N_c/g_1 as an adjustable parameter and observed that they could obtain good fits with Eq. (13) only when unreasonable values of g_1 (of the order of 10, presumably with $m_D^* = 0.070\, m$) were used. However, by including two excited states in their analysis using Eqs. (15) and (17), they obtained good fits to the data with values of $g_1 \approx 2$. The material, in this case, was relatively high purity liquid phase epitaxial GaAs.

From the study of the excited state photoconductivity of shallow donors in GaAs, which will be discussed in Part III, we now know that the energy level structure used in these two analyses is not correct, but the reasons why excited states are required to obtain reasonable fits to the data in some cases and not in others are not understood. It is probably related to the

---

[78] J. C. Carballes, D. Diguet, and J. Lebaily, in "Gallium Arsenide" (*Proc. 2nd Int. Symp., Dallas, Texas, 1968*), p. 28. Inst. Phys. and Phys. Soc., London, 1969.

variation of the scattering factor r_H with temperature, or possibly to the effects of sample inhomogeneity.

All of these analyses indicate that high purity GaAs is characterized by very shallow donor levels. The thermal ionization energy of these donors for the highest purity material is about 5.5 meV and decreases rapidly as the purity decreases. At low temperatures most of the electrons are bound to these shallow donor centers, but for samples with donor concentrations $N_D \gtrsim 10^{15}$ cm^{-3}, the conductivity in an impurity band at 4°K is comparable to that due to free electrons in the conduction band. Oliver[79,80] first observed that electrons could be excited from the impurity levels or the impurity band into the conduction band in GaAs through impact ionization by energetic free carriers. This process was previously observed in Ge.[81,82] The impact ionization of shallow impurities in high purity GaAs was first

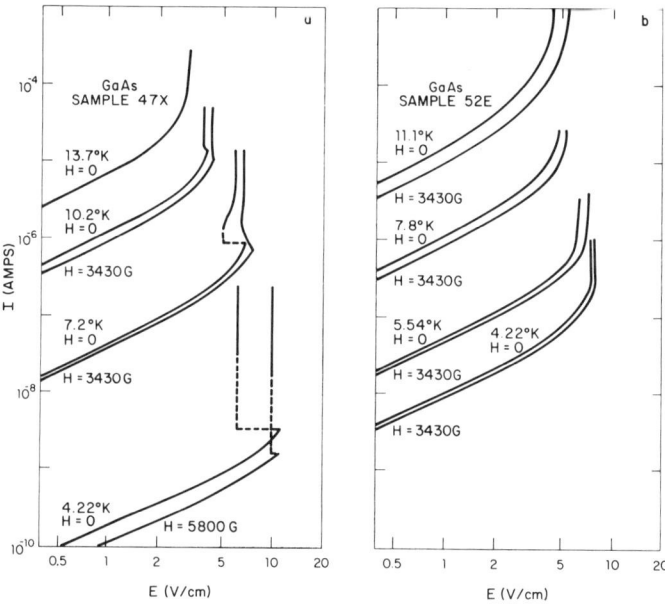

FIG. 15. Current versus electric field curves at several temperatures, with and without a transverse magnetic field. (a) Sample 47X was heavily compensated with $N_D \approx 4.7 \times 10^{14}$ cm^{-3} and $N_A \approx 4.2 \times 10^{14}$ cm^{-3}; (b) sample 52E was characterized by $N_D \approx 7.4 \times 10^{14}$ cm^{-3} and $N_A \approx 2.0 \times 10^{14}$ cm^{-3}. (After Reynolds.[83])

[79] D. J. Oliver, *Phys. Rev.* **127**, 1045 (1962).
[80] D. J. Oliver, *Proc. Int. Conf. Phys. Semicond., Exeter, 1962*, p. 133. Inst. Phys. and Phys. Soc., London, 1962.
[81] N. Sclar and E. Burstein, *J. Phys. Chem. Solids* **2**, 1 (1957).
[82] S. H. Koenig and G. R. Gunther-Mohr, *J. Phys. Chem. Solids* **2**, 268 (1957).

studied in detail by Reynolds[83] and was also observed by Stillman et al.[84] The variation of the current–voltage characteristics for two different samples analyzed by Reynolds are shown in Fig. 15 for several different temperatures both with and without a magnetic field. At low electric fields the current increases linearly with field (Ohmic behavior), but as a critical field strength E_c is approached, the current increases more rapidly with increase in field. The sample in Fig. 15a is heavily compensated, and for this sample the electric field decreases to a lower sustaining value E_s with a small increase in current and then remains constant at this value. This results in a current controlled negative resistance. Applying a magnetic field increases the sustaining electric field while having essentially no effect on the critical electric field for breakdown. This decreases the magnitude of the negative resistance. Similar behavior was observed by Stillman[85] for magnetic fields up to 55 kG. An increase in temperature reduces both E_c and E_s, but E_c decreases more rapidly so that the negative resistance decreases and in fact disappears at $T \approx 10°K$. This negative resistance is similar to that observed in compensated Ge by McWhorter and Rediker.[86,87] Several models were proposed to explain the current controlled negative differential resistance in the earlier work on Ge.[87-91] A new model for high purity GaAs was proposed by Crandall[92] that depends on the screening of ionized impurities by impactionized free carriers and the electron–phonon interaction. In this model an initial increase in carrier density causes an increase in the mobility because of increased screening and a decrease in the electron–phonon interaction because of the reduced scattering. The reduced scattering results in a hotter electron distribution which produces still more impact ionization and a decrease in the electric field results, while the hot distribution maintains the same rate of impact ionization. For the less compensated sample of Fig. 15b there is no indication of the negative differential resistance behavior for any temperature or magnetic field shown, and this is also qualitatively consistent with Crandall's model.[93] Impact ionization and/or non-Ohmic behavior as

---

[83] R. A. Reynolds, *Solid State Electron.* **11**, 385 (1968).
[84] G. E. Stillman, C. M. Wolfe, I. Melngailis, C. D. Parker, P. E. Tannenwald, and J. O. Dimmock, *Appl. Phys. Lett.* **13**, 83 (1968).
[85] G. E. Stillman, unpublished data, 1968.
[86] A. L. McWhorter and R. H. Rediker, *Proc. IRE* **47**, 1207 (1959).
[87] A. L. McWhorter and R. H. Rediker, *Proc. Int. Conf. Phys. Semicond., Prague, 1960*, p. 134. Czech. Acad. Sci., Prague and Academic Press, New York, 1961.
[88] J. Yamashita, *J. Phys. Soc. Japan* **16**, 720 (1961).
[89] A. Zylbersztejn, *J. Phys. Chem. Solids* **23**, 297 (1962).
[90] L. M. Lambert, *J. Phys. Chem. Solids* **23**, 1481 (1962).
[91] L. Kurosawa, *J. Phys. Soc. Japan* **20**, 1405 (1965).
[92] R. S. Crandall, *Phys. Rev. B* **1**, 730 (1970).
[93] R. S. Crandall, *J. Phys. Chem. Solids* **31**, 771 (1970).

well as magnetic freeze-out effects of shallow donors in GaAs have been studied by a number of other workers.[94-98]

b. *Use of 77°K Mobility Measurements*

There is considerable effort involved in taking Hall constant data over the required wide temperature range and then in analyzing the data to determine the donor and acceptor concentrations by the method described above. For routine evaluation of material or in other instances where it is not practical to use this procedure, it is desirable to be able to estimate the sample purity by some simpler method. One method that has been frequently used is to calculate the donor and acceptor concentrations from the Brooks–Herring[99] mobility formula for ionized impurity scattering using the experimental mobility and carrier concentration measured at liquid nitrogen temperature (77°K). The Brooks–Herring equation for the ionized impurity scattering mobility μ_I in an n-type semiconductor is

$$\mu_I = \frac{3.28 \times 10^{15}(m/m^*)^{1/2}\varepsilon_0^2 T^{3/2}}{(2N_A + n)[\ln(b+1) - b/(b+1)]} \quad \text{cm}^2/\text{V-sec}, \tag{18}$$

where

$$b = \frac{1.29 \times 10^{14}(m^*/m)\varepsilon_0 T^2}{n^*}, \tag{19a}$$

and n^* is an effective screening density

$$n^* = n + [(n + N_A)(N_D - N_A - n)/N_D] \quad \text{cm}^{-3}. \tag{19b}$$

The rest of the terms have their usual meaning. For temperatures and/or samples in which donor deionization occurs, it is not possible to solve for N_D and N_A explicitly, and N_D and N_A must be adjusted separately to obtain agreement between the experimental and calculated mobilities. Where donor deionization does not occur, $n = N_D - N_A$, so $n^* = n$. Thus, the acceptor concentration N_A can be calculated directly from the measured mobility and carrier concentration. In following this procedure the effects of lattice scattering on the mobility usually have been neglected, and this leads to

---

[94] I. Akasaki and T. Hara, *Proc. 9th Int. Conf. Phys. Semicond., Moscow, 1968*, Vol. 2, p. 787. Nauka, Leningrad, 1968.
[95] S. Asai, T. Toyabe, and M. Hirao, *Proc. 10th Int. Conf. Phys. Semicond., Cambridge, Massachusetts, 1970*, p. 578. USAEC Div. Tech. Informat., Oak Ridge, Tennessee, 1970 (available as CONF-700801 from Nat. Tech. Informat. Serv., Springfield, Virginia 22151).
[96] L. Halbo and R. J. Sladek, *Phys. Rev.* **173**, 794 (1968).
[97] L. Halbo and R. J. Sladek, *J. Non-Crystalline Solids* **4**, 85 (1970).
[98] T. O. Poehler, *Phys. Rev. B* **4**, 1223 (1971).
[99] H. Brooks, *Advan. Electron. Electron Phys.* **7**, 158 (1955).

significant errors at low and intermediate impurity concentrations. The effect of lattice scattering can be included in an approximate way by assuming a lattice scattering mobility, and simply combining it with the Brooks–Herring mobility. Although this latter procedure is somewhat better than neglecting the lattice scattering, it can still lead to significant errors in the intermediate impurity concentration range. The Brooks–Herring equation has also been used to determine the impurity concentration from the experimental mobility measured at some lower temperature where the effects of lattice scattering should be negligible.[74] However, if the temperature is too low, other scattering mechanisms or impurity banding effects can have an effect on the measured mobility, whereas if the temperature is too high lattice scattering dominates. Therefore, there is some optimum temperature at which the experimental mobility should be measured to make the determination of the impurity concentration in this way most reliable. Such an analysis of the mobility over a wide temperature range has been carried out by Wolfe et al.[100] The temperature variation of the *effective* values of

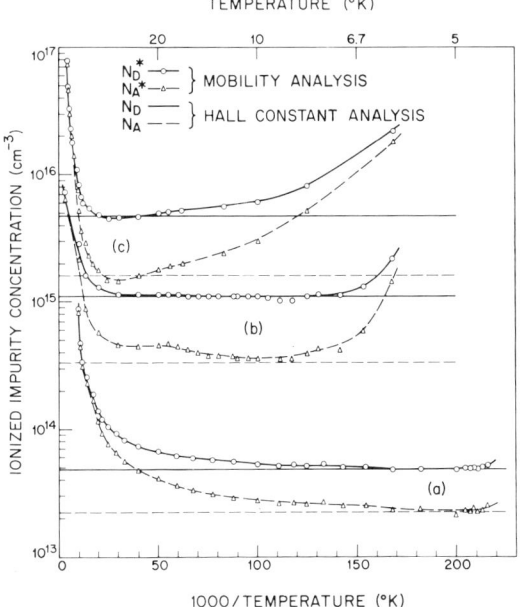

FIG. 16. Temperature variation of effective donor and acceptor concentrations as determined from the Brooks–Herring equation. The minimum effective donor and acceptor densities, N_D^* and N_A^*, are in good agreement with the values of N_D and N_A obtained from the Hall constant analysis over the range where both methods can be used. (After Wolfe et al.[100])

[100] C. M. Wolfe, G. E. Stillman, and J. O. Dimmock, *J. Appl. Phys.* **41**, 504 (1970).

donor and acceptor concentrations, N_D^* and N_A^*, determined in this way from experimental mobility measurements over the temperature range 300–4°K for three different samples is shown in Fig. 16. Also shown in this figure are the values of N_D and N_A determined from analyses of Hall constant versus temperature measurements for these samples. Because of the effects of lattice scattering mechanisms, at higher temperatures the calculated values of N_D^* and N_A^* for these temperatures are much larger than the actual values of N_D and N_A. Similarly, at very low temperatures there is the possibility that impurity banding and/or hopping conduction effects may decrease the experimental mobility so the calculated values of N_D^* and N_A^* may also be too large in this low temperature region. At intermediate temperatures the analysis should give an upper bound on the donor and acceptor concentrations, and in fact it can be seen that there is very good agreement between the minimum values of N_D^* and N_A^* determined in this way and N_D and N_A from the Hall constant analysis. This type of analysis and the

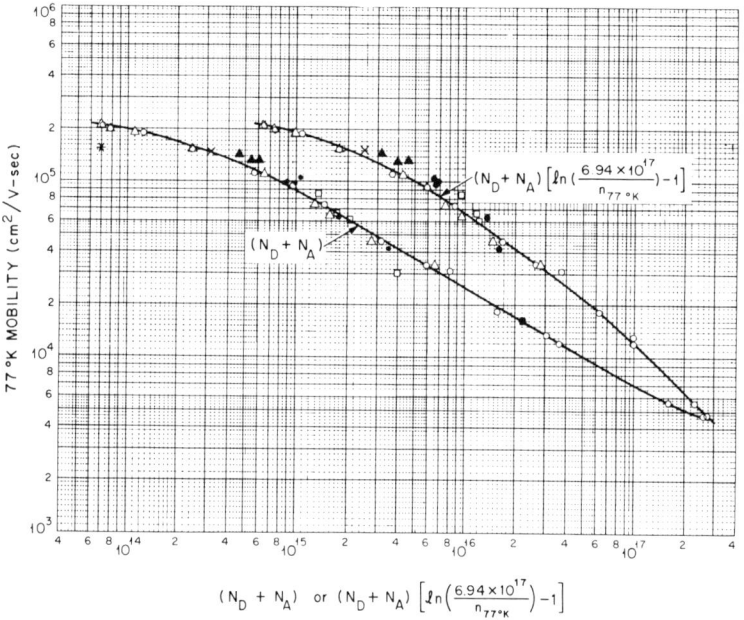

FIG. 17. Empirical curves of the total impurity density to be expected for a given 77°K mobility. The lower curve represents raw data and the upper curve accounts for variations in screening. For example, a sample with a 77°K Hall mobility at 5 kG of 100,000 cm²/V-sec from the lower curve has an $N_D + N_A$ of approximately 8×10^{14} cm^{-3}. At 100,000 cm²/V-sec the abscissa for the upper curve is 5.1×10^{15} cm^{-3}. If the measured $n_{77°K} = 6 \times 10^{14}$ cm^{-3}, then $(N_D + N_A) = 8.4 \times 10^{14}$ cm^{-3}, whereas if $n_{77°K} = 1 \times 10^{14}$ cm^{-3}, then $(N_D + N_A) = 6.5 \times 10^{14}$ cm^{-3}. (After Wolfe et al.[100])

corresponding experimental measurements are not much easier to make than the standard Hall constant measurements and analysis, but it allows an optimum temperature to be selected for this type of analysis and it can be extended to more heavily doped samples which do not show significant carrier freeze-out. The impurity concentrations obtained from this type of analysis as well as Hall constant analyses on a large number of samples were correlated with the experimental 77°K mobility to obtain empirical curves from which the total impurity concentration can be obtained with adequate precision for routine material characterization from a simple measurement of the mobility and carrier concentration at 77°K. The empirical curves are shown in Fig. 17. The curve labeled $(N_D + N_A)$ gives a direct rough estimate of this quantity. A more accurate value of $(N_D + N_A)$ can be determined by reading the abscissa corresponding to the experimental mobility from the other curve and dividing it by the screening factor $[\ln(6.9 \times 10^{17}/n_{77°K}) - 1]$ to account for different compensation ratios. Although only the measurements of mobility and carrier concentration at 77°K are required in order to determine N_D and N_A, the room temperature values are also important in judging the quality of a sample. This is because it is often possible to make good measurements at 77°K when the room temperature mobility is very low or when it is even impossible to make room temperature measurements because of sample inhomogeneity.

c. Inhomogeneity Effects

Many different types of inhomogeneity can have an effect on the measured mobility of semiconductors[101,102] and therefore on the donor and acceptor concentrations determined from mobility measurements. Most of the inhomogeneities considered cause an anomalously low mobility, either as a result of averaging inherent in the Hall constant and resistivity measurements or as a result of additional carrier scattering. These types of inhomogeneity lead to an overestimate of the total impurity concentration and in this way give an indication of a degradation of sample quality. It has recently been shown, however, that a simple conducting inhomogeneity can lead to an anomalously high mobility as determined from Hall effect and resistivity measurements.[103-104a] By introducing conducting homogeneities in high purity epitaxial GaAs it was possible to increase the measured mobility at

[101] A. C. Beer, "Galvanomagnetic Effects in Semiconductors," p. 308. Academic Press, New York, 1963.
[102] R. T. Bate, in "Semiconductors and Semimetals" (R. K. Willardson and A. C. Beer, eds.), Vol. 4, p. 459. Academic Press, New York, 1968.
[103] C. M. Wolfe, and G. E. Stillman, Appl. Phys. Lett. **18**, 205 (1971).
[104] C. M. Wolfe, G. E. Stillman, and J. A. Rossi, J. Electrochem. Soc. **119**, 250 (1972).
[104a] C. M. Wolfe and G. E. Stillman, in "Semiconductors and Semimetals" (R. K. Willardson and A. C. Beer, eds.), Vol. 10, p. 175. Academic Press, New York, 1975.

300°K from 7400 to 24,000 cm^2/V-sec and the 77°K measured mobility from 150,000 to 740,000 cm^2/V-sec. Samples with this type of conducting inhomogeneity can appear to be quite uncompensated when this is not actually the case, so the measurement of a high mobility alone is not a sufficient indication of sample quality. The homogeneity of the material must be established in addition, and this can be done by checking for agreement between the calculated and experimental variation of mobility with temperature, looking for variations of the resistivity ratio with temperature for van der Pauw measurements or in measurements on different sets of contacts for standard Hall samples, and by examining the magnetic field dependencies of the Hall constant.[104a] Although the use of mobility measurements for routine analysis of sample quality remains a very useful method, it is clear that the quality or purity of a sample cannot be established conclusively by a high 77°K mobility alone, particularly when the room temperature value exceeds the theoretical value of about 8000 ± 1000 cm^2/V-sec, or when room temperature measurements cannot be made.

III. Photoconductivity

7. INTRODUCTION

Extrinsic far-infrared photoconductivity was first observed in n-type epitaxial GaAs using DCN and HCN lasers at 195 and 337 μm, respectively.[84] Photoconductivity was also observed at 902 μm using a carcinotron backward wave oscillator. It was proposed that the response at the two shorter wavelengths was due to photoionization of the shallow donor levels, since the long wavelength threshold as calculated from the thermal ionization energy of the shallow donors could occur at wavelengths longer than 400 μm corresponding to donor ionization energies $\lesssim 3$ meV. The response at 902 μm was much less than that at 337 μm and this was considered evidence that the photoionization threshold occurred between these two wavelengths, with the response at 902 μm presumably due to the free carrier photoconductive mechanism characteristic of InSb far-infrared detectors. It was later suggested by Kimmitt[105] that the decreased response at 902 μm may have been due to the fact that the thickness of the sample used for these measurements was only 40 μm and was thus much smaller than the wavelength of the radiation. The far-infrared absorption and photoconductivity spectra were measured on n-type epitaxial GaAs samples with a room temperature carrier concentration of 1×10^{15} cm^{-3} by Bosomworth et al.[106] The measurements were made at 4°K using interferometric techniques. Figure 18 shows the

[105] M. F. Kimmitt, "Far Infrared Techniques," p. 83. Pion Limited, London, 1970.
[106] D. R. Bosomworth, R. S. Crandall, and R. E. Enstrom, *Phys. Lett.* **28A**, 320 (1968).

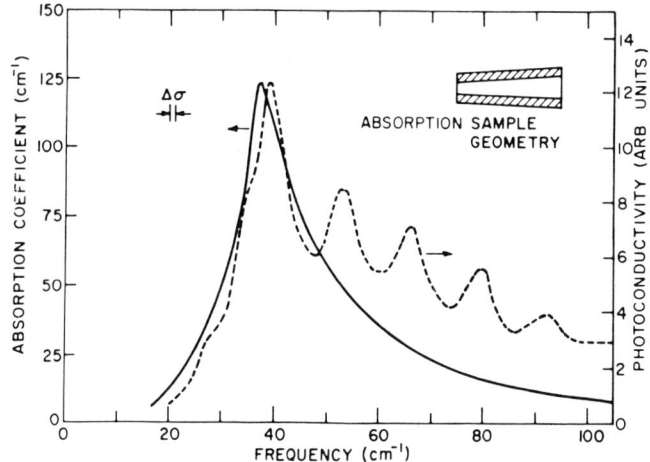

FIG. 18. Far-infrared absorption spectrum of a wedge shaped high resistivity GaAs substrate and two 170 μm thick epitaxial layers shown in the inset (solid curve), and the photoconductivity spectrum of a 75 μm thick layer of epitaxial GaAs (dashed curve). The latter spectrum is normalized to a constant photon number. (After Bosomworth et al.[106])

results they obtained. Both the absorption and photoconductivity spectra have a long wavelength threshold and one dominant peak. The peak of the absorption spectrum at 37 cm^{-1} (4.6 meV) was identified with the donor ionization energy. (The structure in the photoconductivity spectrum is due to multiple reflections in the 75 μm thick plane-parallel sample used for the photoconductivity measurement.) The absence of any structure due to excited state transitions on the low frequency side of the absorption spectrum was explained by the broadening of the excited states into a continuum caused by wave function overlap of neighboring donors. This also explained the ionization energy being smaller than that calculated from the effective mass theory for an isolated donor impurity. Later measurements on higher purity GaAs have shown that the far-infrared absorption/photoconductivity is not as simple as this, but these measurements confirmed the existence of a long wavelength threshold.

a. Optical Absorption by Shallow Impurities

The optical properties of shallow impurities in Si and Ge, as well as the theoretical properties of hydrogen-like centers, have been reviewed by Burstein et al.[107] As discussed in Part I, there are two optical absorption processes that can occur for shallow impurity centers: (1) photoionization

[107] E. Burstein, G. Picus, and N. Sclar, in *Photoconduct. Conf. Atlantic City, 1954* (R. G. Breckenridge, B. R. Russell, and E. E. Hahn, eds.), p. 353. Wiley, New York, 1956.

absorption in which transitions are from the ground state to the conduction band; and (2) absorption in which transitions are from the ground state to the higher excited states. The photoionization absorption should result in a broad absorption continuum at energies higher than the threshold energy while the absorption to the excited states should appear as narrow bands or peaks at energies less than the long wavelength threshold for photoionization absorption. The photoionization absorption cross section for a hydrogenic center with effective mass ratio m^*/m and ionization energy E_I (eV) is given by[108,109]

$$\sigma(\tilde{v}) = \frac{1}{\sqrt{\varepsilon_0}}\left(\frac{\mathscr{E}_e}{\mathscr{E}}\right)^2 \frac{2^8 \pi^2 h e^2}{3m^* c E_I}\left(\frac{E_I}{h\tilde{v}}\right)^4 f(n')$$

$$= \frac{1}{\sqrt{\varepsilon_0}}\left(\frac{\mathscr{E}_e}{\mathscr{E}}\right)^2 \frac{4.685 \times 10^{-15}}{E_I}\left(\frac{m}{m^*}\right)\left(\frac{E_I}{h\tilde{v}}\right)^4 f(n') \quad \text{cm}^2 \quad (20)$$

for $h\tilde{v} \geqslant E_I$, where

$$f(n') = \frac{\exp[-4n' \operatorname{arc cot} n']}{1 - \exp[-2\pi n']} \quad (21)$$

and

$$n' = [(h\tilde{v} - E_I)/E_I]^{-1/2}. \quad (22)$$

The factor $(\mathscr{E}_e/\mathscr{E})$ is the ratio of the effective local field $\mathscr{E}_e$ to the radiation field $\mathscr{E}$, and is ≈ 1 for shallow centers, ε_0 is the dielectric constant. For transitions between discrete levels denoted by b and a, the absorption cross section is given by

$$\sigma(\tilde{v}_{ba}) = \frac{1}{\sqrt{\varepsilon_0}}\left(\frac{\mathscr{E}_e}{\mathscr{E}}\right)^2 \frac{2\pi^2 h e^2}{m^* c} f_{ba}$$

$$= \frac{1}{\sqrt{\varepsilon_0}}\left(\frac{\mathscr{E}_e}{\mathscr{E}}\right)^2 (1.098 \times 10^{-16})\left(\frac{m}{m^*}\right) f_{ba} \quad \text{eV-cm}^2. \quad (23)$$

The oscillator strengths f_{ba} are the same as those for the hydrogen atom, and for the more important transitions are[109]

$$f_{1s \to 2p} = 0.4162, \quad f_{1s \to 3p} = 0.0791,$$

$$f_{1s \to 4p} = 0.0290, \quad f_{1s \to 5p} = 0.0139,$$

$$f_{1s \to 6p} = 0.0078, \quad f_{1s \to 7p} = 0.0048,$$

$$f_{1s \to 8p} = 0.0032, \quad f_{1s \to \text{cont}} = 0.436.$$

[108] M. Lax, in Photoconduct. Conf. Atlantic City, 1954 (R. G. Breckenridge, B. R. Russell, and E. E. Hahn, eds.), p. 111. Wiley, New York, 1956.
[109] H. A. Bethe and E. E. Salpeter, "Quantum Mechanics of One and Two Electron Atoms," p. 265. Academic Press, New York, 1957.

From the above equations and the neutral donor concentration N_D^0 it is possible to calculate the absorption constant for photoionization absorption as $\alpha(\tilde{v}) = \sigma(\tilde{v})N_D^0$. However, information concerning the linewidth and shape is required in order to calculate the absorption constant for transitions between discrete states.

b. Electron-Ionized Donor Recombination

Photoconductivity results from the photoionization absorption of radiation because the ionized electron is free to move throughout the conduction band and thus contribute to the conductivity of the sample until it recombines with an ionized donor center. For a semiconductor in thermal equilibrium, the electron concentration in the conduction band is constant and is determined by the equality of the electron generation and recombination rates. When external radiation of sufficient energy to photoionize the shallow donors is applied to the sample, the generation rate increases and the number of electrons in the conduction band begins to increase. The recombination rate also increases, and if the radiation remains constant, a new steady state is eventually reached in which the electron concentration in the conduction band is larger than the thermal equilibrium value and in which the generation and recombination rates are again equal. If the recombination rate increases, the nonequilibrium electron concentration decreases, and therefore the photoconductive response decreases. Thus, the electron-ionized donor recombination mechanism has a direct influence on the observed photoconductivity.

There are three mechanisms of electron-ionized donor recombination which are identified by the way in which the transition energy is released.

(1) *Auger recombination.* In this type of recombination the electron gives up its energy to another free electron in a collision and then recombines with the ionized donor. This process is not important for the small free carrier concentrations present in samples which exhibit extrinsic photoconductivity, and so will not be considered further.

(2) *Radiative recombination.* In this process the recombination energy is emitted as a photon. Calculation of the radiative recombination rates or cross sections indicate that this is not a very significant process for electron-ionized donor recombination. Emitted radiation corresponding to this mechanism has been observed, however. Far-infrared radiation from Ge corresponding to electron-ionized donor radiative recombination has been observed at very low intensities by Koenig and Brown,[110] Ascarelli and Brown,[111] and Salomon and Fan.[112] In the work of Salomon and Fan an

---

[110] S. H. Koenig and R. D. Brown, *Phys. Rev. Lett.* **4**, 170 (1960).
[111] G. Ascarelli and S. C. Brown, *Phys. Rev.* **120**, 1615 (1960).
[112] S. N. Salomon and H. Y. Fan, *Phys. Rev. B* **1**, 662 (1970).

attempt was made to determine the spectral dependence of the emitted radiation by using various bandpass filters, and these results indicated that the radiation was emitted in two bands, one at ~ 4 meV, corresponding to transitions between the conduction band and (2p, $m = 0$) states and one at ~ 7.5 meV corresponding to transitions between the (3p, $m = 0$) and the ground states. The spectral dependence of the far-infrared recombination radiation from impact-ionized shallow donors in GaAs has been measured[113] and this will be discussed in more detail in Section 15. All of these measurements indicate that only a very small fraction ($\sim 10^{-7}$) of the transitions are radiative.

(3) *Phonon recombination.* The dominant mechanism in electron-ionized donor recombination is that in which the recombination energy is released in the form of phonons. A direct transition from the conduction band to the ground state of the donor could require the simultaneous emission of a large number of phonons—an unlikely process. A more likely process is described by Lax's giant trap model[114,115] in which the electrons are initially captured in highly excited states of the shallow donor centers and, in a cascade process involving successive absorption and emission of single phonons, finally either escape into the conduction band or are trapped in the ground state. This is the dominant process for capture into an attractive Coulomb center in semiconductors.

This model and more recent modifications and/or extensions of it lead to the consideration of a *sticking probability* $P(U)$. The sticking probability is the probability that an electron in a bound state of energy U will eventually enter the ground state before it escapes into the conduction band. It is assumed in the theory that $P(U)$ does not depend on the history of the electron before it enters the state of energy U. The classical calculation of Lax also assumes that there are enough excited states with energies in the vicinity of kT to permit the use of classical methods. His agrument was that the fate of the electron will be decided for states with energies $U \approx kT$, so that even though there are subsequent steps (through discrete lower states) in the capture process which would require a quantum mechanical treatment, the classical treatment should still be valid. The calculation of Lax was modified by Hamann and McWhorter,[116] and the values of $P(U)$ determined in this work are larger because of the inclusion of some additional transitions between bound states which were neglected in Lax's calculation. Another improvement of Lax's calculation for the *electron capture cross*

[113] I. Melngailis, G. E. Stillman, J. O. Dimmock, and C. M. Wolfe, *Phys. Rev. Lett.* **23**, 1111 (1969).
[114] M. Lax, *J. Phys. Chem. Solids* **8**, 66 (1959).
[115] M. Lax, *Phys. Rev.* **119**, 1502 (1960).
[116] D. R. Hamann and A. L. McWhorter, *Phys. Rev.* **134**, A250 (1964).

section has been given by Smith and Landsberg,[117] but the previous sticking probability results were used.

A quantum mechanical calculation of the sticking probability and recombination cross section has been made for shallow donors in Ge and Si.[118,119] In this work it is assumed that the recombination takes place by initial capture of an excited state of the donor, but not necessarily a highly excited state. The sticking probability initially calculated by Ascarelli and Rodriguez[118] was different for As and Sb donors which have the same excited state structure, but slightly different ground state energies (due to different central cell corrections and valley–orbit splittings). However, in the revised calculations of Brown and Rodriguez[119] it is argued that the sticking probability is not significantly affected by the ground state energy for low temperatures. In these papers only the lower s-like states are considered since electron capture in these states is more important. However, the p-like states were included in a calculation by Beleznay and Pataki.[120] They found that the theory of Lax (and Hamann and McWhorter), which neglects the lower lying states, gives a larger estimate of the sticking probability than they obtain, and that the neglect of the p-like states within the framework of their calculation also leads to a considerable overestimate of the sticking probability. The calculation of Beleznay and Pataki has been criticized by Brown and Rodriguez[119] because it included only the longitudinal acoustic phonons. Although it is difficult to quantitatively compare the results of these calculations because of the different approximations, parameter values, etc., the results of Beleznay and Pataki[120] indicate that it is important to include the contribution of the p-like states in the calculation of the sticking probability since the neglect of these states leads to a considerable overestimation of $P(U)$.

8. EXCITED STATES AND PHOTOCONDUCTIVITY

The initial measurements of the absorption and photoconductivity of shallow donors in GaAs[106] were not interpreted in terms of excited states. (See Fig. 18.) However, later measurements on higher-purity material changed the interpretation of these spectra. The extrinsic photoconductivity spectrum of a GaAs sample with a donor concentration nearly two orders of magnitude smaller than the samples used by Bosomworth *et al.*[106] is shown in Fig. 19. The presence of two definite peaks in this spectrum suggests that excited states may be involved. If the energy of the main peak at about 4.40 meV (35.5 cm^{-1}) is identified with the excitation of electrons from the

[117] E. F. Smith and P. T. Landsberg, *J. Phys. Chem. Solids* **27**, 1727 (1966).
[118] G. Ascarelli and S. Rodriguez, *Phys. Rev.* **124**, 1321 (1961).
[119] R. A. Brown and S. Rodriguez, *Phys. Rev.* **153**, 890 (1967).
[120] F. Beleznay and G. Pataki, *Phys. Status Solidi* **13**, 499 (1966).

ground state to the first excited state ($n = 1$ to $n = 2$), the energies of the other discrete transitions from the ground state and the photoionization energy can be calculated using the hydrogenic model. These calculated values are indicated in Fig. 19. (In zero magnetic field, the 2s and 3s states are degenerate with the 2p and 3p states, but optical transitions to the former states are not allowed.) It can be seen that the energy of the smaller peak agrees well with the calculated value for the $n = 1$ to $n = 3$ transition (1s → 3p), although the separation between the two peaks cannot be determined accurately enough to permit a determination of the donor ionization energy as is usually done in Ge and Si. The ionization energy calculated from the (1s → 2p) transition energy is 47.3 cm^{-1} or 5.86 meV.

a. Mechanisms

If the excited states were really bound states, electrons in these states could not contribute to the conductivity of the sample and the excited state absorption would not result in peaks in the photoconductivity spectrum. There are several ways in which the electrons that are excited from the ground state to higher bound excited states can contribute to the conductivity of the sample.

First, if the excited state in question is broadened considerably by interaction with neighboring ionized donor and acceptor centers, there is the possibility that the conductivity of the sample could be increased because of the conductivity of the electrons in this excited state impurity band. For sufficiently high excited states this is just another way of saying that these states are unbound or have merged with the conduction band. For the lower excited states there is the possibility of an excited state impurity band which is not merged with higher excited states or the conduction band. Since the mobility of electrons in a band such as this is expected to be quite low for liquid He temperatures, this mechanism should not be important for the lower excited states, at least in the impurity range of the GaAs sample in Fig. 19.

The other mechanisms for impurity excited state photoconductivity all involve the subsequent transfer of the electron into the conduction band after its excitation to the excited state by the absorption of a photon. The mechanisms that must be considered for this transfer include (1) *field-induced tunneling* from the excited state into the conduction band; (2) *impact ionization* of the electrons in the excited states by energetic free electrons; (3) *thermal ionization* by the absorption of one or more phonons; and (4) *photoionization* by the absorption of a second photon.

All of these mechanisms are difficult to describe theoretically, and with the exception of the third mechanism very little work has been done on them. The third mechanism has been described theoretically by Kogan and

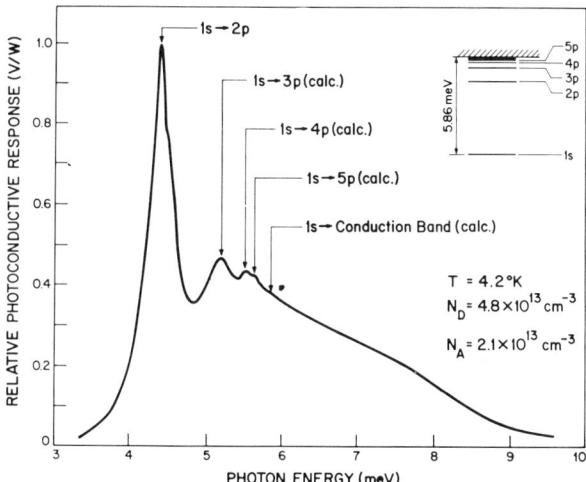

FIG. 19. Far-infrared photoconductivity spectrum of a high purity GaAs sample showing the measured transition energies and those calculated from the hydrogenic model using the (1s → 2p) transition energy. The hydrogenic energy level diagram is shown in the inset. (After Wolfe and Stillman.[61])

Sedunov,[121] who used the term "*photothermal ionization*" for this type of photoconductivity. There are two processes that can contribute to the photothermal ionization:

(a) An electron is first excited to a higher bound state by the absorption of a photon. Then by absorbing and emitting single phonons the bound electron moves up and down through the excited states until it is finally either captured in the ground state, from which thermal ionization is improbable, or released into the conduction band, where it can contribute to the conductivity. The process following the absorption of the photon is identical to that described by the giant trap cascade recombination model, and if the electron is released into the conduction band the process could be called cascade emission instead of cascade capture.

(b) The second possibility for photothermal ionization is the simultaneous or coherent absorption of a photon and phonon(s). Kogan and Sedunov[121] have shown that this is a very unlikely event for the case where the interaction between the impurity center and the lattice is weak.

An electron which is optically excited to the bound state (n, l) of the

[121] Sh. M. Kogan and B. I. Sedunov, *Fiz. Tverd. Tela* **8**, 2382 (1966) [*English Transl.: Sov. Phys.—Solid State* **8**, 1898 (1967)].

impurity center eventually either returns to the ground state or escapes into the conduction band. The probabilities of these two events are just the sticking probability $P_{n,l}$ and the ionization probability $I_{n,l}$, and since one or the other event must happen, $P_{n,l} + I_{n,l} = 1$. It is clear that the p-like states must be considered in determining these probabilities, at least for photothermal ionization, since electrons will be optically excited predominantly from the ground state to these states. A method for the determination of these probabilities from photoconductivity measurements has been described by Lifshits et al.[122] The thermal ionization probability of the state (n, l) can be written as

$$I_{n,l} = \sigma_{i_{n,l}} / \sigma_{(n,l)}, \quad (24)$$

where $\sigma_{i_{n,l}}$ is the photothermal ionization cross section and $\sigma_{(n,l)}$ is the optical absorption cross section from the ground state to the state (n, l). It is reasonable to expect that the electrons thermally ionized from the excited states are within an energy $\sim kT$ from the bottom of the conduction band. Since an electron which arrives in the conduction band in this way is indistinguishable from one which is optically excited from the ground state to this same energy range of the conduction band, the contribution to the photoconductivity should be the same. Thus, the amplitudes of the photoconductivity due to the two mechanisms should have the same ratio as the corresponding ionization cross sections. The amplitude of the photoconductive response due to photothermal ionization of the (n, l) excited state $\mathcal{R}_{n,l}$ can be expressed as

$$\mathcal{R}_{n,l} = k\sigma_{i_{n,l}} = k\sigma_{(n,l)} I_{n,l}. \quad (25)$$

The photoconductive response $\mathcal{R}_c$, due to photoionization of the ground state to an energy greater than the photoionization energy, but within $\approx kT$ from the bottom of the conduction band continuum, can be expressed as

$$\mathcal{R}_c = k\sigma, \quad (26)$$

assuming the escape probability for an electron photoexcited to the bottom of the conduction band is unity. In Eq. (26) σ is the photoionization absorption cross section and the constants of proportionality k in Eqs. (25) and (26) are equal. The ionization probability of the state (n, l) can then be written as

$$I_{n,l} = \frac{\mathcal{R}_{n,l}}{\mathcal{R}_c} \frac{\sigma}{\sigma_{n,l}}. \quad (27)$$

---

[122] T. M. Lifshits, N. P. Likhtman, and V. I. Sidorov, Fiz. Tekh. Poluprov. 2, 782 (1968) [English Transl.: Sov. Phys.—Semicond. 2, 652 (1968)].

Thus, the ionization and sticking probability of the state (n, l) can be determined from the photoconductivity and absorption measurements. These parameters are of interest in the giant trap cascade recombination calculations, but it should be emphasized that the values obtained in this way refer only to the p-like states, while it is the s-like states which have a dominant effect on the electron-ionized donor capture cross sections.[120]

b. *Observation in Other Materials*

Photoconductivity involving transitions to discrete excited states was first observed in Cu doped Ge by Loh and Picus.[123] At low electric fields (20V/cm) there was no photoresponse beyond the 30 μm ionization threshold at liquid-He temperatures, but at higher electric fields the response curve had well-defined shoulders at energies corresponding to excited state absorption peaks. These suggested that the ionization mechanism involved might be field-induced tunneling.

The first observation of excited state photoconductivity due to photothermal ionization was reported by Lifshits and Nad'[11] in Ge doped with Sb and As. Similar measurements were reported for p-type Ge by Sidorov and Lifshits,[124] in which it was noted that photoconductivity was only observed for excited states within about $3kT$ of the edge of the valence band, and that the correspondence between excited state absorption and photoconductivity was only obvious for samples with $N_A \lesssim 3 \times 10^{14}$ cm^{-3}. For higher acceptor concentrations the peaks in the photoconductivity spectra are not well resolved,[125] and this is probably the reason that excited-state photoconductivity was not observed by Shenker et al.[10] in their work on B-doped Ge.

The temperature dependence of the excited state photoconductivity in p-type Ge was studied by Lifshits et al.[122,126] to verify that the mechanism involved was photothermal ionization. These workers selected p-type Ge for these measurements to avoid the complications in n-type material introduced by the temperature dependent occupation of the valley–orbit–split donor ground states. The spectra that they obtained for In-doped Ge are shown in Fig. 20. At 4.2°K there is very little photo response at energies less than the 11.6 meV ionization energy for In acceptors, but for higher temperatures the excited state photoconductivity becomes comparable to the photoionization photoconductivity. Similar behavior was observed for

[123] E. Loh and G. S. Picus, *Bull. Amer. Phys. Soc. Ser. II* **7**, 173 (1962); G. S. Picus, *J. Phys. Chem. Solids* **23**, 1753 (1962).

[124] V. I. Sidorov and T. M. Lifshits, *Fiz. Tverd. Tela* **8**, 2498 (1966) [*English Transl.: Sov. Phys.—Solid State* **8**, 2000 (1967)].

[125] T. M. Lifshits, F. Ya. Nad', and V. I. Sidorov, *Fiz. Tverd. Tela* **8**, 3208 (1966) [*English Transl.: Sov. Phys.—Solid State* **8**, 2567 (1967)].

[126] T. M. Lifshits, N. P. Lichtman, and V. I. Sidorov, *Proc. 9th Int. Phys. Semicond., Moscow, 1968*, Vol. 2, p. 1081. Nauka, Leningrad, 1968.

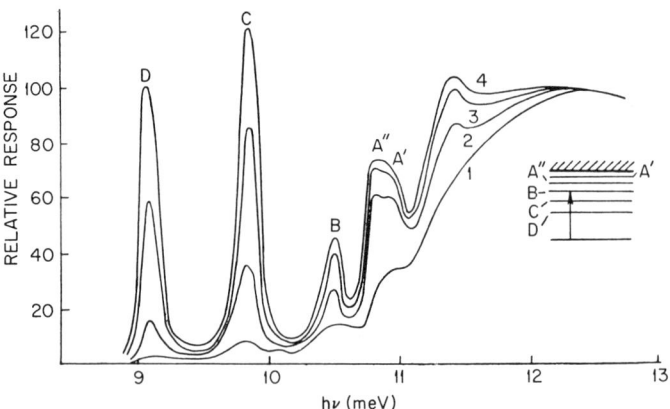

FIG. 20. Variation with temperature of excited state photoconductivity in In-doped Ge. (1) 4.2°K, (2) 6.2°K, (3) 8.4°K, (4) 10.7°K. The energy level diagram of the shallow acceptor center is also shown. (After Lifshits et al.[122,126])

B- and Al-doped Ge. In each case considerable photoconductivity corresponding to excited states within about 3 to $4kT$ of the valence band edge was observed for samples in which the acceptor concentration was no higher than about 5×10^{13} cm^{-3}. The relative intensities of the peaks and the photoconductivity maximum at 4.2°K were nearly independent of electric field. These results indicated that the ionization of electrons from excited impurity states was essentially all due to interaction with phonons and thus confirmed the photothermal ionization process.

Excited state photoconductivity in Sb-doped Ge has also been observed by Nagasaka et al.[127] and was initially attributed to conduction through excited state impurity bands. Later studies of the temperature dependence of the photoconductivity spectra by Nagasaka and Narita[128] indicated that two mechanisms were involved in the photoconductivity at energies less than the photoionization energy: one was the thermal ionization of discrete excited states; the other was an unknown mechanism which had little temperature dependence. The latter produced photoconductivity over a broad spectral range. Spectra which they obtained with a sample doped with 1.5×10^{14} cm^{-3} Sb at several different temperatures are shown in Fig. 21. Peaks that are labeled with the subscript 3 have the triplet ground state as the initial state, and the peaks labeled with the subscript 1 have the lower energy singlet ground state as the initial state. The valley–orbit splitting between the singlet and triplet states for Sb in Ge is only 0.32 meV.[129] Also

[127] K. Nagasaka, Y. Oka, and S. Narita, Solid State Commun. **5**, 333 (1967).
[128] K. Nagasaka and S. Narita, Solid State Commun. **7**, 467 (1969).
[129] J. H. Reuszer and P. Fisher, Phys. Rev. **135**, A1125 (1964).

4. FAR-INFRARED PHOTOCONDUCTIVITY IN HIGH PURITY GaAs

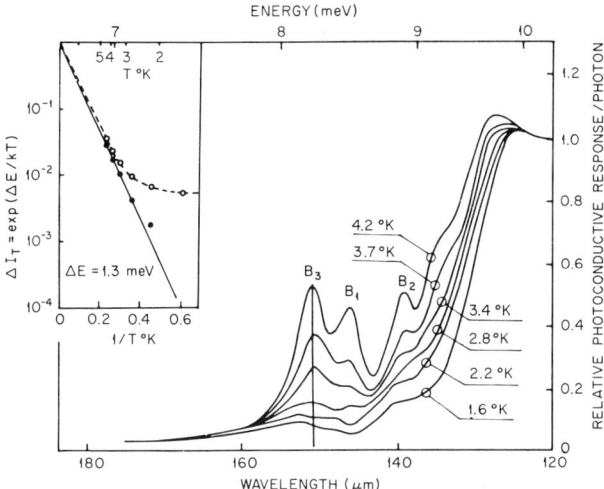

FIG. 21. Temperature-dependent photoconductivity spectra of Ge doped with 1.5×10^{14} cm^{-3} Sb. In the inset the dotted curve shows the logarithmic plot of the photoconductivity response at the peak B_3 at various temperatures against $1/T$ (°K), and the solid straight line shows the logarithmic plot of the difference between the response at B_3 at the various temperatures and the response at 1.6°K. (After Nagasaka and Narita.[128])

shown in the inset of the figure is the logarithmic variation with reciprocal temperature of the magnitude of the photoresponse at peak B_3 (dashed line) and the difference between the response at peak B_3 at a given temperature and the response at the same energy at 1.6°K (straight line). For the latter case a straight line is obtained, and this supports the existence of a temperature independent residual photoconductivity. The slope of this straight line gives an activation energy of about 1.3 meV for the photothermal ionization process, which is reduced to 1.0 to 1.2 meV when the temperature dependent population of the triplet state is included. The strongest support for the existence of the residual photoconductivity comes from higher resolution measurements of the photoresponse shown in Fig. 22. The peaks in the curve for 4.2°K due to photothermal ionization of the donors show a direct correspondence to dips in the spectrum for 1.6°K. This indicates that absorption to the excited states at the lower temperature detracts from the residual photoconductivity, and proves the existence of two different mechanisms. These authors suggest that the residual photoconductivity may result from photoionization of impurity atoms in which the ionization energy of the ground state is reduced due to clustering or inhomogeneous distribution of the impurity atoms. In this model the long tail and continuous nature of the residual photoconductivity would result from a continuous distribution of ionization energies. However, no direct support was given for this model.

Photothermal ionization of deeper or nonhydrogenic impurity levels has also been observed. The photothermal ionization of acceptors in semiconducting diamond has been studied extensively by Collins and Lightowlers.[130] The excited state photoconductivity, in this case, occurred in the energy range 300–400 meV and was studied over a much wider temperature range (5–150°K) than is possible with shallow impurity states. Analysis of the variation of the photoconductivity spectrum with temperature established that the mechanism for the excited state photoconductivity was indeed photothermal ionization. Excited state photoconductivity due to photothermal ionization was also recently reported for the doubly charged S-donor in Si by Sah et al.[131] These workers also reported the lack of excited state photoconductivity for the neutral Au center in Si, which was consistent with the lack of excited states for the neutral impurity potential.

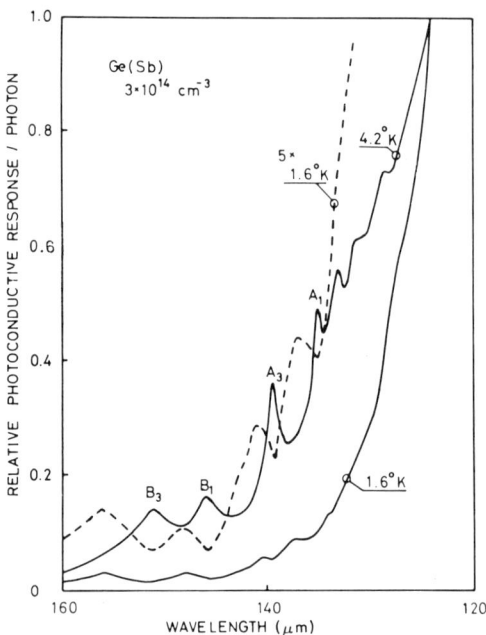

FIG. 22. High resolution far-infrared photoconductivity spectra of Sb-doped Ge at two temperatures. The peaks in the solid curve for 4.2°K correspond to the minima in the dashed curve for 1.6°K. (The dashed curve is a factor of five enlargement of the lower solid curve for 1.6°K.) (After Nagasaka and Narita.[128])

[130] A. T. Collins and E. C. Lightowlers, *Phys. Rev.* **171**, 843 (1968).
[131] C. T. Sah, T. H. Ning, L. L. Rosier, and L. Forbes, *Solid State Commun.* **9**, 917 (1971).

The main application of the photothermal ionization mechanism so far has been in the study of impurity excited states, since this method permits the observation of these states in much purer samples than can be studied by absorption measurements. Lifshits et al.[122,126] have pointed out that this method is very sensitive for the observation of impurity centers and that the intensity of the photoconductivity lines actually increases noticeably when the impurity concentration is decreased, i.e., the lines become narrower and higher. This is because the field effect of neighboring ionized impurities decreases. They note that in Ge with impurity concentrations of 10^{13} to 10^{14} cm^{-3} the lines are observed only at temperatures a few degrees higher than liquid He, whereas in Ge with impurity concentrations of only 10^{11} to 10^{12} cm^{-3}, the lines are clearly visible at 4°K. This high sensitivity at low impurity concentrations makes the observation of excited-state photoconductivity one of the best means of identifying residual impurities or dopants in Ge. The photoconductivity spectrum of a Ge sample with $N_A - N_D = 3 \times 10^{11}$ cm^{-3} containing two different acceptor impurities is shown in

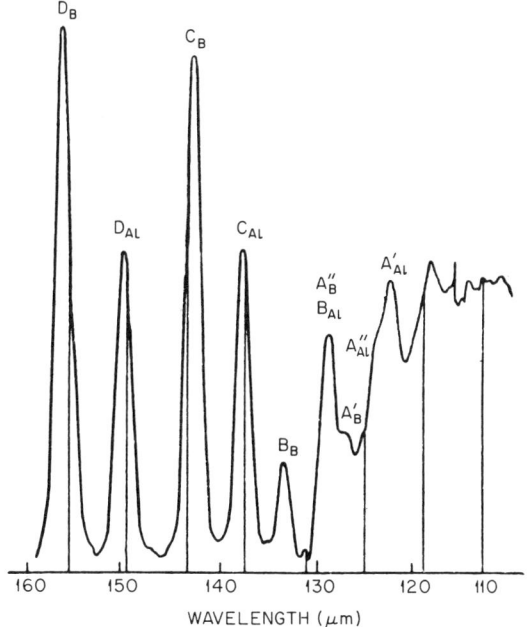

FIG. 23. Far-infrared photoconductivity spectrum of a high purity p-type Ge sample with a net acceptor concentration $N_A - N_D = 3 \times 10^{11}$ cm^{-3} containing both B and Al acceptors. The transitions labeled A_B, A_{Al}, B_B, B_{Al}, etc. correspond to the inset of Fig. 20 for boron and aluminium acceptors, respectively, which have different central cell shifted ground states, but similar excited state energies. (After Lifshits et al.[126])

Fig. 23. This is the most sensitive method of determining the identity of residual impurities, and since the total number of acceptors can be determined by the analysis of Hall constant measurements, it is possible to determine the concentration of each different impurity species.

c. *Behavior in GaAs*

Because of the success of the photothermal ionization process in explaining the experimental results in Ge and the decrease in the relative photoresponse of the dominant peak in the GaAs far-infrared photoconductivity spectrum with decreasing temperature,[66,132,133] it was assumed that this mechanism was also dominant in the excited state photoconductivity in GaAs. The donor levels in GaAs are so shallow that the photothermal ionization process should be important at much lower temperatures than in Ge, but some of the other processes may also be more important than in Ge.

(1) *Experimental method.* Most of the experimental measurements of impurity absorption and excited state photoconductivity in Ge have been made with far-infrared grating spectrometers, and this method has also been used for some measurements on GaAs in the 100–500 μm wavelength range. The problems of stray and higher order radiation are very severe for this part of the spectrum, however, and a method that has found much wider use for this wavelength range is Fourier transform spectroscopy. Both of these spectroscopic methods have been discussed in some detail and compared by Martin.[134] The schematic diagram of a Michelson interferometer is shown in Fig. 24. The collimated light from the source, which is usually a high pressure mercury lamp, is divided into two beams by the beam splitter. One beam travels a constant distance from the beam splitter to a fixed mirror and back, while the other travels a variable distance to the movable mirror and back. The beams are recombined at the beam splitter and directed either to a Golay cell detector or to a light pipe which directs the radiation to a photoconductive sample or a cryogenic detector. When the path difference from the beam splitter to the movable mirror is the same as the path difference to the fixed mirror, the two beams have the same phase for all wavelengths and the intensity of the combined beam will be a maximum. If the two beams are monochromatic with wavenumber $\tilde{v}$ (cm^{-1}) and of equal intensity $S_{\tilde{v}}$, the intensity of the combined beam $S(x)$ as a

---

[132] G. E. Stillman, C. M. Wolfe and J. O. Dimmock, *in* "Gallium Arsenide and Related Compounds" *(Proc. 3rd Int. Symp., Aachen, 1970)*, p. 212. Inst., of Phys. and Phys. Soc., London, 1971.

[133] G. E. Stillman, C. M. Wolfe, and J. O. Dimmock, *Proc. Symp. Submillimeter Waves, New York, 1970*, p. 345. Polytechnic Press, Brooklyn, 1971.

[134] D. H. Martin, *in* "Spectroscopic Techniques" (D. H. Martin, ed.), p. 3. North–Holland Publ., Amsterdam, 1967.

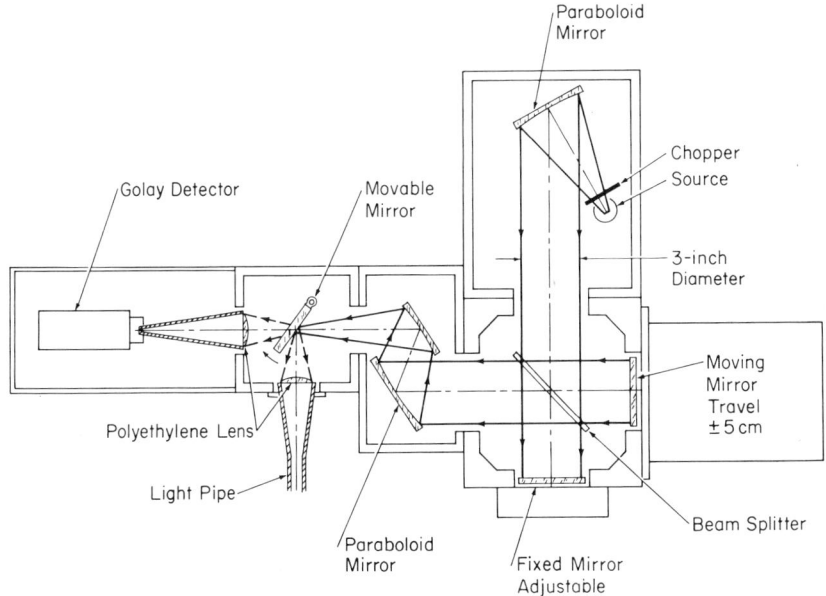

FIG. 24. Schematic diagram of a Michelson Fourier spectrophotometer (Beckman Scientific Instruments Division, Model FS-720).

function of the path difference between the two beams x (cm) is

$$S(x) = S_{\tilde{v}}[1 + \cos(2\pi\tilde{v}x)]. \tag{28}$$

Now, if the responsivity of the detector at wavenumber $\tilde{v}$ is $\mathcal{R}(\tilde{v})$ in amperes per watt or volts per watt, and if there are other filters or samples in the path of the recombined beam with total transmission at wavenumber $\tilde{v}$ of $T_{\tilde{v}}$, the signal detected by the detector due to the chopped monochromatic light as a function of path difference will be

$$I_{\tilde{v}}(x) = \mathcal{R}(\tilde{v})\{S_{\tilde{v}} \cdot T_{\tilde{v}} \cdot [1 + \cos(2\pi\tilde{v}x)]\}. \tag{29}$$

For radiation from the mercury lamp, which has a continuous spectral distribution $S(\tilde{v})$, the signal detected at a given path difference x (the interferogram) will be given by the integral over all wavenumbers as

$$I(x) = \int_0^\infty \mathcal{R}(\tilde{v})S(\tilde{v})T(\tilde{v})[1 + \cos(2\pi\tilde{v}x)]\,d\tilde{v}$$

$$= \tfrac{1}{2}I(0) + \int_0^\infty \mathcal{R}(\tilde{v})S(\tilde{v})T(\tilde{v})\cos(2\pi\tilde{v}x)\,d\tilde{v} \tag{30}$$

where $I(0)$ is the signal for zero path difference. The spectral dependence of the transmission of the filters (or samples) is $T(\tilde{\nu})$, and the variation of the beam splitter efficiency with $\tilde{\nu}$ can be included in the spectral distribution of the source, $S(\tilde{\nu})$. This equation can be inverted by the Fourier cosine transform to give

$$\mathscr{R}(\tilde{\nu})S(\tilde{\nu})T(\tilde{\nu}) = 4 \int_0^\infty [I(x) - \tfrac{1}{2}I(0)] \cos(2\pi\tilde{\nu}x)\, dx. \tag{31}$$

With a given source and detector, it is possible to determine $T(\tilde{\nu})$, the transmission of the filters (or samples) by two measurements: one with and one without the filters (or samples). Similarly, with a given source and filter combination it is possible to determine $\mathscr{R}(\tilde{\nu})$, the response of a photodetector, by two measurements: one with this detector and one with a detector of known or flat spectral response. Fourier transform spectroscopy is now a well-established spectroscopic method, and the techniques for obtaining the interferogram in a form from which its transform can be calculated, as well as problems of finite path difference, apodization, resolution, phase error, digital calculation, etc. are described in the literature.[105,134-140]

(2) *Variation of spectral response with temperature.* The temperature variation of the excited state photoconductivity for a GaAs sample with $N_D = 4.8 \times 10^{13}$ cm^{-3}, $N_A = 2.1 \times 10^{13}$ cm^{-3} and $\mu_{300°K} = 8600$ cm^2/V-sec, $\mu_{77°K} = 210{,}000$ cm^2/V-sec is shown in Fig. 25.[140] The calculated spectral resolution for these spectra was about 0.16 cm^{-1}, and all of the spectra were normalized to have the same photoresponse in the 75–85 cm^{-1} range for comparison purposes only. The applied electric field for all of these spectra was well below that required for impact ionization of the electrons from the ground state of the shallow donor level. There is very little change in the spectra as the temperature is decreased from 4.2°K down to about 3°K, but below 2.5°K there is a significant reduction in the excited state photoconductivity for a further decrease in temperature. These photoresponse spectra are are plotted on a per photon basis and, because the sample used for these

[135] E. V. Lowenstein, *Appl. Opt.* **5**, 845 (1966).
[136] P. L. Richards, in "Spectroscopic Techniques" (D. H. Martin, ed.), p. 31. North–Holland Publ., Amsterdam, 1967.
[137] L. Mertz, "Transformations in Optics." Wiley, New York, 1966.
[138] J. Connes, *Rev. Opt. (France)* **40**, 45, 116, 171, 231 (1961) [*English Transl.*: in NAVWEPS Rep. No. 8099, U.S. Naval Ordnance Test Station, China Lake, California].
[139] K. D. Moller and W. G. Rothschild, "Far Infrared Spectroscopy," p. 128. Wiley (Interscience), New York, 1971.
[139a] R. J. Bell, "Introductory Fourier Transform Spectroscopy." Academic Press, New York, 1972.
[140] G. E. Stillman, C. M. Wolfe, and D. M. Korn, in *Proc. 11th Conf. Phys. Semicond. Warsaw, Poland, 1972,* p. 863. Polish Sci. Publ., Warsaw, 1973.

4. FAR-INFRARED PHOTOCONDUCTIVITY IN HIGH PURITY GaAs

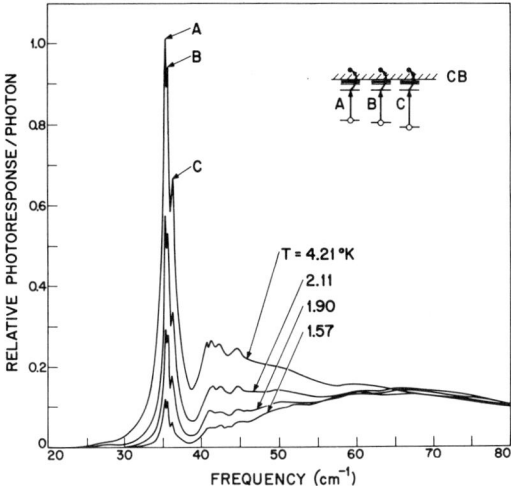

FIG. 25. Variation with temperature of the excited state photoconductivity in high purity GaAs. (The fine structure apparent in these spectra is discussed in Section 10.)

measurements was thin and the absorption constant small, the higher temperature photoresponse in the continuum should be similar to the spectral variation of the absorption. At the lowest temperature of 1.2°K the photoresponse resulting from the excitation of electrons from the ground state of the shallow donors to the first excited state is about as large as that at the energy corresponding to the photoionization of the shallow donors (~ 48 cm^{-1}). At this temperature there is also a significant amount of photoresponse, due to higher excited states, at energies just below the energy corresponding to the photoionization energy. For the excited states the photothermal ionization spectrum is related to the absorption spectrum and the ionization probability of the excited states through Eq. (24). Since the concentration of neutral donors is essentially constant, the small change in the photoconductivity spectrum as the temperature is decreased from 5.5°K to about 3°K indicates that the thermal ionization probability of the excited states is not changing significantly in this temperature range. Therefore, in this temperature range the shape of the excited state photoconductivity spectrum should be the same as the absorption spectrum, and the thermal ionization probability of the p-like excited states is about unity (i.e., the sticking probability of these excited states is approximately zero). At still lower temperatures the excited state photoconductivity is reduced, indicating that the ionization probability of the excited states is decreasing with decreasing temperature below about 2.5°K. From Fig. 25 it appears that at low temperatures electrons excited higher into the conduction band

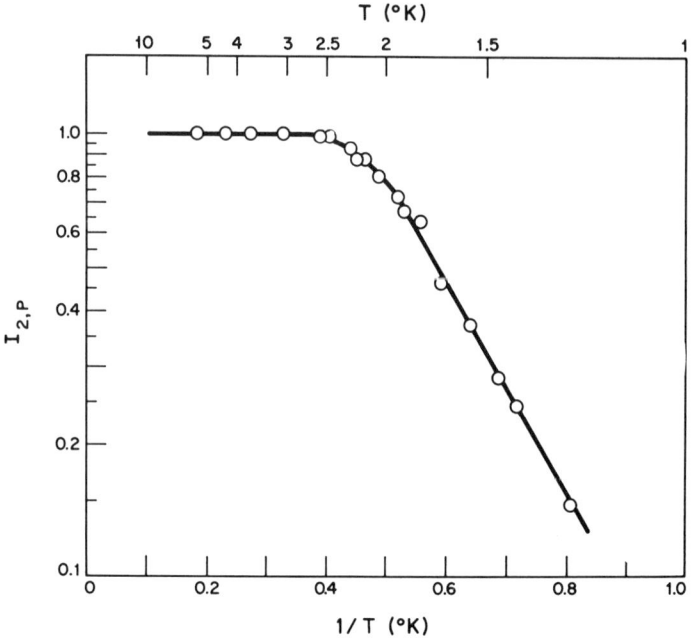

FIG. 26. Temperature variation of the ionization probability of the 2p excited states of shallow donors in GaAs.

contribute relatively more photoresponse than those excited to within kT of the bottom of the band. This may be due to the fact that the electron mobility is limited by ionized impurity scattering, and that the mobility of the carriers excited higher in the band is significantly higher than for those at the bottom of the conduction band. However, since the thermally excited electrons should be excited to within about kT of the bottom of the conduction band, the photoresponse within this energy range should be used in determining the ionization probability with Eq. (27). The variation of the ionization probability determined in this way is shown in Fig. 26. At first glance the exponential decrease in I_{2p} for temperatures less than about 2.5°K would seem to be good support for the assumption that the main mechanism responsible for the excited state photoconductivity is photothermal ionization.[129] The activation energy for the ionization probability determined from this data is $\Delta E = 0.47$ meV, and in the simple picture this would correspond to the energy separation between the 2p excited states and the effective continuum of the higher overlapping excited states. From the thermal ionization energy of $E_{D_t} = 5.52$ meV (determined from the analysis of Hall constant data for this sample), this energy separation can be estimated to be about 1.12 meV, considerably larger than the value of $\Delta E = 0.47$ meV. However, it is clear from the work of Beleznay and Pataki[120] that there is more involved in the photothermal ionization than just a thermal activation energy from a given excited state into the conduction band, since electrons in the 2s and 2p states have the same energy separation from the continuum, but significantly different sticking (or ionization) probabilities. Although the calculated ionization probabilities of the 2p and 3p states in Ge[121] cannot be compared quantitatively with the experimental results for GaAs, the calculated values in Ge are also much larger than expected at a given temperature from the separation of these states from the continuum.

(3) *Variation of spectral response with electric field.* At 4.2°K there is essentially no dependence of the excited state spectra on the magnitude of the applied electric field. This is consistent with the conclusion that at this temperature the ionization probability of an electron in the 2p excited state is essentially unity, even at low electric fields. At the lower temperatures for which $I_{2p} < 1$, there is only a small increase in the photoconductive response of the peak as the electric field is increased nearly two orders of magnitude. The small increase observed at 1.2°K corresponds to an increase in the ionization probability of the 2p state from about 0.14 to about 0.15.

The contribution of the absorption of a second photon to the excited state photoconductivity is apparently insignificant, since the relative photoresponse at the peak does not increase as the intensity of the radiation is increased. Therefore, the weak dependence of the spectral response on the electric field and the decrease in the excited state photoconductivity with decreasing temperature below 2.5°K indicate that photothermal ionization is the dominant mechanism.

9. Dependence of Spectral Response on Doping

The thermal ionization energy of shallow donors in GaAs is smaller in samples with higher donor concentrations, and for samples with $N_D \gtrsim 10^{16}$, $E_{D_t} \approx 0$. Thus, it was initially assumed[106] that the absorption and photoconductivity threshold would occur at lower energies in GaAs samples with larger donor concentrations and that it would be possible to vary the long wavelength threshold somewhat by adjusting the doping level. However, because the long wavelength photoconductivity is due to excited state transitions, it was found that the threshold could be changed only slightly by varying the doping level.

a. Experimental Results

The variation of the spectral response of the photoconductivity with doping was studied by Stillman *et al.*[66] Photoconductivity spectra for five different

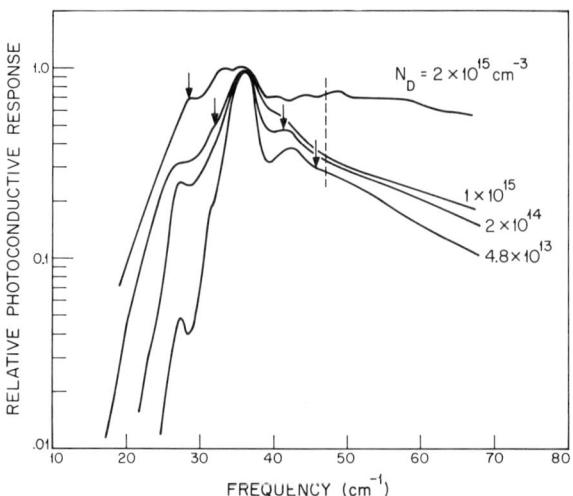

FIG. 27. Spectra showing the change in the photoconductive response of GaAs with increasing donor concentration. The vertical arrows indicate the thermal ionization energies for the samples and the photoionization energy is indicated by the dashed line.

GaAs samples with donor concentrations in the range 4.8×10^{13}–2.1×10^{15} cm^{-3} are shown in Fig. 27. These samples are the same as are listed in Table V. To show the change in response with doping level, the spectra were normalized to have the same amplitude at the peak. As the donor concentration is increased from the lowest value of $N_D = 4.8 \times 10^{13}$ cm^{-3}, the dominant peak corresponding to (1s → 2p) transitions at 35.5 cm^{-1} remains at the same frequency, but broadens considerably, while the smaller peak at about 42.2 cm^{-1} becomes less distinct and finally disappears. For the purest sample there is some structure on the low energy side of the dominant peak in this logarithmic plot, although it is hardly noticeable on a linear scale. The amplitude of this structure increases with increasing donor concentration and in the sample with the largest donor concentration shown there are two peaks evident on the low energy side of the (1s → 2p) peak. The possible origin of this structure is discussed in the next subsection. A photoconductivity spectrum could not be obtained for a sample with $N_D = 4.72 \times 10^{15}$ (sample 1 of Table V), possibly because the mobility in the impurity band was comparable to that in the conduction band and the excitation of electrons into the conduction band in this case would not significantly increase the sample conductivity. The frequencies corresponding to the thermal ionization energies determined for these samples are also shown in Fig. 27, along with the donor binding energy or photoionization energy of 47.3 cm^{-1} (5.86 meV) calculated from the (1s → 2p) transition energy in the hydrogenic approximation. It was felt that the calculation of the photoionization energy in this way was justified, since the calculated (1s → 3p) energy of 42.1 cm^{-1} was in good agreement with the smaller peak observed in the purest sample at 42.2 cm^{-1}. The photoionization energy calculated in this way is the same for all of these samples because the (1s → 2p) transition energy does not change over the range of donor concentrations investigated.

b. Relationship of Spectral Broadening to Thermal Ionization Energy

The variation of the donor ionization energy determined from Hall constant analyses is shown in Fig. 28, along with the photoionization energy calculated from the (1s → 2p) transition. The decrease in thermal ionization energy with increasing donor concentration has been known since the earliest work on transport properties of semiconductors. One of the first theoretical treatments of the problem was by Shifrin[76,141] who suggested that the lowering of the ionization energy with increasing impurity concentration was due to conduction in the overlapping excited states of the ionized impurity atoms. He reasoned that starting from some excited state N with energy E_N and Bohr radius a_N the distance between the bound outer electron and the impurity atom would be so large that this electron might just as well belong to some neighboring *ionized* impurity atom, so that electrons having an energy less than E_N below the conduction band belong to the continuum. The excited state which marks the boundary between the discrete excited states and the continuum was determined from the condition that the separation between the *ionized* donors be about twice the corresponding Bohr radius.

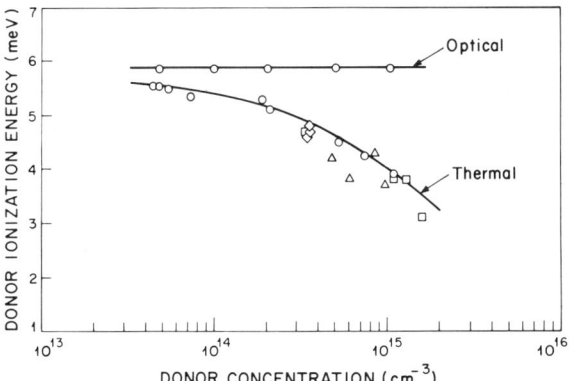

FIG. 28. Variation of donor ionization energy with donor concentration N_D, as determined from the (1s → 2p) transition energy in photoconductivity and from analyses of Hall coefficient data. ○, Stillman et al.[66]; △, Bolger et al.[71]; □, Eddolls et al.[75]; ◇, Maruyama et al.[73] (After Wolfe and Stillman.[61])

[141] K. Shifrin, *J. Phys. (USSR)* **8**, 242 (1944).

$$a_N = a_B N^2 \lesssim \tfrac{1}{2}(1/N_D^+)^{1/3} \tag{32}$$

so

$$N \sim [2a_B(N_D^+)^{1/3}]^{-1/2}. \tag{33}$$

With this model the thermal ionization energy E_D depends on the ionized donor concentration $N_D^+ = n + N_A$. Since N_D^+ is a function of temperature, E_{D_t} is also a function of temperature, and is given by

$$E_{D_t} \approx E_I\left[1 - \frac{1}{N^2}\right] \approx E_I\{1 - 2a_B[n(T) + N_A]^{1/3}\}. \tag{34}$$

Broadening of the impurity excited levels similar to that considered by Shifrin was proposed by Erginsoy[142] to explain the behavior of the resistivity and Hall constant of Ge at low temperatures. He calculated [143] the energy of the lower edge of the impurity bands formed by broadening of the 1s, 2s,

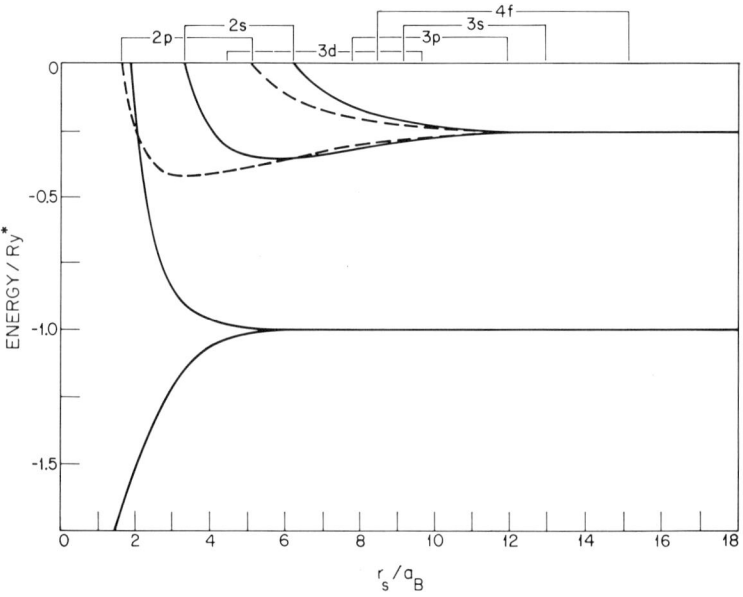

FIG. 29. Broadening of the 1s, 2s, and 2p donor impurity levels with increasing donor concentration. The solid curves are the edges of the 1s and 2s bands and the dashed curves are the edges of the 2p band. The abscissa is the ratio of $r_s = (3/4\pi N)^{1/3}$, where N is the donor concentration, to a_B, the Bohr radius of the donors. Also shown are the values of r_s/a_B, for which the edges of some of the higher bands cross the value $E = 0$. (After Baltensperger.[144])

[142] C. Erginsoy, Phys. Rev. **80**, 1104 (1950).
[143] C. Erginsoy, Phys. Rev. **88**, 893 (1952).

and 3s states as a function of the ratio of the radius of a Wigner–Seitz sphere r_s, given by $4\pi r_s^3/3 = 1/N_D$, where N_D is the donor concentration. Although his results predicted significant broadening of the excited states at moderate impurity concentrations, he concluded that this mechanism could not explain the decrease of E_{D_t} to zero, because the rise of the bottom of the 2s band with increasing impurity concentration was so steep that it was unlikely that the 1s and 2s bands could overlap. Baltensperger[144] calculated the edges of the 1s, 2s, and 2p bands in the same approximation as Erginsoy[143] and the results which he obtained are shown in Fig. 29. It can be seen that the 1s and 2p bands do overlap at sufficiently high donor concentrations and therefore this mechanism can explain the decrease of E_{D_t} to zero. The abscissa is the ratio of the Wigner–Seitz sphere for the donor impurities to the Bohr radius r_s/a_B, so his calculation predicts that broadening of the ground state will start to occur when this ratio is about 5 and that the 1s and 2p bands will overlap for $r_s/a_B \approx 2$ or for $N_D^{1/3} a_B \approx 0.31$. For GaAs this corresponds to $N_D \approx 2 \times 10^{16}$ cm^{-3}. The broadening of the 2p (and 2s) levels begins for $r_s/a_H \approx 12$. It should be pointed out that these calculations were made in the approximation that the impurity atoms were regularly spaced and the model corresponds to metallic hydrogen. This type of calculation was extended to higher impurity concentrations by Stern and Talley.[145] The effect of a random distribution of impurity centers was considered by Aigrain.[146] In this model, although the thermal activation energy decreases because of the overlap of the excited states as in the model proposed by Shifrin,[76] the decrease in E_{D_t} depends on the *total donor concentration rather than on just the ionized donor concentration.*

The decrease in E_{D_t} with increasing impurity concentration was observed in p-type Si by Pearson and Bardeen.[147] They found that the variation depended on the cube root of the acceptor concentration in B-doped, p-type Si which would correspond to the donor concentration N_D in an n-type semiconductor. This conclusion was incorrect because the ionization energy determined for their purest sample was not due to B, but resulted from some other residual impurity.[9] The decrease of E_{D_t} in n-type Ge was studied by Debye and Conwell[148] who found that the activation energy could be fitted by the expression

$$E_D = E_I - \alpha(N_D^+)^{1/3}, \qquad (35)$$

where N_D^+ represents an average density of ionized donors over the

---

[144] W. Baltensperger, *Phil. Mag.* **44**, 1355 (1953).
[145] F. Stern and R. M. Talley, *Phys. Rev.* **100**, 1638 (1955).
[146] P. Aigrain, *Physica* **20**, 978 (1954).
[147] G. L. Pearson and J. Bardeen, *Phys. Rev.* **75**, 865 (1949).
[148] P. P. Debye and E. M. Conwell, *Phys. Rev.* **93**, 693 (1954).

temperature range that was used to determine E_D. The functional form of this equation is the same as Eq. (34), but the mechanism used to justify this empirical result was not the same as that considered by Shifrin.[76] This type of variation would also be expected if the decrease in E_D was due to the potential energy of attraction between the ionized donors and free electrons.[149]

Another mechanism which has been widely used to explain the decrease of E_D with increasing donor concentration is that of the screening of the impurity centers by free carriers. Mott[150,151] has considered a crystalline array of hydrogenic atoms and shown that at absolute zero temperature there should be a sharp transition from a state with zero conductivity to a state with finite conductivity (i. e. the metallic state) as the spacing between the hydrogenic atoms was decreased. He concluded that it was only possible to have a finite conductivity at $T = 0°K$ if there was a considerable concentration of free carriers. In this case, the Coulomb potential is replaced by the Debye–Hückel screened potential

$$V(r) = -(e^2/\varepsilon_0 r)e^{-qr}, \qquad (36)$$

where q is a screening constant, given in the Thomas–Fermi approximation by

$$q^2 = 4m^*e^2n^{1/3}/\varepsilon_0\hbar^2 = 4n^{1/3}/a_B. \qquad (37)$$

Mott used the condition for the nonexistence of bound states for this potential

$$q > 1/(\hbar^2\varepsilon_0/m^*e^2) = 1/a_B \qquad (38)$$

to estimate the value of n for which the transition occurs. Combining Eqs. (37) and (38) we obtain the well-known equation for the "Mott transition,"

$$n^{1/3}a_B \approx 0.25. \qquad (39)$$

In this relation n is the free carrier concentration, but this has been sometimes changed to N_D or $N_D - N_A$ since the criterion is derived for the case of metallic conduction where freezeout does not occur. This equation predicts the carrier concentration (and therefore the net donor concentration) for which $E_{D_t} = 0$, and the experimental results in Ge and Si are in good agreement with this prediction. The carrier concentration predicted for GaAs at which $E_{D_t} = 0$ is $n \approx 1.6 \times 10^{16}$ cm^{-3}. Since experimentally E_{D_t} decreases to zero for N_D between about 1 and 2×10^{16} cm^{-3}, there is also good

---

[149] G. W. Castellan and F. Seitz, in "Semiconducting Materials" (*Proc. Int. Conf., Reading, 1950*), p. 8. Butterworth, London and Washington, D. C., and Academic Press, New York, 1951.

[150] N. F. Mott, *Proc. Phys. Soc. London* **A62**, 416 (1949).

[151] N. F. Mott, *Phil. Mag.* **6**, 287 (1961).

agreement for GaAs. The modifications of this picture required for a random array of centers have been discussed by Mott and Davis.[152]

The energy levels of a bound particle in a screened Coulomb or Debye–Hückel potential have been considered by a number of authors.[153,154] This model has been particularly attractive in trying to explain the lack of an ionization energy in high purity InSb,[155] as well as the magnetic freezeout of carriers observed in this material.[156–158] A variational calculation for this potential in GaAs has been performed by Katana et al.[159] They obtain an expression for the energy of the ground state as a function of a screening radius which predicts that the donor merges with the conduction band in GaAs for a free carrier concentration n (or net donor concentration $N_D - N_A$) of 2×10^{16} cm^{-3} in good agreement with the observed value. These workers also calculated the variation of the ground state and first excited state energies with n, and their results are given in Table VI. From these results the two lowest bound states of a screened Coulomb potential both approach the conduction band as the free carrier concentration increases, and the separation between these states also decreases. This behavior is not in agreement with the experimental observation[66,132] that the energy

TABLE VI

RESULTS OF VARIATIONAL CALCULATIONS FOR THE GROUND STATE ENERGY $-E_1$ AND FIRST EXCITED STATE ENERGY $-E_2$ OF SHALLOW DONORS IN GaAs WITH A SCREENED COULOMB POTENTIAL FOR DIFFERENT CARRIER CONCENTRATIONS, n^a

n (cm^{-3})	$-E_1$ (meV)	$-E_2$ (meV)	n (cm^{-3})	$-E_1$ (meV)	$-E_2$ (meV)
2.8×10^{13}	5.37	0.892	5.0×10^{15}	0.963	—
8.3×10^{13}	4.91	0.576	7.3×10^{15}	0.633	—
2.2×10^{14}	4.26	0.237	9.0×10^{15}	0.466	—
9.3×10^{14}	2.72	0	1.0×10^{16}	0.354	—
2.6×10^{15}	1.66	—	2.0×10^{16}	0	—

^{a} After Katana et al.[159]

[152] N. F. Mott and E. A. Davis, *Phil. Mag.* **17**, 1269 (1968).
[153] V. L. Bonch-Bruevich and V. B. Glasko, *Opt. Spektrosk.* **14**, 495 (1963) [*English Transl.: Opt. Spectrosc.* **14**, 264 (1963)].
[154] K. Colbow and D. Dunn, *Phil. Mag.* **22**, 237 (1970).
[155] S. P. Li, W. F. Love, and S. C. Miller, *Phys. Rev.* **162**, 728 (1967).
[156] E. W. Fenton and R. R. Haering, *Phys. Rev.* **159**, 593 (1967).
[157] J. Durban and N. H. March, *J. Phys. C (Proc. Phys. Soc.)* **1**, 1118 (1968).
[158] P. E. Hanley and E. H. Rhoderick, *J. Phys. C (Solid State Phys.)* **2**, 365 (1969).
[159] P. K. Katana, S. D. Tiron, and A. G. Cheban, *Fiz. Tekh. Poluprov.* **4**, 260 (1970) [*English Transl.: Sov. Phys.—Semicond.* **4**, 210 (1970)].

of the (1s → 2p) transition does not change for samples with donor concentrations in the range from about 4.7×10^{13} to 1.4×10^{15} cm^{-3}, even when allowance is made for the fact that the carrier concentration is considerably lower at the temperature at which the photoconductivity measurements are made.

All of the mechanisms described above predict the donor or carrier concentration for which the thermal ionization energy descreases to zero quite accurately. However, only some form of banding of the impurity excited states seems to explain the broadening of the excited state photoconductivity spectra *and* the gradual decrease in the thermal ionization energy with increasing donor concentration, although the mechanism for the banding is still in doubt. Summers *et al.*[160] have pointed out that the spectral resolution of the far-infrared donor excited state spectra is very dependent on the banding of the $n = 2$ and higher states. They extended the Mott "broadening" parameter to include excited states by writing it as $\beta = \alpha N_D^{1/3} \geqslant 0.25$, where α is the Bohr radius of the (excited) state in question and N_D is the donor concentration. Using this relation they conclude that the $n = 2$ states will be banded in samples with $N_D > 1 \times 10^{14}$ cm^{-3} and that the $n = 3$ states will be banded for $N_D > 1 \times 10^{13}$ cm^{-3}. It is clear from the prior discussion that this extension to excited states *is not* valid within the framework of screening by free carriers for which it was originally derived. However, the conclusion that banding in excited states affects the spectral resolution *is* correct for samples with higher donor concentrations. Although they attribute the 5 cm^{-1} half-widths of the peaks corresponding to the (1s → 2p) transitions in their samples to this banding, there are other complications which must be considered and these will be discussed below. It can be concluded, however, that the broadening of the (1s → 2p) peak with increasing donor concentration shown in Fig. 27 is related to the broadening or banding of the 2p levels, and the broadening of this and the other excited states as well as the ground state can also qualitatively explain the decrease in the thermal ionization energy with increasing donor concentration shown in Fig. 28. The lack of a decrease in the (1s → 2p) transition energy with increasing donor concentration indicates that screening by free carriers cannot explain the observed decrease in the donor thermal ionization energy. Thus, the good agreement between the experimental donor concentration for which $E_{D_t} = 0$ and that predicted by the equation for the Mott transition [Eq. (39)] must be regarded as accidental.[161] Whether the broadening of the excited states and thus the decrease in E_{D_t} is due to a banding effect which varies with the neutral donor concentration as in the metallic hydrogen model or is simply due to the overlap of excited state wave functions with

[160] C. J. Summers, R. Dingle, and D. E. Hill, *Phys. Rev. B* **1**, 1603 (1970).
[161] N. F. Mott, *Comments Solid State Phys.* **2**, 183 (1970).

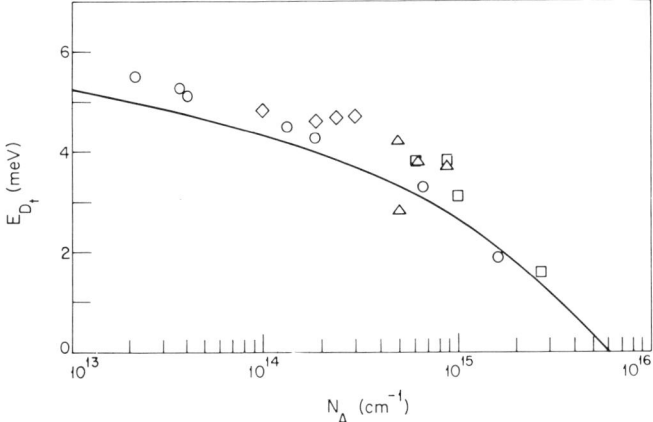

FIG. 30. Variation of donor ionization energy with acceptor concentration. At the low temperature at which E_{D_t} is determined, the acceptor concentration is approximately equal to the ionized impurity concentration. The curve was calculated from the relation $E_{D_t} = E_I (1 - \alpha N_A^{1/3})$. ○, Stillman et al.[66]; Δ, Bolger et al.[71]; □, Eddolls[74]; ◇, Maruyama et al.[73]

neighboring ionized donors which varies with the ionized donor concentration or acceptor concentration is open to question. The decrease in E_{D_t} is shown in Fig. 30 as a function of the acceptor concentration, which is approximately the ionized donor concentration for the low temperatures at which E_{D_t} was determined. The curve in this figure is calculated from a relation corresponding to Eq. (34), $E_{D_t} = E_I(1 - \alpha N_A^{1/3})$ where α was chosen to make $E_{D_t} = 0$ for $N_A \approx 6 \times 10^{15}$ cm^{-3}. E_I is the ionization energy at infinite dilution and 5.86 meV was used for this value in the calculated curve. The reasonable agreement between this calculated curve and the experimental points, particularly for acceptor concentration $> 10^{14}$ cm^{-3}, could be regarded as support for the hypothesis of overlap with neighboring ionized donors. However, the calculation of Baltensperger[144] (Fig. 29) also gives good agreement with experiment for donor concentrations $\gtrsim 10^{15}$ cm^{-3}, if the thermal ionization energy is identified with the separation between the bottom of the 2p band and the top of the 1s band.

c. *Intentionally Doped Crystals*

The variation of the spectral response described above has presumably all been due to variation in the *concentrations* of chemical donors and acceptors, since these samples were not intentionally doped. As discussed previously in Section 5, the carrier concentration and the donor and acceptor concentrations of samples grown in a particular growth reactor can be varied over a wide range by simply varying the growth conditions and/or the crystalline orientation of the substrate. This is because the segregation coefficients of impurities always present in the system are different for different growth conditions. However, the peak corresponding to the (1s → 2p) transition at 35.5 cm^{-1} observed in the spectra of Fig. 27 does not occur at the same energy as the corresponding transition at 37 cm^{-1} in the absorption spectrum reported by Bosomworth et al.[106] (Fig. 18) for their vapor phase epitaxial material. Other workers have observed the peak in the photoconductivity spectrum at ~37 cm^{-1} in liquid phase epitaxial material,[162] and at ~35.5 cm^{-1} in vapor[163] and liquid[164] epitaxial material prepared at other laboratories, so it appears that some other mechanism must be involved in the shift of the (1s → 2p) peak in addition to the general broadening described above. Summers et al.[160] have studied the far-infrared absorption and photoconductivity of n-type GaAs samples that were grown with the deliberate addition of particular chemical dopants. The properties of some of the samples which they studied are given in Table VII, along with some of the results determined from the absorption and photoconductivity measurements. The spectra obtained for these samples are shown in Fig. 31. The three Ge doped samples were grown in an AsCl$_3$–Ga–H$_2$ vapor phase reactor which, without the deliberate introduction of Ge, produced epitaxial layers that were either high resistivity or p-type with $p \approx 10^{13}$ cm^{-3}. No estimates were made of the donor and acceptor concentrations of this undoped material so that the residual impurity concentrations are not known. The three samples designated M1, M2, and M3 were grown in a similar, but different system which produced n-type layers with growth conditions such that the residual carrier concentrations were about 10^{12} cm^{-3} when no intentional dopant was added. The dopants S and Se were added by introducing H$_2$S and H$_2$Se in the gas stream. The sample M1, designated as Si doped, was grown without intentional doping, but with the growth conditions adjusted so that the carrier concentration was about 1×10^{15} cm^{-3}. It was assumed that the most likely impurity in this case was Si because of the fused silica used in the growth reactor. The (1s → 2p) peak for this sample at 35.2 ± 0.2 cm^{-1} agrees with the value of 35.5 cm^{-1} for the spectra of Fig. 27. It should be noted, however, that Bosomworth et al.[106] also speculated that the dominant residual donor, which produced the peak at ~37 cm^{-1} in their spectrum, was Si. The donor and acceptor concentrations calculated using Fig. 17 and the 77°K mobilities reported by Summers et al.[160] are also given in Table VII. The spectrum for sample NF 177 is characterized by a strong absorption at ~37.4 cm^{-1}, a shoulder at ~44 cm^{-1}, and an increase in the absorption beginning at about 48

[162] J. Pipher, private communication, 1970.
[163] R. Kaplan, private communication, 1970.
[164] R. A. Stradling, private communication, 1970.

TABLE VII

PHYSICAL, ELECTRICAL, AND OPTICAL PROPERTIES OF THE SAMPLES FOR THE ABSORPTION AND PHOTOCONDUCTIVITY SPECTRA OF FIG. 31[a]

Sample	Dopant	Temp. (°K)	n (cm^{-3})	μ (cm^2/V-sec)	t (μm)	N_D^d (cm^{-3})	N_A^d (cm^{-3})	(1s → 2p) (cm^{-1})	$E_I - E_H^f$ (meV)	(1s → cont.) = E_I^f (cm^{-1})
								Transition energies and $E_I - E_H$ from spectral measurements		
				Electrical and physical properties						
NF177[b]	Ge	300	7.7×10^{14}	7500	40	1.8×10^{15}	1.0×10^{15}	37.4 ± 0.2	0.29	48.0 ± 1.0
		77	7.7×10^{14}	48,000					(0.46)	(50.0)
AM94D[b]	Ge	300	2.2×10^{15}	7100	11.2	4.3×10^{15}	2.4×10^{15}	37.3 ± 0.2	0.29	48.2 ± 0.5
		77	1.9×10^{15}	30,000					(0.32)	(48.9)
NF99[b]	Ge	300	1.9×10^{15}	6640	65	—[e]	—[e]	37.4 ± 0.2	0.29	48.2 ± 0.5
									(0.46)	(50.0)
M1[b,c]	Si	300	1×10^{15}	6300	20.5	1.1×10^{15}	0.2×10^{15}	35.2 ± 0.2	0.02	46.5 ± 0.5
		77	9×10^{14}	80,000					(0.06)	(46.8)
M2[c]	Se	300	2.2×10^{14}	6560	24	—[e]	—[e]	35.9 ± 0.2	0.10	47 ± 0.5
									(0.15)	(47.5)
M3[c]	S	300	2.9×10^{14}	8250	4.5	1.3×10^{15}	1.0×10^{15}	37.5 ± 1	0.31	47 ± 0.5
		77		50,000					(0.35)	(49.1)

[a] After Summers et al.[160]
[b] Absorption measurements.
[c] Photoconductivity measurements.
[d] These values were calculated from Fig. 17 using 77°K mobility and carrier concentration when given.
[e] Could not be determined because the 77°K mobility was not given.
[f] The values in parentheses were determined using $R_Y^* = 46.10$ cm^{-1} + nonparabolic corrections of Stillman et al.[57] instead of $R_Y^* \approx 46.7$ cm^{-1}.

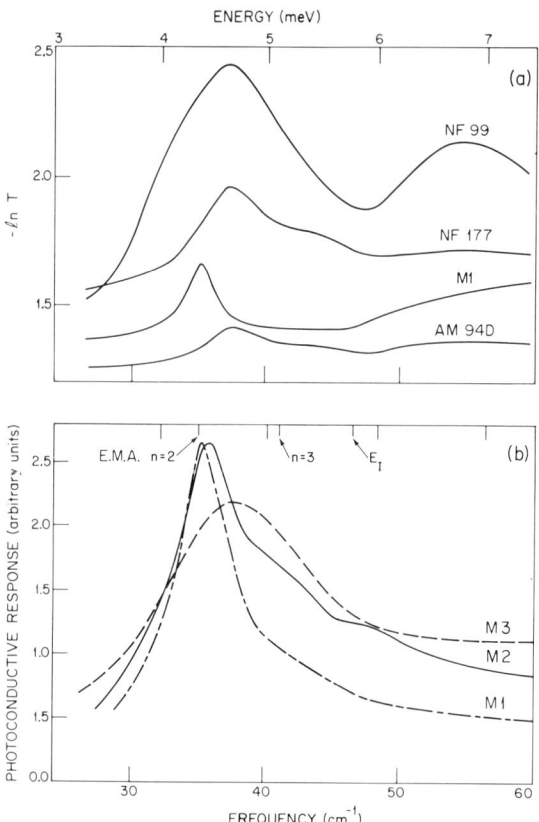

FIG. 31. Far-infrared donor absorption (a) and photoconductivity (b) spectra for the doped GaAs samples listed in Table VII. (After Summers et al.[160])

cm^{-1} which they suggested marked the ionization threshold. The energies of the dominant absorption peak for each sample were identified with (1s → 2p) transitions for different chemical donors in which the 1s ground state energies are slightly different because of small central cell corrections. By using the measured (1s → 2p) transition energies and the energies E_H, calculated from the hydrogenic model with $m^*/m = 0.0665$ and $\varepsilon_0 = 12.5$, they determined the central cell corrections ($E_I - E_H$) and the ionization energies given in Table VII. The values for $E_I - E_H$ and E_I shown in parentheses were obtained using the measured (1s → 2p) transition energies and the more accurate value for E_H of 46.10 cm^{-1}.[57] It is interesting to note that the absorption spectrum of Bosomworth et al.[106] shown in Fig. 18 does not show any indication of an increase in absorption in the spectral region corresponding to the ionization threshold. The absorption calculated from the hydrogenic model (Eq. 20) also decreases monotonically for energies above the ionization threshold so that the spectra for samples NF 99 and M1 are particularly unusual. The reasons for the differences in the measured spectra are not understood. In view of the high resolution measurements of the excited state photoconductivity due to residual donors discussed in Section 10, it is clear that the ionization energies obtained by Summers et al.[160] for the particular dopants studied do not apply to the corresponding donors. More experiments will have to be done on higher purity intentionally doped material to make an accurate identification of the residual donors in GaAs and an accurate determination of the central cell corrections and ionization energies of specific donor species. High resolution measurements of the type described in Section 10 should be particularly useful for these experiments.

d. Long Wavelength Response

There is no adequate explanation of the photoconductive response on the low energy side of the (1s → 2p) peak apparent in the spectra of Fig. 27, but what is known about this response will be described, and one possible model will be suggested. Similar transitions have been seen in high purity GaAs prepared by liquid phase epitaxial techniques[162,164] as well as in that prepared by vapor phase epitaxial techniques, so the long wavelength response does not seem to be dependent on the method of material preparation. From Fig. 27 it is evident that the long wavelength response consists of a discrete transition at ~ 26.5 cm^{-1} and a shoulder at ~ 32 cm^{-1}. The photoresponse from these transitions relative to that of the (1s → 2p) transition increases with increasing donor concentration, and in the sample with the highest donor concentration the shoulder at ~ 32 cm^{-1} has become comparable to the (1s → 2p) peak. The observation of the discrete peaks at the low temperature used for these measurements leads us to the conclusion that the initial state involved in these transitions must either be the ground state of the donor center or at least be nearly degenerate with the ground state energy. The temperature dependence of the long wavelength photoconductivity is also consistent with the photothermal ionization of some state between the ground state and first excited state of the donor center.

A possible model which can explain these results is shown in Fig. 32. This model consists of the usual shallow donors with their excited states which contribute to the photoresponse through photothermal ionization in the usual way, and, in addition, a large concentration of electrically inactive or isoelectronic-type centers which can serve as intermediate states for optically excited electrons. The origin of these centers could possibly be N or some complex involving O,[165] or pairs of C atoms situated on

[165] M. E. Weiner and A. S. Jordan, *J. Appl. Phys.* **43**, 1767 (1972).

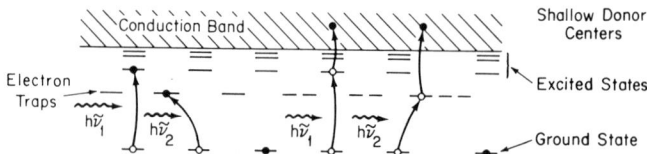

FIG. 32. Electron trap model to explain long wavelength photoconductivity. The usual photoconductivity results from the excitation of electrons from the ground states to the excited states of neutral shallow donors, but in addition the transitions from the donor ground states to the ground states of the electron traps produce photoconductive response at longer wavelengths.

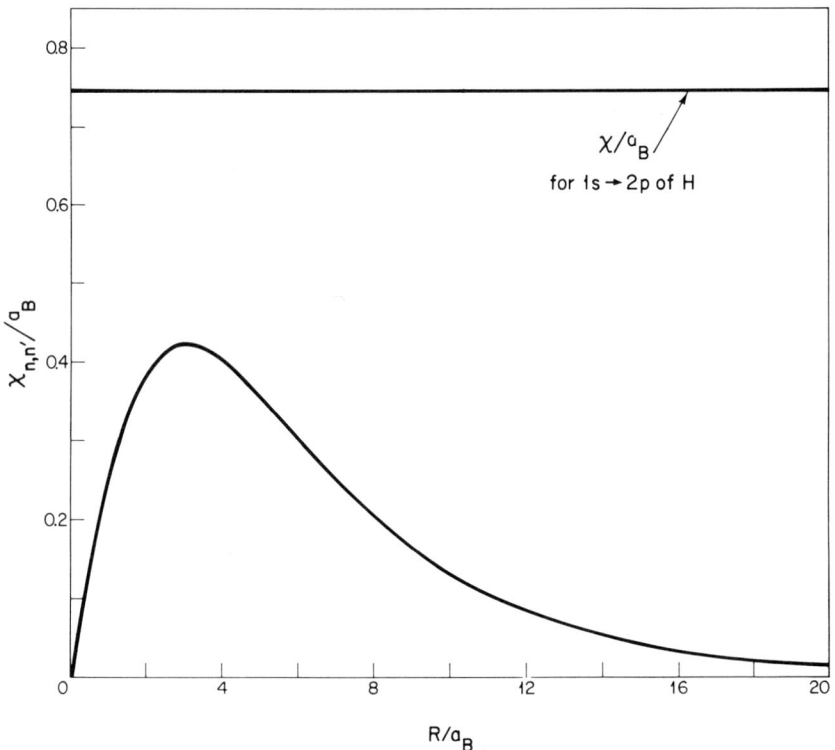

FIG. 33. Calculated matrix element for the transition of an electron from the donor ground state to the isoelectronic trap as a function of the separation between the neutral shallow donors and isoelectronic traps. The matrix element and separation have been normalized by the Bohr radius of the donors a_B. Also shown is the value of the matrix element for the (1s → 2p) transition of hydrogenic donors.

adjacent Ga and As sites, since high purity GaAs contains a large amount of electrically inactive C.[166] Although such centers may not be strictly isoelectronic they have the same properties and act as electron traps. The long wavelength transitions in this model correspond to transitions between the ground states of the shallow donor centers and the ground states of the electron trap centers ($1s_D \rightarrow 1s_I$), followed by subsequent thermal excitation into the conduction band. The ratio of the calculated matrix element for the transition from the donor ground state to the electron trap and a_B, $\chi_{n,n'}/a_B$, is shown in Fig. 33 as a function of the ratio of the distance R between the two types of centers to a_B, R/a_B. A δ-function potential was assumed for the electron trap, and a_B is the Bohr radius of the donor. The maximum occurs for $R/a_B = 3$ and is over one-half of that for the (1s $\rightarrow$ 2p) donor transition. The corresponding value of the matrix element for the (1s $\rightarrow$ 2p) transition of the hydrogenic donor is also shown for comparison. If we assume that the number of these isoelectronic centers is essentially constant, the relative increase of the peak corresponding to the ($1s_D \rightarrow 1s_I$) transition with increasing donor concentration would be explained by the occurrence of a larger number of neutral donor centers within about three Bohr radii of the electron trap centers. This behavior is in agreement with the observation that the long wavelength peaks are more prominent in less compensated samples with the same total donor concentration. The occurrence of more than one long wavelength peak is due to other different types of electron traps with different energy levels.

In a magnetic field, the long wavelength peaks do not split, but remain essentially constant in energy. As the magnetic field is increased to relatively large values (~ 20 kG) the amplitudes of the long wavelength peaks decrease, which is qualitatively consistent with the shrinking of the donor ground state wave function and a corresponding decrease in the oscillator strength (matrix element). The lack of a shift or splitting in a magnetic field is consistent with the transition between two s-like states. (Since the two states occur on different centers, the usual selection rule $\Delta l = \pm 1$ does not hold.)

10. Magnetic Field Effects

The occurrence of excited state photoconductivity in GaAs was suggested by the observation of discrete peaks in the photoconductivity spectrum and by the good agreement between the observed transition energies and those calculated from the hydrogenic model.[38] The identification of this mechanism was confirmed by the observation of the Zeeman splitting of the (1s $\rightarrow$ 2p) transition and these and other measurements in a magnetic field have yielded a considerable amount of information about the shallow donors in GaAs.

[166] C. M. Wolfe, G. E. Stillman, and E. B. Owens, *J. Electrochem. Soc.* **117**, 129 (1970).

a. Zeeman Effect

Zeeman effect measurements were made on Sample 6 of Table V,[38,66] and the results are shown in Fig. 34. These spectra were taken at a resolution of about 1 cm^{-1}. The sample was mounted in a light pipe at the center of a superconducting solenoid in effectively a combined Faraday–Voigt configuration so that all transitions from the 1s ground state to higher p-states were allowed. In a magnetic field the dominant peak splits into three distinct components corresponding to the transitions from the 1s to the (2p, m = ± 1, 0) states. At low magnetic fields a considerable amount of structure exists at the higher energies which probably results from transitions to higher donor

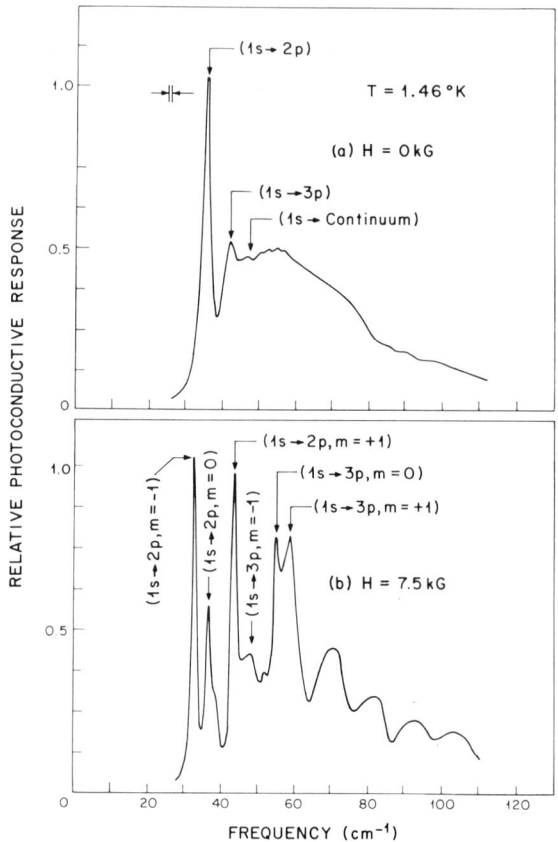

FIG. 34. Magnetic field dependence of the photoconductivity spectra of sample 6 in Table V at 1.46°K and (a) $H = 0$, (b) $H = 7.5$ kG, (c) $H = 15.0$ kG, and (d) $H = 29.9$ kG. The transitions designated as (1s → 3p, m = 0, ± 1) are now known to not correspond to these states over the whole magnetic field range. (After Stillman et al.[66])

excited states and to Landau states in the conduction band continuum. At higher magnetic fields three smaller peaks at shorter wavelengths become distinct, and these peaks were tentatively identified as due to (1s → 3p, m = 0, ±1) transitions. More recent work indicates that the peak identified as (1s → 3p, m = 0) does not correspond to a transition with $\Delta m = 0$,[167] and this will be discussed below. The theory of Hasegawa and Howard[168] for the behavior of a hydrogenic atom in high magnetic fields predicts that at high magnetic fields the peak corresponding to the (1s → 2p, m = ±1) transition should become dominant since this is the only transition which has finite oscillator strength in the limit of infinite magnetic field. In addition, the amplitude of the (1s → 2p, m = −1,0) transition may decrease because of

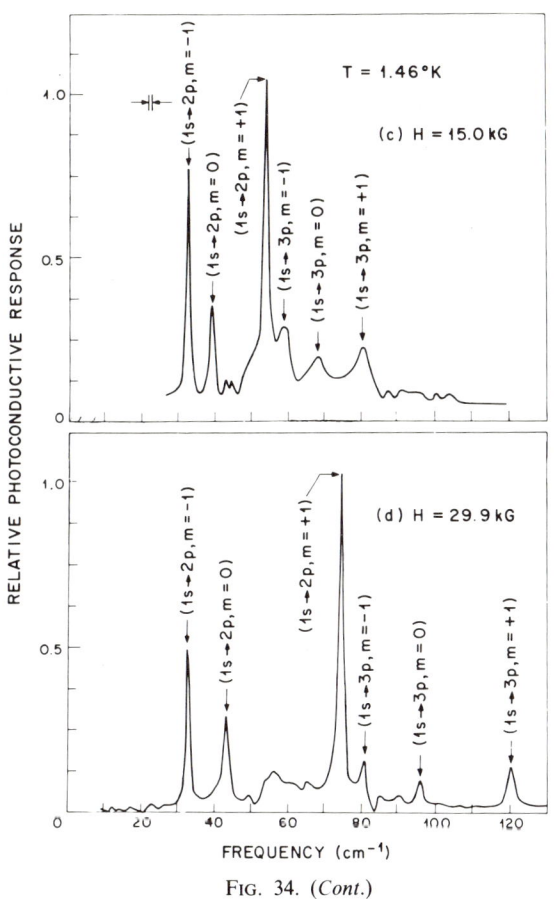

Fig. 34. (Cont.)

[167] S. Narita and M. Miyao, Solid State Commun. 9, 2161 (1971).
[168] H. Hasegawa and R. E. Howard, J. Phys. Chem. Solids 21, 179 (1961).

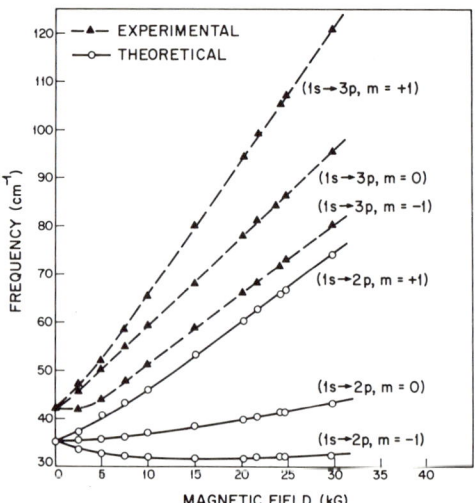

FIG. 35. Energies of the (1s → 2p) and higher donor transitions as a function of magnetic field. The solid curves are from the variational calculations of Larsen and the data points are from the experimental measurements. The higher transitions were initially identified as (1s → 3p) transitions. (After Stillman et al.[66])

the decrease in the photothermal ionization probability of the (1s → 2p, m = −1, 0) transitions due to the increase in the binding energy of the (2p, m = −1) and (2p, m = 0) states with increasing magnetic field. The same decrease is not expected for the (2p, m = +1) state because it becomes degenerate with the continuum of the lowest Landau level.

The variations of the experimental energies of the (1s → 2p) and (1s → 3p) transitions with magnetic field are shown in Fig. 35. The peaks labeled (1s → 2p) and (1s → 3p) in Fig. 34a each split into three components, and the splitting of both the (1s → 2p, m = ±1) and the (1s → 3p, m = ±1) transitions are equal and linear in magnetic field. The normal Zeeman splitting of the m = ±1 states of a hydrogenic donor is given by

$$\Delta E_{+1,-1} = E(2p_+) - E(2p_-) = e\hbar H/m_0^* c, \tag{40}$$

where m_0^* is the electron effective mass at the bottom of the conduction band. Using this equation and a least squares straight line fit to the data shown in Fig. 35, an effective mass of $m_0^*/m = 0.0665 \pm 0.0005$ was determined. The same value of the effective mass has been obtained from Zeeman measurements on less pure samples. The energies of the ground state (1s) and the two excited states (2p, m = ±1) of a hydrogenic impurity in an arbitrarily large magnetic field have been calculated variationally by

Larsen,[169] and more recently the energy of the (2p, m = 0) state has also been calculated using the same approach.[170] The solid curves in Fig. 35 are the results of Larsen's calculation applied to GaAs using the above value for the effective mass. It can be seen that there is extremely good agreement between theory and experiment.

The photoconductivity spectra of less pure, but uncharacterized samples of GaAs were measured by Kaplan et al.[171] in magnetic fields up to about 50 kG. In this material the dominant peak at zero magnetic field also split into three components when the magnetic field was applied. Only one additional more diffuse line appearing at higher energies could be followed consistently as the magnetic field was increased. No attempt was made to determine the effective mass from the Zeeman splitting. Nevertheless, good agreement was obtained between the experimental energy levels and those calculated from Larsen's theory using the value $m^* = 0.0675$ m. Kaplan et al. used the high magnetic field quantum numbers (NMλ) instead of the usual quantum numbers for hydrogenic atoms (n/m) to describe the energy levels. The high magnetic field limit is defined by $\gamma \gg 1$, and the low magnetic field limit by $\gamma \ll 1$, where $\gamma = \hbar\omega_c/2$ R$\hat{y}$, ω_c is the cyclotron frequency, and R$\hat{y}$ is the effective Rydberg for the hydrogenic donors. In general, there is no convincing answer as to how the low field hydrogenic levels (n/m) correspond to the high field states (NMλ), but there is general agreement that the 2p (m = +1), 2p(m = 0), and 2p(m = −1) states correspond to the high field states (110), (001), and (0$\bar{1}$0), respectively. The problem of the correspondence of the low field and high field states has been considered by Kleiner,[172] and by Elliott and Loudon[173] using the concept of "nodal-surface conservation" and also by Boyle and Howard.[174] The correspondences obtained are given in Table VIII. The variational calculations of the energies of states higher than the 2p states using the method of Larsen is difficult, but recent calculations of Larsen[170] indicate that the concept of nodal-surface conservation is not valid. The energies of some of the higher states have been calculated in a simplified way by Narita and Miyao,[175] and the results of this calculation are shown in Fig. 36. The dashed curves were computed using the Kohn–Luttinger (KL)-type[176] trial functions, while the solid curves were obtained using Yafet–Keyes–Adams (YKA)-type[177]

[169] D. M. Larsen, J. Phys. Chem. Solids 29, 271 (1968).
[170] D. M. Larsen, private communication, 1970.
[171] R. Kaplan, M. A. Kinch, and W. C. Scott, Solid State Commun. 7, 883 (1969).
[172] W. H. Kleiner, Lincoln Lab., M.I.T., Progr. Rep., February 1958.
[173] R. J. Elliott and R. Loudon, J. Phys. Chem. Solids 15, 196 (1960).
[174] W. S. Boyle and R. E. Howard, J. Phys. Chem. Solids 19, 181 (1961).
[175] S. Narita and M. Miyao, Solid State Commun. 9, 2161 (1971).
[176] W. Kohn and J. M. Luttinger, Phys. Rev. 98, 915 (1955).
[177] Y. Yafet, R. W. Keyes, and E. N. Adams, J. Phys. Chem. Solids 1, 137 (1956).

TABLE VIII

CORRESPONDENCE BETWEEN THE LOW-FIELD
HYDROGEN-ATOMIC LEVELS (n/m) AND THE
HIGH-FIELD LEVELS (NMλ)[a]

(n/m)	(NMλ)[b]	(NMλ)[c]
3p (m = +1)	210	112
3p (m = 0)	101	003
3p (m = −1)	1$\bar{1}$0	0$\bar{1}$2
2p (m = +1)	110	110
2p (m = 0)	001	001
2p (m = −1)	0$\bar{1}$0	0$\bar{1}$0
1s	000	000

[a] After Narita and Miyao.[175]
[b] Kleiner[172] and Elliott and Loudon.[173]
[c] Boyle and Howard.[174]

trial functions. The true energy values should be closer to the lower of the two calculated curves, and the lower curves agree well with the energies for the (000) state from Larsen's calculation[169] and for the (2p, m = ±1) state from the calculations of Stillman et al.[38] shown as closed circles in Fig. 36. The overly simplified wave functions used by Narita and Miyao[175] can be criticized on theoretical grounds. Nevertheless, the agreement between the calculated and experimental values shown in Fig. 36b is very good. The middle transition of the three higher energy lines agrees with the calculated curve for the state (112): a state with m = 1 (in n/m) instead of with the state (101) as previously identified.[38] Comparisons of the photoconductivity spectra in predominantly Voigt and predominantly Faraday configurations[175] confirmed that the previous indentification of this transition as (1s → 3p, m = 0) was incorrect. Recent calculations by Larsen[170] indicate that the (1s → 3p, m = ±1) transitions were also incorrectly identified and that the (3p, m = −1, 0) states occur at energies between the (2p, m = 0) and (2p, m = +1) states. They also show that the transition previously identified as (1s → 3p, m = 0) actually corresponds to the (1s → 3p, m = +1) transition. Some of these transitions are labeled in the spectra of Fig. 37. (The origin of the structure apparent in the $2p_{-1}$, $2p_0$, and $2p_{+1}$ transitions in these spectra is discussed in the next section.) The first 14 energy levels of a hydrogenic atom have also recently been calculated by Praddaude,[177a] and, except for the 3s and 3d (m = 0) levels, his results agree with those of Larsen. A thorough study of the higher energy transitions could help clarify the correspondence between the high and low field states.

[177a] H. C. Praddaude, *Phys. Rev. A* **6**, 1321 (1972).

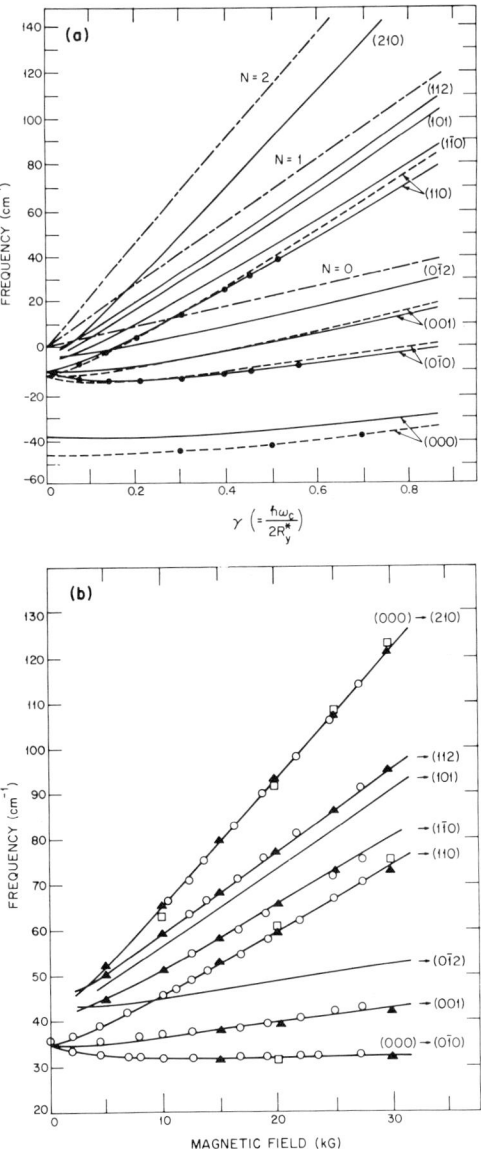

FIG. 36. (a) Computed energies for donor energy levels in a magnetic field. Solid curves and dashed curves are from variational calculations using YKA and KL type trial functions, respectively. Closed circles are from Stillman et al.[66] (After Narita and Miyao.[175]) (b) Calculated transition energies from the variational calculation of Narita and Miyao and experimental transition energies measured by Narita and Miyao, ○ ; Kaplan et al.,[171] □ ; and Stillman et al.,[66] ▲ . (After Narita and Miyao.[175])

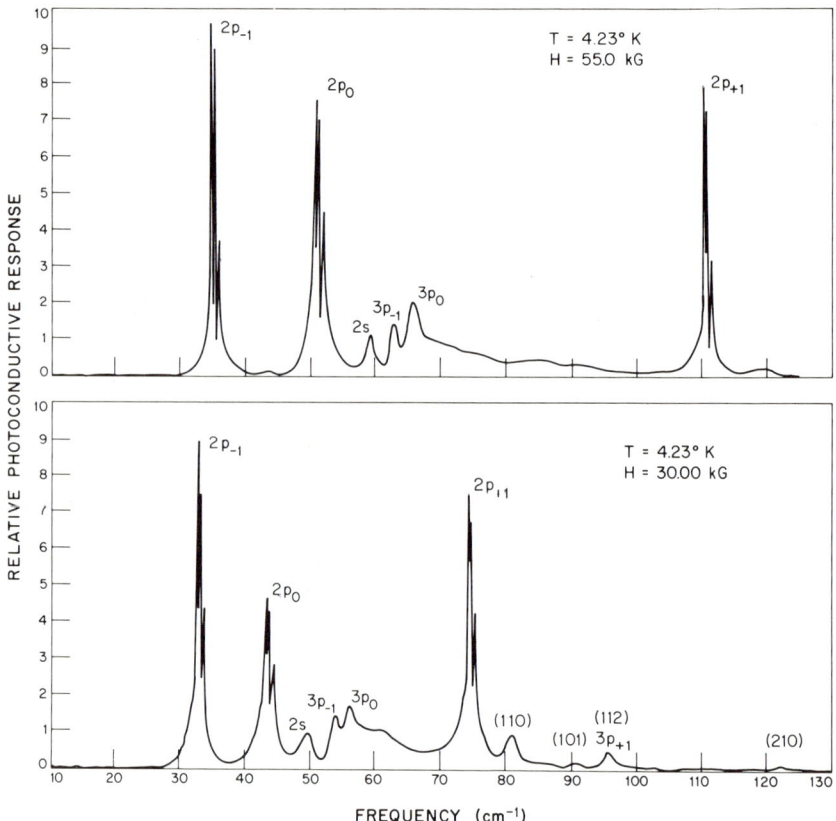

FIG. 37. High resolution (∼ 0.16 cm⁻¹) photoconductivity spectra of a high purity GaAs sample at 30.0 and 55.0 kG showing the identification of some of the peaks resulting from transitions from the ground state to some of the higher energy levels.

b. High Resolution Measurements

The observation of a small shoulder on the *short* wavelength side of the (1s → 2p) transition in high purity GaAs,[61] along with the observation of different energies for the (1s → 2p) transition peak in material doped with specific donors,[160] indicated that there was some deviation of the behavior of shallow donors in GaAs from that predicted by the hydrogenic model. Under high resolution the single (1s → 2p) transition observed at lower resolution is seen to consist of several different, previously unresolved transitions, each presumably arising from the (1s → 2p) transition of a different residual donor species. The first high resolution measurements were made by Fetterman *et al.*[178] using a far-infrared laser for both photoconductivity and absorption in a magnetic field. The magnetic field was precisely deter-

[178] H. R. Fetterman, D. M. Larsen, G. E. Stillman, P. E. Tannenwald, and J. Waldman, *Phys. Rev. Lett.* **26**, 975 (1971).

4. FAR-INFRARED PHOTOCONDUCTIVITY IN HIGH PURITY GaAs   249

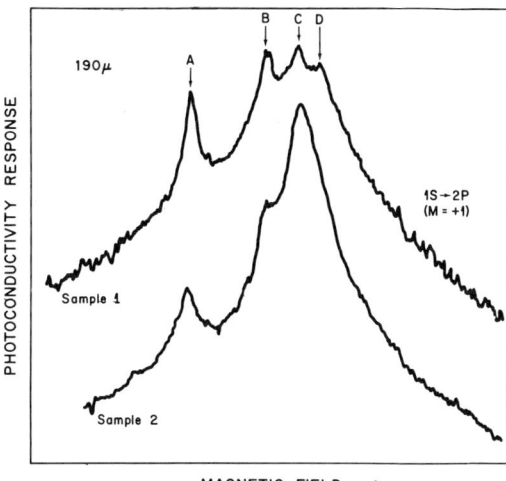

FIG. 38. The photoconductivity due to the (1s → 2p, m = + 1) transition as a function of magnetic field for two different samples when excited by 190 μm laser radiation. For sample 1, $N_D = 2.0 \times 10^{14}$ cm^{-3}, $N_A = 4.0 \times 10^{13}$ cm^{-3}, $\mu_{77°K} = 153{,}000$ cm^2/V-sec. For sample 2, $N_D = 4.3 \times 10^{13}$ cm^{-3}, $N_A = 2.5 \times 10^{13}$ cm^{-3}, and $\mu_{77°K} = 180{,}000$ cm^2/V-sec. (After Fetterman et al.[178])

mined by use of NMR techniques (∼ ±15 G) and the transitions were observed at the same magnetic fields in both photoconductivity and transmission, indicating that there is no shift of the excited state photoconductivity peaks from the corresponding absorption maxima, as observed in some of the early work on Ge.[11,127] The variation of the photoconductivity corresponding to the (1s → 2p, m = + 1) transition with a magnetic field is shown in Fig. 38 for two different high purity GaAs samples. The laser frequency for these spectra was 52.65 cm^{-1} (190 μm) and the magnetic field shown covers a range of a little more than 1 kG with 14.63 kG corresponding to the field at the dominant peak. These transitions as well as the (1s → 2p, m = 0) transitions were studied with other laser frequencies from 51.36 cm^{-1} (195 μm) to 127.48 cm^{-1} (78.4 μm) and corresponding magnetic fields in the range 13–67 kG. The different transitions labeled A, B, C, and D each correspond to a different donor species, and variation of the amplitude of these transitions in the different samples results from different relative concentrations of the particular donor species. The separation between the different peaks in a given transition group (1s → 2p, m = +1 or 1s → 2p, m = 0) increased with increasing magnetic field and this increase was interpreted as a magnetic field dependence of the small central cell corrections responsible for the different transitions observed for different donor species. By treating the central cell potential, $v_j(\mathbf{r})$, as a weak short-range perturbation on the Coulomb potential of the donor species j, and writing the effective mass wave function Ψ_{em} as the product of the band edge Bloch functions

$U_1(\mathbf{r})$, and the envelope function for the ground state of a hydrogenic donor in a magnetic field $\varphi_1(\mathbf{r})$, the central cell energy correction was calculated as

$$\langle \Psi_{em} | v_j(\mathbf{r}) | \Psi_{em} \rangle = \kappa_j |\varphi_1(0)|^2, \tag{41}$$

where κ_j is $\langle U_1(\mathbf{r}) | v_j(\mathbf{r}) | U_1(\mathbf{r}) \rangle$. Although κ_j is different for different donor species, it is independent of magnetic field. The difference between the central cell corrections $\Delta E_{ji}(H)$ for two different donor species j and i at magnetic field H can be determined from the experimental energy separations. In terms of Eq. (41) this energy difference is given by

$$\Delta E_{ji}(H) = E_j - E_i = (\kappa_j - \kappa_i)|\varphi_1(0)|^2 \equiv K_{ji}|\varphi_1(0)|^2, \tag{42}$$

where E_j is the energy of the transition associated with donor j. The value of $|\varphi_1(0)|^2$ was calculated using the variational, parabolic band, ground state wave functions of Larsen,[169] and the value of K_{ji} was then determined using the experimental values for $\Delta E_{ji}(H)$. Although the energy separation of peaks labeled A and C in Fig. 38 increased from 1 to 1.42 cm^{-1} (a 42% increase) as the magnetic field was increased from about 13 to 67 kG, the value of K_{ji} remained essentially constant, confirming that the splitting observed in the (1s → 2p) transition was due to central cell corrections and that the increase in the splitting with increasing magnetic field was due to the shrinking of the wave function of the bound electron into the central cell by the magnetic field.

c. *Determination of Effective Rydberg for GaAs*

Although the electron effective mass is known very precisely for GaAs, there is, as indicated in Section 4, considerable uncertainty in the static dielectric constant, and this makes it impossible to accurately calculate the hydrogenic binding energy or effective Rydberg for shallow donors in GaAs from Eq. (1). The usual way of determining the Rydberg experimentally is to measure accurately the energy between discrete excited states (the Balmer series of the donors) and then calculate the Rydberg using the hydrogenic model, since the p-like excited states observed in optical measurements are not affected by central cell corrections. In an attempt to use this procedure in GaAs by making high resolution measurements of the (1s → 2p) and (1s → 3p) transitions it was found that the transitions to the 3p states could not be resolved sufficiently well to obtain accurate results. The spectra obtained for two different high purity samples are shown in Fig. 39. Although the structure observed in the (1s → 3p) transition is similar to that in the (1s → 2p) transition (which results from the different residual donors), the difference in the spectra of these two samples of comparable purity indicates that other effects are involved. Thus, it was impossible to accurately determine the Rydberg in this way. However, the effective Rydberg for GaAs

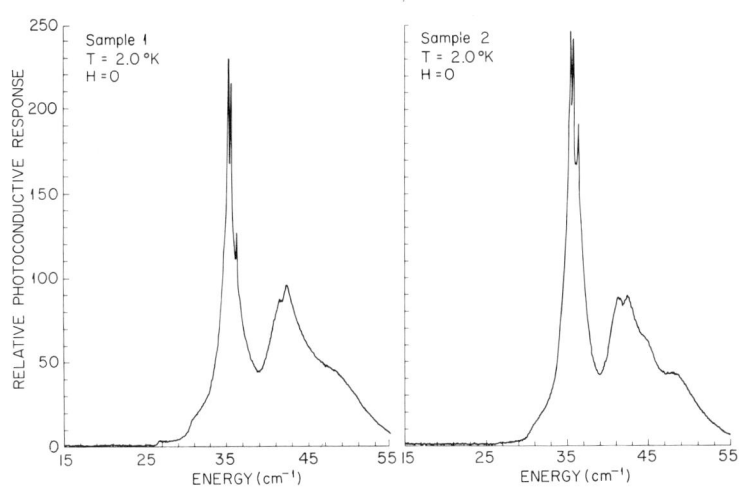

FIG. 39. High resolution photoconductivity spectra of two high purity epitaxial GaAs samples at 20°K and zero magnetic field. The decrease in the response for energies greater than 45 cm^{-1} in these spectra is due to filters used for these measurements.

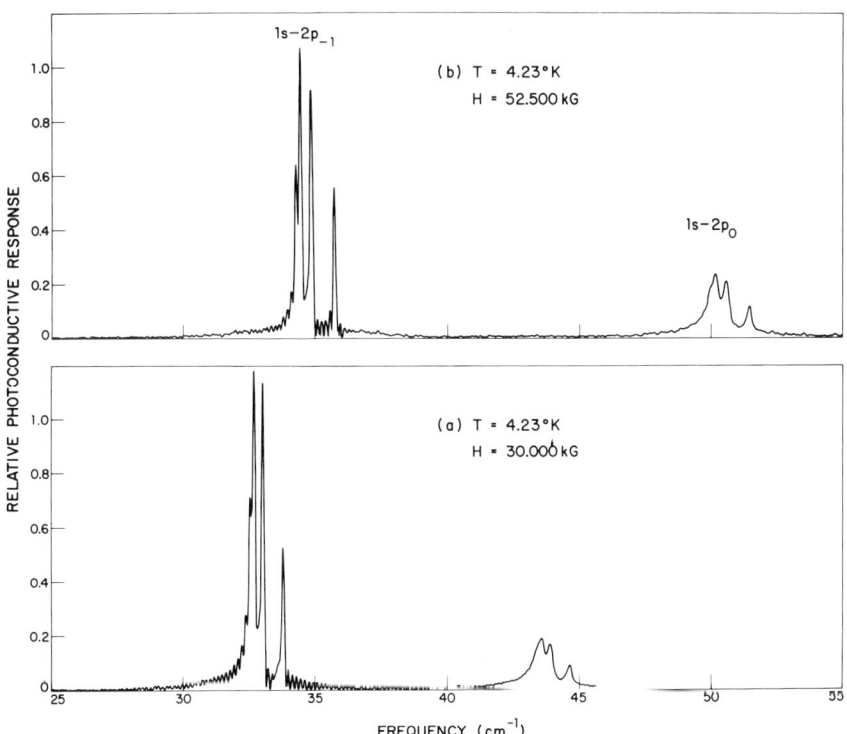

FIG. 40. High resolution (0.076 cm^{-1}) photoconductivity spectra at magnetic fields of (a) 30.000 kG and (b) 52.500 kG for an unintentionally doped high purity GaAs sample (sample 6 of Table V) with four different donor species. (After Stillman et al.[57])

donors has been determined by analyzing the Zeeman splitting of the 2p levels.[57]

Typical high resolution measurements of the photoconductivity spectra in a magnetic field are shown in Fig. 40. The spectra in this figure were measured at 4.23°K and magnetic fields of 30.000 and 52.500 kG on the sample listed as sample 6 in Table V, but measurements were made throughout the range 0–55 kG. The calculated resolution of these spectra was 0.076 cm^{-1}, and the spectral bandwidth was limited to less than 70 cm^{-1}. The (1s → 2p$_{+1}$) transitions where the +1 indicates the value of the quantum number m, occur out of this spectral range for these magnetic fields. Nuclear magnetic resonance techniques were used to measure the magnetic field to an accuracy of about 0.02%. The structure in the lower energy group of (1s → 2p$_{-1}$) transitions is similar to that in the higher energy group of (1s → 2p$_0$) transitions and is due to transitions from the ground states of the four different donor species present in this high purity GaAs. The transitions occur at slightly different energies because of small, but different central cell shifts of the ground states.[178] From measurements such as this it was possible to determine the energy difference between the 2p$_0$ and 2p$_{-1}$ levels to an accuracy of about 0.01 cm^{-1} for the three deepest ground states. Although four different (1s → 2p$_{-1}$) transitions were well-resolved at high magnetic fields, indicating that there are four donors present, the (1s → 2p$_0$) transition for the shallowest residual donor was not well-resolved, so the (2p$_0$ → 2p$_{-1}$) energy difference could not be determined for this donor species. No dependence of the (2p$_0$ → 2p$_{-1}$) energy difference on donor species was observed.

The energy difference between the two lowest energy 2p levels can be written as

$$\Delta E_{0,-1} \equiv E(2p_0) - E(2p_{-1}) \equiv f(\gamma) R\overset{*}{y} \qquad (43)$$

where $E(2p_0)$ and $E(2p_{-1})$ are the energies of the states which approach the corresponding zero magnetic field hydrogenic 2p states as $H \to 0$,[57] and $R\overset{*}{y}$ is the effective hydrogenic Rydberg as given by Eq. (1). The function $f(\gamma)$, where as before $\gamma = \hbar\omega_c/2R\overset{*}{y}$ with $\omega_c = eH/m^*c$, can be calculated quite accurately using variational methods to determine the energy eigenvalues $E(2p_0)$ and $E(2p_{-1})$ in the hydrogenic *effective mass approximation*. The (2p$_0$ → 2p$_{-1}$) energy splitting depends on the effective Rydberg (or on m_0^* and ε_0) both directly through $R\overset{*}{y}$, and indirectly through $f(\gamma)$, in contrast to the splitting of the (2p, m = ± 1) levels which depends only on the electron effective mass m_0^*. Thus, by accurately measuring the experimental energy splitting $\Delta E_{0,-1}$ and adjusting the value of $R\overset{*}{y}$ for best agreement in Eq. (43) using the calculated value for $f(\gamma)$, it is possible to determine the effective Rydberg for donors in GaAs.

The results of the experimental determination of $\Delta E_{0,-1}$ and the theoretical

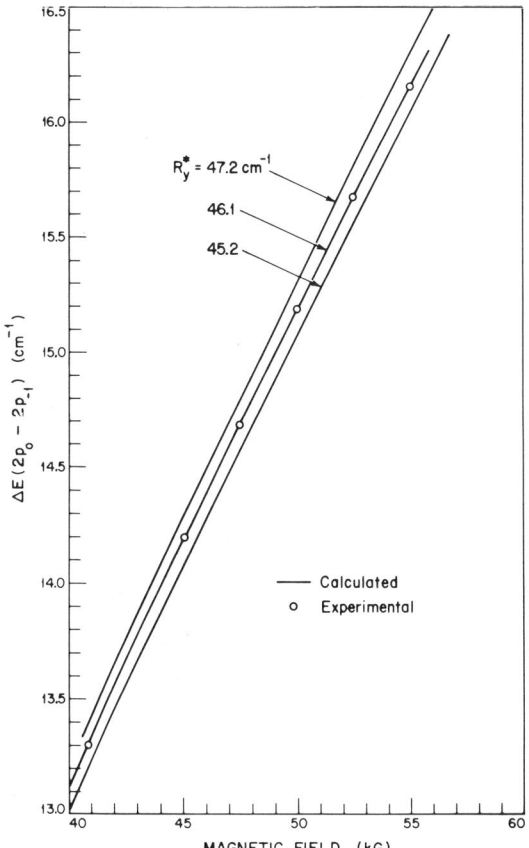

FIG. 41. Theoretical (effective mass approximation) and experimental $(2p_0 - 2p_{-1})$ energy differences for shallow donors in GaAs as a function of magnetic field. The only adjustable parameter in the calculated values is $R\overset{*}{y}$. (After Stillman et al.[57])

calculations for three different values of $R\overset{*}{y}$ are shown in Fig. 41. Although the results in Fig. 41 are only shown over the magnetic field range 40–55 kG so that an expanded scale could be used, the measurements and calculations were actually made over the range 20–55 kG and agreement between theory and experiment is better over the entire range than can be shown in the figure. The theoretical fit is never worse than ± 0.015 cm^{-1} in this range for $R\overset{*}{y} = 46.10$ cm^{-1} and this corresponds to a precision in $R\overset{*}{y}$ of $\sim \pm 0.15$ cm^{-1}. Using the value $m_0^*/m = 0.06650 \pm 0.00005$, this value of $R\overset{*}{y}$ corresponds to $\varepsilon_0 = 12.56 \pm 0.04$ for the static dielectric constant.

Central cell corrections as well as the binding energy of the different residual donor species can be determined using this value of $R\overset{*}{y}$. The energy

TABLE IX

CENTRAL CELL CORRECTIONS TO THE PARABOLIC BAND R_y^* ($= 46.10$ cm^{-1}), FOR FOUR RESIDUAL DONORS IN HIGH PURITY GaAs[a]

Donor	Central cell correction	Ionization energy[b,c]
1	0.064 meV (0.52 cm^{-1})	5.800 meV (46.79 cm^{-1})
2	0.081 meV (0.65 cm^{-1})	5.817 meV (46.92 cm^{-1})
3	0.117 meV (0.94 cm^{-1})	5.854 meV (47.22 cm^{-1})
4	0.200 meV (1.61 cm^{-1})	5.937 meV (47.89 cm^{-1})

[a] After Stillman et al.[57]
[b] The ionization energies include the deepening of the 1s state due to both central cell shifts and conduction band nonparabolicity.
[c] The magnitudes of the ionization energies are accurate within about ± 0.15 cm^{-1}, while the differences between the energies are known to about ± 0.01 cm^{-1}.

difference between the (1s) ground state and the (2p) excited states in the effective mass approximation is $(0.75 + 8.12 \times 10^{-5} R_y^*) R_y^*$, where the term $8.12 \times 10^{-5} (R_y^*)^2$ is the nonparabolic correction to the (1s → 2p) transition, most of which results from the deepening of the ground state. The central cell corrections can be determined by subtracting this energy difference from the experimental (1s → 2p) transition energy, and the ionization energies can be determined by adding the central cell corrections and the nonparabolic corrections to the parabolic band value of 46.1 cm^{-1}. The values obtained in this way are given in Table IX. The observation of the extremely sharp lines due to four residual donors in this high purity GaAs casts doubt on the identification of the broad peaks observed in earlier measurements with particular chemical donors and on the previous determination of the ionization energies of particular donor species.[160]

d. *Effects of Ionized Impurities and Internal Electric Fields*

The high resolution photoconductivity measurements in GaAs were examined carefully in the low magnetic field range ($\lesssim 20$ kG) where nonparabolic corrections are negligible[179] to obtain an independent verification of the cyclotron effective mass $m_0^*/m = 0.06650 \pm 0.00005$.[36] With the use

[179] G. E. Stillman, D. M. Larsen, and C. M. Wolfe, *Phys. Rev. Lett.* **27**, 989 (1971).

4. FAR-INFRARED PHOTOCONDUCTIVITY IN HIGH PURITY GaAs

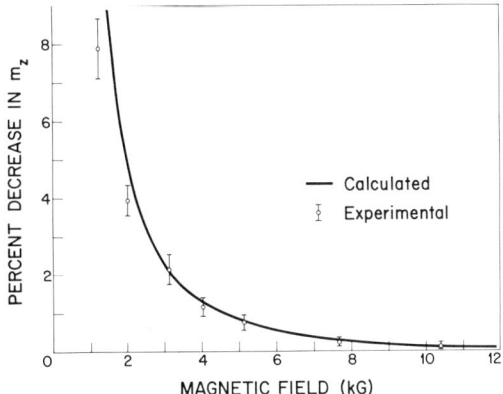

FIG. 42. Percent deviation of the Zeeman mass from the cyclotron effective mass, $[(m_0^* - m_z^*)/m_0^*] \times 100$, as a function of magnetic field for a high purity GaAs sample with $N_D = 2.5 \times 10^{13}$ cm^{-3} and $N_A = 2.2 \times 10^{13}$ cm^{-3}. (After Stillman et al.[179])

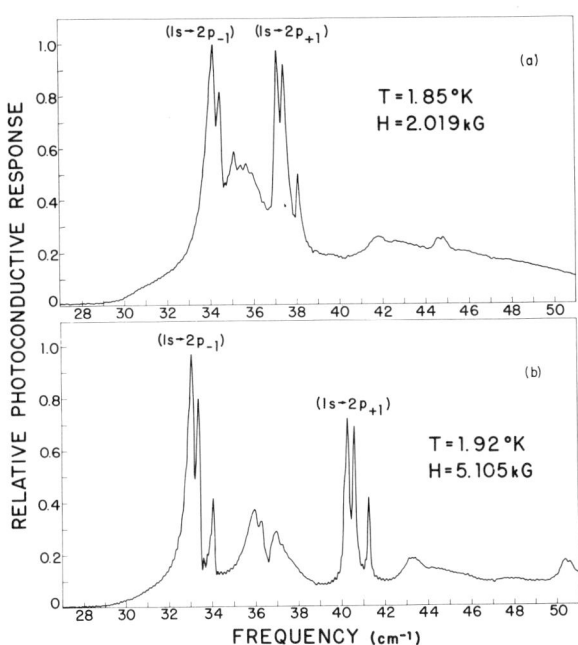

FIG. 43. Photoconductivity spectra at two low magnetic fields for the sample used in Fig. 42. The groups of peaks labeled (1s → 2p$_{-1}$) and (1s → 2p$_{+1}$) result from transitions from the slightly different ground state energies of four different residual donor species present in this unintentionally doped high purity GaAs. The calculated resolution for these spectra was 0.076 cm^{-1}, and only transitions from three of the donor species were well resolved. (After Stillman et al.[179])

of NMR techniques to determine the magnetic field very accurately, the high resolution measurements of the Zeeman splitting of the (1s → $2p_{\pm 1}$) transitions in magnetic fields between about 10 and 20 kG yielded effective mass values in agreement with the cyclotron mass value. However, at magnetic fields less than about 10 kG, the "Zeeman effective mass" not only did not agree with the cyclotron resonance value, but was also magnetic field dependent. The deviation of the Zeeman mass from the cyclotron resonance value over the magnetic field range from about 1 to 10 kG is shown in Fig. 42. The experimental points in this figure were calculated using the value of the Zeeman effective mass determined from Eq. (40) and the value of $\Delta E_{+1,-1}$ obtained from the (1s → $2p_{+1}$) and (1s → $2p_{-1}$) donor transition lines. Two typical spectra used for this determination are shown in Fig. 43. As the magnetic field is decreased in this low magnetic field range, the $2p_{-1}$ and the $2p_{+1}$ lines become broader and more asymmetric with significant tails to lower and higher energies, respectively. The structure between these two lines may cause the measured values for the energy splitting, $\Delta E_{+1,-1}$ to be too small, but the error bars shown indicate only the uncertainties in the measurements of the magnetic field and in the determination of the peak energies.

This variation in the Zeeman mass, or, equivalently, the energy perturbation of the $2p_{\pm 1}$ states was interpreted as being due to the Stark effect on the shallow donor states resulting from the internal electric fields. The internal electric fields due to point charges in semiconductors were first considered by Redfield.[180] To explain the magnetic field variation of the Zeeman mass shown in Fig. 42, the internal electric field in the sample resulting from ionized donors and acceptors was treated as uniform over the region of space occupied by the 2p states, since for this sample the mean spacing between the donors and acceptors was $r_0 \approx (N_D + N_A)^{-1/3} \approx 2400$ Å, while $a_B \approx 100$ Å. The energy splitting of the $2p_{\pm 1}$ states was written as

$$\Delta E_{\pm 1} = \hbar\omega_c [1 + |M|^2 / \{[E(2p_{+1}) - E(2s)][E(2s) - E(2p_{-1})]\}], \quad (44)$$

where

$$M = e \langle \varphi_{2s} | \mathscr{E}_x x + \mathscr{E}_y y | \varphi_{2p \pm 1} \rangle, \quad (45)$$

and $\mathscr{E}_x$ and $\mathscr{E}_y$ are the electric field components perpendicular to the direction of the applied magnetic field at the donor in question. The donor eigenfunctions *in a magnetic field* were represented by φ_{2s} and $\varphi_{2p \pm 1}$ and were calculated variationally. The solid curve in Fig. 42 is that calculated from Eq. (44) using the variational wave functions and the single adjustable parameter $\mathscr{E}_\perp^2$. The value of $\mathscr{E}_\perp^2$ used for the fit shown was 195 V^2-cm^{-2},

---

[180] D. Redfield, *Phys. Rev.* **130**, 914 (1963).

compared to the estimated value of 205 V^2-cm^{-2} for the sample used. In this model $\mathscr{E}_\perp^2$ is the most probable value of the statistical distribution of the quantity $(\mathscr{E}_x^2 + \mathscr{E}_y^2)$. Although the agreement between the experimental points and the calculated curve was remarkably good, a reexamination by Larsen[181] of the general problem of line shapes and broadening of donor transitions in semiconductors has shown that the treatment described above is not completely correct.

The early work on line shape and broadening of optical lines has been reviewed by Breene,[182] and the Holtsmark theory[183] for Stark broadening has been summarized by Garbuny.[184] The Holtsmark theory of Stark broadening assumes that the radiating or absorbing atoms (corresponding to the donor impurities in our case) are in the electric fields of either (1) ions, (2) dipoles, or (3) quadrupoles. These electric fields then perturb the spectral emission of the radiating atoms. In the case of the linear Stark effect the emission line is split into two or more lines, while in the case of the quadratic Stark effect the line is shifted in energy. Although any given atom (donor) experiences a well-defined electric field at a given instant, the distribution of electric fields over all of the atoms in conjunction with the linear or quadratic Stark effect results in a broadened line. Holtsmark[183] has shown that for hydrogen-like spectra, resulting from a large number of atoms which experience an average field F_D from surrounding dipoles, the half-width of the broadened line is given by

$$\Delta v_H = aF_D, \quad (46)$$

where a is the coefficient for the linear Stark shift, and according to this theory the line shape is Lorentzian. The average field F_D depends on the concentration of dipoles N, and on the dipole moment X, so that the half-width of a dipole broadened line can be written as

$$\Delta v_H = 4.54aNX. \quad (47)$$

The line shapes for the cases of ions and quadrupoles could not be evaluated analytically in the Holtsmark theory, but were determined graphically. These results indicated that for quadrupoles the line shape was essentially Lorentzian, while for ions the line had considerably smaller tails on either side of the center frequency for the same half-width. The half-width determined graphically for ions of concentration N and charge e was

$$\Delta v_H = 1.25aF_I = 3.25aeN^{2/3} \quad (48)$$

---

[181] D. Larsen, private communication; *Phys. Rev.* **B13**, 1681 (1976).
[182] R. G. Breene, *Rev. Mod. Phys.* **29**, 94 (1957).
[183] J. Holtsmark, *Ann. Phys.* **58**, 577 (1919); *Phys. Z.* **20**, 162 (1919).
[184] M. Garbuny, "Optical Physics," p. 130. Academic Press, New York, 1965.

and for quadrupoles of moment Q and concentration N was

$$\Delta v_H = 0.67 a F_Q = 5.53 a Q N^{4/3}. \tag{49}$$

For the quadratic Stark effect there is not a splitting of the line but a shift, which is proportional to the square of the electric field. This is of interest for hydrogenic donors in a magenetic field since in this case the degeneracy of the various l-states, particularly the $p_{\pm 1}$ states, is removed. The statistical distribution of the electric field in this case results in a broadened and shifted line.

Although the Holtsmark theory has been widely used in describing ionized gases, it is not directly applicable to the case of ionized donors and acceptors in a semiconductor. Larsen[181] has treated the problem of line broadening in GaAs in a magnetic field by expanding the potential at a neutral donor due to the Coulomb fields produced by distant charged donors and acceptors $\Phi(\mathbf{r})$ as a Taylor series

$$\Phi(\mathbf{r}) = \Phi(0) + \nabla\Phi|_{r=0} \cdot \mathbf{r} + \tfrac{1}{2} \sum_{i=1}^{3} x_i^2 \, \partial^2\Phi/\partial x_i^2 |_{r=0} + \cdots. \tag{50}$$

If the potential varies slowly over distances of the order of the Bohr radius for the state in question, only the first few terms of this expansion are required. Using this expansion the energy shift of the lth donor level with wave function $\varphi_l(\mathbf{r})$ was calculated in first order perturbation theory as

$$\Delta E_l^{(1)} = e\langle \varphi_l | \Phi(\mathbf{r}) | \varphi_l \rangle = e\Phi(0) + \frac{e}{2} \sum_{i=1}^{3} \langle \varphi_l | x_i^2 | \varphi_l \rangle \frac{\partial^2 \Phi}{\partial x_i^2}\bigg|_{r=0}$$

$$= e\Phi(0) + \frac{e}{4} \langle \varphi_l | 3z^2 - r^2 | \varphi_l \rangle \frac{\partial^2 \Phi}{\partial z^2} \equiv \tfrac{1}{4} Q_l \frac{\partial \mathcal{E}_z}{\partial z} + e\Phi(0), \tag{51}$$

where the electric field is given by $\bar{\mathcal{E}} = \nabla\Phi$, with the assumption that φ_l is cylindrically symmetric about the z axis (direction of magnetic field). Here Q_l is the quadrupole moment of the charge distribution of the electron in state φ_l. The energy shift of the optical transition from level 0 to level 1 at a particular donor is then

$$\Delta E_{0,l}^{(1)} = \tfrac{1}{4}(Q_l - Q_0)\, \partial \mathcal{E}_z / \partial z|_{r=0}. \tag{52}$$

If the charged impurities are randomly distributed, the statistical distribution of $\partial \mathcal{E}_z / \partial z$ is Lorentzian and the resulting line shape is an unshifted Lorentzian with half-width given by

$$2.53 \frac{|Q_l - Q_0|}{a_B^2} \, \text{R\mathring{y}} \, (N a_B^3), \tag{53}$$

where N is the concentration of ionized impurities ($\approx 2N_A$), a_B is the Bohr radius, and R$\mathring{y}$ is the hydrogenic Rydberg. It is this quadrupole broadening which is the most significant broadening mechanism in Larsen's treatment.

With increasing magnetic field from 10 to 55 kG, the widths of the $2p_{-1}$ lines decrease significantly, while the widths of the $2p_0$ lines remain essentially constant. An indication of this variation can be seen in the spectra of Figs. 40 and 43. Larsen has shown that this behavior can be understood by considering the variation with magnetic field of the quadrupole moment Q_{2p} for the charge distribution of each of these states. In zero magnetic field the $2p_{\pm 1}$ wave functions are pancake-shaped and a charge distribution of this shape yields a negative quadrupole moment. With increasing magnetic field the $2p_{\pm 1}$ wave functions are elongated and at high magnetic fields have a cigar shape. The quadrupole moment of this shape charge distribution is positive.

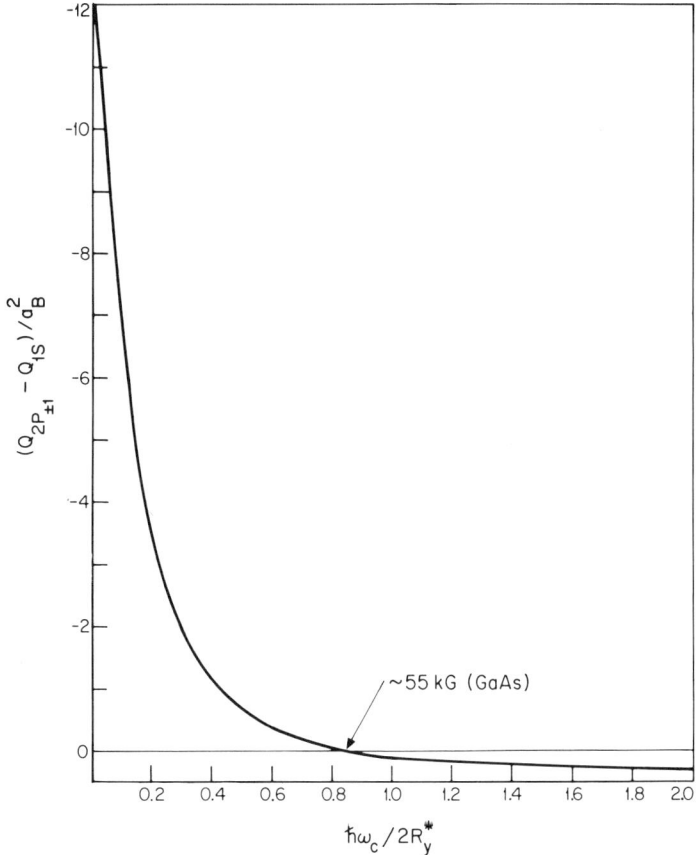

FIG. 44. Variation with magnetic field of the change in the quadrupole moment for a ($1s \rightarrow 2p_{-1}$) electron transition. For a magnetic field of about 55 kG, the change in quadrupole moment is zero. (After Larsen.[181])

Thus, as the magnetic field is increased from zero, the quadrupole moment of the $2p_{\pm 1}$ states should increase from a negative value, pass through zero, and become positive. Larsen[181] has calculated the variation of the quadrupole moment for the $(1s \rightarrow 2p_{\pm 1})$ transition with magnetic field and his results are shown in Fig. 44. It can be seen that in the low magnetic field range the magnitude of the quadrupole moment for these transitions decreases very rapidly with increasing magnetic field and this decrease results in a decrease in the half-width of the lines as indicated in Eq. (53). At ~55 kG the quadrupole moment for this transition becomes zero and the broadening due to this mechanism should disappear.

In second order perturbation theory the second term in the Taylor series expansion in Eq. (50), $\nabla\Phi|_{r=0} \cdot \mathbf{r}$ (the quadratic Stark shift), gives a significant contribution, and this contribution is formally only a factor $\sim (a_B/r_0)$ smaller

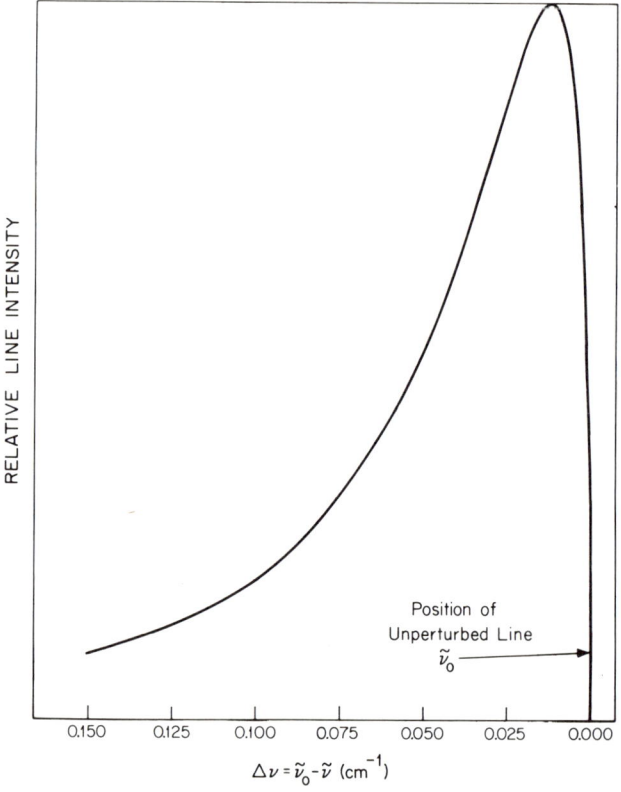

FIG. 45. Calculated line shape of the $(1s \rightarrow 2p_{-1})$ transition in GaAs at a magnetic field of about 55 kG where there is zero quadrupole broadening. (After Larsen.[181])

than the quadrupole contribution to the energy shift. The Stark shift of the state l (assuming φ_l is nondegenerate in the absence of electric fields) is given by

$$\Delta E_l^{(2)} = e^2 \sum_i |\langle \varphi_i | \mathscr{E} \cdot \mathbf{r} | \varphi_l \rangle|^2 / (E_l - E_i). \tag{54}$$

This expression cannot be evaluated exactly, but Larsen has shown that the shift of the $2p_{-1}$ state for a magnetic field of 55 kG, where the quadrupole moment vanished, is approximately

$$\Delta E_{2p-1}^{(2)} \approx 4 \, \mathrm{R} \overset{*}{\mathrm{y}} \, (N a_\mathrm{B}^3)^{4/3} (5.8) [\mathscr{E}_x^2 + \mathscr{E}_y^2 + 0.8 \mathscr{E}_z^2], \tag{55}$$

where the $\mathscr{E}_i$'s are the components of the field at the donor. By calculating the statistical distribution of the quantity in brackets (for $N = 4.4 \times 10^{13}$ and $\mathrm{R}\overset{*}{\mathrm{y}} = 46.1 \text{ cm}^{-1}$) he obtained the asymmetric line shape shown in Fig. 45. Although the asymmetry is very similar to that observed experimentally, the calculated linewidths are narrower than the resolution of the photo-

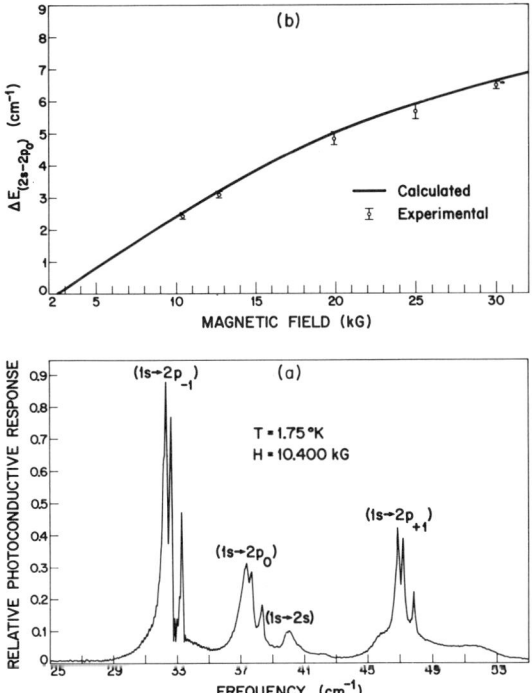

FIG. 46. (a) Typical photoconductivity spectrum showing the peak(s) identified as (1s → 2s) transitions, for the same sample as in Figs. 42 and 43, and (b) Variation of the (2s → 2p$_0$) energy separation with magnetic field. (After Stillman et al.[179])

conductivity spectra, so a direct comparison of theory and experiment cannot be made. At low magnetic fields both broadening mechanisms will be important, and the quadratic Stark term is probably responsible for the shift in the (1s → 2p) transitions and, thus, the apparent variation of the effective mass shown in Fig. 42. At zero magnetic field the linear Stark effect will be important in determining the lineshapes, since in this case the 2s and 2p levels will all be degenerate in the absence of electric fields.

Another example of the influence of internal electric fields is shown in Fig. 46. These electric fields cause a mixing of the 2s and $2p_0$ states so that there is some dipole oscillator strength in the resulting mixed 2s state. The weak photoconductive peaks shown on the high energy side of the (1s → $2p_0$) peaks, Fig. 46a, result from (1s → 2s) transitions. The variation with magnetic field of the separation between the 2s and 2p peaks as calculated variationally and measured experimentally is shown in Fig. 46b.

Although much work remains to be done on understanding the lineshapes and broadening observed in photoconductivity spectra of high purity GaAs, it is clear that the internal elecric fields exert a considerable influence on these spectra.

IV. Detector Performance

11. Introduction

Research in the far-infrared part of the spectrum has been limited in the past by the low intensity of the available sources and by the lack of sensitive detectors. However, a large number of laser transitions are now known in the spectral range from 50 μm to 1 mm and a large assortment of detectors is now available. The Golay cell has been the standard room temperature detector in the past and is still widely used. Another room temperature detector which has recently been developed is the pyroelectric detector[184a] and with further advances it may surpass the performance of the Golay cell. It already has two advantages in that it is less subject to microphonic noise due to vibration, and it can also be used with a fast response time. However, it is not yet as sensitive as the Golay cell. The most sensitive detectors for the far-infrared all require liquid He temperatures for their operation. Until recently, Ge and Si bolometers were probably the most sensitive practical detectors, but because of their fragility, slow response time, and the usual requirement for pumped He temperatures, there are many applications where these detectors are not satisfactory. The far-infrared detectors with fast response times are the InSb free electron bolometers[185] and the doped

---

[184a] See E. H. Putley, Chapter 7, this volume.
[185] For a discussion of this type of detector, see E. H. Putley, this volume, Chapter 3.

4. FAR-INFRARED PHOTOCONDUCTIVITY IN HIGH PURITY GaAs

Ge and Si extrinsic photoconductive detectors.[186] The InSb detectors have their optimum response at wavelengths beyond 1 mm and fall off roughly as λ^2 at shorter wavelengths while the Ge extrinsic photodetectors have a long wavelength limit of about 120 μm set by the ionization energy of shallow donors and acceptors in Ge. Thus, there is a region between about 120–500 μm where the sensitivity of these detectors is low. It is in this region that the GaAs extrinsic photodetectors have their maximum responsivity.

The photoconductivity mechanisms described in Part III are utilized in the GaAs far-infrared detectors. In this part our interest will be in the factors affecting the detector responsivity, noise performance, response time, etc., rather than in the physical mechanisms themselves.

12. Analysis of Detector Performance

The typical circuit generally used for photoconductive detectors consists of a battery and load resistor in series with the detector, as shown in Fig. 47. The preamplifier can be either connected across the detector as shown, or equivalently across the load resistor with no change in performance. The signal voltage e_s across the detector in this circuit can be written as

$$e_s = V \frac{g_s}{g_L + g} = V \frac{(g_s/g)}{1 + (g_L/g)}, \quad (56)$$

where V is the dc bias voltage across the detector and g_s, g_L, and g are the signal, load, and detector conductances, respectively. This equation is valid as long as the detector is Ohmic and the susceptance due to the stray capacitance at the operating frequency is small relative to the conductance $(g_L + g)$. Thus, it would seem that the signal voltage could be increased idefinitely by increasing V. However, because of the impact ionization discussed in Part II, there is a maximum practical value of V which depends on the sample parameters and the noise generated by the breakdown mechanism. Within this limitation it can be seen that to maximize the signal voltage and

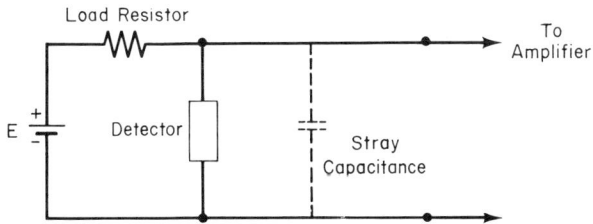

FIG. 47. Bias circuit for photoconductive detectors.

[186] For a discussion of this type of detector see P. Bratt, this volume, Chapter 2.

thus the responsivity, we want to make g_s as large as possible and g and g_L as small as possible.

a. Factors Affecting Responsivity

The signal conductance results from the generation of free electrons by the incident radiation, and the magnitude of this conductance depends on the average length of time that the free electrons spend in the conduction band as well as the efficiency with which they are generated by the incident photons.

The kinetics of the generation and recombination of free carriers have been described for the case of Ge which also exhibits impact ionization at relatively low electric fields,[107,123,187,188] and this same analysis can be applied to GaAs extrinsic detectors. The equation for the rate of change of the free electron concentration n is written as

$$dn/dt = A_T(N_D - N_A - n) + A_I n(N_D - N_A - n) + A_P(N_D - N_A - n)$$
$$- B_T n(N_A + n) - B_I n^2(N_A + n) - B_P n(N_A + n), \quad (57)$$

where the first three terms in the right member of this equation represent generation processes and the last three terms represent recombination processes. The subscripts T, I, and P refer to thermal, impact, and photo processes, respectively. All of the generation rates for the different processes are proportional to the number of neutral donors $(N_D - N_A - n)$, and the generation due to the impact ionization process is also proportional to the number of free carriers n. The term $A_T(N_D - N_A - n)$ represents the thermal generation rate of free electrons by the ionization of neutral donors through interaction with both lattice vibrations *and* that part of the blackbody radiation of the background which is in thermal equilibrium with the sample. The coefficient A_T increases rapidly with increasing temperature. The term $A_I n(N_D - N_A - n)$ is the generation rate due to impact ionization of the neutral donors $(N_D - N_A - n)$, by the free electrons n, and the impact ionization coefficient A_I depends strongly on the electric field. The term $A_P(N_D - N_A - n)$ is the rate of generation of free carriers by the photons incident on the sample, both due to the signal radiation as well as any background radiation not in thermal equilibrium with the sample. The coefficient A_P depends on the quantum efficiency of the sample and the number of incident photons. This term includes the contribution of the photothermal ionization process for excited state photoconductivity in GaAs. The recombination rates for the different processes are all proportional to the product of the free carrier and ionized donor concentrations,

[187] S. H. Koenig, R. D. Brown, and W. Schillinger, *Phys. Rev.* **128**, 1668 (1962).
[188] R. F. Wallis and H. Shenker, Naval Res. Lab. Rep. 5996, 1963.

$n(N_A + n)$. Because the Auger recombination process depends on the transfer of the excess energy and momentum of the recombining electron to other free carriers, this recombination rate varies as $n^2(N_A + n)$, and for the practical cases of interest in GaAs this term is negligible. As discussed in Section 7, only a small fraction of the recombination transitions are radiative and the dominant recombination mechanism in GaAs is that due to the phonon or lattice interaction. Thus, in Eq. (57), B_T is much greater than B_P.

At the liquid He temperatures required for the operation of GaAs detectors, $n \ll N_D - N_A$ and $n \ll N_A$. Thus, Eq. (57) simplifies to

$$dn/dt = A_I n(N_D - N_A) + (A_T + A_P)(N_D - N_A) - B_T n N_A$$
$$= (A_T + A_P)(N_D - N_A) - n[B_T N_A - A_I(N_D - N_A)]. \quad (58)$$

The increase in the free carrier concentration Δn resulting from a step increase in the photogeneration rate due to an applied signal $G_s = \Delta A_P(N_D - N_A)$ at $t = 0$, can be obtained from the solution of Eq. (58) as

$$\Delta n = \Delta A_P(N_D - N_A)\tau[1 - e^{-t/\tau}], \quad (59)$$

where

$$\tau = [B_T N_A - A_I(N_D - N_A)]^{-1}. \quad (60)$$

For $t \gg \tau$ a new steady state carrier concentration is attained which is larger than that before the signal was applied by the amount

$$\Delta n = G_s \tau. \quad (61)$$

The photogeneration rate G_s is related to the intensity of the radiation and the quantum efficiency which in turn depends on the reflectivity, the absorption constant, and the thermal ionization probability of the donor excited states when they are involved. In general G_s will depend on position in the sample so that the absorption must be integrated over the sample thickness d. For practical thicknesses attainable in GaAs detectors $\alpha d < 1$, so that we can treat the absorption as constant, and the quantum efficiency η for photoionization, neglecting multiple reflections, is given by

$$\eta = (1 - R)\alpha d, \quad (62)$$

where R is the reflectivity and α is the absorption constant. Since the reflectivity is determined by the material properties ($R \approx |(1 - \sqrt{\varepsilon_0})/(1 + \sqrt{\varepsilon_0})| \approx 0.314$), η can be maximized by making α as large as possible and the sample as thick as possible. The increase in the carrier concentration due to G_s causes an increase in the sample conductance given by

$$g_s = \Delta n e \mu A/l, \quad (63)$$

where A is the cross-sectional area of the detector, l is the distance between contacts, and μ is the carrier mobility. In this equation it is assumed that the average mobility of the photogenerated carriers is the same as that of the carriers present before the signal was applied, and that the carriers are generated uniformly throughout the sample. The steady state free electron concentration before the signal was applied can be written as $n = G\tau$, where G is the total generation rate per unit volume due to background radiation and impact ionization of the neutral shallow donors (and thermal generation if the temperature is not low enough), and the corresponding steady state conductance is $g = ne\mu A/l$. Then, the ratio of the signal conductance g_s to g is just

$$g_s/g = \Delta n/n = G_s/G \qquad (64)$$

and from Eq. (56) it can be seen that the photoconductive signal is proportional to this ratio.

b. Non-Ohmic Detector Characteristics

Without an external signal the steady state carrier concentration can be written in terms of the generation and recombination coefficients of Eq. (58) as

$$n = \frac{(A_T + A_P)(N_D - N_A)}{[B_T N_A - A_I(N_D - N_A)]}. \qquad (65)$$

Since the coefficient A_I increases with increasing electric field, an increase in the detector voltage will result in an increase in n, a corresponding increase in g, and thus a non-Ohmic detector characteristic. Equation (56) for the photoconductive signal does not apply in this case, but the analysis of a non-Ohmic detector by Putley[1] or Wallis and Shenker[188] can be used. The total change in the sample conductance for such a non-Ohmic detector when a signal e_s is observed can be written as

$$\delta g = (\partial g/\partial G)_V G_s - (\partial g/\partial V)_G e_s = g_s - (\partial g/\partial V)_G e_s, \qquad (66)$$

where the negative sign on the second term results because e_s is defined to be positive for decreasing g, and where e_s and δg are related by

$$e_s = -(\partial V/\partial g)\,\delta g. \qquad (67)$$

Since $V = g_L E/(g_L + g)$ for the circuit of Fig. 47,

$$\partial V/\partial g = -g_L E/(g_L + g)^2 = -V/(g_L + g)$$

and combining Eqs. (66) and (67),

$$e_s = \frac{V g_s}{g + g_L + (\partial g/\partial V)_G V} = \frac{V(g_s/g)}{1 + (g_L/g) + 1/g(\partial g/\partial V)_G V}. \qquad (68)$$

By comparing this expression with Eq. (56), it is clear that the photoconductive signal will be smaller than for the Ohmic case even when $g_L \ll g$ since $(\partial g/\partial V)_G$ is positive. For $g_L \gg g + (\partial g/\partial V)_G V$ the non-Ohmic characteristic will have a negligible effect. The signal voltage and thus the responsivity should vary linearly with the detector voltage until the variation of g becomes significant.

c. Effect of Temperature and Background Radiation

The responsivity of a photoconductive detector is determined primarily by the ratio $g_s/g = \Delta n/n$. The change in this ratio with temperature and/or background radiation is due mainly to the changes in n, unless the mobility of the signal generated carriers is significantly different from that of the other free electrons or the mobility of all the carriers is changed by the presence of the signal generated carriers, except in the case of photothermal generation of carriers. For the usual case of interest where $n \ll (N_D - N_A)$ and $n \ll N_A$, the thermal equilibrium value of the carrier concentration with only equilibrium background radiation can be calculated from Eq. (13). From this equation it can be seen that n decreases exponentially with decreasing temperature. In the approximation used here, the number of neutral donors $N_D - N_A - n \approx N_D - N_A$ does not change significantly so the absorption is essentially constant and therefore Δn is also approximately constant, provided τ is constant. Hence, the ratio g_s/g should also increase exponentially, and it would seem that the photoconductive responsivity could be increased arbitrarily by simply decreasing the temperature as long as the conductance of the load resistor is negligible compared to that of the sample. Practically, this does not occur even if there is no generation due to background radiation in the case of GaAs. This is because at some low temperature the conductivity in an impurity band or the conductivity due to hopping between impurity centers can be comparable to that in the conduction band. Thus, a further decrease in temperature does not necessarily decrease the conductance.

When the generation of free electrons by the background radiation is high enough that the carrier concentration calculated from Eq. (13) is much less than the observed carrier concentration, the sample conductance will not be significantly reduced by a further decrease in temperature, although it may decrease slightly because of a decrease in the free electron mobility (or a decrease in the photothermal ionization probability). Thus, there is no advantage to reducing the temperature below the value where the thermal generation rate is small in comparison with the background (and impact ionization) generation rates in the detector.

13. NOISE MECHANISMS

There are several sources of noise that can influence the sensitivity of a photoconductive detector.[1,189-191] The Johnson noise and thermal generation–recombination noise are inherent properties of the detector, while the photon generation–recombination noise depends on the detector and its environment. Other sources of noise include contacts, inhomogeneities, and surface states, all of which can hopefully be eliminated by careful sample selection. In addition, amplifier noise and the noise of the associated load resistor may have to be considered.

a. Johnson Noise

Johnson noise, which arises from the fluctuation of the thermal velocities of the free charge carriers, occurs in any resistive material and appears even when no current is flowing in the circuit. The noise voltage due to this mechanism is given by

$$V_N = (4kTR\,\Delta f)^{1/2} \qquad (69)$$

for practical frequencies. The Johnson noise for two resistors at different temperatures[192] in series is

$$V_N = [4k(R_1T_1 + R_2T_2)\,\Delta f]^{1/2}, \qquad (70)$$

and for two resistors in parallel is

$$V_N = [4k(R_1T_2 + R_2T_1)\,\Delta f[R_1R_2/(R_1+R_2)^2]]^{1/2}. \qquad (71)$$

When the detector is non-Ohmic, the Johnson noise calculated using the lattice temperature of the detector only gives a lower limit for this noise source, since the free carrier temperature is actually somewhat higher.[193]

b. Generation–Recombination Noise

Generation–recombination noise, which arises from fluctuations in the number of free electrons, can be divided into three different components for GaAs detectors according to the generation–recombination mechanism involved.

(1) *Fluctuation in background radiation.* For most extrinsic photoconductive detectors, the background is at a higher temperature than the detector. The statistical fluctuation in the number of photons reaching the

[189] K. M. van Vliet, *Proc. IRE* **46**, 1004 (1958).
[190] P. W. Kruse, L. D. McGlauchlin, and R. B. McQuistan, "Elements of Infrared Technology," p. 235. Wiley, New York, 1962.
[191] J. A. Jamieson, R. H. McFee, G. N. Plass, R. H. Grube, and R. G. Richards, "Infrared Physics and Engineering," p. 357. McGraw-Hill, New York, 1963.
[192] R. A. Smith, F. E. Jones, and R. P. Chasmar, "The Detection and Measurement of Infrared Radiation," p. 187. Oxford Univ. Press (Clarendon), London and New York, 1968.
[193] W. J. Moore and H. Shenker, *Infrared Phys.* **5**, 99 (1965).

detector from the background causes a fluctuation in the generation rate of the free electrons. There is also a fluctuation in the recombination rate and, even though the detector and the background are not in thermal equilibrium, it has been shown that the recombination fluctuation is the same as that of the generation process.[1,189] The total fluctuation corresponding to a background photon flux J is

$$\Delta N_J = 2\tau(\eta J)^{1/2}(1 + \omega^2\tau^2)^{-1/2}(\Delta f)^{1/2}, \quad (72)$$

where ΔN_J is the RMS fluctuation in total number of carriers due to this source, η is the quantum efficiency, τ is the carrier lifetime, $\omega \, (= 2\pi f)$ is the angular frequency, and Δf the frequency bandwidth for which the fluctuation is observed.

(2) *Fluctuation in thermally generated electrons.* The thermal generation of free carriers and the various recombination mechanisms determine the equilibrium carrier concentration in the absence of nonequilibrium background radiation, and there are fluctuations in these processes which result in thermal generation–recombination noise. Putley[1] has shown that the RMS fluctuation due to this process can be written as

$$\Delta N_G = 2\tau G^{1/2}(1 + \omega^2\tau^2)^{-1/2}(\Delta f)^{1/2}. \quad (73)$$

(3) *Fluctuation in impact ionization rates.* The results for nonequilibrium background radiation can be extended to the generation of free carriers by impact ionization as

$$\Delta N_I = 2\tau(G_I)^{1/2}(1 + \omega^2\tau^2)^{-1/2}(\Delta f)^{1/2}, \quad (74)$$

where G_I is the generation rate due to impact ionization.

Generation–recombination noise only exists when current flows in the detector circuit, and for a non-Ohmic detector the G–R noise is given by

$$\Delta V_{GR} = \frac{V(\Delta N/N)}{1 + (g_L/g) + g^{-1}(\partial g/\partial V)_G V} \quad (75)$$

since $(\Delta N/N)$ is the same as the ratio of the "noise conductance" to the mean conductance (see Eq. 68). When two or more of these generation–recombination processes are comparable they can be combined to give

$$\Delta V_{G+J+I} = \left[\frac{2V(\eta J + G + G_I)^{-1/2}}{1 + (g_L/g) + 1/g(\partial g/\partial V)_G V}\right]\left[\frac{(\Delta f)^{1/2}}{(1 + \omega^2\tau^2)^{1/2}}\right], \quad (76)$$

where $1/\tau$ is given by the combination of the time constants for each mechanism considered separately,

$$1/\tau = (1/\tau_G) + (1/\tau_J) + (1/\tau_I), \quad \Delta N = (\Delta N_J^2 + \Delta N_G^2 + \Delta N_I^2)^{1/2},$$

and N, the total number of free carriers in the sample, is given by $(G + \eta J + G_I)\tau$.[191]

14. GaAs Extrinsic Photodetectors

a. Description and Operation of Detectors

The GaAs detectors are generally epitaxial layers grown on semi-insulating GaAs substrates. The resistivity of the substrate is of the order of 10^7–10^9 Ω-cm at room temperature. Because the high resistivity is due to compensating centers near the middle of the GaAs bandgap, there is no observed far-infrared absorption in the substrate and it serves as a passive support for

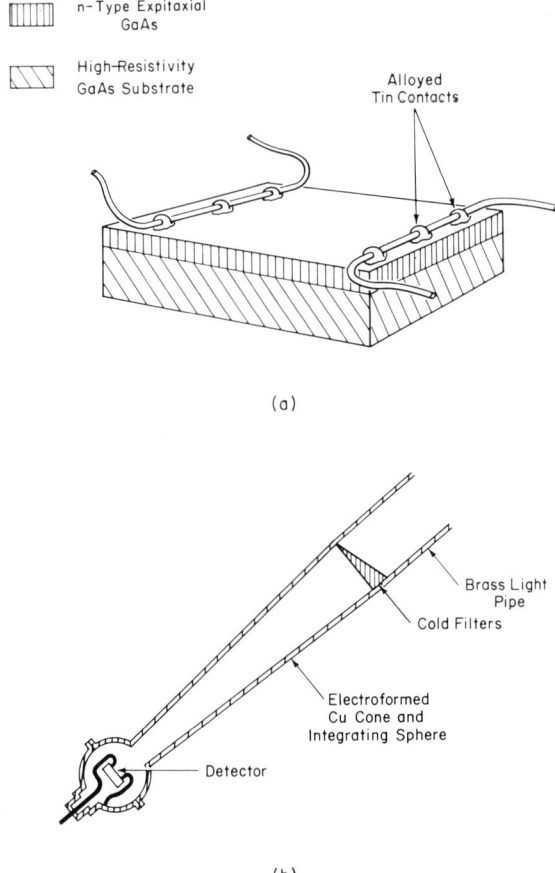

FIG. 48. Schematic diagrams of (a) a GaAs detector sample showing the insulating substrate, and (b) the copper cone and integrating sphere commonly used with these detector elements.

the epitaxial layer. A sketch of a GaAs detector is shown in Fig. 48a. The substrates are typically 250 μm thick while the epitaxial layers can be from 100 to over 400 μm in thickness. As discussed in Section 12 because of the small absorption, the thickness of this layer should be as large as possible to make the responsivity as high as possible. For a typical detector with $N_D - N_A \approx 2 \times 10^{13}$ cm^{-3}, the calculated absorption coefficient at the ionization energy is ~ 25 cm^{-1}.

The contacts shown on the sample in the diagram are tin contacts which are alloyed into the epitaxial layer in a reducing atmosphere at a temperature of about 300–350°C. This procedure has been found to give reproducible, Ohmic, low-noise contacts. Alloyed In,[106] evaporated Au–Ge eutectic,[162] and Sn-doped liquid phase epitaxial grown contacts[194] have also been used.

Because of the small absorption and the small active crystal volume, the detectors are often mounted in an integrating sphere[195] as shown in Fig. 48b to increase the responsivity. The use of this type integrating chamber and the condensing cone increases the responsivity for typical detectors by a factor of 2 to 4, depending on the sample volume and the area of the entrance aperture of the chamber. The integrating sphere also serves to eliminate the interference effect that is sometimes observed in photoconductivity spectra for detectors with parallel faces.[106] The actual spectral response of the detector is not changed significantly by operating them in integrating spheres, and this indicates that even in this mode of operation only a small amount of the incident radiation is absorbed.

b. *Variation of Responsivity with Doping*

Increasing the uncompensated donor concentration should increase the responsivity by increasing the absorption constant and thus the quantum efficiency. However, as discussed previously the donor concentration cannot be increased to very high values because of the decrease in the thermal ionization energy which limits the carrier freezeout and because of the impurity-band conduction at high donor concentrations.

The responsivity spectra for five GaAs samples with different donor concentrations are shown in Fig. 49. The spectral variation of the responsivity was determined from interferometric measurements that were corrected for the energy distribution of the source and instrument by ratioing the measured spectra with the spectra obtained with a Golay cell which was assumed to have a constant responsivity over the wavelength range of interest. The magnitude of the responsivity was determined from measurements of the detector signal in response to calibrated and filtered blackbody radiation.

[194] D. Woodward, private communication, 1970.
[195] G. A. Morton, M. L. Schutz, and W. E. Harty, *RCA Rev.* **20**, 599 (1959).

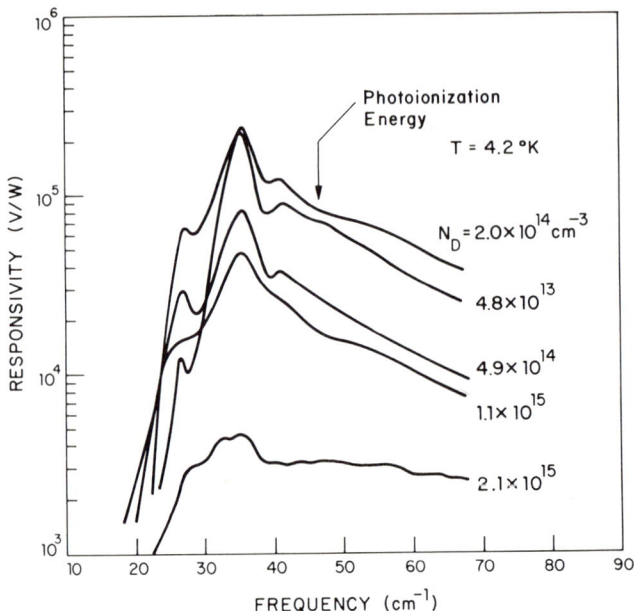

FIG. 49. Responsivity spectra at 4.2°K for five GaAs photodetectors with the donor concentrations as shown. (After Stillman et al.[133])

The responsivity of the detector can be expressed as

$$\mathscr{R}(\tilde{v}) = R_P r(\tilde{v}), \tag{77}$$

where R_P is the responsivity at the peak of the relative response curve and $r(\tilde{v})$ is the relative spectral response normalized to unity at the peak. Then, the signal $V(T_{BB})$, due to radiation from a blackbody at temperature T_{BB}, can be calculated as

$$V(T_{BB}) = R_P \int_0^\infty r(\tilde{v}) W(\tilde{v}, T_{BB}) \, d\tilde{v}, \tag{78}$$

where $W(\tilde{v}, T_{BB})$ is the spectral flux (W/cm^{-1}) incident on the light-pipe aperture with the blackbody at temperature T_{BB}. $W(\tilde{v}, T_{BB})$ was corrected for the emissivity of the chopper blades and the transmission of a cold long wavelength pass filter which was included for the calibration. R_P was determined from the measured blackbody signal by dividing $V(T_{BB})$ by the integral in Eq. (78) which was evaluated numerically. The resulting spectral responsivity $\mathscr{R}(\tilde{v}) = R_P r(\tilde{v})$ determined for each of the five samples is shown in Fig. 49. The background radiation for these calibration measurements corresponded to a background temperature of about 80°K. Although the effects of the different sample sizes and epitaxial layer thicknesses were minimized by enclosing the detectors in an integrating sphere, these differences create some uncertainty in the optimum donor concentration. However, we can conclude that the donor concentration should be in the range between 4×10^{13} and 5×10^{14} cm^{-3}. The donor concentrations for the samples in Fig. 49 are given in that figure.

c. V–I Characteristics

The V–I characteristics for a typical GaAs extrinsic detector at 4.2 and 1.93°K under normal and reduced background conditions are shown in Fig. 50. The reduced background conditions were obtained with a cold long wavelength pass filter which had a sharp cutoff at about 150 μm. For these measurements the detector was mounted in an integrating sphere at the end of a lightpipe (Fig. 48b) and was immersed in liquid He. The bias circuit used was that shown in Fig. 47. The load resistor and the detector were at the same temperature with $R_L = 20$ MΩ at 4.2°K and $R_L = 26$ MΩ at 1.93°K. The characteristics shown in Fig. 50 are typical of those observed for other detectors. The general features of all the curves are similar. At low currents

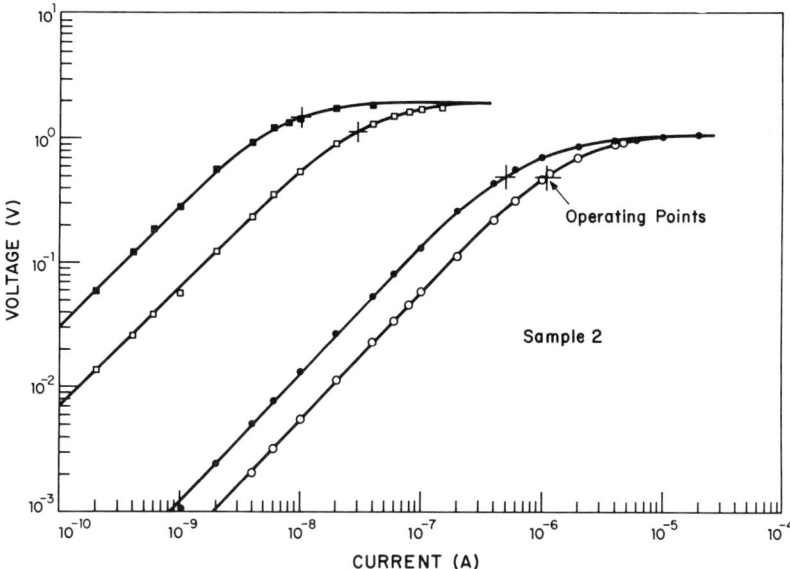

FIG. 50. V–I characteristics of a GaAs extrinsic photodetector for two background conditions at 4.2 and 1.93°K. The operating points shown were the bias values for maximum responsivity. Other characteristics and properties of this detector are listed under sample 2 of Table X. ○, 4.2°K, normal background; ●, 4.2°K, reduced background; □, 1.93°K, normal background; ■, 1.93°K, reduced background. (After Stillman et al.[133])

and voltages the curves are linear, indicating Ohmic behavior, but at higher currents the voltage does not increase as rapidly as the current. This decrease in sample resistance results from the increase in the impact ionization coefficient A_1 of Eq. (58). At a critical field, which depends on the electron mobility and the compensation ratio, the current increases several orders of magnitude, while the field remains relatively constant. The critical or breakdown field for this detector was about 1.5 V/cm at 4.2°K. In detectors which are more heavily compensated, the breakdown field is larger than the sustaining field so that a negative resistance region appears, as discussed in Section 6. It is clear from Fig. 50 that the background radiation has a negligible effect on the breakdown voltage, and that there is a slight increase in the breakdown voltage as the temperature is decreased from 4.2 to 1.93°K. This increase is due to the corresponding decrease in the electron mobility.

The resistance of this detector in the Ohmic region at 4.2°K is about 600 KΩ for normal background conditions, 1.4 MΩ for reduced background, and 8.6 MΩ in the dark. At 1.93°K the corresponding Ohmic resistances are 6.5, 18.5, and 330 MΩ. These values indicate that for the normal and reduced background conditions the sample resistance is limited by the generation of free carriers by the background radiation. The background optical generation rate should not change significantly for normal extrinsic photoconductivity, so that change in resistance as the temperature is decreased from 4.2 to 1.93°K should be just due to the change in mobility. However, the mobility for this sample only decreases by about a factor of 2 between 4.2 and 1.93°K, while the resistance increases by more than a factor of 10. The major reason for this discrepancy is that, although the optical absorption does not change significantly as the temperature is lowered, the number of free carriers produced by this absorption decreases, because fewer of the electrons which have been optically excited into the excited states are thermally transferred into the conduction band. That is, the photothermal ionization probability of the shallow donor levels decreases as the temperature is reduced. Thus, decreasing the temperature produces a decrease in g [Eq. (68)] even though g is limited by the background radiation, but the responsivity at the longer wavelengths is not higher because g_s also decreases. This decrease in the photothermal ionization probability causes a considerable decrease in the response of the detectors at wavelengths longer than the ionization threshold as the temperature is decreased below about 2.5°K. The decrease in response at these wavelengths also causes the larger relative increase in detector resistance shown in Fig. 50 between normal and reduced background conditions at 1.93°K than between normal and reduced background conditions at 4.2°K. The effective decrease in the generation of free carriers by the background radiation occurs because the cold filter blocks a *larger percentage* of the effective carrier-producing background radiation at the lower temperature than it blocks at 4.2°K.

d. Responsivity and Noise Characteristics

The variations of the signal and noise voltages and the signal-to-noise ratio with sample current for the detector whose $V-I$ characteristics are given in Fig. 50 are shown in Fig. 51. The signal increases linearly with current for low bias values, but then levels off and finally decreases for higher current. This behavior can be understood in terms of Eq. (68) and the $V-I$ characteristics of Fig. 50. The dashed curve is the variation of the signal voltage calculated from Eq. (68) using the values of g and $(\partial g/\partial V)_G$ determined from the $V-I$ characteristics and assuming a constant value for g_s. The linear increase in signal with sample current occurs in the current range where the detector is Ohmic and is due to the linear increase in sample voltage V. At currents of about 1×10^{-7} A the signal begins to increase more slowly with increasing current, due to a slower increase in V and the increase in the sample conductance g. For currents higher than about 5×10^{-7} A the signal actually decreases with a further increase in current, and this is due to the large increase of the term with $(\partial g/\partial V)_G$ in Eq. (68) in this current range.

The variation of the noise voltage with bias current shown in Fig. 51 can be understood in terms of the Johnson noise of the parallel combination of the detector and load resistor and the G–R noise due to the combination of

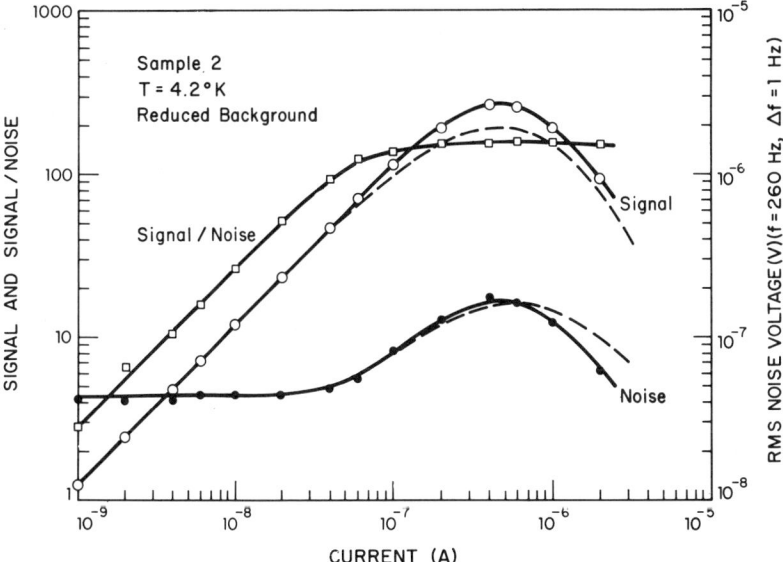

FIG. 51. Signal, noise, and signal/noise dependence on bias current for sample 2 in Table X. The dashed curves are calculated using Eqs. (68) and (76) and the $V-I$ characteristics of Fig. 50 as described in the text.

TABLE X

CHARACTERISTICS OF THREE GaAs FAR-INFRARED DETECTORS[a]

Sample	Material properties	Operating conditions ($T = 4.2°$K)	Peak responsivity $R_{282\mu m}$ (V/W)	$\text{NEP}_{282\mu}$ (W Hz$^{-1/2}$) ($f = 260$ Hz)	Detector dimensions L (mm) XW (mm) xt (μm)
1	$N_D = 1.2 \times 10^{14}$ cm^{-3} $N_A = 1.0 \times 10^{14}$ cm^{-3} $E_{D_t} = 5.4$ meV	Normal background[b]	4.0×10^4	2.0×10^{-12}	$4.3 \times 4.3 \times 295$
		Reduced background[c]	1.4×10^5	9.0×10^{-13}	
2	$N_D = 2.0 \times 10^{14}$ cm^{-3} $N_A = 0.4 \times 10^{14}$ cm^{-3} $E_{D_t} = 5.1$ meV	Normal background[b]	4.0×10^4	1.3×10^{-12}	$6.3 \times 5.5 \times 73$
		Reduced background[c]	2.4×10^5	5.3×10^{-13}	
3	$N_D = 4.9 \times 10^{14}$ cm^{-3} $N_A = 1.0 \times 10^{14}$ cm^{-3} $E_{D_t} = 4.6$ meV	Normal background[b]	3.0×10^4	2.8×10^{-12}	$3.8 \times 3.0 \times 57$
		Reduced background	9.9×10^4	1.1×10^{-12}	

[a] After Stillman et al.[133]

[b] The detector measurements for normal background conditions were made with the detector mounted in an integrating sphere at the end of a brass light pipe about 3 ft. long with warm black polyethylene and cold crystal quartz filters.

[c] The reduced background conditions were the same as for normal background conditions using an additional cold long wavelength pass filter with a cut on wavelength of 150 μm.

4. FAR-INFRARED PHOTOCONDUCTIVITY IN HIGH PURITY GaAs

background radiation and impact ionization processes. The nearly constant noise at low currents is due to the combination of preamplifier and Johnson noise. At higher currents the noise increases and reaches a maximum at about the same current which produces the maximum signal. The increase in noise is due to an increase in the G–R noise. The variation of the total calculated noise is also shown in this figure. The G–R noise was calculated from Eq. (76) using the total generation rate G, determined from the carrier concentration and the sample volume assuming a constant value for τ, and the total noise was determined as the rms of the Johnson noise, the amplifier noise, and the G–R noise. It can be seen that the qualitative agreement is good. The nearly constant value of the signal/noise ratio over a large current range is characteristic of G–R noise. At still larger current values than shown in the figure the noise increases several orders of magnitude and in this range is apparently associated with the breakdown mechanism.

Measurements of the variation of the noise with frequency show a $1/f$ dependence at frequencies below 200 Hz, with a less rapid decrease with increasing frequency above 200 Hz. Thus, some of the excess noise in the 260 Hz measurements shown in Fig. 51 may be due to contact noise, sample inhomogeneities, surfaces, or other fabrication problems.

The characteristics of three GaAs detectors at 4.2°K for two different background conditions are summarized in Table X. The responsivity curve for sample 2 is that labeled $N_D = 2.0 \times 10^{14}$ cm^{-3} in Fig. 49. The sensitivity in this table is given in terms of NEP instead of D^*, since these measurements were made with detectors mounted in an integrating sphere at the end of a lightpipe and in this case the area of the lightpipe aperture is perhaps of more importance than the area of the detector.

15. Performance in Reduced Background Conditions

The increase in the responsivity of the detectors in reduced background conditions shown in Table X indicates that the responsivity is a strong function of the background radiation as expected. Since one of the intended applications of the GaAs extrinsic photodetectors is in far-infrared rocket-borne observations of the low temperature cosmic background, a study was made of the variation of the detector responsivity with background temperature. This was done using the far-infrared recombination radiation from impact ionized GaAs as a calibrated signal source, while the background radiation was set by simultaneously exposing the detector to a cold blackbody at a controlled temperature. The experimental arrangement used for these measurements is shown in Fig. 52. The rotating turret, positioned between the bottom end of the lightpipe and the entrance to the detector cone, was used to either insert cold filters into the light path or block the path with the conical blackbody. A vacuum sealed crystal quartz

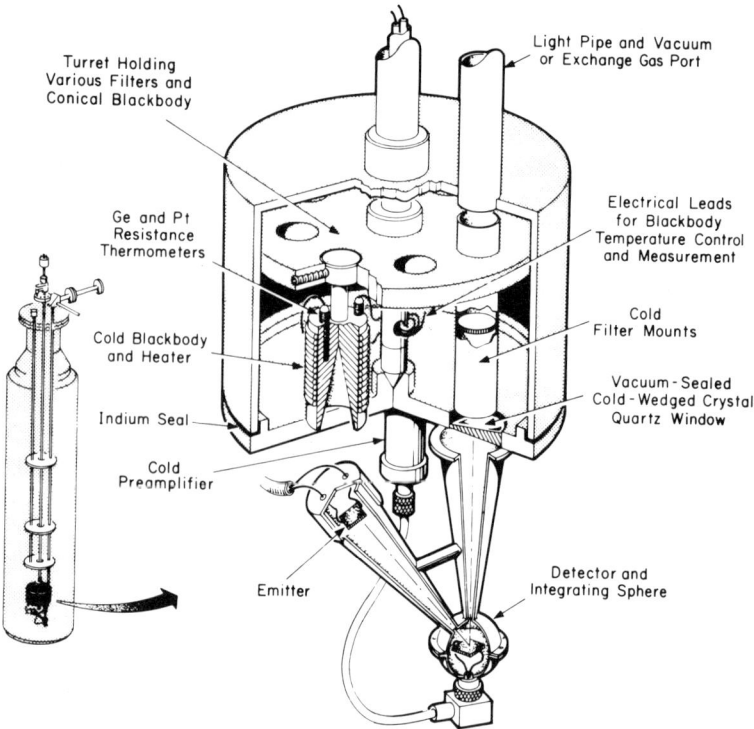

FIG. 52. Experimental arrangement for reduced background measurements.

window was at the entrance to the detector. The chamber containing the turret could either be evacuated or backfilled with He exchange gas. In this manner the blackbody temperature could be maintained constant at any value between about 4.2 and 300°K. The emitter sample was mounted in the other cone attached to the detector integrating sphere. In this way the detector could be exposed to the radiation from the emitter and the radiation from either the blackbody or an external source simultaneously. A small cold preamplifier[196] with a 37 MΩ load resistor was used to reduce the effects of shunt capacitance because of the high detector resistance.

a. GaAs Far-Infrared Emitter

The impact ionization of neutral donors in GaAs was described briefly in Section 6 and the recombination mechanisms in Section 7. A small fraction of the electron-ionized donor recombination transitions that take place are radiative and the far-infrared recombination radiation from this

[196] P. D. Feldman and D. P. McNutt, *Appl. Opt.* **8**, 2205 (1969).

process has been studied in GaAs by Melngailis et al.[113] Because of the current controlled negative resistance observed in the $I-V$ characteristic of compensated GaAs samples at low temperatures (see Fig. 15), the current becomes filamentary in this part of the $I-V$ characteristic, with the steady-state concentration of neutral donors and free carriers within the filament remaining essentially constant. The increasing conductance at constant voltage in the breakdown region is due mainly to the spreading of this

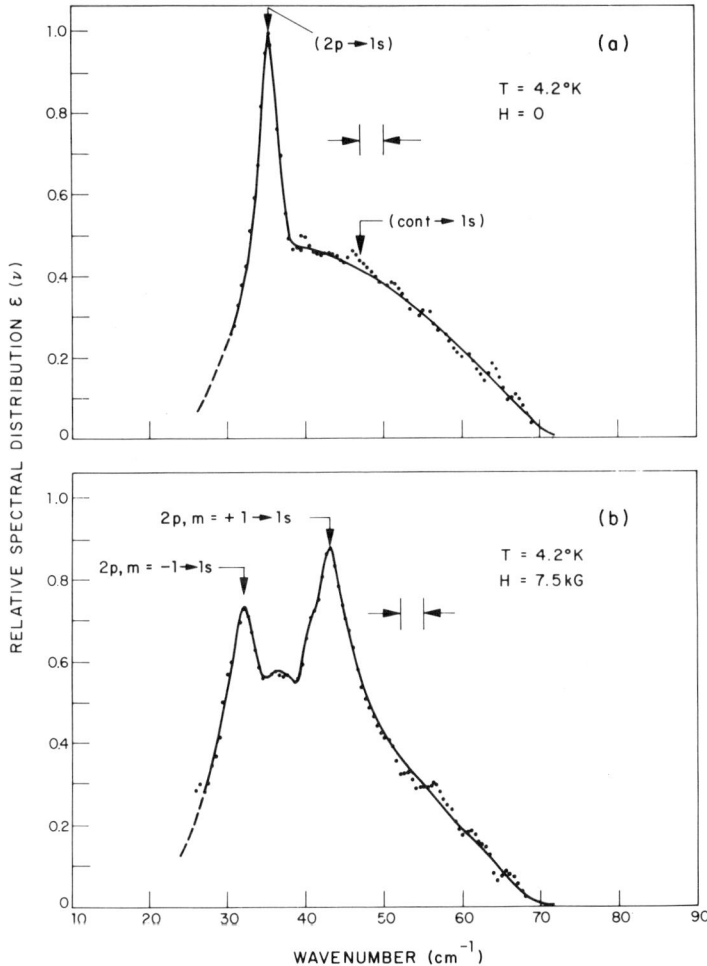

FIG. 53. Relative spectral emission due to impact ionization and radiative electron-ionized donor recombination for a GaAs sample at (a) zero magnetic field and (b) a magnetic field of $H = 7.5$ kG. (After Stillman et al.[132])

filament throughout the volume of the sample, which results in a larger number of both ionized donors and free electrons. When the filament has spread throughout the sample volume, the voltage again increases for increasing currents and the conductance approaches a constant value.

Within the current filament the generation and recombination rates of free electrons are constant. The relative spectral emission resulting from the small fraction of those recombination transitions which are radiative is shown in Fig. 53a for a sample at 4.2°K in zero magnetic field.[132] This spectrum was measured using an interferometer with a GaAs detector and was corrected for the relative spectral sensitivity of the detector. The dominant peak occurs at the same energy as the dominant peak observed in photoconductivity and is due to radiative transitions of electrons from the first excited state to the ground state of the donor. At higher energies and zero magnetic field the emission results from radiative transitions from higher excited states, the quasicontinuum, and states higher in the conduction band. The (cont. → 1s) energy shown in Fig. 53a denotes the conduction band edge. The relative emission spectrum does not change significantly with electric fields up to ~ 8 V/cm, or with temperature down to about 1.5°K. Figure 53b shows that in a magnetic field this peak splits into two major components corresponding to the (2p, m = ±1 → 1s) transitions. The observed splitting is in good agreement with that observed in photoconductivity.

The far-infrared radiation from the emitter was measured by first calibrating the detector with an external blackbody source as described in Section 14. Then, with the same background condition the detector signal was measured as a function of emitter current. The emitter current was a square wave current pulse with a frequency of 260 Hz, and the resulting signal was synchronously detected with a lock-in voltmeter. The spectral emitter power at current I and wavenumber $\tilde{v}$ can be written as $\mathscr{E}(I, \tilde{v}) = \mathscr{E}_P(I)\mathscr{E}(\tilde{v})$, since the relative spectral distribution $\mathscr{E}(\tilde{v})$ does not change in the range of currents which were used here. The total emitted power is then

$$\mathscr{E}_T(I) = \int_0^\infty \mathscr{E}_P(I)\,\mathscr{E}(\tilde{v})\,d\tilde{v} \tag{79}$$

and the peak spectral radiant intensity $\mathscr{E}_P(I)$ can be determined from

$$\mathscr{E}_P(I) = V(I,f)\Big/\mathscr{R}_P \int_0^\infty r(\tilde{v})\mathscr{E}(\tilde{v})\,d\tilde{v}, \tag{80}$$

where $V(I,f)$ is the detector signal for specified emitter current and frequency, and $\mathscr{R}_P$ is the peak spectral responsivity for the detector under the same background condition.

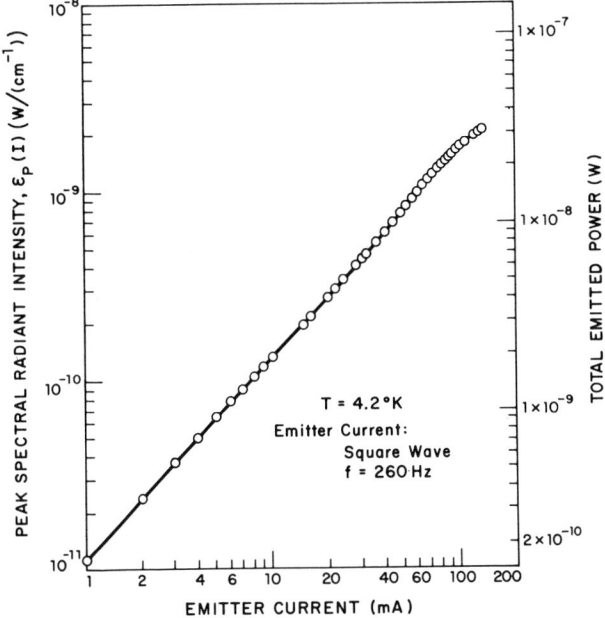

FIG. 54. Peak spectral radiant intensity $\mathscr{E}_p(I)$ in the relation $\mathscr{E}(I, v) = \mathscr{E}_p(I) \cdot \mathscr{E}(\tilde{v})$ as a function of current for a GaAs far-infrared emitter.

The results of this calibration are shown in Fig. 54. The day-to-day repeatability of this calibration was about ±1%, even though the detector responsivity varied somewhat. The detector responsivity variation was apparently caused by warming of the lightpipe and an associated increase in background radiation at lower liquid helium levels. The total emitted power could be varied from about 10^{-10} to about 3×10^{-8} W for the configuration used here without evidence of heating.

b. *V–I Characteristics in Reduced Background*

The variation of the $V-I$ characteristics of a GaAs detector for several different blackbody temperatures as well as for the "reduced" and "normal" 300°K background conditions mentioned earlier is shown in Fig. 55. The curves all have essentially the same shape, as observed previously. They are all Ohmic at low currents and voltages, and they all show the characteristic increase in conductance near the breakdown voltage which is independent of the background radiation. For the 4.2°K background the Ohmic resistance is about 1×10^8 Ω. For a 10°K background the Ohmic resistance has only decreased to 9.4×10^7 Ω. This indicates that at these background levels the thermal generation of carriers from the shallow donor levels is dominant.

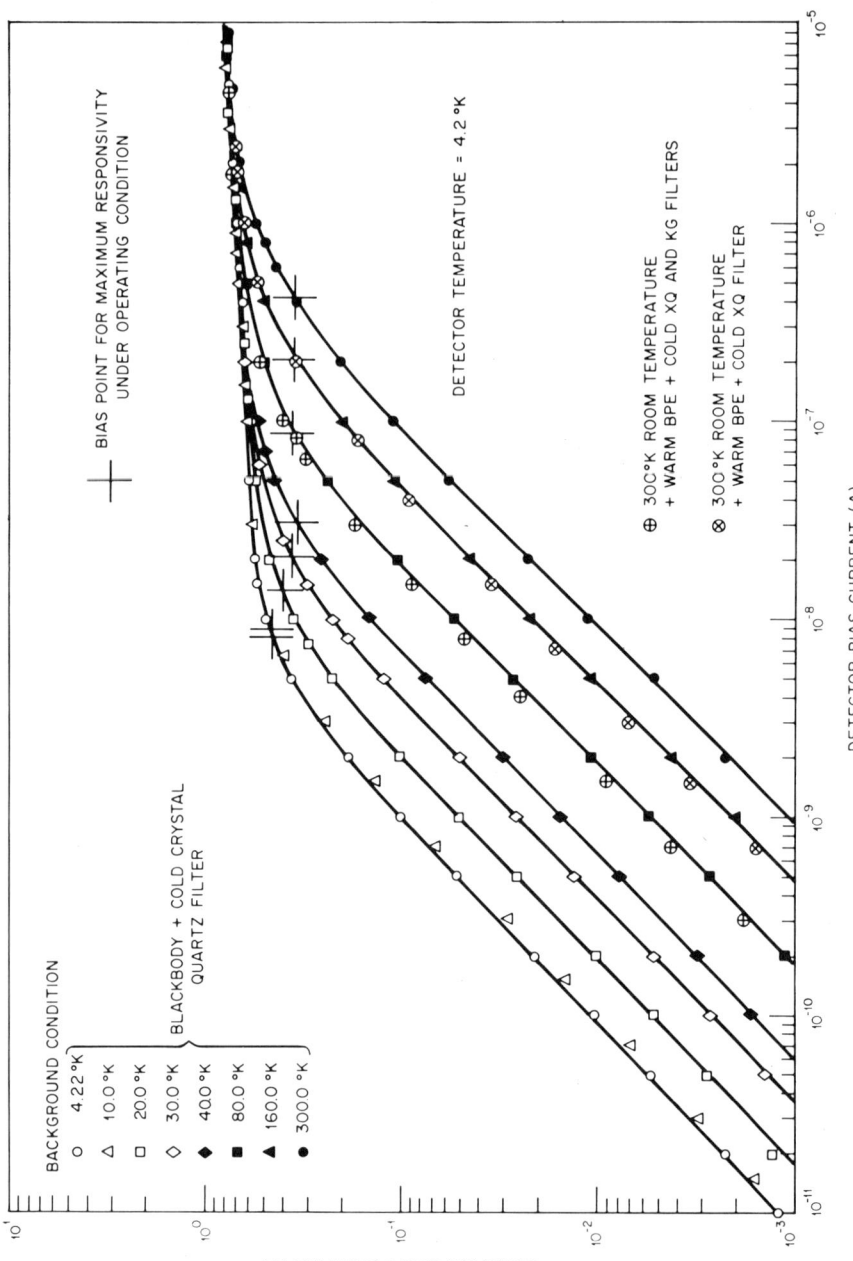

FIG. 55. V–I characteristics of a GaAs detector for different background temperatures.

For the 20°K blackbody background, however, the Ohmic resistance has decreased to about $5.1 \times 10^7 \, \Omega$. This indicates that for this 20°K background and a detector temperature of 4.2°K the thermal and background carrier generation rates are approximately equal. The Ohmic sample resistance for a 300°K blackbody background is just two orders of magnitude lower than that for the 4.2°K background. The "normal" and "reduced" 300°K background conditions correspond to blackbody backgrounds of about 160 and 80°K, respectively. The lower resistance for the 300°K background indicates that the cold light-pipe attenuates the room temperature background by about a factor of two.

The operating points for maximum responsivity are also shown in Fig. 55. The higher bias voltages at background temperatures below 40°K are probably due to loading effects of the 37 MΩ resistor. That is, at the higher bias voltages the sample resistance is lower and more closely matched to the load resistor.

c. Responsivity Variation

The variation of the peak responsivity of the detector with blackbody

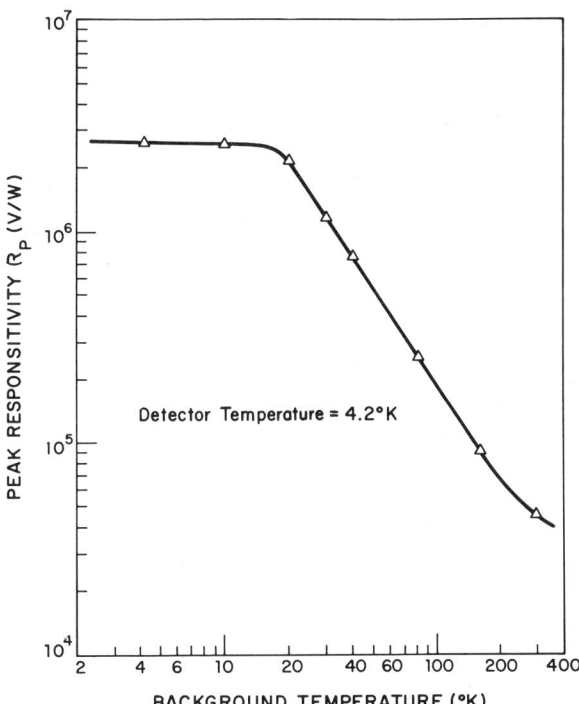

FIG. 56. Variation of detector peak responsivity with blackbody background temperature.

background temperature is shown in Fig. 56. The responsivity was determined by using the relation

$$\mathscr{R}_P(T_{BG}) = V(\mathscr{E}_1, T_{BG})/\mathscr{E}_P(I) \int_0^\infty r(v)\mathscr{E}(\tilde{v}) \, d\tilde{v}, \qquad (81)$$

where $\mathscr{R}_P(T_{BG})$ is the peak detector responsivity (volts per watt) for the blackbody background temperature T_{BG}, and $V(\mathscr{E}_1, T_{BG})$ is the signal measured for the calibrated emitter peak spectral radiant intensity $\mathscr{E}_P(I)$.

There is little change in the responsivity for background temperatures between 4.2 and 10°K, another indication that in this background range the detector conductance is limited mainly by thermal generation of carriers. However, for background temperatures of 20°K and higher the responsivity decreases rapidly with increasing background temperature. The major variation of the signal with background radiation stems from the variation of the sample conductance in Eq. (68), since the signal conductance g_s is expected to remain relatively constant. The sample conductance at the operating point decreased by a factor of 67 as the background temperature was decreased from 300 to 4.2°K. The peak responsivity increased by a factor of 57 over this same range.

Although a detailed analysis of the noise mechanisms was not made for these conditions, the NEP of this detector at the wavelength of peak responsivity and for $f = 260$ Hz decreased from about 1×10^{-12} W-Hz$^{-1/2}$ for the "normal" background condition (equivalent to blackbody $T_{BG} = 160°$K) to about 1×10^{-13} W-Hz$^{-1/2}$ for the 4.2°K background condition.

16. Response at Millimeter and Centimeter Wavelengths

In the initial work on the extrinsic photoconductivity of GaAs,[84] response was observed at 902 μm using radiation from a carcinotron, and response has since been observed at still longer wavelengths using other microwave sources. This response was at much lower energies than that required for photothermal ionization of the shallow donors, but in the interferometric measurements no response was observed between the long wavelength threshold and about 1 mm (10 cm^{-1}). There appear to be two different mechanisms responsible for this long wavelength response.

The response of GaAs detectors to 4 mm radiation has been studied by Fetterman et al.[197] The frequency response of the detectors at this wavelength was measured by a method similar to that used by Whalen and Westgate[198] in InSb, which consists of mixing the radiation from two different 4 mm sources in the GaAs detectors at 4.2°K. These results are shown in

[197] H. Fetterman, P. E. Tannenwald, and C. D. Parker, *Proc. Symp. Submillimeter Waves, New York, 1970*, p. 591. Polytechnic Press, Brooklyn, 1971.
[198] J. Whalen and C. Westgate, *Appl. Phys. Lett.* **15**, 292 (1969).

4. FAR-INFRARED PHOTOCONDUCTIVITY IN HIGH PURITY GaAs

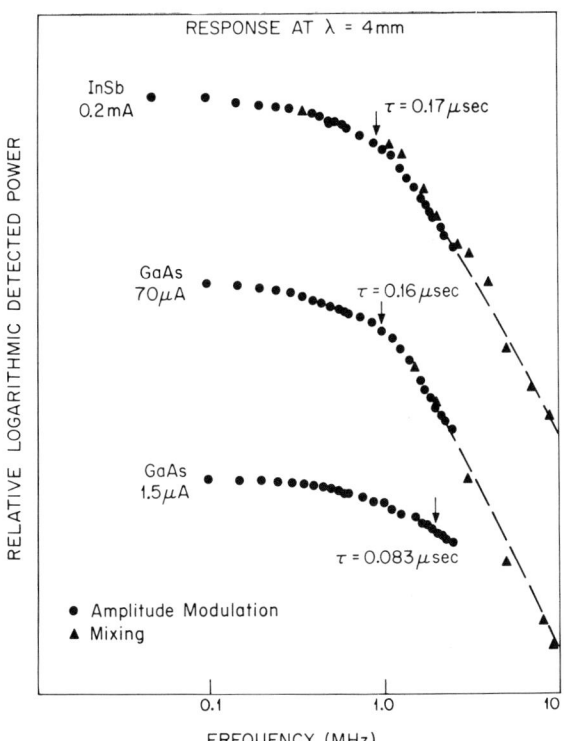

FIG. 57. Frequency response of InSb and GaAs far-infrared detectors excited by millimeter radiation ($\lambda = 4$ mm). The response of the GaAs detector is shown for two bias currents. The data points were obtained by two methods; ●, by amplitude modulation of 4 mm carcinotron radiation; and ▲, by mixing the 4 mm radiation of a carcinotron and a klystron. (After Fetterman et al.[197])

Fig. 57, along with results for frequencies from about 1 to 2.5 MHz obtained by amplitude modulating the radiation from a single carcinotron. The time constants determined for InSb and GaAs are nearly equal at high bias levels, and the time constant for GaAs decreases with decreasing bias. This behavior is consistent with the free electron bolometer mechanism important in InSb, since the time constant for this mechanism increases with increasing fields. The lowest bias current level of 1.5 μA for the GaAs sample shown in Fig. 57 results in too small an electric field to cause impact ionization of the neutral shallow donors. Thus, the free electron concentration should be quite low. However, the origin of the carriers responsible for the signal at low applied dc fields was explained by impact ionization of the neutral shallow donors by the sum of the microwave field and the applied dc field. The variation of the I–V characteristics of the GaAs detector with incident

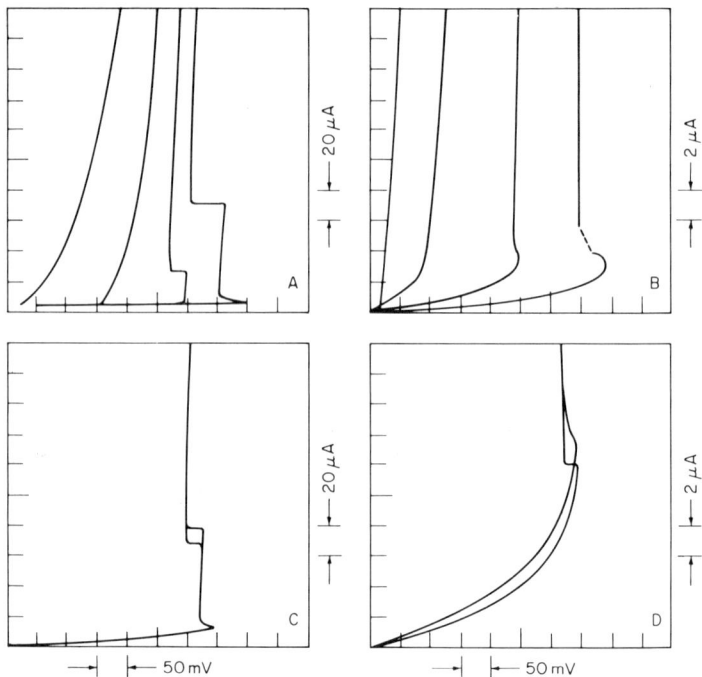

FIG. 58. I–V curves of GaAs with incident far-infrared radiation: A and B, 4 mm radiation at (left to right) 10, 5, 2. 5, and 0 mW for two different scales; and C and D, 337 μm radiation at about 3 mW for the same two current scales. The 4 mm radiation causes a change in the breakdown voltage while the 337 μm has little effect on the breakdown voltage, but causes considerable photoconductivity. (After Fetterman et al.[197])

4 mm microwave power is shown in Fig. 58A and B. It can be seen that the dc bias voltage at breakdown decreases with incident microwave power. An incident power of 1 mW/cm^2 corresponds to a microwave field of about 1 V/cm, and Fetterman et al.[197] concluded that the total electric field (dc + microwave) required for breakdown remains essentially constant. Thus, the 10 mW of microwave power used for the mixing and amplitude modulation was sufficient to generate free carriers by impact ionization even at low bias currents. The microwave power incident on these samples was essentially constant, however, since even for the amplitude modulation experiments the depth of modulation was only about 10%. The behavior of the I–V characteristics of the same GaAs detector with 3 mW of incident 337 μm radiation from an HCN laser is shown in parts C and D of Fig. 58. It is evident that the laser radiation produces significant photoconductivity, but essentially no change in the breakdown voltage. The reason for the

difference between the effect of the 337 μm radiation and 4 mm radiation is presumably that the frequency of the 4 mm radiation, 75 GHz, is low relative to the collision frequency of the electrons ($1/\langle \tau \rangle \approx 530$ GHz for $\mu \approx 50{,}000$ cm^2/V-sec), so that the energy of the carriers can be increased sufficiently by the microwave field to cause impact ionization while the frequency of the 337 μm radiation, 890 GHz, is higher than the collision frequency so that the carriers are not heated sufficiently to cause impact ionization. The problem of far-infrared mixing in high purity GaAs has been studied theoretically by Lao and Litvak.[199]

High purity GaAs has also been used for the detection of still lower frequency pulsed microwave radiation by Baukus and Ballantyne[200] in a different mode of operation. They found that by operating the detector at bias currents well above breakdown, the response time to the microwave radiation was of the order of 7 nsec. The I–V and pulse response characteristics of their detector operated in this mode are shown in Fig. 59. They measured the responsivity and NEP as 6.5 V/W and 6.0×10^{-10} W/Hz$^{1/2}$, respectively, at a microwave frequency of 26 GHz and a temperature of 4.2°K. The detector response was linear with X-band power over four orders of magnitude. More recent studies by Ballantyne et al.[201] have shown that this mode of operation results from the modulation of the breakdown voltage by the incident pulsed microwave field, as discussed in connection with Fig. 58. They called this mode of operation the LMB (light modulated breakdown) mode. Noise-equivalent-power values of less than 10^{-10} W/Hz$^{1/2}$

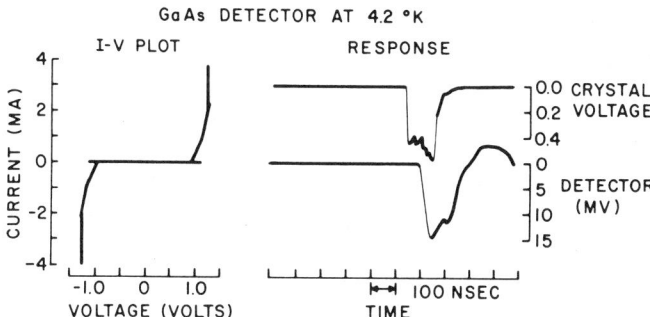

FIG. 59. I–V characteristics and pulse response of a GaAs detector to X-band microwave radiation when operated in the LMB mode. (After Baukus and Ballantyne.[200])

[199] B. Y. Lao and M. M. Litvak, *J. Appl. Phys.* **42**, 3357 (1971).
[200] J. Baukus and J. M. Ballantyne, *Proc. Symp. Submillimeter Waves, New York, 1970*, p. 583. Polytechnic Press, Brooklyn, 1971.
[201] J. M. Ballantyne, J. P. Baukus, and J. M. Lavin, *Int. Electron Devices Meeting, Washington, 1971*, paper 8.5; *Appl. Opt.* **12**, 2486 (1973).

with response times ≈ 16 nsec were measured in this work and it was shown that this detector has significant advantages over other competitors in the millimeter spectral region.

17. Response Time

The response time of the GaAs detectors operated in the usual extrinsic photoconductive mode was initially estimated to be about 10 nsec from measurements of the recombination time of the impact ionized free carriers.[84] Response times from as long as 250 nsec in samples with low acceptor concentrations to as short as 23 nsec in samples with higher acceptor concentrations have been measured by Holcomb et al.[202] They observed the decay of photoconductivity generated in response to 337 μm or 195 μm HCN laser radiation modulated with a Ge impact ionization modulator.[203] The measured response time was the same for the two wavelengths, indicating that as expected the photothermal ionization mechanism does not affect the response time. The response time should be described by the time constant of Eq. (60), and since the measurements by Holcomb et al.[202] were made at relatively high bias voltages where τ should be longer, their values only represent an upper limit for the response times.

The photoconductive response time was also measured by Fetterman et al.[197] by mixing radiation at slightly different frequencies from two different 337 μm (890 GHz) HCN lasers and detecting the difference frequency. Because of the limited spontaneous HCN linewidth, the lasers could only be tuned apart in frequency about 4 to 5 MHz. However, at low bias currents the frequency response of the GaAs detector was flat to beyond 4 MHz ($\tau < 40$ nsec), the limit of the measurements. The variation of the response time with applied voltage was not studied.

All of these measurements indicate that the response time of the GaAs detectors is quite short. However, the relatively high resistance of the detectors requires careful consideration of circuit capacitance, and special bootstrapping, or current-mode amplifier techniques to utilize the full responsivity of the GaAs detectors at response times shorter than 1 μ sec.

V. Summary

In summary, n-type epitaxial GaAs of sufficient purity to study extrinsic far-infrared photoconductivity and to optimize detector performance is now available, and the physics of the photoconductive processes and the performance of the detectors are reasonably well understood. The GaAs

---

[202] T. Holcomb, I. Melngailis, L. J. Belanger, and C. D. Parker, Solid State Res. Rep., Lincoln Lab., M.I.T., p. 6, 1970: 4.
[203] I. Melngailis and P. E. Tannenwald, *Proc. IEEE* **57**, 806 (1969).

4. FAR-INFRARED PHOTOCONDUCTIVITY IN HIGH PURITY GaAs

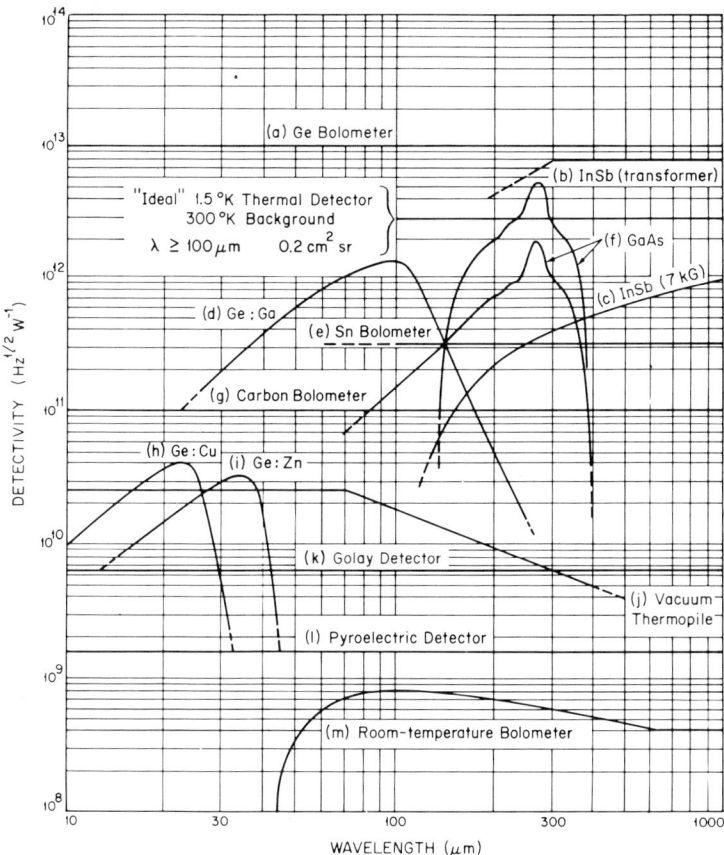

FIG. 60. Detectivity of far-infrared detectors. (After Kimmitt.[105])

detector is a useful addition to the selection of fast far-infrared detectors that is now available. Because the performance of cooled far-infrared detectors is very dependent on the actual background radiation present and on the particular operating conditions, it is difficult to make a meaningful comparison of the various detectors characterized under different or unknown conditions. However, Fig. 60 and Table XI give the spectral detectivities and corresponding measurement conditions for far-infrared detectors as recently summarized by Kimmitt[105] to which the best performance attained with GaAs extrinsic photodetectors under two measurement conditions has been added. It is clear that in the spectral range from about 100 μ to about 400 μ the GaAs detectors should be very useful when high sensitivity and speed of response are required.

TABLE XI

Measurement Conditions for Spectral Detectivities Given in Fig. 60[a]

Detector	Area (mm^2)	Operating temperature (°K)	Approximate response time (μsec)
(a) Ge bolometer	1	2	10^4
(b) InSb (transformer)	2	1.2	10^2
(c) InSb (7 kG)	10	1.5	10^{-1}
(d) Ge:Ga	6	4.2	10^{-2}
(e) Sn bolometer	6	3.7	10^4
(f) GaAs[b]	4	4.2	10^{-2}
(g) Carbon bolometer	20	1.8	10^3
(h) Ge:Cu	10	4.2	10^{-2}
(i) Ge:Zn	10	4.2	10^{-2}
(j) Vacuum thermopile	3	300	10^4
(k) Golay detector	7	300	10^4
(l) Pyroelectric detector	7	300	10^4
(m) Room temperature bolometer	20	300	35

[a] After Kimmitt.[105]

[b] The two GaAs curves are for the same detector, but the curve with the higher peak detectivity was obtained with a cold long wavelength pass filter with a cut-on wavelength of about 150 μm. (See Kimmitt[105] for further details on the other detectors.)

Acknowledgments

The authors would like to express their appreciation to D. M. Larsen for many helpful discussions during the preparation of this chapter, and to Miss Peggy Southard for her careful typing of the manuscript.

Note Added in Proof

Recent measurements of the ordering of the lower conduction bands in GaAs have indicated that the minima at the L_1 points of the Brillouin zone are 0.176 eV below the minima at X_1 and that the $\Gamma_1 - X_1$ energy separation is 0.462 eV, in contrast to the values given in the discussion on pp. 178–179.[204] This ordering is consistent with a reinterpretation of previous experimental data,[205] and also with recent velocity-field calculations and Gunn threshold measurements under uniaxial stress.[206] These changes do not influence any of the results presented in this chapter, but the reader is referred to the references below for a discussion of these recent developments.

[204] D. E. Aspnes, C. G. Olsen, and D. W. Lynch, *Phys. Rev. Lett.* **37**, 766 (1976).
[205] D. E. Aspnes, *Phys. Rev. B* **14**, 5331 (1976).
[206] P. J. Vinson, C. Pickering, A. R. Adams, W. Fawcett, and G. D. Pitt, *Proc. Int. Conf. Phys. Semicond., Rome, 1976*, to be published.

CHAPTER 5

Avalanche Photodiodes*

G. E. Stillman and C. M. Wolfe

I.	INTRODUCTION	291
	1. Performance Limits of PIN Photodiodes	292
	2. Performance Limits of Avalanche Photodiodes . . .	297
II.	AVALANCHE GAIN MECHANISM	300
	3. Low Frequency Avalanche Gain	301
	4. Gain-Bandwidth-Product Limitations	308
	5. Gain Saturation Effects	312
III.	MULTIPLICATION NOISE	314
	6. Excess-Noise Factor	314
	7. Avalanche Gain Statistics	321
IV.	ELECTRON AND HOLE IONIZATION COEFFICIENTS	325
	8. General Discussion	325
	9. Measurement Methods	330
	10. Experimental Results	337
V.	AVALANCHE PHOTODIODE DETECTORS	350
	11. Device Structure	351
	12. Si Avalanche Photodiodes	361
	13. Ge Avalanche Photodiodes	371
	14. Avalanche Photodiodes in Other Materials	373
VI.	ELECTROABSORPTION AVALANCHE PHOTODIODE DETECTORS . .	380
	15. Experimental Results	380
	16. Franz–Keldysh Effect and Quantum Efficiency . . .	381
	17. Avalanche Gain	386
VII.	SUMMARY AND CONCLUSIONS	391

I. Introduction

The widespread development of optical fiber and other optical communication and data processing systems in the visible and near-infrared spectral regions has been responsible for renewed interest in solid-state avalanche and *PIN* photodiodes. In many cases this interest has been stimulated by expectations of a small, durable solid-state device, which has a higher quantum efficiency and a larger bandwidth than a photomultiplier tube. While such a superior solid-state photodiode may never be achieved, there are many applications where the inherent advantages of solid-state devices are

*This work was sponsored by the Defense Advanced Research Projects Agency and by the Department of the Air Force.

particularly attractive. For some of these applications commercially available Si or Ge avalanche photodiodes can significantly improve the system signal-to-noise ratio over that obtainable with nonavalanche *PIN* photodiodes. For other applications avalanche devices fabricated from materials other than Si or Ge may offer even greater advantages. However, there are also applications where avalanche photodiodes cannot improve the system performance obtainable with *PIN* detectors. The decision as to whether to use a *PIN* photodiode or an avalanche photodiode for a given application depends upon the specific requirements of the system.

In this chapter we will first describe the performance limitations of Si *PIN* detectors and avalanche photodiodes so that a decision as to which device to employ can be based on the system requirements. For those applications that require an avalanche photodiode, we will discuss in detail the basic properties of these devices. This discussion should enable the reader to select or design an avalanche photodiode that will provide the best possible performance of the system under consideration. Finally, we will summarize the performance of avalanche photodiodes in various materials and point out the areas where further research would be particularly fruitful.

1. PERFORMANCE LIMITS OF *PIN* PHOTODIODES

The generalized photodetection process for a *PIN* photodiode[1,2] is shown schematically in Fig. 1. The block diagram in Fig. 1a illustrates the separate elements of the detection process. An optical signal and background radiation are absorbed in the diode, where they produce electrical currents by the internal photoelectric effect. That is, the radiation gives up its energy to electrons in the valence band of the semiconductor, exciting the electrons into the conduction band and leaving holes in the valence band. These electrons and holes are then separated and collected by the electric field of the device to produce an electric current. The quantum efficiency $\eta(\lambda, \omega)$ of this conversion depends upon the wavelength λ and the modulation frequency ω of the incident radiation, the properties of the material used to form the junction, and the geometry of the detector. Some of these considerations will be discussed in detail later.

To determine the current generated by this photoelectric process, let us consider an intensity-modulated optical signal given by

$$P(\omega) = P_o[1 + m \cos(\omega t)], \qquad (1)$$

where P_o is the average optical signal power, m is the modulation index,

[1] L. K. Anderson and B. J. McNurtry, *Proc. IEEE* **54**, 1335 (1966); L. K. Anderson, M. DiDomenico Jr., and M. B. Fisher, *Advan. Microwaves* **5**, (1970).

[2] H. Melchior, in "Laser Handbook" (F. T. Arecchi and E. O. Schutz-Dubois, eds.), Vol. 4, p. 725. North-Holland Publ., Amsterdam, 1972.

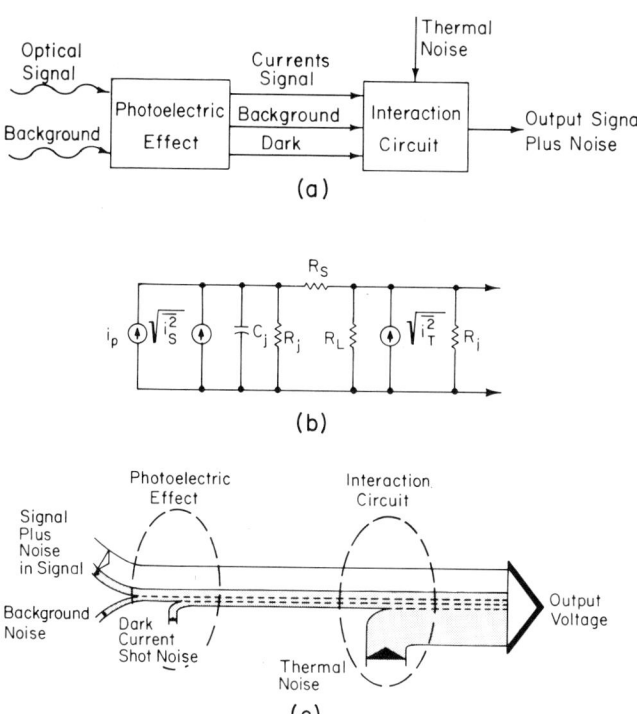

FIG. 1. Generalized photodetection process for *PIN* detectors. (a) Block diagram separating elements of detection process. (b) Equivalent circuit of *PIN* detector. The current sources shown are described by Eqs. (2), (5), and (9). (c) Variation of signal and noise currents through different stages of detection process. (After Anderson et al.[1])

and ω is the modulation frequency. Since the rms optical signal power incident on the detector is $mP_o/\sqrt{2}$, the rms signal current can be written as

$$i_p(\lambda, \omega) = q\eta(\lambda, \omega)mP_o/hv\sqrt{2}, \qquad (2)$$

where q is the electronic charge and hv is the photon energy corresponding to radiation of wavelength λ. For well-designed diodes $\eta(\lambda, \omega)$ can be close to unity and essentially independent of wavelength and frequency over wide operating ranges. In addition to the modulated signal current, there is an *average* current due to the optical signal which is given by

$$I_P = q\eta(\lambda, 0)P_o/hv, \qquad (3)$$

and a current due to the background radiation,

$$I_B = q\int_\lambda \{[\eta(\lambda', 0)P_B(\lambda')]/hv\}\, d\lambda'. \qquad (4)$$

In this equation $P_B(\lambda')$ is the background power per unit wavelength. Even without any optical signal or nonequilibrium background radiation absorbed in the detector, there will be some "dark current" I_D, due to thermal excitation of electron–hole pairs in the depletion region of the device and/or surface leakage currents. Because of the randomness of the generation of all of these currents, they contribute shot-noise fluctuations which can be expressed as a mean-square shot-noise curent,

$$\langle i_S^2 \rangle = 2q(I_P + I_B + I_D)B, \tag{5}$$

where B is the system bandwidth.

The signal and shot-noise currents interact with some form of circuit to deliver a signal and noise output to a load. This circuit, represented by the box labeled "interaction circuit" in Fig. 1a, also acts as a source of thermal noise. In the simplest case this circuit would consist of an external load resistor through which the signal and noise currents flow to develop the total output voltage of the detector.

The equivalent circuit of a PIN diode with a more realistic interaction circuit is shown in Fig. 1b. Here i_P, $\sqrt{\langle i_S^2 \rangle}$, C_j, R_j, and R_s are associated with the PIN detector, R_L is an external load resistor, and R_i is the input resistance of the following amplifier. All the resistances in this equivalent circuit contribute additional thermal noise so that the output of the overall detector consists of the desired signal plus the various noise contributions that result from the average optical signal, the optical background power, the dark current, and the thermal-noise voltage of the interaction circuit. These currents are represented by the current sources shown in Fig. 1b. The variation of the signal and noise currents throughout the detection process is shown schematically in Fig. 1c, and it is clear that the signal-to-noise ratio is continually degraded throughout the detection process.

For a well-designed PIN detector the series resistance R_s can be small compared to the other resistances and can be neglected. The mean-square thermal-noise currents for the other resistances in the interaction circuit are given by

$$\langle i_j^2 \rangle = 4kT_D(1/R_j)B \tag{6}$$

for the shunt resistance R_j of the detector at temperature T_D,

$$\langle i_L^2 \rangle = 4kT_L(1/R_L)B \tag{7}$$

for the load resistor R_L at temperature T_L, and

$$\langle i_A^2 \rangle = 4kT_A(1/R_A)B \tag{8}$$

for the amplifier, where k is Boltzmann's constant. In this case T_A is not the actual physical temperature, but rather an effective temperature which is related to the noise figure of the following amplifier when operated with the source resistance, $1/R_A = 1/R_j + 1/R_L$. For convenience all of these thermal-noise sources as well as the excess noise of the following amplifier can be included in a single equivalent thermal-noise source,

$$\langle i_T^2 \rangle = 4kT_{\text{eff}}(1/R_{\text{eq}})B, \tag{9}$$

where $1/R_{\text{eq}} = (1/R_j) + (1/R_L) + (1/R_i)$.

For a 100% modulated signal with average power P_o, the signal-to-noise power ratio of a PIN detector can be written as

$$\frac{S}{N} = \frac{i_P^2(\lambda, \omega)R_{\text{eq}}}{\langle i_N^2 \rangle R_{\text{eq}}} = \frac{\frac{1}{2}(q\eta P_o/hv)^2}{(\langle i_S^2 \rangle + \langle i_T^2 \rangle)}$$

$$= \frac{\frac{1}{2}(q\eta P_o/hv)^2}{2q(I_P + I_B + I_D)B + (4kT_{\text{eff}}B/R_{\text{eq}})}. \tag{10}$$

From Eqs. (10) and (3), P_o, the minimum, average, 100% modulated, optical power required to obtain a given signal-to-noise ratio is

$$P_o = \frac{2hvB}{\eta}\left(\frac{S}{N}\right)\left\{1 + \left[1 + \frac{I_{\text{eq}}}{qB(S/N)}\right]^{1/2}\right\}, \tag{11}$$

where

$$I_{\text{eq}} = I_B + I_D + (2kT_{\text{eff}}/qR_{\text{eq}}). \tag{12}$$

Let us examine Eq. (11) for P_o under two limiting regimes. First, when $I_{\text{eq}}/qB(S/N)$ is much less than one, the minimum optical power P_o corresponds to a situation where the only limiting factor is the quantum noise associated with the optical signal itself. In practice, for bandwidths up to about 50 MHz and (S/N) ratios of 40 to 50 dB, this limit is realized. However, this limit cannot be obtained for larger bandwidths because amplifier noise increases (T_{eff} increases and R_{eq} decreases) or for applications at this bandwidth where (S/N) is smaller. The other limiting regime is obtained when $I_{\text{eq}}/qB(S/N)$ is much greater than one. This corresponds to applications where either the background radiation or the thermal noise of the equivalent load resistor is high. Under these conditions a commonly used figure-of-merit for detectors is the rms optical signal power that produces an output signal current equal to the rms noise current. The wavelength and modulation frequency of the optical signal are usually specified and, when normalized to a bandwidth of 1 Hz, this figure-of-merit is called the noise-equivalent-power. Since the noise-equivalent-power is just the incident rms optical power required to produce a power signal-to-noise ratio of one in a 1 Hz bandwidth, the NEP of the *PIN* detector is given by

$$\text{NEP} = \sqrt{2}(hv/\eta)[I_{\text{eq}}/q]^{1/2}. \tag{13}$$

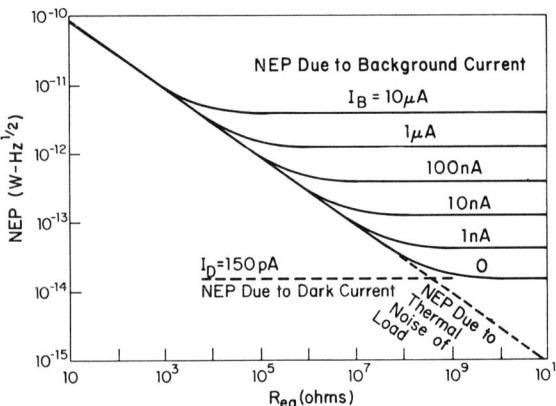

FIG. 2. Variation of NEP of PIN diode (with $\eta = 75\%$ at $\lambda = 0.77$ μm) with load resistance R_{eq} for 150 pA dark current and various background currents.

From Eqs. (12) and (13) the ultimate sensitivity of a *PIN* detector with a quantum efficiency η and dark current I_D can be obtained by making R_{eq} sufficiently large and by reducing the background radiation so that $2kT_{eff}/qR_{eq}$ and I_B are negligible with respect to I_D. In Fig. 2 we have plotted the NEP calculated for a typical Si *PIN* detector, with a quantum efficiency of 75% at a wavelength of 0.77 μm and a dark current $I_D = 150$ pA, as a function of R_{eq} for various values of I_B. These results show that high values of R_{eq} must be used to achieve an NEP limited by the dark current or background current shot noise, even with relatively large background photocurrents. This high load resistance required to achieve the dark-current- or background-current-limited NEP can impose a bandwidth or high frequency limitation which is unacceptable in many applications. The parallel combination of R_{eq} and the junction capacitance C_j, shown in Fig. 1b, determines the base-bandwidth and upper cutoff frequency, although in most applications additional stray capacitance and the input capacitance of the following amplifier must also be taken into account.

Figure 3 shows the variation of the 3 dB cutoff frequency with R_{eq} for three values of C_j. For example, from Fig. 2, R_{eq} must be greater than 10^4 ohms to obtain background-limited operation, even for $I_B = 10$ μA. For $R_{eq} \sim 10^4$ ohm, Fig. 3 indicates a cutoff frequency of less than 10 MHz for a typical junction capacitance of 2 pF. To obtain background-limited operation for smaller values of background current, R_{eq} must be larger than 10^4 ohm, which gives a cutoff frequency even less than 10 MHz. To obtain an NEP limited by the 150 pA dark current, the equivalent load resistance must be greater than 10^9 ohms with the bandwidth thus limited to a few

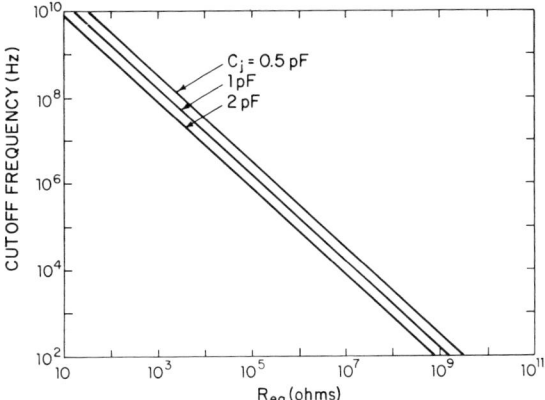

FIG. 3. Variation of RC cutoff frequency or base bandwidth with load resistance R_{eq} for three typical values of capacitance.

hundred hertz. Hence, *PIN* diodes are best suited to applications involving relatively high backgrounds and low bandwidths.

To obtain lower values of NEP over wider bandwidths, some internal current gain mechanism at the detector is needed which can reduce the relative importance of the thermal noise of the interaction circuit. This is true even if the following amplifier has zero noise (a noise figure of 1.0) and zero input capacitance, as long as the detector and the amplifier are operated at finite temperatures. Internal current gain[3] can be provided by avalanche multiplication even at microwave frequencies,[4,5] and the performance of avalanche photodiode detectors is considered in the next section.

2. PERFORMANCE LIMITS OF AVALANCHE PHOTODIODES

The generalized photodetection process for avalanche photodiodes, including the internal current gain, is shown schematically in Fig. 4. The current gain mechanism multiplies not only the signal current, but also the background current and that part of the dark current which flows through the region of the device where multiplication occurs. The gain mechanism also generates extra noise so that the signal-to-noise ratio *prior to the interaction circuit* is actually degraded further by the multiplication process. However, if the overall signal-to-noise ratio without current gain is determined mainly by the thermal noise to the interaction circuit, the internal gain can produce a significant increase in the overall signal-to-noise ratio.

[3] K. G. McKay and K. B. McAfee, *Phys. Rev.* **91**, 1079 (1953).
[4] K. M. Johnson, *IEEE Trans. Electron. Devices* **ED-12**, 55 (1965).
[5] L. K. Anderson, P. G. McMullin, L. A. D'Asaro, and A. Goetzberger, *Appl. Phys. Lett.* **6**, 62 (1965).

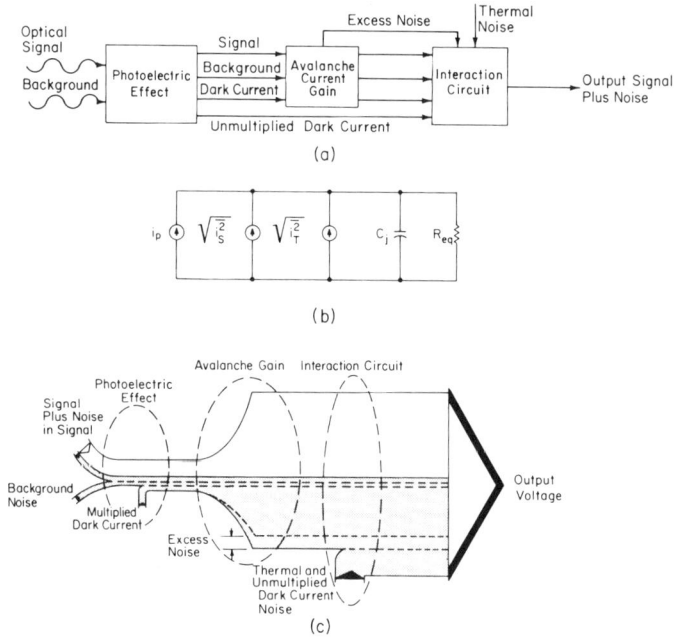

FIG. 4. Generalized photodetection process for an avalanche photodiode detector. (a) Block diagram of detection process with avalanche gain. (b) Equivalent circuit of avalanche photodiode detector. The current sources are described by Eqs. (14), (16), and (9). (c) Variation of signal and noise currents through different stages of detection process showing effect of avalanche gain and excess noise on S/N ratio. (After Anderson et al.[1])

The equivalent circuit for an avalanche photodiode is shown schematically in Fig. 4b. The multiplied rms photocurrent is given by

$$i_P = q\eta(\lambda, \omega)mP_oM(\lambda, \omega)/hv\sqrt{2}, \qquad (14)$$

where $M(\lambda, \omega)$ is the average multiplication or avalanche gain which can be both wavelength and modulation-frequency dependent. The other elements of the equivalent circuit are the same as for the PIN detector equivalent circuit of Fig. 1b. The mean-square shot-noise current after multiplication is given by

$$\langle i_S^2 \rangle = 2q(I_P M_P^2 F_P + I_B M_B^2 F_B + I_{D_B} M_D^2 F_D + I_{D_S})B, \qquad (15)$$

where M_P and F_P, M_B and F_B, and M_D and F_D are the average multiplication and excess-noise factors for the average photocurrent, background current, and multiplied dark current, respectively. Because of the wavelength and frequency dependence of the average multiplication, M_P, M_B, and M_D are not necessarily the same as $M(\lambda, \omega)$ given in Eq. (14). The excess-noise factors F_P, F_B, and F_D also are not necessarily equal. For convenience, even when the multiplication factors are not equal, we can redefine the excess-noise factors so that we can write

$$\langle i_S^2 \rangle = 2q[(I_P F_P + I_B F_B + I_{D_B} F_D)M^2(\lambda, \omega) + I_{D_S}]B. \qquad (16)$$

In Eqs. (15) and (16) the dark current has been separated into two components: I_{D_B}, the bulk dark current which is multiplied in the avalanche region, and I_{D_S}, the surface leakage dark current which is not multiplied, but nevertheless contributes shot noise. The distinction between the two currents is necessary since in many practical avalanche photodiodes I_{D_S} can be orders of magnitude larger than I_{D_B}.

The thermal noise of the interaction circuit, including the excess amplifier noise, is the same as for the PIN detector [Eq. (9)], and is represented in the equivalent circuit of Fig. 4b by the $\sqrt{\langle i_T^2 \rangle}$ current source. The signal-to-noise power ratio for the avalanche photodiode is then,

$$\frac{S}{N} = \frac{\frac{1}{2}(q\eta P_o/h\nu)^2 M^2(\lambda, \omega)}{2q[(I_P F_P + I_B F_B + I_{D_B} F_D)M^2(\lambda, \omega) + I_{D_S}]B + (4kT_{eff}B/R_{eq})}$$

$$= \frac{\frac{1}{2}(q\eta P_o/h\nu)^2}{2q(I_P F_P + I_B F_B + I_{D_B} F_D)B + [2qI_{D_S}B/M^2(\lambda, \omega)] + [4kT_{eff}B/R_{eq}M^2(\lambda, \omega)]}. \qquad (17)$$

The variation of the signal and noise currents throughout the detection and gain process is shown schematically[1] in Fig. 4c. From Eq. (17) we see that avalanche gain can only increase the signal-to-noise ratio by reducing the importance of the last two terms in the denominator of Eq. (17). In addition, because the excess-noise factors in Eq. (17) are always equal to or greater than unity and also increase monotonically with multiplication, there is some optimum value of multiplication which will produce the maximum signal-to-noise ratio for a given optical power. This optimum multiplication is roughly that value of $M(\lambda, \omega)$ for which the first term in the denominator of Eq. (17) is approximately equal to the sum of the other two terms in the denominator. Equation (17) can be solved for the minimum, average, 100% modulated, optical power P_o required to produce a given signal-to-noise ratio with avalanche gain. The resulting expression is,

$$P_o = \frac{2h\nu B F_P}{\eta}\left(\frac{S}{N}\right)\left\{1 + \left[1 + \frac{I_{eq}}{qBF_P^2(S/N)}\right]^{1/2}\right\}, \qquad (18)$$

where

$$I_{eq} = I_B F_B + I_{D_B} F_D + \frac{I_{D_S}}{M^2(\lambda, \omega)} + \frac{2kT_{eff}}{qR_{eq}M^2(\lambda, \omega)}. \qquad (19)$$

Under conditions such that $I_{eq}/qBF_p{}^2(S/N)$ is much less than one, the minimum detectable power is again limited by the quantum noise in the signal itself, and the ratio F_p/η (which has a minimum value of one) should be as small as possible for the lowest value of P_o. Without avalanche gain the excess-noise factor is one, so *in the quantum-noise limit* the minimum detectable power is determined by $1/\eta$ [cf. Eq. (11)]. This minimum detectable power would be smaller than that for a detector of the same quantum efficiency with avalanche gain [cf. Eq. (18)]. However, because of the thermal noise of the load resistor and the excess noise of currently available wide-bandwidth amplifiers, it is not possible to attain the quantum-noise limit in wide-bandwidth detectors without some internal gain mechanism. Even for avalanche photodiodes where sufficient multiplication can be obtained to make the last two terms of I_{eq} in Eq. (19) insignificant compared to the first two terms, it is only possible to attain the quantum-noise limit if the background current I_B and the multiplied dark current I_{DB} are sufficiently small and the bandwidth and signal-to-noise ratio are sufficiently large.

Thus, for detection at high frequencies and large bandwidths, avalanche photodiodes can have a significant advantage over *PIN* diodes. This is simply because the minimum detectable power at high frequencies and large bandwidths is limited by the thermal noise of the load and the noise figure of the following amplifier stage. The parameters such as the excess-noise factors and $M(\lambda, \omega)$ which are required to make a direct comparison of *PIN* and avalanche photodiodes in a particular systems application are discussed in the following sections.

II. Avalanche Gain Mechanism

In avalanche photodiodes, current gain or multiplication is obtained when the photogenerated or other primary free carriers gain sufficient energy from the electric field to generate additional (secondary) free carriers by impact ionization of the valence electrons into the conduction band, leaving free holes in the valence band. Secondary carriers that are generated in this way can in turn be accelerated by the electric field and generate more secondary carriers when they impact-ionize other valence electrons. The generation of electron–hole pairs and the current gain or multiplication of avalanche photodiodes can be described in terms of impact-ionization coefficients which are electric-field dependent. These phenomenological coefficients, which have dimensions of cm^{-1} and are often referred to as ionization *rates* since for a given saturated carrier velocity they determine the carrier generation rate, are the reciprocal of the average distance a carrier will travel at a given electric field before impact ionization generates an additional electron–hole pair. In general, the ionization coefficients of electrons and holes are not the same and are represented by $\alpha_n(E)$ and $\beta_p(E)$, respectively. The magnitude and field variation of $\alpha_n(E)$ and $\beta_p(E)$, which are determined by the appropriate carrier scattering mechanisms for a given semiconductor, together with the actual diode structure, have a strong influence on the low frequency avalanche gain, the gain-bandwidth-product limitations, and the excess-noise factor associated with the avalanche multiplication process.

3. Low Frequency Avalanche Gain

The general features of the gain process can be seen by considering the high-field depletion region of an avalanche photodiode shown schematically in the top part of Fig. 5. The spatial variation of the field in this region does not need to be specified, but the direction of the field is assumed to be as shown, so that electrons within this region travel in the positive x direction and holes travel in the negative x direction. Thus, the direction of current flow, whether due to electrons or holes, is in the same direction as the electric

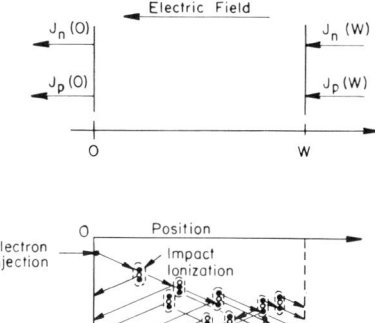

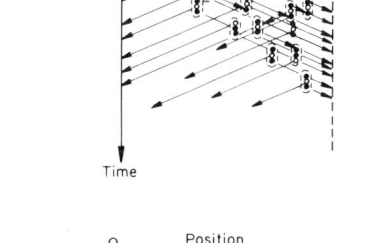

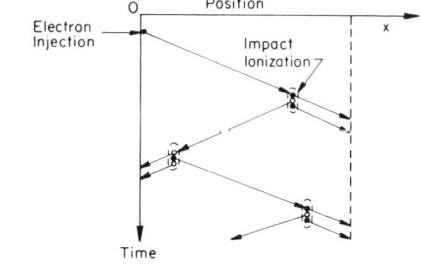

FIG. 5. Schematic diagram of avalanche multiplication process. Top: definition of boundary conditions and direction of electric field. Middle: avalanche gain process for $\beta_p = 0$. Bottom: avalanche gain process for $\alpha_n = \beta_p$.

field, but the electron current increases with increasing x while the hole current decreases with increasing x. The variation of the electron current in the high-field depletion region can be written as

$$\frac{d}{dx} J_n(x) = \alpha_n(x) J_n(x) + \beta_p(x) J_p(x) + qG(x), \tag{20}$$

where q is the magnitude of the electronic charge, and $G(x)$ is the generation rate of electron–hole pairs by absorbed photons at position x in the depletion region. (The variation of the ionization coefficients with x can be obtained from the spatial dependence of the electric field as discussed below.) The variation of the hole current is given by

$$-\frac{d}{dx} J_p(x) = \alpha_n(x) J_n(x) + \beta_p(x) J_p(x) + qG(x). \tag{21}$$

The total current J is the sum of the electron and hole currents and under dc conditions, $J = J_n(x) + J_p(x) = \text{constant}$. The differential equations for the electron and hole currents can then be written in terms of the total current J as

$$\frac{d}{dx} J_n(x) = [\alpha_n(x) - \beta_p(x)] J_n(x) + \beta_p(x) J + qG(x) \tag{22}$$

and

$$\frac{d}{dx} J_p(x) = [\alpha_n(x) - \beta_p(x)] J_p(x) - \alpha_n(x) J - qG(x). \tag{23}$$

These expressions can be solved by using the integrating factor $\exp[-\int_0^x (\alpha_n - \beta_p) dx'] \equiv \exp[-\varphi(x)]$ and integrating from $x = 0$ to W to obtain,

$$J = \frac{J_p(W) + J_n(0) \exp[\varphi(W)] + q \exp[\varphi(W)] \int_0^W G(x) \exp[-\varphi(x)] dx}{1 - \int_0^W \beta_p(x) \exp\left[\int_x^W (\alpha_n - \beta_p) dx'\right] dx} \tag{24}$$

from Eq. (22) and

$$J = \frac{J_p(W) \exp[-\varphi(W)] + J_n(0) + q \int_0^W G(x) \exp[-\varphi(x)] dx}{1 - \int_0^W \alpha_n(x) \exp\left[-\int_0^x (\alpha_n - \beta_p) dx'\right] dx} \tag{25}$$

from Eq. (23). The equivalence of Eqs. (24) and (25) can be demonstrated by using the identity

$$-\int_0^W (\alpha_n - \beta_p) \exp\left[-\int_0^x (\alpha_n - \beta_p) \, dx'\right] dx = \exp\left[-\int_0^W (\alpha_n - \beta_p) \, dx\right] - 1. \tag{26}$$

If the space-charge generation is taken to be zero and hole injection at $x = W$ and electron injection at $x = 0$ are considered separately, the usual multiplication factors for electrons and holes can be obtained from Eqs. (24) and (25). Thus,

$$M_p = \frac{J}{J_p(W)} = \frac{1}{1 - \int_0^W \beta_p \exp\left[\int_x^W (\alpha_n - \beta_p) \, dx'\right] dx}$$

$$= \frac{\exp\left[-\int_0^W (\alpha_n - \beta_p) \, dx\right]}{1 - \int_0^W \alpha_n \exp\left[-\int_0^x (\alpha_n - \beta_p) \, dx'\right] dx} \tag{27}$$

and

$$M_n = \frac{J}{J_n(0)} = \frac{\exp\left[\int_0^W (\alpha_n - \beta_p) \, dx\right]}{1 - \int_0^W \beta_p \exp\left[\int_x^W (\alpha_n - \beta_p) \, dx'\right] dx}$$

$$= \frac{1}{1 - \int_0^W \alpha_n \exp\left[-\int_0^x (\alpha_n - \beta_p) \, dx'\right] dx}. \tag{28}$$

These relations are the same as those derived by Howard[6] and Lee et al.,[7] and differ from the equations derived by Moll[8] only because of the assumed direction of the electric field. The avalanche breakdown voltage is defined as that voltage for which the multiplication becomes infinite and thus is the

---

[6] N. R. Howard, J. Electron. Contr. 13, 537 (1962).
[7] C. A. Lee, R. A. Logan, R. L. Batdorf, J. J. Kleimack, and W. Wiegmann, Phys. Rev. 134, A761 (1964).
[8] J. L. Moll, "Physics of Semiconductors," p. 225. McGraw-Hill, New York, 1964.

voltage for which

$$\int_0^W \beta_p \exp\left[\int_x^W (\alpha_n - \beta_p)\,dx'\right]dx = 1 \tag{29}$$

with hole injection at $x = W$, or for which

$$\int_0^W \alpha_n \exp\left[-\int_0^x (\alpha_n - \beta_p)\,dx'\right]dx = 1 \tag{30}$$

with electron injection at $x = 0$. With the aid of Eq. (26) it can be shown that these two conditions are identical. When both electrons and holes are injected simultaneously and electron–hole pairs are photogenerated in the space-charge region, the breakdown voltage is that voltage for which the total current J in Eqs. (24) and (25) becomes infinite. This voltage is the same as that given by Eqs. (29) and (30). Thus, the breakdown voltage is not influenced by the *type* of carrier injection.

The equations presented above are, in general, quite complicated because of the variation of $\alpha_n(x)$ and $\beta_p(x)$ with position. However, to demonstrate the effects of the ionization coefficients on the avalanche gain mechanism, we will consider two special cases, both assuming a uniform field in the depletion region: (1) β_p (or α_n) = 0, which will later be shown to be the desired case for the highest performance avalanche photodiode; and (2) $\alpha_n = \beta_p$, the worst case for the performance of an avalanche photodiode.

For $\beta_p = 0$ and a uniform field, the multiplication factor for electrons injected at $x = 0$ is simply [see Eq. (28)],

$$M_n = \exp\left(\int_0^W \alpha_n\,dx\right) = \exp(\alpha_n W). \tag{31}$$

Under these conditions there is no avalanche breakdown and M_n just continues to increase exponentially with increasing values of $\alpha_n W$. The buildup of the avalanche gain process with time in this special case is shown schematically in the center part of Fig. 5. The current pulse which results from the single injected electron increases during a time period which is essentially that of the transit time through the high-field region of the initially injected electron. The current pulse then decreases to zero in approximately the hole transit time. Thus, the current pulse lasts about twice as long with avalanche gain as it would without avalanche gain. Since the pulse-width is independent of the amount of multiplication, there is no gain-bandwidth-product limitation for avalanche gain when β_p (or α_n) = 0. It can also be seen that for high gain and dc steady-state conditions there are a large number of ionizing carriers (electrons for $\beta_p = 0$) in the high-field region at a given time. Thus, the statistical variation of the impact-ionization process will cause only a small fluctuation in the total number of carriers in the high-field region and the avalanche gain process should create little excess noise.

The behavior is drastically different, however, when $\alpha_n = \beta_p$ and the field is uniform. For this case [see Eqs. (27) and (28)],

$$M_n = M_p = \frac{1}{1 - \int_0^W \alpha_n \, dx} = \frac{1}{1 - \alpha_n W} \qquad (32)$$

and a true avalanche breakdown is obtained. The breakdown voltage corresponds to the situation where $\alpha_n W = 1$; that is, when each injected carrier on the average generates one electron–hole pair during its transit through the depletion region. The buildup of the avalanche gain process with time is shown schematically in Fig. 5c for electron injection with $\alpha_n = \beta_p$. The gain or multiplication can be very high, but when the gain is high the pulse width is also very long, so that there is a definite gain–bandwidth relationship. (When the multiplication is infinite, the current pulse length is infinite and we have avalanche breakdown.) It can be seen from this diagram that even at high gain there are relatively few carriers in the depletion region at any given time. Thus, statistical variations in the impact-ionization process can cause large fluctuations in the gain or multiplication and contribute considerable excess noise.

From a practical point of view, in most semiconductors both the electrons and holes contribute to the ionization process and the ionization coefficients are not equal, so that real multiplication processes are somewhere between the two extremes just considered. In the general case, given arbitrary (constant) values of α_n and β_p that are independent of position (an ideal PIN avalanche photodiode), the multiplication of electrons injected into the high-field region at $x = 0$ is [see Eq. (28)],

$$M_n = \frac{[1 - (\beta_p/\alpha_n)] \exp\{\alpha_n W[1 - (\beta_p/\alpha_n)]\}}{1 - (\beta_p/\alpha_n) \exp\{\alpha_n W[1 - (\beta_p/\alpha_n)]\}}. \qquad (33)$$

It has been pointed out[1] that, for small values of β_p/α_n, the positive feedback characteristic of the avalanche breakdown becomes apparent by rewriting Eq. (33) as

$$M_n \approx M_1/[1 - (\beta_p/\alpha_n) M_1], \qquad (34)$$

where $M_1 = \exp(\alpha_n W)$ is the electron multiplication [Eq. (31)] that would apply for $\beta_p = 0$ (the unilateral gain), and β_p/α_n is the positive feedback factor.

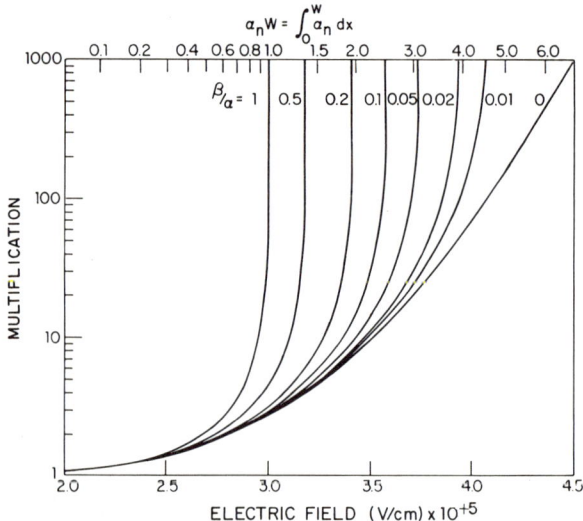

FIG. 6. Calculated electron multiplication as a function of electric field for a 1 μm wide PIN diode with $\alpha_n = 3.36 \times 10^6 \exp(-1.75 \times 10^6/|E|)$ for various ratios β_p/α_n. (After Webb et al.[9])

The electron multiplication calculated from Eq. (33) as a function of $\alpha_n W$ for various ratios of β_p/α_n is shown in Fig. 6. These results are plotted in terms of the electric field in a 1 μm wide "*I*" region of a PIN avalanche photodiode for which the electron ionization coefficient is given by

$$\alpha_n = 3.36 \times 10^6 \exp(-1.75 \times 10^6/|E|).$$

(This is the variation of α_n with electric field obtained in Si by Lee et al.[7]) The corresponding reverse-bias voltage is 1×10^{-4} cm times the electric field and thus ranges from 20 to 45 V in this figure. It is clear from Fig. 6 that, when the ratio of β_p/α_n is near unity, there will be—in addition to the lower gain-bandwidth-product and higher excess noise mentioned above— severe practical limitations on the uniformity and stability of the avalanche gain process for high values of multiplication.[9] Webb et al.[9] have noted that, for $\beta_p/\alpha_n = 0.01$, at a reverse bias of 100 V, a 0.5% variation in the electric field would result in a 20% variation in multiplication. However, for $\beta_p/\alpha_n = 1$, the same variation in electric field would result in a completely unusable 320% variation in multiplication. Since in real avalanche photodiodes such local variations in electric field could easily be caused by small inhomogeneities in the semiconductor doping level, these results show that there will be severe (probably insurmountable) technological problems in

[9] P. P. Webb, R. J. McIntyre, and J. Conradi, RCA Rev. 35, 234 (1974).

fabricating uniform high-gain avalanche photodiodes from semiconductors in which α_n and β_p are nearly equal.

For devices fabricated from material in which the ionization coefficients are significantly different, it is important that the device structure be designed to take advantage of this difference. That is, when α_n and β_p are not equal, the avalanche gain will depend on where the optically excited (primary) carriers are generated in or injected into the high-field multiplication region. To demonstrate this point, let us consider the multiplication or gain that results from the injection (or generation) of an electron–hole pair at position x_0 in the depletion region. This situation can be obtained from Eqs. (24) and (25) by setting $J_n(0) = J_p(W) = 0$ and letting $G(x) = G\delta(x - x_0)$, where $\delta(x - x_0)$ is the Dirac delta function. This gives

$$M(x_0) = \frac{J}{qG} = \frac{\exp\left[\int_{x_0}^{W}(\alpha_n - \beta_p)\,dx\right]}{1 - \int_0^W \beta_p \exp\left[\int_x^W (\alpha_n - \beta_p)\,dx'\right] dx}$$

$$= \frac{\exp\left[-\int_0^{x_0}(\alpha_n - \beta_p)\,dx\right]}{1 - \int_0^W \alpha_n \exp\left[-\int_0^x (\alpha_n - \beta_p)\,dx'\right] dx}. \quad (35)$$

Thus, the actual multiplication depends on where the electron–hole pair is is injected (that is, on the value of x_0) as well as on the variation of α_n and β_p with position throughout the depletion region.

A qualitative picture of the dependence of multiplication on where an electron–hole pair is generated or injected in the depletion region is shown in Fig. 7. The carriers that are injected in the P region for $x < 0$ and the N region for $x > W$ must first diffuse to the high-field region before they are multiplied. For $\alpha_n = \beta_p$ it makes no difference where the electron–hole pairs are injected since $M(x) = $ constant, independent of x. However, for $\beta_p > \alpha_n$ holes should be injected from the N side of the high-field region so that they can travel the entire width of the depletion region. For $\alpha_n > \beta_p$ electrons should be injected from the P side so that they can travel the entire width of the depletion region. In this manner the optimum multiplication can be obtained.

We will see in a later section that it is also important to have the carrier with the higher ionization coefficient injected into the high-field region to obtain the lowest possible excess-noise factor. The dependence of the multiplication on the position at which the electron–hole pairs are injected

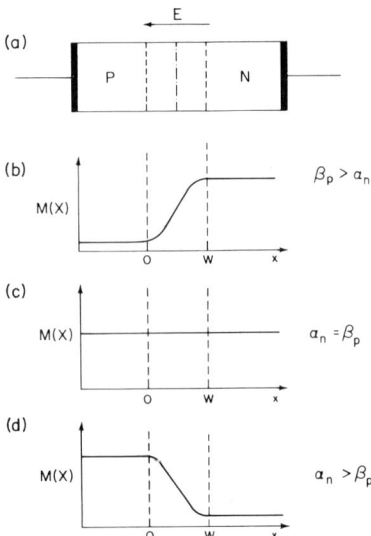

FIG. 7. Average multiplication of electron–hole pairs injected at position x in PN junction avalanche photodiode for different relative values of the electron and hole ionization coefficients.

in the high-field region is the source of the wavelength (λ) dependence of the average multiplication $M(\lambda, \omega)$ in Eq. (14). The frequency (ω) dependence of $M(\lambda, \omega)$ will be considered in the next section.

4. GAIN-BANDWIDTH-PRODUCT LIMITATIONS

In the discussion of the impact-ionization process given above, it was demonstrated that, for $\alpha_n = \beta_p$, the length of a current pulse increases as the multiplication increases and that there is a gain-bandwidth-product limitation when both of the ionization coefficients are finite (nonzero). This multiplication-bandwidth relationship in a PIN avalanche photodiode was first examined by Emmons and Lucovsky,[10] while Chang[11] and then Emmons[12] studied this interaction for arbitrary values of β_p/α_n. To examine this dependence of multiplication on frequency, we write the time-dependent transport equations corresponding to Eqs. (20) and (21). For the field configuration and current definitions given in Fig. 5, these equations have the

[10] R. B. Emmons and G. Lucovsky, *Proc. IEEE* **52**, 869 (1964).
[11] J. J. Chang, *IEEE Trans. Electron. Devices* **ED-14**, 139 (1967).
[12] R. B. Emmons, *J. Appl. Phys.* **38**, 3705 (1967).

form

$$\frac{1}{v_n}\frac{\partial J_n(x,t)}{\partial t} = \frac{\partial J_n(x,t)}{\partial x} - \alpha_n J_n(x,t) - \beta_p J_p(x,t) - qG(x,t) \qquad (36)$$

and

$$\frac{1}{v_p}\frac{\partial J_p(x,t)}{\partial t} = \frac{\partial J_p(x,t)}{\partial x} + \alpha_n J_n(x,t) + \beta_p J_p(x,t) + qG(x,t), \qquad (37)$$

where v_n and v_p are the electron and hole drift velocities, respectively. Assuming a constant dc electric field and an optical photocurrent with a dc and an ac component, the solutions of Eqs. (36) and (37) can be written as the sum of a dc and an ac component as

$$J_n(x,t) = J_{n_0}(x) + J_{n_1}(x)e^{i\omega t} \qquad (38)$$

and

$$J_p(x,t) = J_{p_0}(x) + J_{p_1}(x)e^{i\omega t}. \qquad (39)$$

Substituting Eqs. (38) and (39) into Eqs. (36) and (37), we obtain a pair of equations for the ac currents, J_{n_1} and J_{p_1}, and the pair of dc equations previously discussed.

The differential equations for J_{n_1} and J_{p_1} are linear if v_n, α_n, v_p, and β_p are independent of x. Under normal operating conditions the average electric field in the depletion region of an avalanche photodiode is greater than 10^4 V/cm, so it is reasonable to assume that the electron and hole velocities are saturated and thus constant, independent of both x and t. Both Chang[11] and Emmons[12] show that it is also a good approximation to treat α_n and β_p as independent of x and t in *PIN* avalanche photodiodes under normal operating conditions, even though both α_n and β_p are strong functions of the electric field. The total ac current in the device must include the displacement current and thus is given by

$$J_{ac} = J_{n_1} + J_{p_1} + \varepsilon\varepsilon_0(\partial E/\partial t), \qquad (40)$$

where ε is the static dielectric constant, and ε_0 is the permittivity of free space. Read[13] has shown that for the short-circuit case the external ac current, including the displacement current in the depletion region, is given by

$$J_{ac} = \frac{1}{W}\int_0^W (J_{n_1} + J_{p_1})\,dx. \qquad (41)$$

The normalized frequency response $F(\omega)$ of the multiplication of electrons

[13] W. T. Read, Jr., *Bell Syst. Tech. J.* **37**, 401 (1958).

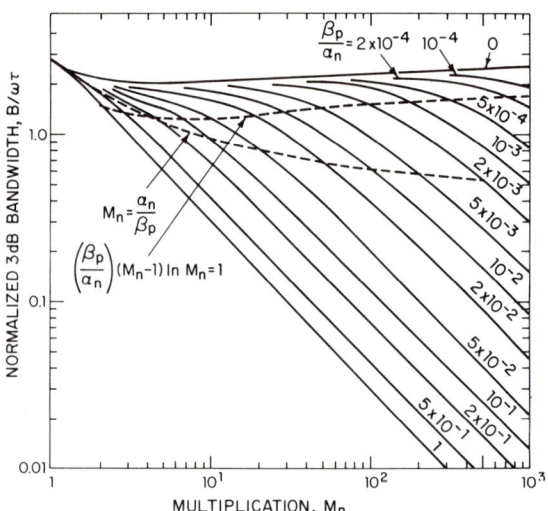

FIG. 8. Calculated 3 dB bandwidth of $F(\omega)$ [Eq. (41a)], normalized to $\omega\tau$, as a function of the dc multiplication of electrons injected at $x = 0$ for various values of β_p/α_n in a uniform electric field avalanche photodiode. (After Emmons.[12])

injected into the high-field region at $x = 0$ is given by

$$F(\omega) = |J_{ac}|/J = |J_{ac}|/M_n J_n(0), \qquad (41a)$$

where J is the low frequency or dc multiplied current. Emmons[12] has calculated $F(\omega)$ as a function of the normalized frequency $\omega\tau$, where $\tau = W/v_n = W/v_p$ (that is, the electron and hole saturated velocities are assumed equal) for various values of β_p/α_n and M_n. The results of these calculations are shown in Fig. 8, where the 3 dB bandwidth of $F(\omega)$ normalized to $\omega\tau$ is given as a function of the dc multiplication M_n for various values of β_p/α_n. For $M_n > \alpha_n/\beta_p$, the frequency variation of the multiplication can be represented by an equation of the form

$$M(\omega) = M_n/(1 + \omega^2 M_n^2 \tau_1^2)^{1/2}, \qquad (42)$$

where τ_1 is an effective transit time.[12] This effective transit time is approximately equal to $N(\beta_p/\alpha_n)\tau$, where τ is the actual carrier transit time given above, and N is a slowly varying number that is $\frac{1}{3}$ for $\beta_p/\alpha_n = 1$ and is 2 for $\beta_p/\alpha_n = 10^{-3}$. Anderson et al.[1] have shown that for average values of multiplication M_n such that

$$(\beta_p/\alpha_n)(M_n - 1)\ln(M_n) \leqslant 1,$$

the multiplication is essentially independent of frequency. Also, under this condition the bandwidth and response time are limited simply by the transit time. These results are the same as for a nonavalanche PIN detector (neglecting RC limitations).

The gain-bandwidth product for values of $M_n > \alpha_n/\beta_p$ and high frequencies is constant and is given by [see Eq. (42)]

$$M(\omega)\omega = \frac{1}{\tau_1} = \frac{1}{N\tau(\beta_p/\alpha_n)} = \frac{1}{N(W/v_n)(\beta_p/\alpha_n)}, \quad (43)$$

Thus, to obtain large gain-bandwidth products, v_n should be large and β_p/α_n and W should be small. It should be noted, however, that these parameters are not independent, and for Si, a decrease in W results in an increase in β_p/α_n. Emmons' results[12] were obtained for injection of electrons into the depletion region at $x = 0$ [cf. the top part of Fig. 5]. Because of the symmetry of the problem, the results given in Fig. 8 also apply to the injection of holes at $x = W$ if the ratio of the electron and hole ionization coefficients, α_n/β_p is used in place of the value given for the ratio β_p/α_n, and hole multiplication M_p is used in place of M_n.

The frequency response of the multiplication $M(\omega)$ for electron velocities equal to and a factor of two greater than the hole velocities has been presented by Chang.[11] His calculations were performed for three different photon absorption (electron–hole pair generation) conditions described by $\exp(-\alpha x)$ with $\alpha = \infty$, $\alpha W = 1$, and $\alpha = 0$, where α is the optical absorption constant and the photons are incident from the left in Fig. 5. For $\alpha = \infty$ the situation is the same as that considered by Emmons[12] and corresponds to electron injection at $x = 0$. The results of Chang's analysis indicate that the cutoff or 3 dB frequency of $M(\omega)$—that is, the quantity $f_{co} = \omega_{co}/2\pi$—is essentially independent of the absorption coefficient α, but increases with the ratio α_n/β_p, and decreases with an effective transit time given by $W(v_n + v_p)/v_n v_p$. Nevertheless, the gain-bandwidth-product is larger for high α since the average low-frequency multiplication is larger when α is high and α_n/β_p is large (see Fig. 7).

These results show that, although the actual frequency response of the multiplication $M(\omega)$ is essentially independent of the spatial generation of the primary electron–hole pairs, it is important that the device structure be chosen to give an optimum low-frequency multiplication. In this manner the highest possible gain-bandwidth-product can be obtained. The actual frequency response of an avalanche photodiode can be limited by either the frequency response of $M(\omega)$, including transit time effects, or by the $R_{eq}C_j$ cutoff. Which of these factors limits the frequency response will be determined not only by the semiconductor material parameters, but also by the device structure, the operating wavelength, and the avalanche gain.

5. Gain Saturation Effects

In the discussion presented above, it was assumed that the gain-bandwidth product is the only limitation to the maximum multiplication obtainable with an avalanche photodiode. In real devices, however, the maximum achievable multiplication at high multiplied-current levels can be limited, even at low frequencies, by several factors: (1) the voltage drop across the load resistor, across the series resistance of the contacts, and across the undepleted bulk material; (2) space-charge effects in which carriers drifting through the depletion region reduce the electric field[14]; (3) thermal resistance which causes a heating of the junction with a subsequent reduction in the electron and hole ionization coefficients (and a corresponding increase in the breakdown voltage); and (4) microscopic defects which limit the electric field that can be maintained over the active area of the device.[15]

The first three of these multiplication-limiting factors can be treated analytically by combining their effects into one effective series resistance, R. Using this approach, Melchior and Lynch[16] have extended an empirical relationship of Miller[17,18] (for the variation of the multiplication with bias voltage) to describe the multiplied photocurrent in the prebreakdown region and account for gain saturation. The multiplication for photogenerated carriers is written as

$$M_{Ph} = (I - I_{MD})/(I_P - I_D) = 1 \bigg/ \left[1 - \left(\frac{V - IR}{V_B} \right)^n \right], \quad (44)$$

where I is the total multiplied current, I_P is the total primary (unmultiplied) current, and I_D and I_{MD} are the primary and multiplied dark currents, respectively. V is the reverse-bias voltage and V_B is the breakdown voltage. The exponent n is an adjustable parameter that is used to fit Eq. (44) to the experimental data. This exponent depends on the material parameters, the device structure, and the wavelength of the incident radiation. Using this relation, Melchior and Lynch demonstrated that, for high light intensities ($I_P \gg I_D$), the maximum photomultiplication which occurs at $V = V_B$ is given by

$$(M_{Ph})_{max} = (V_B/nI_{Ph}R)^{1/2} \equiv I_{max}/I_{Ph}. \quad (45)$$

Thus, the maximum multiplication varies inversely with the square root of the primary photocurrent, $I_{Ph} = I_P - I_D \approx I_P$.

---

[14] W. Shockley, *Solid-State Electron.* **2**, 36 (1961).
[15] H. Kressel, *RCA Rev.* **28**, 175 (1967).
[16] H. Melchior and W. T. Lynch, *IEEE Trans. Electron. Devices* **ED-13**, 829 (1966).
[17] S. M. Miller, *Phys. Rev.* **99**, 1234 (1955).
[18] S. M. Miller, *Phys. Rev.* **105**, 1246 (1957).

When the photocurrent is smaller than the dark current, the maximum multiplication is limited by the dark current and is given by

$$(M_{Ph})_{max} = (V_B/nI_D R)^{1/2}. \tag{46}$$

Thus, it is important that the dark current be as low as possible so as not to limit either the maximum attainable multiplication, because of gain saturation, or the minimum detectable power [as given in Eq. (19)], because of the shot noise associated with the dark current. The high dark current of room-temperature Ge avalanche photodiodes is primarily responsible for the lower maximum multiplication observed for these devices as compared to similar Si avalanche photodiodes.[16]

The exponent n in the modified Miller equation [Eq. (44)] has been shown to be closely related to the electron and hole ionization coefficients α_n and β_p.[19] This exponent attains its smallest value when the carrier with the highest ionization coefficient is injected into the high-field region. Thus, the conditions that give the largest low-frequency multiplication and highest gain-bandwidth product also give the highest possible saturation-limited multiplication.

The saturation of the multiplied photocurrent as limited by the first three factors listed above has recently been considered in a more general way, without resorting to the empirical Miller equation, by Webb et al.[9] They obtain an expression for the photocurrent which shows how the current saturates with multiplication. This expression can be used to estimate the maximum primary photocurrent in the linear range. Their results also indicate that for large values of multiplication and primary photocurrent, the multiplied photocurrent I_{max} only increases with the square root of I_{Ph}, in agreement with Eq. (45).

The fourth factor given above which can limit the maximum achievable multiplication in practical avalanche photodiodes is associated with crystalline imperfections and/or doping nonuniformities which are present in the semiconductor starting material or introduced during device fabrication. This is not a fundamental limitation, but for high-performance avalanche photodiodes these problems must be minimized. The success that has been attained for various semiconductor materials and device structures will be discussed later.

To summarize the results of this section, for the optimum low-frequency multiplication, the highest gain-bandwidth product, and the highest multiplication before gain saturation becomes important, a semiconductor material with greatly different electron and hole ionization coefficients is required. In addition, the device structure must be designed so that the carrier with the higher ionization coefficient is injected into the multiplying

[19] J. Urgell and J. R. Leguerre, *Solid State Electron.* **17**, 239 (1974).

region. In the next section we will see that these same conditions must also be fulfilled to obtain the lowest possible excess-noise factor associated with the avalanche gain process.

III. Multiplication Noise

The electrons and holes that pass through or are generated in the depletion region of a semiconductor diode are collected (emerge from the depletion region) randomly. Since the collection of an individual carrier is independent of the rest of the carriers, this random collection of the carriers follows a Poisson distribution. Therefore, the average current I, which flows through the diode, contributes the mean-square shot-noise current

$$\langle i_s^2 \rangle = 2qIB, \qquad (47)$$

in a bandwidth B.[20] With an ideal or noise-free current multiplication M, the shot-noise current would be multiplied by M and the mean-square shot-noise current would be multiplied by M^2. In an avalanche photodiode the multiplication is not ideal, however, and the impact-ionization process must be described statistically. The multiplication or gain calculated [Eq. (35)] for an electron–hole pair injected at position x_0 in the depletion region of an avalanche photodiode represents the average gain of the injected electron–hole pair. The statistical variation of the multiplication from this average is responsible for the increased noise resulting from avalanche current gain. In this section we will first review the procedures that have been used to describe the excess shot noise due to the multiplication process, summarize the results of these procedures, and then discuss the implications of these results for the design of avalanche photodiode detectors.

6. Excess-Noise Factor

In the early work on the statistical properties of the avalanche gain process, Tager[21] pointed out that the positive feedback characteristic which exists when both α_n and β_p are finite can greatly amplify any initial current fluctuations. He also found that for the same average value of M, the mean-square fluctuation when $\alpha_n = \beta_p$ is about M times greater than when $\beta_p = 0$.

McIntyre[22] extended this analysis to arbitrary values of the ionization coefficients. The procedure he used was to consider a small element dx in the avalanche region where the incremental increase in the electron and hole currents can be described by equations similar to Eqs. (20) and (21).

---

[20] B. M. Oliver, *Proc. IEEE* **53**, 436 (1965).
[21] A. S. Tager, *Fiz. Tverd. Tela* **6**, 2418 (1964) [*English Transl.: Sov. Phys.—Solid State* **6**, 1919 (1965)].
[22] R. J. McIntyre, *IEEE Trans. Electron. Devices* **ED-13**, 164 (1966).

That is,

$$dI_n = (\alpha_n I_n + \beta_p I_p + qG)\,dx, \quad (48)$$

where the current I_n rather than the current density J_n was used. Since each impact-ionization by a given particle is a random event, the probability that a given electron or hole entering the element dx will produce a certain number of secondary electron–hole pairs is described by Poisson statistics. Thus, the current generated in the element dx should contribute the shot noise $2q\,dI_n\,B$ in a bandwidth B. This noise current generated at x will be multiplied in the same manner as dI_n, which is also generated at x.

The resulting change in the noise spectral density of the total current flowing through the diode φ due to multiplication at x in the element dx is given by

$$d\varphi(x) = 2qM^2(x)\,dI_n(x), \quad (49)$$

where $M(x)$ is given by Eq. (35). Integrating over the depletion width,

$$\varphi = 2q\left[I_n(0)M_n^2 + I_p(W)M_p^2 + \int_0^W (dI_n/dx)M^2(x)\,dx\right], \quad (50)$$

where $I_n(0)$ and $I_p(W)$ are the electron and hole currents entering the depletion region at $x=0$ and $x=W$, respectively [cf. Fig. 5], and $M_n = M(0)$ and $M_p = M(W)$ are given by Eqs. (27) and (28). If we use the solution to Eq. (48) and integrate the last term in Eq. (50) by parts, φ can be written as

$$\varphi = 2q\Big\{2[I_n(0)M_n^2 + I_p(W)M_p^2 + \int_0^W GM^2(x)\,dx]$$
$$+ I[2\int_0^W \alpha_m M^2(x)\,dx - M_p^2]\Big\}, \quad (51)$$

where I, the total current (i.e., including multiplication) flowing through the diode, is given by

$$I = I_n(0)M_n + I_p(W)M_p + q\int_0^W GM(x)\,dx. \quad (51a)$$

Thus, if $\alpha_n(E)$, $\beta_p(E)$, and $E(x)$ are known, the noise spectral density can be calculated for any type of carrier injection or spatial generation of electron–hole pairs. Using this result, McIntyre[22] considered several special cases, two of which will be discussed here.

(1) $\alpha_n = \beta_p$. As discussed previously, M is independent of x so $M(x) = M_n = M_p = M =$ constant and using the solution of Eq. (48), Eq. (51) reduces

to
$$\varphi = 2qI_t M^3, \qquad (52)$$

where I_t is the total injected current given by

$$I_t = I_n(0) + I_p(W) + q\int_0^W G(x)\,dx \qquad (52a)$$

and M is given by

$$M = 1/\left[1 - \int_0^W \alpha_n\,dx\right] \qquad (52b)$$

as in Eq. (32). This expression for the noise spectral density is the same as the low frequency limit of the result obtained by Tager.[21] Thus, for $\alpha_n = \beta_p$ the noise spectral density only depends on the total injected current and not on where the current is injected or generated. Since the shot-noise spectral density for ideal (that is, noise free) multiplication is given by $2qI_t M^2$, we can define an excess-noise factor F such that the shot noise with non ideal multiplication is given by F times the shot noise with noise-free multiplication $2qI_t M^2 F$. Thus, for equal ionization coefficients the excess-noise factor is just the average multiplication $F = M$ and can be very large.

(2) $\beta_p = k\alpha_n$. The second special case, considered by McIntyre because of its mathematical simplicity and its applicability to PIN avalanche diodes with constant electric field, is for a constant ratio of the electron and hole ionization coefficients, $k = \beta_p/\alpha_n$. With this condition the excess-noise factor is dependent on where the current is injected or generated and, in fact, can be either greater or smaller than the excess-noise factor for $\alpha_n = \beta_p$. For electron injection into the high field region at $x = 0$ [see Fig. 5], the excess-noise factor for electrons is given by

$$F_n = M_n\{(1 - (1-k)[(M_n - 1)/M_n]^2)\}. \qquad (53)$$

For hole injection at $x = W$, the excess-noise factor for holes is given by

$$F_p = M_p\{1 - [1 - (1/k)][(M_p - 1)/M_p]^2\}, \qquad (54)$$

where M_n and M_p are the electron and hole multiplication factors given by Eqs. (28) and (27), respectively.

The excess-noise factors calculated from these expressions are plotted in Fig. 9 for various values of $k = \beta_p/\alpha_n$. Because of the symmetry of these equations in n, p, k, and $1/k$, only one set of curves is required. In this figure, it can be seen that for a low excess-noise factor the electron and hole ionization coefficients must be greatly different. In addition, the device structure must be designed so that the carrier with the highest ionization coefficient is injected into the high-field region. If this is not the case, the excess-noise factor will actually be worse than that for a device with equal ionization

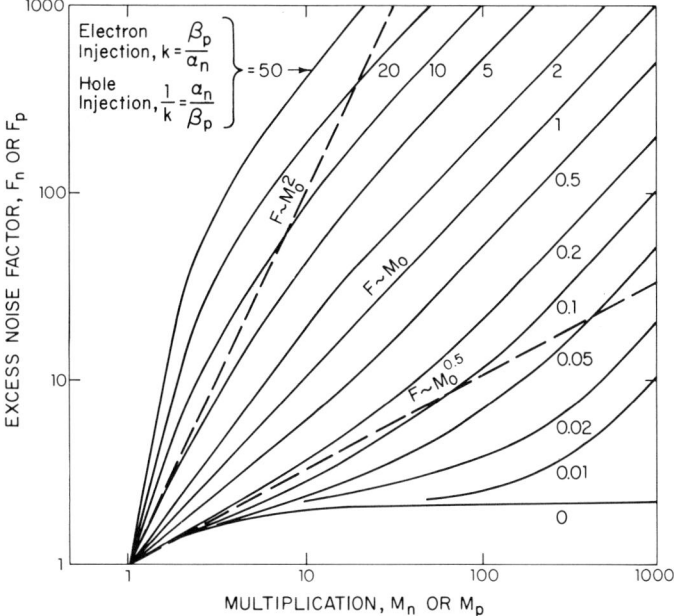

FIG. 9. Excess-noise factors calculated from Eqs. (53) and (54) for various values of multiplication and ratios of the electron and hole ionization coefficients.

coefficients. These results apply to an ideal *PIN* avalanche photodiode, in which the electric field is constant, or to a device fabricated from a semiconductor material with properties such that the ratio of the electron and hole ionization coefficients is constant. As we will see in the next section, there are very few semiconductors for which this is true.

Since $\beta_p/\alpha_n = k =$ constant was such a poor approximation for Si avalanche detectors, McIntyre[23] also examined the validity of the constant k approximation by defining the quantities,

$$k_1 = \int_0^W \beta_p M(x)\,dx \bigg/ \int_0^W \alpha_n M(x)\,dx \qquad (55)$$

and

$$k_2 = \int_0^W \beta_p M^2(x)\,dx \bigg/ \int_0^W \alpha_n M^2(x)\,dx. \qquad (56)$$

In these equations $M(x)$ is the position-dependent multiplication of

[23] R. J. McIntyre, *IEEE Trans. Electron. Devices* **ED-19**, 703 (1972).

electron–hole pairs injected at position x given by Eq. (35). Using these quantities and Eq. (51), the expressions for the excess-noise factors corresponding to Eqs. (53) and (54) can be written as[9]

$$F_e = \frac{k_2 - k_1^2}{1 - k_2} M_n + 2\left[1 - \frac{k_1(1 - k_1)}{1 - k_2}\right] - \frac{(1 - k_1)^2}{M_n(1 - k_2)}$$

$$= k_{\text{eff}} M_n + \left(2 - \frac{1}{M_n}\right)(1 - k_{\text{eff}}) \qquad (57)$$

and

$$F_h = \frac{k_2 - k_1^2}{k_1^2(1 - k_2)} M_p + 2\left[1 - \frac{k_2(1 - k_1)}{k_1^2(1 - k_2)}\right] + \frac{k_2(1 - k_1)^2}{M_p k_1^2(1 - k_2)}$$

$$= k'_{\text{eff}} M_p + \left(2 - \frac{1}{M_p}\right)(1 - k'_{\text{eff}}), \qquad (58)$$

where

$$k_{\text{eff}} = (k_2 - k_1^2)/(1 - k_2) \qquad (59)$$

and

$$k'_{\text{eff}} = k_{\text{eff}}/k_1^2. \qquad (60)$$

Both sets of equations [Eqs. (53), (54) and Eqs. (57), (58)] have been used to analyze the noise performance of avalanche photodiodes and some of these results will be discussed in Part V.

The derivation of the shot-noise spectral density given by McIntyre[22] is valid only in the low frequency limit, since the time dependence of the multiplication as described by Eq. (42) was neglected. A derivation that does include the time dependence of the multiplication process has been given by Naqvi.[24] When $\beta_p = k\alpha_n$, Naqvi's result for the total avalanche shot noise in terms defined above is

$$\langle i_S^2 \rangle = \frac{qI_p(W)M_p^2 F_p B}{1 + \omega^2 \bar{M}^2 \tau^2} + \frac{2qI_n(0)M_n^2 F_n B}{1 + \omega^2 \bar{M}^2 \tau^2}$$

$$+ \frac{2qB}{1 + \omega^2 \bar{M}^2 \tau^2}\left[2\int_0^W GM^2(x)\,dx + \frac{kM_n^2 - M_p^2}{1 - k}\int_0^W GM(x)\,dx\right]. \qquad (61)$$

The first two terms in this equation are the shot noise resulting from the injected hole and electron currents, respectively, and the last term is the shot noise resulting from the space-charge generated current. Each term

---

[24] I. M. Naqvi, *Solid-State Electron.* **16**, 19 (1973).

includes the time dependence of the multiplication process where

$$\bar{M} = (M_n I_n(0) + M_p I_p(W))/I_t \tag{61a}$$

and τ is as defined by Kuvas and Lee.[25]

Figure 10a shows a normalized plot of Eq. (61), neglecting the space-charge generation term, for different fractions of the electron and hole injection currents at a frequency of 30 MHz. The shot noise is normalized to the shot noise which would result for $\alpha_n = \beta_p$. As a means of comparison, the low frequency shot noise for either pure electron or hole injection calculated from Eqs. (53) and (54) for different values of $k = \beta_p/\alpha_n$ is given in Fig. 10b. The curves for $k = 0.1$ in Fig. 10b can be compared directly with the 100% hole current and 100% electron current curves in Fig. 10a. It can be seen that, even for a frequency as low as 30 MHz, the time dependence of the multiplication process decreases the effective shot noise for multiplication values higher than about 100. However, the noise due to the avalanche process will never be underestimated by using the excess-noise factors of Eqs. (53) and (54).

The effects of mixed injection indicated in Fig. 10a can be seen more explicitly by neglecting the multiplication time and space-charge generation in Eq. (61) and writing an effective noise factor F_{eff} defined by

$$\langle i_s^2 \rangle = 2qI_t \bar{M}^2 F_{\text{eff}} B. \tag{62}$$

In this equation

$$I_t = I_p(W) + I_n(0) \quad \text{and} \quad \bar{M} = (I_n(0)M_n + I_p(W)M_p)/I_t.$$

If the fraction of the total injected current due to electrons is $f = I_n(0)/I_t$, the effective noise factor is given by

$$F_{\text{eff}} = [fM_n^2 F_n + (1-f)M_p^2 F_p]/[fM_n + (1-f)M_p]^2. \tag{63}$$

The variation of F_{eff} with M for $k = \beta_p/\alpha_n = 0.005$ and several values of f is shown in Fig. 11. This effective noise factor for mixed injection is always higher than the factor for pure electron injection at the same average gain. Thus, we see once again that it is not enough to have a semiconductor material with greatly different electron and hole ionization coefficients to obtain low noise in an avalanche photodiode, but that, in addition, the device structure must be designed so that the carriers with the higher ionization coefficient are injected into the high-field region.

In the early work on avalanche photodiodes it was recognized that for noise-free multiplication the shot noise would increase as M^2, while for $\alpha_n = \beta_p$ the shot noise would increase as M^3. It was therefore common to describe the multiplied shot noise by $2qI_t M^d$ rather than by $2qI_t M^2 F$. It

---

[25] R. Kuvas and C. A. Lee, *J. Appl. Phys.* **41**, 1743 (1970).

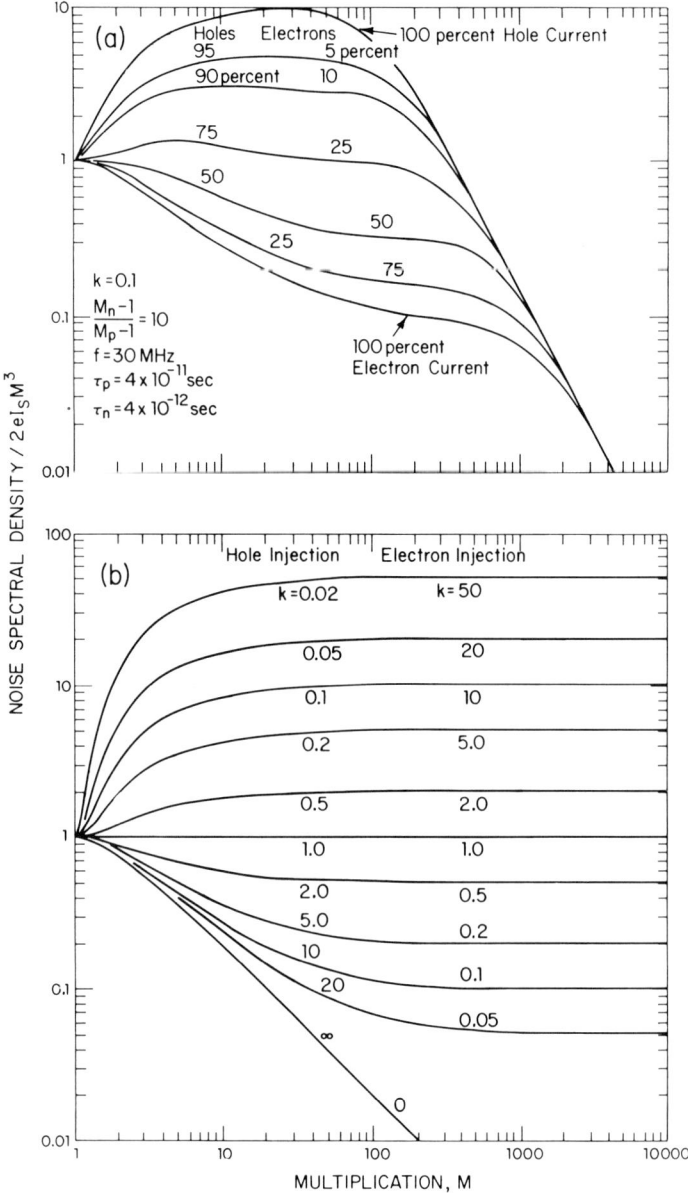

FIG. 10. Calculated avalanche shot noise normalized to the multiplied shot noise for $\alpha_n = \beta_p$, $2eI_sM^3$. (a) For $k = \beta_p/\alpha_n = 0.1$, including the time dependence of the multiplication process for various fractions of electron and hole current. (After Naqvi.[24]) (b) Normalized low-frequency multiplication noise for various values of $k = \beta_p/\alpha_n$ and either electron or hole injection. (After McIntyre.[22])

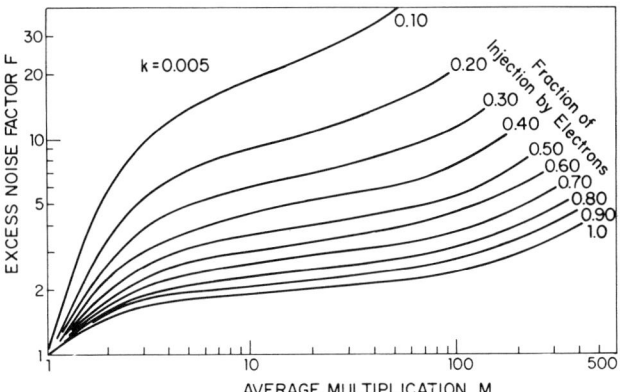

FIG. 11. Low-frequency excess-noise factor as a function of multiplication calculated from Eq. (63) for $k = \beta_p/\alpha_n = 0.005$ for several values of f, the fraction of total injected current due to electrons. (After Webb et al.[9])

can be seen from the above discussion and Figs. 9 and 11 that, unless $\alpha_n = \beta_p$, such an expression can, at best, only apply over a very small multiplication range and then only under certain special circumstances. Unfortunately, it has been generally accepted that the exponent d is a characteristic of the semiconductor material (2.3 to 2.5 for Si, for example). Thus, the expression $2qI_tM^d$ has been used for estimating the performance of optical detection systems using avalanche photodiodes, when in fact the actual excess-noise factor also depends on the device structure and most likely the wavelength of operation. Analyses using this description of the shot noise will not be reliable over a very wide range of operating conditions.

7. Avalanche Gain Statistics

In the introduction, the signal-to-noise ratio of an avalanche photodiode was derived for amplitude-modulated optical power incident on the detector. Using the multiplication and excess-noise factors of the previous sections, the (S/N) ratio of a specific avalanche photodiode structure can be evaluated for a given incident wavelength, modulation frequency, and set of operating conditions. In many of the applications for avalanche photodiodes, however, the optical signal will not be amplitude modulated, but will consist of some form of pulse-coded modulation. For this type of modulation the factor of interest rather than the (S/N) ratio is the error probability P_e; that is, the probability that a pulse will not be detected (a miss) or the probability that a pulse will be detected when none is present (a false alarm).

It can be shown that, for detectors in which the collection of electron–hole pairs follows a Poisson distribution, the error probability can be calculated directly from the power (S/N) ratio[26]: $P_e = \exp[-(S/N)/4]$. However, we have just seen that, although the probability of a given electron or hole injected into the depletion region producing a certain number of electron–hole pairs is described by Poisson statistics, the total number of electron–hole pairs that will be collected is not. This results in a probability distribution of avalanche gain considerably different from the Gaussian distribution that would result if the total number of collected electron–hole pairs were Poisson distributed. The actual gain distribution in avalanche photodiodes has been studied in two different ways by Personick[27,28] and McIntyre[23] with essentially equivalent results. The results of this analysis can be applied to the error rates of pulse-coded detection systems. Unfortunately, the error-rate analysis is much more complicated than the signal-to-noise analysis given in Part I, so that the results cannot be described in any general sense.

McIntyre[9,29] has considered a special case and compared the results obtained for his gain probability distribution with the results that would be obtained for a Gaussian distribution. The actual gain probability distribution was approximated by

$$P(x) = \frac{1}{(2\pi)^{1/2}} \frac{1}{[1+(x/\lambda)]^{3/2}} \exp\left[-\frac{x^2}{2[1+(x/\lambda)]}\right] \quad (64)$$

for

$$-[(F_e - 1)/F_e][(M-1)/M] < x/\lambda < \infty,$$

where $x = (m - n_e M)/\sigma$, $\sigma^2 = n_e M^2 F_e$, $\lambda = (n_e F_e)^{1/2}/(F_e - 1)$. The number of electrons detected is m, while the number of electrons injected is n_e. Equation (64) applies to pure electron injection which obeys a Poisson distribution. The corresponding Gaussian probability distribution is

$$P(x) = (2\pi)^{-1/2} \exp(-\tfrac{1}{2}x^2). \quad (65)$$

It can be seen that for large values of λ, Eq. (64) is approximately the Gaussian distribution of Eq. (65). This regime can be approached for large values of n_e, the number of electrons injected. For small values of n_e, however, the two distributions are significantly different. Conradi has experimentally confirmed the probability distribution given by Eq. (64).[29a]

[26] R. S. Kennedy and E. V. Hoversten, unpublished (1970); see also W. K. Pratt, *in* "Laser Communication Systems," p. 209. Wiley, New York, 1969.
[27] S. D. Personick, *Bell Syst. Tech. J.* **50**, 167 (1971).
[28] S. D. Personick, *Bell Syst. Tech. J.* **50**, 3075 (1971).
[29] R. J. McIntyre, *IEEE Int. Electron. Devices Meeting, Washington, D.C., 1973*. Tech. Digest, p. 213, 1973.
[29a] J. Conradi, *IEEE Trans. Electron. Devices* **ED-19**, 713 (1972).

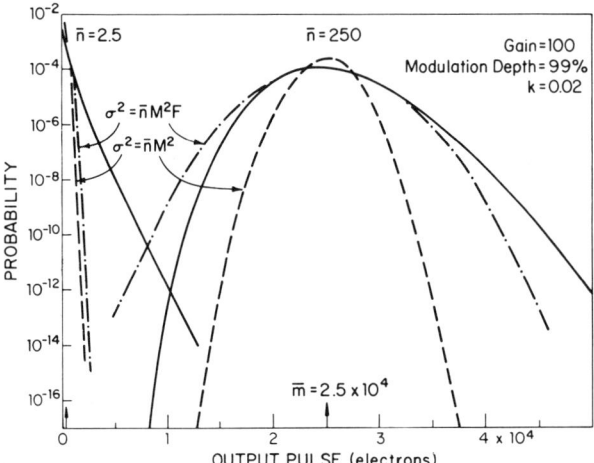

FIG. 12. Gain distributions of on and off pulses for 99% modulation with 250 primary photoelectrons in the on pulse, as calculated from Eq. (64). The dashed curves are Gaussian distributions with and without excess noise. (After Webb et al.[9])

To demonstrate the difference in the two distributions, Webb et al.[9] have calculated, using Eqs. (64) and (65), the amplitude distribution of gain for two pulses with 250 and 2.5 primary electrons. These pulses could conceivably correspond to the "on" and "off" pulses in a pulse-coded modulation system with a modulation depth of 99%. The calculations were performed for a diode operating at a gain of 100 with $k_{\text{eff}} = 0.02$. The probability distributions of the collected electrons for the noise-free Gaussian distribution ($F_e = 1$), the Gaussian distribution, and the correct distribution are shown in Fig. 12 for each pulse. It can be seen that the use of a Gaussian distribution to determine the discriminator setting in a pulse-coded detection system would result in a much higher error rate than would be necessary with the correct discriminator setting.

To show how the minimum number of injected electrons required to give a certain error rate varies with $k = \beta_p/\alpha_n$, Webb et al.[9,29] calculated the data shown in Table I. This table gives the minimum number of primary electrons n_{min} and the corresponding gain M_{opt}, which are required to achieve a bit-error-rate of 10^{-10} for both the "on" and "off" pulses given a modulation depth of 99%. The calculations were made for two settings of an amplifier discriminator set to pass only pulses containing more than m electrons. In a practical system the higher the thermal and excess noise of the amplifier is, the higher m must be. The number of primary photoelectrons required to obtain the same bit-error-rate with a noise-free detector ($F_e = 1$), as determined from the Gaussian distribution of Eq. (65) and the distribution of

TABLE I

Comparison of Minimum Number of Primary Electrons n_{min} Required from an "On" Pulse of a PCM Optical Communications System to Give a 10^{-10} Bit Error Rate[a]

	Ideal detector ($F = 1$)		Aval. photodiode		
	Poisson dist.	Eq. (64)	$k = 0$	$k = 0.02$	$k = 1$
(a) $m = 2 \times 10^4$					
n_{min}	46	49	115	328	1820
M_{opt}	large	—	550	150	25
F_e	1	1	2	5.0	25
n_{min}/F_e	46	49	57	65	73
(b) $m = 4 \times 10^3$					
n_{min}	46	49	115	198	793
M_{opt}	large	—	110	54	11.0
F_e	1	1	2	3.0	11.0
n_{min}/F_e	46	49	57	66	72

[a] Modulation depth is 99%, for two different threshold settings of m electrons: (a) $m = 2 \times 10^4$ electrons; (b) $m = 4 \times 10^3$ electrons. The minimum value of m depends on the amplifier noise. (After Webb et al.[9])

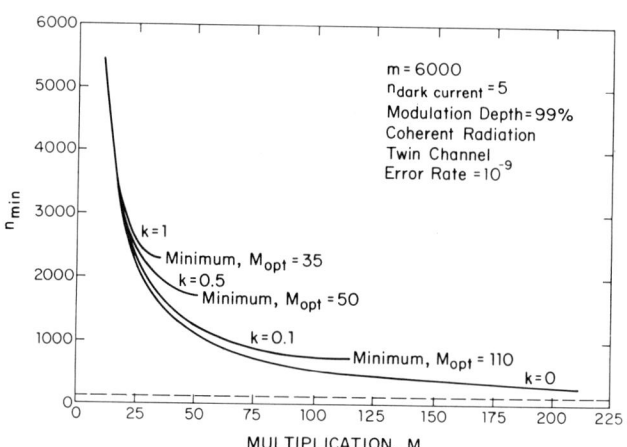

FIG. 13. Calculated minimum number of primary photoelectrons, n_{min}, required for an error rate of less than 10^{-9} in a 99% modulated twin-channel system in which the thermal noise of the amplifier is assumed to be $m = 6 \times 10^3$ electrons. The required multiplication for the smallest value of n_{min} is given for the indicated values of $k = \beta_p/\alpha_n$. (After Personick.[28])

Eq. (64), are also given in Table I. These results show that the minimum number of photoelectrons increases rapidly and the corresponding optimum gain or multiplication M_{opt} decreases as $k = \beta_p/\alpha_n$ increases from 0 to 1.

Another example calculated by Personick[28] is shown in Fig. 13. Here the minimum number of photoelectrons required for a 99% pulse-coded modulated twin-channel system is plotted as a function of avalanche gain. For this example a value of $m = 6 \times 10^3$ was used for the thermal noise of the amplifier. These results show the same behavior as indicated in Table I; that is, for any $k \neq 0$ there is an optimum gain which permits the detection of the smallest number of photoelectrons. The value of M_{opt} increases with decreasing k. Personick[28] has pointed out that, with $k = 0.1$ and $M_{opt} = 100$, the required energy per pulse is within 10 dB of that required with $k = 0$ and gains of several thousand.

IV. Electron and Hole Ionization Coefficients

In the previous two sections, it was shown that the ratio of the electron and hole ionization coefficients is the most significant factor in determining the ultimate sensitivity of a properly designed avalanche photodiode, whether for use in an amplitude or pulse-coded modulation detection system. Since the ionization coefficients are so important in the performance and design of practical detectors, in this part we will first describe ionization coefficients from a theoretical point of view. We will then discuss methods of measuring ionization coefficients and the experimental results of these measurements for particularly interesting semiconductors.

8. GENERAL DISCUSSION

The general problem of impact ionization in semiconductors has recently been reviewed by Chynoweth in a previous volume of this series[30] and also by Mönch.[31] Since these reviews were completed, however, there has been a considerable amount of theoretical and experimental work reported in the literature. Thus, after a brief review of the earlier theoretical work, we will discuss in detail some of these more recent results.

When an electric field is applied to a semiconductor, the free carriers (electrons or holes) gain energy from the field as they drift through the lattice and then lose energy in collisions with (or scattering by) phonons, impurities, or other crystal defects. At low fields there is a balance between the energy gained from the field and that lost by collisions, so that the free carriers are in equilibrium with the lattice. At high electric fields the free carriers

[30] A. G. Chynoweth, in "Semiconductors and Semimetals" (R. K. Willardson and A. C. Beer, eds.), Vol. 4, p. 263. Academic Press, New York, 1968.
[31] W. Mönch, *Phys. Status Solidi* **36**, 9 (1969).

gain energy from the field faster than they lose it, until they have sufficient energy to generate an additional electron–hole pair through impact ionization. Once a carrier has acquired the threshold energy for impact ionization, it has one mean-free-path to create an electron–hole pair before it suffers a collision that reduces its energy below the threshold energy. Impact ionization reduces the energy of the initial free carrier to essentially zero. The initial carrier plus the generated electron–hole pair then gain energy from the electric field to produce additional pairs as they move through the lattice. This results in avalanche multiplication and, at high enough electric fields, avalanche breakdown.

The electron and hole impact-ionization coefficients are used to describe the average distance a carrier will travel in an electric field before generating an electron–hole pair by impact ionization. The general procedure for calculating the carrier ionization coefficient is to determine the distribution of carriers (with velocity v and energy $\mathscr{E}$) as a function of electric field. Various approximations and assumptions have been used to determine this distribution from the Boltzmann transport equation, but once the distribution is known, the impact-ionization coefficient can be obtained by calculating the reciprocal of the average distance between ionizing collisions.

In the first theoretical work on impact-ionization coefficients, Wolff[32] assumed that electrons experience only two types of collisions: collisions with optical phonons and collisions with the lattice that produce impact ionization. In his solution of the Boltzmann equation Wolff argued that the mean-free-path for ionizing collisions would be much smaller than that for phonon collisions. This would tend to keep the electron velocity distribution almost spherically symmetric (that is, a displaced Maxwellian distribution), especially at high electric fields. With this assumption the ionization coefficient was calculated to have the form

$$\alpha(E) = A \exp(-b/E). \tag{66}$$

Refinements by Moll and co-workers[33,34] in the procedure used by Wolff have resulted in essentially the same field variation of $\alpha(E)$.

Another approximate form of the distribution function was used by Shockley.[35] He assumed that the only electrons which can gain sufficient energy to cause impact ionization are the few "lucky" ones that are accelerated to the ionization threshold energy in one free path. This results in a high-velocity spike in the distribution function in the direction of the electric field. The treatment by Shockley results in ionization coefficients of the

[32] P. A. Wolff, *Phys. Rev.* **95**, 1415 (1954).
[33] J. L. Moll and R. van Overstraeten, *Solid State Electron.* **6**, 147 (1963).
[34] D. J. Bartelink, J. L. Moll, and N. Meyer, *Phys. Rev.* **130**, 972 (1963).
[35] W. Shockley, *Solid State Electron.* **2**, 35 (1961).

form

$$\alpha(E) = A \exp[-(b/E)^2], \quad (67)$$

which should apply in the low electric-field range. Moll and Meyer[36] have corrected Shockley's treatment to allow for electrons that suffer just one collision while being accelerated to the threshold energy. Their results change the form of Eq. (67) significantly for finite values of the electric field.

The most general theory of the ionization rates was given by Baraff.[37,38] He solved the Boltzmann equation for a distribution function which was assumed to consist of a spherical part and a spike-shaped part. The assumptions used by Baraff to evaluate this solution numerically were (1) the energy bands in momentum space are parabolic; (2) only scattering by optical phonons is important (except for impact ionization), and the lattice temperature is so low that only emission of optical phonons is important. This results in an energy loss equal to the optical phonon energy $\mathscr{E}_R$; (3) the mean-free-path for optical phonon emission is λ_R which is independent of energy; (4) for electron energies greater than the threshold ionization energy $\mathscr{E}_i$, the mean-free-path for impact ionization λ_i is constant; (5) the scattering is spherically symmetric. The results obtained numerically by Baraff using the additional assumptions that

$$r(\mathscr{E}) = \sigma_i(\mathscr{E})/\sigma_T(\mathscr{E}) = 0 \quad \text{for} \quad \mathscr{E} < \mathscr{E}_i,$$

where $\sigma_i(E)$ is the ionization collision cross section and $\sigma_T(\mathscr{E})$ is the total collision at energy $\mathscr{E}$, and that $\sigma_i = \sigma_R$, where σ_R is the cross section for optical phonon emission, for $\mathscr{E} > \mathscr{E}_i$ (or equivalently, since $\sigma_T = \sigma_i + \sigma_R$, that $r = 0.5$ for $\mathscr{E} > \mathscr{E}_i$), are shown in Fig. 14. The calculated curves are normalized and presented in the form $\alpha\lambda_R$ versus $\mathscr{E}_i/eE\lambda_R$ for various values of $\mathscr{E}_R/\mathscr{E}_i$, the ratio of the phonon energy loss per collision to the threshold energy. The assumption $r = 0.5$ for $\mathscr{E} > \mathscr{E}_i$ was not based on any physical evidence, but corresponds to a situation where the ionization cross section rises abruptly at energy $\mathscr{E}_i$ to the same values as σ_R. However, further calculations for various values of r at a fixed value of $\mathscr{E}_R/\mathscr{E}_i$ indicated that the ionization rate was only weakly dependent on r. Baraff concluded that the curves in Fig. 14 would be useful for $0.25 \leq r \leq 1.0$. Neglecting the uncertainty in r, for a given material the parameter $\mathscr{E}_R/\mathscr{E}_i$ is in general different for electrons or holes since in general $\mathscr{E}_i$ is different for electrons and holes. Thus, there should be one curve applicable to electrons and another to holes. These curves have been used with experimentally determined variations of $\alpha(E)$ with field to obtain λ_R and $\mathscr{E}_i$.

[36] J. L. Moll and N. Meyer, *Solid State Electron.* **3**, 155 (1961).
[37] G. A. Baraff, *Phys. Rev.* **128**, 2507 (1962).
[38] G. A. Baraff, *Phys. Rev.* **133**, A26 (1964).

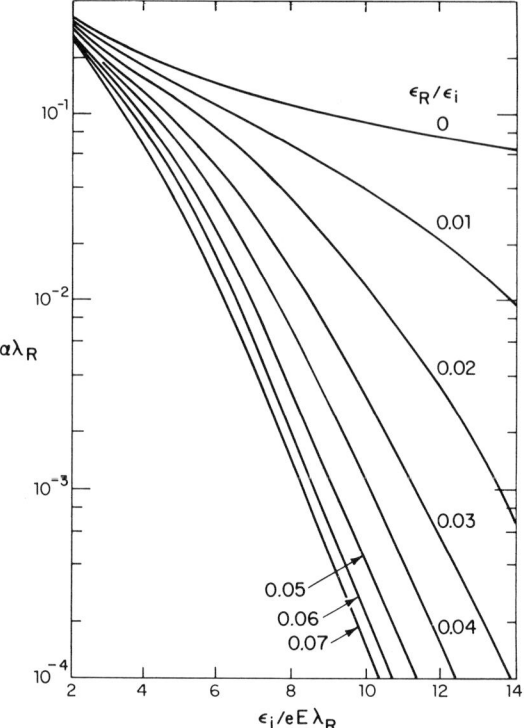

FIG. 14. Normalized universal Baraff curves of the ionization rates versus reciprocal electric field for different ratios of the phonon energy to the impact-ionization energy $\mathscr{E}_R/\mathscr{E}_i$. (After Baraff.[37])

In a study of the temperature dependence of avalanche multiplication, Crowell and Sze[39] pointed out that the effects of optical phonon absorption as well as optical phonon emission or generation should be included, particularly for Ge and GaAs which have optical phonon energies of the order of kT at room temperature. They suggested that in place of $\mathscr{E}_R$ an average energy lost per collision $\langle \mathscr{E}_R \rangle$ could be used in Baraff's theory. They also give an expression for the variation of $\langle \mathscr{E}_R \rangle$ with temperature,

$$\langle \mathscr{E}_R \rangle / \mathscr{E}_R = 1/(2N+1) = \lambda_R/\lambda_o, \tag{68}$$

where

$$N = [\exp(\mathscr{E}_R/kT) - 1]^{-1}$$

and λ_o is the mean-free-path for optical phonon generation in the low-

[39] C. R. Crowell and S. M. Sze, *Appl. Phys. Lett.* **9**, 242 (1966).

temperature limit. The Baraff curves of Fig. 14 were approximated (within $\pm 2\%$ for $0.01 < p < 0.06$ and $5 < x < 16$) by

$$\alpha\lambda_R = \exp[(11.5p^2 - 1.17p + 3.9 \times 10^{-4})x^2$$
$$+ (46p^2 - 11.9p + 1.75 \times 10^{-2})x - (757p^2 - 75.5p + 1.92)], \qquad (69)$$

where

$$p = \langle \mathscr{E}_R \rangle / \mathscr{E}_i, \qquad x = \mathscr{E}_i/eE\lambda_R.$$

With Eqs. (68) and (69) the results of Baraff's calculation can be applied to the analysis of impact-ionization coefficients at different temperatures.

Keldysh[40] has used a general analytic solution of the Boltzmann equation to calculate ionization coefficients. The field dependence of the ionization coefficients he obtained has been compared with Baraff's results by Mönch.[31]

The treatments discussed above all assumed parabolic energy bands and considered the isotropic scattering by optical phonons to be the only energy loss mechanism (except for the impact-ionization process). Dumke[41] has developed a theory of avalanche breakdown, appropriate to InSb and InAs, in which allowance is made for nonparabolic conduction bands and in which the dominant scattering mechanism is anisotropic polar-mode scattering. In polar-mode scattering the preference for small-angle scattering causes the electron distribution to be peaked in the direction of the electric field, similar to the distribution assumed by Shockley.[35] Dumke's results are presented in terms of electron generation rates which can be converted to ionization coefficient data by using the drift velocity corresponding to a given electric field.

The threshold energy for impact ionization $\mathscr{E}_i$ can be calculated from the conduction and valence band structure for a given material using energy and momentum conservation. For parabolic bands with equal electron and hole effective masses, $\mathscr{E}_i = 3E_g/2$. In the early work $\mathscr{E}_i$ was either used as an adjustable parameter to fit experimental results or calculated from elementary considerations of the energy band structure.[42] From these results it was erroneously concluded that the hole ionization energy was always larger than the electron ionization energy. Recently, calculations have been obtained for more detailed band structure.[43-45] The results of Anderson

[40] L. V. Keldysh, *Zh. Eksp. Teor. Fiz.* **48**, 1692 (1965) [*English Transl.: Sov. Phys.—JETP* **21**, 1135 (1965)].
[41] W. P. Dumke, *Phys. Rev.* **167**, 783 (1968).
[42] J. R. Hauser, *J. Appl. Phys.* **37**, 507 (1966).
[43] C. L. Anderson and C. R. Crowell, *Phys. Rev. B* **5**, 2267 (1972).
[44] D. L. Camphausen and C. J. Hearn, *Phys. Status Solidi (b)* **50**, K139 (1972).
[45] R. A. Balliager, K. G. Major, and J. R. Mallenson, *J. Phys. C: Solid State Phys.* **6**, 2573 (1973).

and Crowell[43] show that the idea of an ionization threshold must be used cautiously. There are several different thresholds for electrons or holes traveling in the same crystalline direction, and the thresholds for the corresponding ionization processes are different for different crystalline directions. In constrast to the conclusions of Hauser,[31,42] they also find that the hole ionization energy is not always lower than that of the electrons; that is, for GaAs in all three principal crystalline directions holes have the lowest ionization threshold, while for Ge electrons have the lowest threshold energy in the $\langle 100 \rangle$ and $\langle 111 \rangle$ directions and the holes have the lowest threshold in the $\langle 110 \rangle$ direction. Although no theoretical calculations for λ_R have been made, it seems improbable from these results and from the calculations of Baraff[37] and Dumke[41] that the electron and hole ionization coefficients would fortuitously have the same magnitude and field dependence.

In the following section we will discuss next the methods used to measure the electron and hole ionization coefficients.

9. Measurement Methods

Several procedures have been used to obtain the ionization rates of semiconductors from experimental measurements, including analyses of photomultiplication measurements, breakdown voltage data, and microwave device performance characteristics. Since photomultiplication measurements have been used most extensively, this method will be discussed in some detail.

From the results in Section 3 it is clear that it is a relatively simple matter to calculate the multiplication for an arbitrary electric field variation and distribution of injected carriers, when the electron and hole impact-ionization coefficients are known as a function of electric field. Graphical[45a] and analytical[45b] methods are also available to determine the multiplication factors with various approximations. The inverse of this procedure, determining the magnitude and field variation of α_n and β_p from the variation of multiplication of injected current with applied voltage, is not as straightforward. This is the method, however, that has generally been used most successfully.

In the earliest measurements of ionization coefficients[3,46] either light or α particles were used to inject carriers into the junctions. Either of these techniques resulted in the injection of both electrons and holes into the multiplication region. The resulting multiplication data were analyzed by assuming that $\alpha_n = \beta_p$ [Eq. (32)] and using an effective barrier width W and either an effective "uniform" electric field[3] or a linear or a parabolic electric-field variation in the depletion region.[46]

[45a] C. D. Bulucea and D. C. Prisecaru, *IEEE Trans. Electron. Devices* **ED-20**, 692 (1973).
[45b] P. Spirito, *IEEE Trans. Electron. Devices* **ED-21**, 226 (1974).
[46] K. G. McKay, *Phys. Rev.* **94**, 877 (1954).

The first attempt to determine α_n and β_p separately was made by Miller[17] who used PNP and NPN transistors to obtain pure hole or electron injection, respectively, from the emitter junctions. Subsequently, Miller[18] used complementary P^+N and N^+P junctions, injecting carriers from the high-resistivity side of the junction with light, to obtain α_n and β_p.

The first measurements in which α_n and β_p were separately determined from photomultiplication measurements on the *same* junction were reported by Wul and Shotov.[47] They used specially prepared junctions in which both the P-side and the N-side of the junction could be illuminated with short-wavelength light to obtain either pure electron or pure hole injection, respectively. With this method the uncertainties in obtaining results on different junctions (different possible electric field configurations and scattering environments) were avoided. The analysis of Wul and Shotov[47] assumed a constant value of the ratio $k = \beta_p/\alpha_n$.

Lee et al.[7] showed that if the measurements of the electron and hole multiplication, M_n and M_p, are made in the same junction, there is a simple relationship between the measured values of multiplication and the quantity

$$\int_0^W (\alpha_n - \beta_p)\, dx,$$

which can be used to solve the integral equations exactly, without any assumptions regarding the relative magnitudes of α_n and β_p. From Eqs. (27) and (28) this relationship is

$$\varphi(W) = \int_0^W (\alpha_n - \beta_p)\, dx = \ln(M_n/M_p). \tag{70}$$

Because of the difficulty in extracting the ionization rates from multiplication measurements and because the reliability of the ionization rates obtained from these measurements depends critically on the electric field distribution and the type of current injection, it is worthwhile to carefully design the experiment to simplify the subsequent analysis. Some of the conditions that the ideal experiment should meet are as follows[30]:

(1) Pure electron and pure hole injection should be obtained in the same junction rather than in complementary junctions. Pure injected currents are required to accurately determine α_n and β_p when the ionization rates are significantly different. If the measurements of M_n and M_p are made in the same junction where the electric field configuration and scattering environment are identical, Eqs. (27) and (28) can be combined to determine α_n and β_p.

(2) The magnitude of the electric field and the field profile for the device

[47] B. M. Wul and A. P. Shotov, *Solid State Phys. Electron. Telecommun.* **1**, 491 (1960).

should be accurately known, since the ionization coefficients can be very sensitive to variations of electric field. In addition, the analysis can be considerably simplified if the field profile is constant, linear, or parabolic.

(3) The experiment should be designed so that the quantum efficiency and/or photocurrent without avalanche gain can be accurately determined. In particular, the quantum efficiency without avalanche gain or the collection efficiency should be independent of the reverse-bias voltage applied to the device. This is important because the low multiplication values (where the difference in M_n and M_p is largest) are most useful for determining α_n and and β_p separately [see Eq. (70)]. Also, small errors in the determination of the photocurrent without multiplication can cause significant errors in the determination of M_n and M_p.

(4) The high-field region of the junction should be of sufficient width so that the injected carriers in any given region can be considered to have the energy distribution characteristic of the electric field in that region. If the variation of the electric field with position in the junction is too rapid this is not a valid assumption, and corrections must be made in the analysis of the experimental data for the "dead space" or distance an injected carrier must travel before it acquires sufficient energy to cause impact ionization. This dead-space correction depends on the ionization threshold $\mathscr{E}_i$ and several methods involving end corrections have been used to estimate $\mathscr{E}_i$.[30,33]

(5) The multiplication across the active area of the device must be uniform; that is, if measurements are to be made close to the true breakdown voltage of the junction, devices with microplasmas should not be used. However, it has been shown[48] that for voltages below the "turn-on voltage" of microplasmas, the multiplication factors are not influenced by the presence of the microplasmas. Thus, the ionization rates derived from measurements made under these conditions should be reliable.

Neglecting the details of how a given device structure can be fabricated in a particular semiconductor material, several experimental arrangements, which meet the above requirements with varying degrees of success, are discussed below.

a. PIN Junctions

The simplest configuration for the measurement and the analysis of photomultiplication measurements is the *PIN* junction shown in Fig. 15a. With this arrangement it is possible to obtain very pure electron and hole injection by completely absorbing high energy photons ($h\nu \gg E_g$) in the P^+ and N^+ regions. If the carrier diffusion lengths are sufficiently long or if

---

[48] R. van Overstraeten and H. DeMan, *Solid-State Electron.* **13**, 583 (1970).

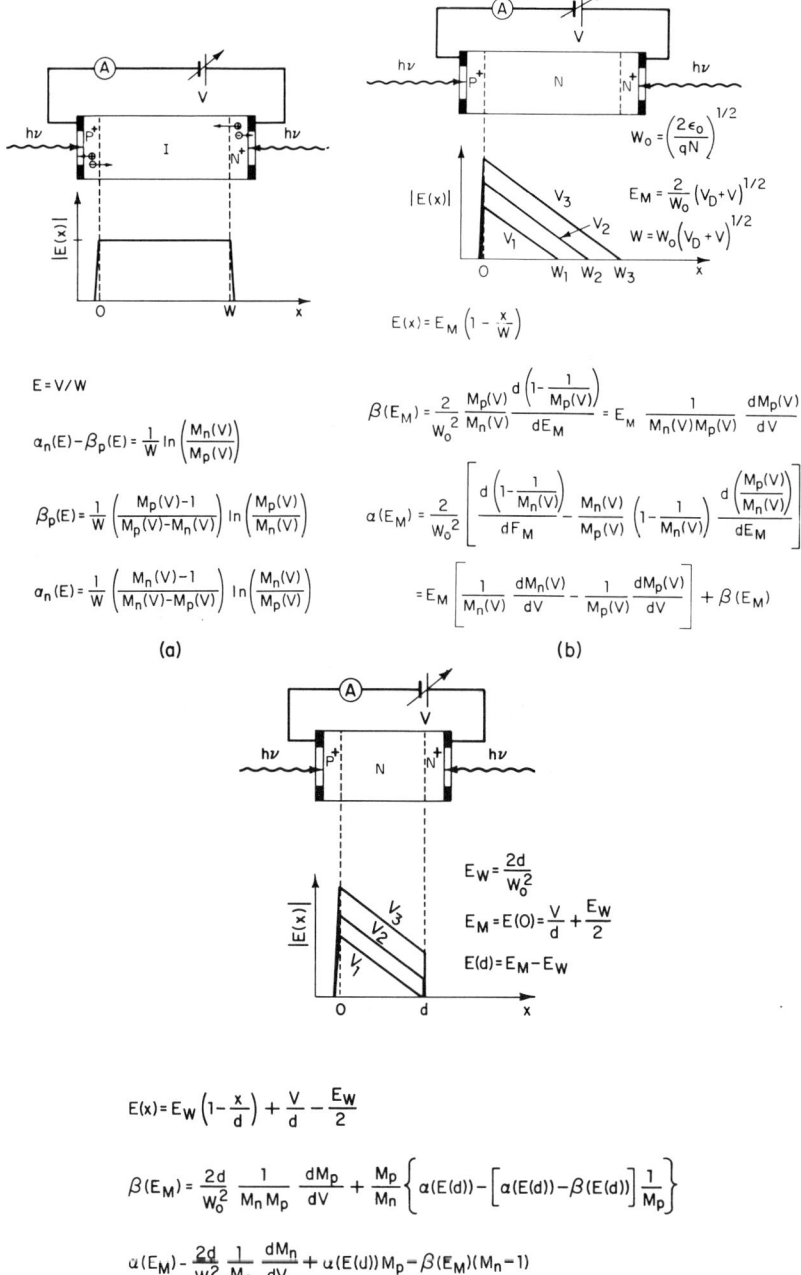

FIG. 15. Experimental arrangements and equations for determining electron and hole ionization coefficients from photomultiplication measurements. (a) *PIN* junction. (b) Abrupt *PN* junction. (c) Abrupt *PN* junction at punch-through.

the P^+ and N^+ regions are sufficiently heavily doped, the small increase in the depletion width of the P^+ and N^+ regions with bias voltage will cause a negligible change in the collection efficiency. Since the electric field is constant in this structure ($E = V/W$, where V is the total voltage across the intrinsic region), the ionization coefficients are constant, and Eqs. (27), (28), and (70) can be solved to obtain the ionization rates given in Fig. 15a. Aside from the technological problems of fabricating this device structure, it is probably the most desirable configuration for determining the ionization coefficients.

b. Abrupt PN Junctions

The structure that next to the PIN junction permits the easiest analysis is the abrupt PN junction shown in Fig. 15b. In this structure, the electric field varies linearly with distance in the depletion region, and with this field variation the values of $\alpha(E_m)$ and $\beta(E_m)$ can be determined from Eqs. (27) and (28) by taking the derivatives with respect to the maximum electric field E_m. The resulting expressions for $\alpha(E_m)$ and $\beta(E_m)$ are shown in Fig. 15b and are only slightly more complicated than those for the PIN structure. In these expressions, V_D is the diffusion voltage for the device being considered, and the applied voltage V is positive for reverse bias.

The abrupt PN junction configuration is not as suitable for the determination of $\alpha(E_m)$ and $\beta(E_m)$ as the PIN structure for two reasons. The first and most important is that the collection efficiency for the injection of holes from the N^+ contact depends on the width of the depletion region and therefore on the applied bias voltage. This requires that corrections for variations in collection efficiency be made when calculating the hole multiplication.[49] The uncertainties in this correction make the lower values of M_p (which are important in the separate determination of α_n and β_p) less reliable. The second less desirable feature of the abrupt PN junction configuration is that the analysis requires the differentiation of the experimental multiplication data. Careful experimental techniques and sophisticated methods of determining the derivatives are necessary to minimize both the noise in the experimental data and that introduced by the numerical differentiation.

c. Abrupt PN Junction at Punch-Through

The problem of the variation in collection efficiency with voltage for the abrupt PN junction case can be eliminated by fabricating devices so that the depletion region punches through or reaches through to the N^+ layer as shown in Fig. 15c. In this configuration the conditions for constant collection efficiency are the same as in the PIN junction. By writing the

[49] M. H. Woods, W. C. Johnson, and M. A. Lampert, *Solid-State Electron.* **16**, 381 (1973).

multiplication integrals of Eqs. (27) and (28) in terms of the electric field and then taking the derivatives with respect to E_m as before, we can obtain the expressions for $\alpha(E_m)$ and $\beta(E_m)$ shown in Fig. 15c. In this figure E_w is the maximum electric field at the punch-through voltage, E_m is the maximum electric field at $x = 0$ and $E(d)$ is the electric field at the edge of the N^+ region for a given applied voltage V. For bias voltages less than that required to achieve punch-through, the relations given in Fig. 15b should be used. For bias voltages greater than the punch-through voltage the expressions shown in Fig. 15c should be used.

By continuing the measurements from voltages less than punch-through to voltages greater than punch-through, the values of $\alpha[E(d)]$ and $\beta[E(d)]$ required in the determination of $\alpha(E_m)$ and $\beta(E_m)$ can be obtained. For applied voltages where multiplication just begins to occur, $\alpha[E(d)]$ and $\beta[E(d)]$ will be negligible when the slope of $E(x)$ is large enough, and the lowest values of $\alpha(E_m)$ and $\beta(E_m)$ can be easily determined. Where d and/or the slope of $E(x)$ are so small that the difference between E_m and $E(d)$ lead to small differences between $\alpha(E_m)$ and $\alpha[E(d)]$ and $\beta(E_m)$ and $\beta[E(d)]$, this analysis is less appropriate and it is better to use the PIN approximation.

d. Linearly Graded PN Junction

In many cases of practical interest, PN junctions are prepared by impurity diffusion and under certain conditions the resulting junctions have an approximately linear net impurity concentration gradient, which results in a parabolic field distribution. This field configuration was first used in the determination of ionization coefficients by McKay[50] with the assumption that $\alpha_n = \beta_p$. Lee et al.[7] showed that for the parabolic field variation, Eqs. (27) and (28) can be written as

$$1 - \frac{1}{M_p} = 2 \exp\left[\int_0^{E_m} (\alpha - \beta)\, dy\right] \times \int_0^{E_m} \beta(E) \cosh\left[\int_E^{E_m} (\alpha - \beta)\, dy'\right] dy, \quad (71)$$

and

$$1 - \frac{1}{M_n} = 2 \exp\left[-\int_0^{E_m} (\alpha - \beta)\, dy\right] \times \int_0^{E_m} \alpha(E) \cosh\left[\int_E^{E_m} (\alpha - \beta)\, dy'\right] dy, \quad (72)$$

where

$$y = [-(w/2)[1 - (E/E_m)]^{1/2}].$$

By restricting the range of multiplication values used in their analyses, Lee et al.[7] were able to approximate Eqs. (71) and (72) in the form of Abel's integral equation which gave solutions for $\alpha(E_m)$ and $\beta(E_m)$ in the form of the derivatives of numerically evaluated integrals.

[50] K. G. McKay, *Phys. Rev.* **94**, 877 (1954).

e. Other Structures

The experimental arrangements just described all permit the determination of α_n and β_p without any assumptions regarding the relative values of the ionization coefficients. The simplicity of the analyses in these structures results from the particular form of the electric field dependence on x and the use of multiplication data determined with pure electron and pure hole injection. However, the requirement of uniform junctions for the multiplication measurements has generally precluded the fabrication of structures with field distribution like those described above in which pure hole and electron injection can be obtained by irradiating the N^+ and P^+ regions, respectively, with strongly absorbed radiation.

A structure that is used in many practical avalanche photodiode detectors and consequently has been used in many of the experimental measurements of ionization coefficients is shown schematically in Fig. 16. The P^+ region is formed by diffusion, and for detector applications the diffused region is thin so that the quantum efficiency at shorter wavelengths will be high. For electron multiplication measurements the wavelength of the incident radiation is chosen so that the photon energy is much greater than the bandgap and the radiation is completely absorbed in the P^+ region. Thus, the electrons which are generated within a diffusion length of the edge of the depletion region will be collected and will result in pure electron injection. The electron multiplication factor can then be determined as accurately as in the idealized cases shown in Fig. 15.

Hole multiplication measurements are made by using long wavelength (low energy) radiation that is weakly absorbed. The weakly absorbed radiation generates electron–hole pairs throughout the P^+ region, the depletion region, and part or all of the N region, depending upon the absorption

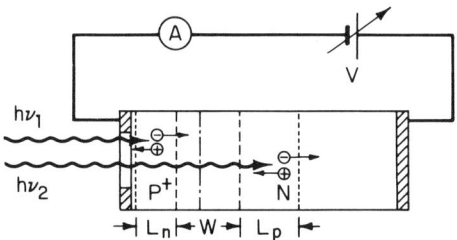

FIG. 16. Schematic diagram of technique commonly used in multiplication measurements to obtain electron and hole injection separately in the same device by using strongly absorbed (hv_1) and weakly absorbed (hv_2) radiation. The broken line indicates the boundary between the P^+ and N regions, the coarse dashed lines indicate the boundaries of the depletion region, and the fine dashed lines indicate the limits of the regions in which minority carriers are collected by diffusion.

coefficient and the dimensions involved. The holes that are generated within a diffusion length L_p of the edge of the depletion region in the N-type material are collected and result in hole injection. In addition, the radiation which is absorbed in the depletion region results in the injection of both electrons and holes. The radiation that is absorbed within a diffusion length of the edge of the depletion region in the P-type material results in electron injection. Thus, with this configuration there is no way of obtaining pure hole injection to accurately determine M_p, and this configuration can lead to large errors if the electron and hole ionization rates are significantly different. Specifically, for the configuration of Fig. 16 if $\alpha_n \gg \beta_p$, the relatively small fraction of the injected current due to electrons when M_p is being measured can lead to large errors in the values of M_p. In addition, the values of α_n and β_p determined from such measurements cannot be used to evaluate the effects of electron contamination on the hole injection current. This is because the experimentally determined values of α_n and β_p are strongly influenced by the electron contamination of the hole current when the true value of α_n is much greater than β_p. Thus, for the determination of α_n and β_p with the configuration shown in Fig. 16, it is important that the injection current for the carrier with the lower ionization rate be extremely pure. This implies that the radiation must be incident on a thin P^+ region if $\beta_p \gg \alpha_n$ and on a thin N^+ region if $\alpha_n \gg \beta_p$. In both cases the space-charge generation of carriers may also need to be considered. If the field profile in the actual device is not one of the simple forms described above, the analysis is more complex, but if the electric field is known, there are methods which can be used to determine α_n and β_p from measurements of M_n and M_p.[50a]

10. EXPERIMENTAL RESULTS

There have been many experimental measurements of ionization rates reported in the literature. Although most of these measurements have been on Si (because of its practical importance), ionization rates have also been reported for GaAs as well as for Ge, GaP, InSb, and InAs. Since the ratio of the electron and hole ionization coefficients is very important in determining the ultimate performance capabilities of avalanche photodiodes, the results of some of these measurements will be discussed here. It will be seen that even for Si, which has been most thoroughly studied, there is still considerable disagreement concerning the relative values and the field dependence of the ionization coefficients.

a. Ionization Rates in Si

The early published ionization-rate data on Si either assumed that α_n

---

[50a] See, for example, W. N. Grant, *Solid-State Electron.* **16**, 1189 (1973).

and β_p were equal[46] or used complementary $P^+ N$ and $N^+ P$ junctions to determine α_n and β_p separately.[18,33] The first measurements which determined α_n and β_p from results on the same junction were reported by Wul and Shotov,[47] but these authors assumed that the ratio α_n/β_p was constant. Subsequently, Lee et al.[7] used the configuration shown in Fig. 16 to determine the ionization coefficients from measurements on the same junction. The results obtained by Lee et al. were significantly different from the earlier results. They found much larger field variations in both α_n and β_p, higher absolute values of α_n, and smaller values of β_p than were previously obtained. Because of the stronger field dependence they were able to fit their ionization-rate data to the Baraff theory with reasonable parameters.[7] This success was

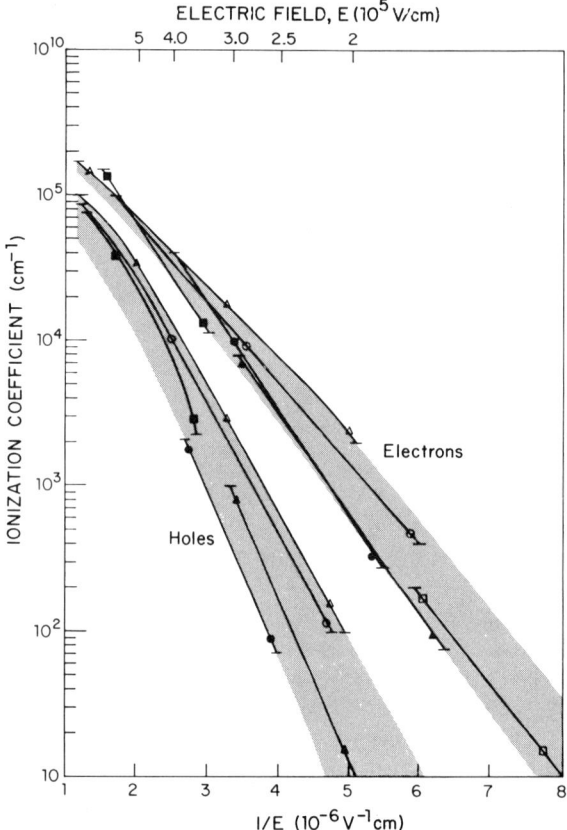

FIG. 17. Experimental ionization coefficients for Si at room temperature; ●, Lee et al.[7]; ▲Ruegg,[53] ■, Moll and van Overstraeten[33]; ○, van Overstraeten and DeMan[48]; △, Grant[50a]; □, Ogawa.[51] (The data points only serve to identify the curves and are not actual calculated points.)

attributed to: (1) the use of microplasma-free junctions; (2) a better method of analysis for measurements made on the same PN junction; and (3) precise control of the purity of electron and hole injection currents used for determining M_n and M_p.

The results of Lee et al.[7] are probably the most widely quoted ionization rates for Si, even though there have been a number of more recent results.[48-53] None of the more recent measurements, however, have been able to reproduce the low values of the hole ionization coefficient obtained by Lee et al. For comparison, the various data are shown in Fig. 17 where it is clear that the hole ionization rates of Lee et al. are significantly lower than any of the other measurements. The lower values of the hole ionization rates have been criticized on several grounds. Kokosa and Davies[54] and Sunshine and Assour[55] have shown that the experimental breakdown voltages of Si diodes cannot be explained using the ionization-rate data of Lee et al. Sunshine and Assour[55] have also shown that much better agreement between experimental and calculated breakdown voltages is obtained by using the ionization-rate data of van Overstraeten and DeMan.[48] van Overstraeten and DeMan[48] also point out that no corrections were made by Lee et al.[7] for the ionization threshold in the measurements on low breakdown voltage diodes. The multiplication data of Lee et al.[7] were reanalyzed by van Overstraeten and DeMan[48] using their method of analysis and they obtained results that do not agree with the electron ionization rates obtained by Lee et al. The differences were attributed to inaccuracies in the numerical solution of Abel's integral equation in the analysis of Lee et al.[7]

Another criticism has been made by Woods et al.[49] In their determination of the Si ionization coefficients using Schottky barrier diodes, Woods et al. made careful corrections for the effect of the depletion width variation with voltage on the collection efficiency of holes in the determination of the hole multiplication factor M_p. The hole ionization coefficients they obtained are somewhat higher than the results of Lee et al.,[7] but when Woods et al.[49] reanalyzed their data, using a linear extrapolation of the photocurrent at low voltages to determine the hole multiplication factor (the procedure used by Lee et al.), the results for β_p were in fairly good agreement with the results obtained by Lee et al. This suggested that the neglect of the variation of the collection efficiency with voltage by Lee et al. in their analysis was the reason for the discrepancy between their results and those obtained in the more recent measurements.

[51] T. Ogawa, *Jap. J. Appl. Phys.* **4**, 473 (1965).
[52] R. D. Baertsch, *IEEE Trans. Electron. Devices* **ED-13**, 987 (1966).
[53] H. W. Ruegg, *IEEE Trans. Electron. Devices* **ED-14**, 239 (1967).
[54] R. A. Kokosa and R. L. Davies, *IEEE Trans. Electron. Devices* **ED-13**, 874 (1966).
[55] R. A. Sunshine and J. Assour, *Solid-State Electron.* **16**, 459 (1973).

It should also be pointed out that Lee et al.[7] only presented hole ionization-coefficient data for one device and they observed a large variation in the calculated ionization rates from device-to-device. In fact, the electron ionization-coefficient data for one sample was essentially equal to the hole ionization-coefficient data criticized in the above references. Although Lee et al. attribute this variation from device-to-device to different scattering environments,[7] one would not expect optical phonon scattering to be strongly dependent on the presence of impurities or imperfections.[30,56]

Finally, an additional conflict arises from recent calculations of the performance of microwave IMPATT devices.[57] These results indicate that better agreement with experimental results is obtained using the more recent ionization-coefficient data.[50a]

In support of the results of Lee et al.,[7] it must be pointed out that, in contrast to some of the recent results,[50a] their experiment was arranged so that there was negligible electron contamination of the hole injection current. This is very important because, as discussed above, if $\alpha_n \gg \beta_p$ even a small electron contamination of the hole current can significantly increase the measured values of M_p and thus result in values of β_p which are too large. In addition, recent calculations[58] of the gain-voltage curves of reach-through avalanche photodiodes have shown that much better agreement with the experiment is obtained with the ionization-rate data of Lee et al.[7] than with that of van Overstraeten and DeMan.[48] Also, calculations of the excess-noise factor of avalanche photodiodes are in better agreement with the experiment when the ionization-rate data of Lee et al. are used. Although the breakdown voltages calculated using the ionization-rate data of Lee et al. are higher than the experimental values,[50a,54,55] it can be argued that the experimental values could be lower than the true breakdown voltages because of edge effects, imperfections, etc. Thus, this discrepancy may not reflect on the validity of the results of Lee et al.

In summary, there is still considerable uncertainty about the magnitude and field variation of α_n and β_p in Si, but there is general agreement that $\alpha_n > \beta_p$ and that the ratio α_n/β_p is larger for low electric fields.[59]

b. Ionization Rates in Ge

The first measurements of the electron and hole ionization coefficients in Ge were made by Miller[17] using complementary transistors, and essentially equivalent results were obtained by Wul and Shotov[47] using electron and hole multiplication measurements made on the same *PN* junctions. These measurements indicated that β_p was about a factor of two larger than α_n

[56] J. L. Moll, *Phys. Rev.* **137**, A938 (1965).
[57] D. R. Decker and C. N. Dunn, *I. Electron. Mat.* **4**, 527 (1975).
[58] J. Conradi, *Solid-State Electron.* **17**, 99 (1974).
[59] W. E. Sayle and P. O. Lauritzen, *IEEE Trans. Electron. Devices* **ED-18**, 58 (1971).

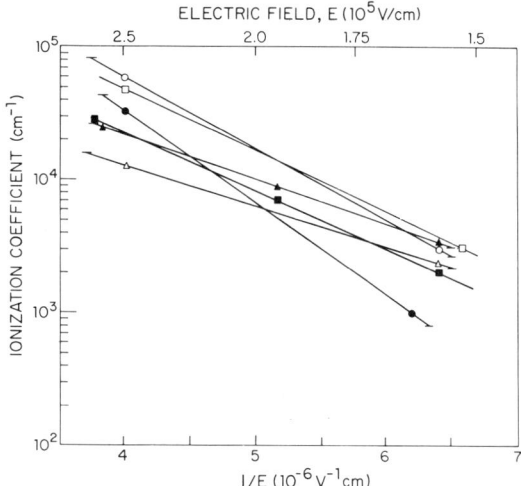

FIG. 18. Experimental ionization coefficients for Ge at room temperature. The solid data points identify the electron ionization coefficient curves and the open data points identify the hole ionization coefficient curves: ▲, △, Decker and Dunn[61]; ●, ○, Miller[17]; ■, □, Wul and Shotov.[47]

over the range of electric fields employed (~ 1.5–2.5×10^5 V/cm). Subsequent measurements using both the method depicted in Fig. 16 and complementary PN junctions[60] supported these earlier results. However, a recent determination of the Ge ionization coefficients from an analysis of IMPATT diode characteristics gave considerably different results.[61] This IMPATT diode analysis indicated that α_n was greater than β_p, just the opposite of the other measurements. The results for the ionization rates in Ge are summarized in Fig. 18.

Although there could be considerable uncertainty in the values of the ionization rates determined by Miller,[17] Wul and Shotov,[47] and Logan and Sze[60] due to uncertainties in the electric field and possible errors in the analyses, the conclusion of these workers that β_p is greater than α_n is probably correct. There appears to be a much greater possibility of error in the analysis of the IMPATT diode small signal admittance characteristics. This is due to the difficulties in obtaining reliable device admittance measurements over a wide frequency range (because of the effects of the device package, etc.) and the indirect and more model-dependent method for determining the ionization coefficients. Thus, for Ge, it is likely that β_p is greater than α_n and that the relative difference between the two is not very large.

[60] R. A. Logan and S. M. Sze, *Proc. Int. Conf. Phys. Semicond., Kyoto, 1966* (*Suppl. J. Phys. Soc. Japan.* **21**, 434). Phys. Soc. Japan, Tokyo, 1966.
[61] D. R. Decker and C. N. Dunn, *IEEE Trans. Electron. Devices* **ED-17**, 290 (1970).

However, the absolute magnitudes of the coefficients and their field variation have not been well-established experimentally.

c. Ionization Rates in GaAs

As with Si and Ge, the first ionization-coefficient measurements on GaAs[62] assumed that the electron and hole ionization coefficients were equal. The measurements were made on very narrow junctions and no corrections were made in the analysis for the ionization threshold energy. Thus, the results are not representative of the true values of the ionization coefficients in GaAs. However, these first ionization-coefficient data agreed with the theory of Baraff[37] and were linear in $1/E^2$ over the range of electric fields used in the measurements. Later measurements by Kressel and Kupsky[63] on abrupt PN junctions with higher breakdown voltages, which were also analyzed with the assumption that $\alpha_n = \beta_p$, indicated that the ionization coefficients were linear in $1/E$ (cf. Williams[64]).

The first measurements on GaAs which attempted to determine α_n and β_p separately were made by Logan and Sze[60] using diffused P^+N and N^+P junctions. These authors used the procedure that had been followed previously for Si: nonpenetrating light was employed to measure the multiplication characteristics of the minority carriers of the diffused regions (M_n for P^+N and M_p for N^+P junctions) and penetrating light was used to determine the multiplication factor M_{np} for multiplication initiated by both electrons and holes. The results of these measurements indicated that, to within experimental error $M_n \simeq M_{np} \simeq M_p$. Therefore, Logan and Sze[60] concluded that $\alpha_n = \beta_p$ and again obtained good agreement with Baraff's theory and a $1/E^2$ dependence of the ionization rates.

Subsequent ionization-coefficient measurements and analyses reported for GaAs assumed that $\alpha_n = \beta_p$, either because of the work of Logan and Sze[60] or simply for ease of analysis.[65-70a] Hall and Leck[65,66] made measurements on P^+N diffused junctions over the temperature range from -20 to $+80°C$. Their results also gave a $1/E^2$ dependence of the ionization rates, but they were unable to obtain good agreement with Baraff's theory using

[62] R. A. Logan, A. G. Chynoweth, and G. G. Cohen, *Phys. Rev.* **128**, 2518 (1962).
[63] H. Kressel and G. Kupsky, *Int. J. Electron.* **20**, 535 (1966).
[64] R. Williams, *RCA Rev.* **27**, 336 (1966).
[65] R. Hall and J. H. Leck, *Int. J. Electron.* **25**, 529 (1968).
[66] R. Hall and J. H. Leck, *Int. J. Electron.* **25**, 539 (1968).
[67] Y. J. Chang and S. M. Sze, *J. Appl. Phys.* **40**, 5392 (1969).
[68] S. N. Shabde and C. Yeh, *J. Appl. Phys.* **41**, 4743 (1970).
[69] G. Salmer, J. Pribetich, A. Farrayre, and B. Kramer, *J. Appl. Phys.* **44**, 314 (1973).
[70] G. H. Glover, *J. Appl. Phys.* **44**, 3253 (1973).
[70a] D. McCarthy and C. A. Lee, unpublished (see Tech. Rep. RADC-TR-72-351, January 1973, Rome Air Develop. Center, Griffiss Air Force Base, New York).

reasonable values for the adjustable constants. Hall and Leck[65] suggested that this discrepancy could be accounted for by the energy band structure of GaAs for which, as suggested by Conwell,[71,72] the dominant energy-loss mechanism may not be optical phonon scattering. The breakdown voltages calculated for abrupt and linearly graded *PN* junctions using the ionization-coefficient data of Hall and Leck[65] are in better agreement with experimental values than those calculated from the data of Logan and Sze.[60] Chang and Sze[67] also measured the temperature dependence of the ionization coefficients of GaAs with the assumption that $\alpha_n = \beta_p$ and found that for room temperature and below there was a $1/E$ dependence of the ionization rates. The temperature dependence of the ionization coefficients obtained by Chang

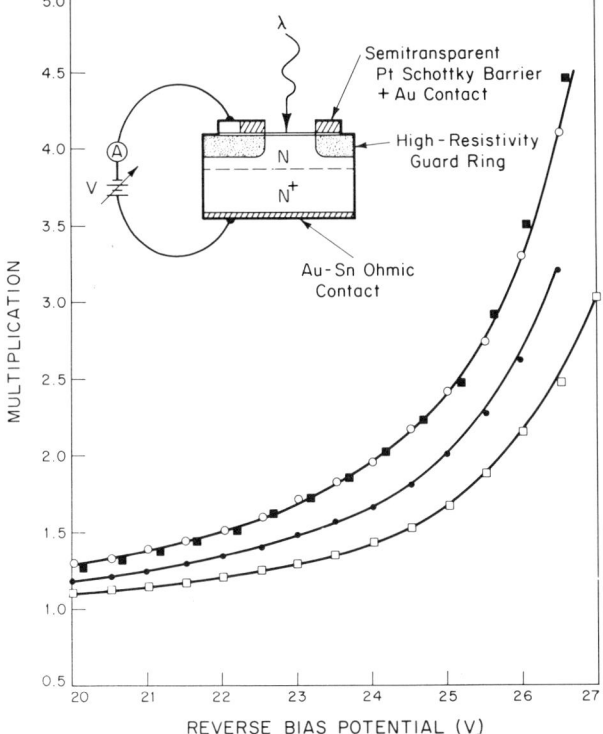

FIG. 19. Variation of multiplication of a GaAs Schottky barrier avalanche photodiode with bias voltage for different wavelengths of incident radiation: ○, $\lambda = 0.85$ μm, ●, $\lambda = 0.65$ μm, □, $\lambda = 0.50$ μm; ■, microscope lamp. The inset shows the device configuration. (After Stillman et al.[73])

[71] E. M. Conwell, *Appl. Phys. Lett.* **9**, 383 (1966).
[72] E. M. Conwell, comment in discussion of Logan and Sze.[60]

and Sze were adequately described by Baraff's theory. Shabde and Yeh[68] made measurements of the ionization coefficients of $Al_xGa_{1-x}As$ and obtained a field dependence of $1/E^2$ throughout the alloy composition range examined ($x = 0$ to 0.24). Their results for GaAs ($x = 0$) also agreed with the results of Logan and Sze.[60]

Multiplication measurements on GaAs Schottky barriers by Glover,[70] using white light and the assumption that $\alpha_n = \beta_p$, gave ionization coefficients which were in good agreement with those of Logan and Sze.[60] However, more recent measurements of the multiplication in GaAs Schottky barrier diodes have shown that the multiplication depends on the wavelength of the incident radiation.[73] For example, Fig. 19 shows the multiplication characteristics obtained with a GaAs Schottky barrier diode for white light from a microscope lamp and for radiation at three different wavelengths. The longest wavelength (0.85 μm) used for these measurements was chosen to be above the absorption edge so that shifts of the absorption edge with bias voltage due to the Franz–Keldysh effect[74,75] would have little effect on the results. The multiplication data indicate that higher values of multiplication are obtained for the more penetrating light. These results show that α_n and β_p are not equal in GaAs and, because the Schottky barrier is on N-type material, imply that β_p is greater than α_n.

There have been several other indications that α_n and β_p are not equal in GaAs. As mentioned in Section 8, calculations of the ionization threshold energies in GaAs by Anderson and Crowell[43] indicate that the lowest threshold energy is not the same for holes and electrons and that the lowest ionization threshold energy for holes is always lower than that for electrons. The noise performance[76] and gain-bandwidth product[77] of some GaAs and $In_xGa_{1-x}As$ avalanche photodiodes also indicate a large asymmetry in the electron and hole ionization coefficients. However, it was not feasible to determine α_n and β_p from the measurements shown in Fig. 19 because of the mixed injection which results for all three discrete wavelengths and for the white light of the microscope lamp. As discussed in the previous section, the experiment should be designed so that pure electron- and pure hole-initiated multiplication can be achieved in the same device.

The technique used by Stillman et al.[77] in an attempt to obtain pure electron and pure hole injection in the same Schottky barrier device is

---

[73] G. E. Stillman, C. M. Wolfe, J. A. Rossi, and A. G. Foyt, *Appl. Phys. Lett.* **24**, 471 (1974).
[74] W. Franz, *Z. Naturforsch.* **13A**, 494 (1958).
[75] L. V. Keldysh, *Zh. Eksp. Teor. Fiz.* **34**, 1138 (1958) [*English Transl.: Sov. Phys.—JETP* **34**, 788 (1958)].
[76] W. T. Lindley, R. J. Phelan, Jr., C. M. Wolfe, and A. G. Foyt, *Appl. Phys. Lett.* **14**, 197 (1969).
[77] G. E. Stillman, C. M. Wolfe, A. G. Foyt, and W. T. Lindley, *Appl. Phys. Lett.* **24**, 8 (1974).

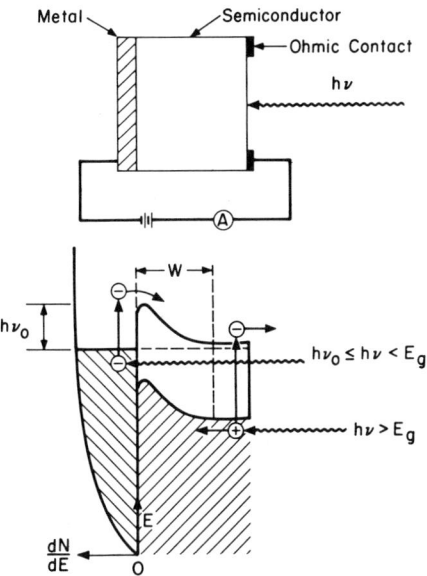

FIG. 20. Schematic diagram of technique used to obtain electron and hole injection in the same Schottky barrier device.

shown schematically in Fig. 20. (The same technique was used earlier by Woods et al.[49] for ionization-coefficient measurements on Si Schottky barriers.) With this technique pure electron injection can be obtained by irradiating the metal Schottky barrier with light of sufficient energy to excite electrons from the metal over the potential barrier (internal photoemission), but of insufficient energy to excite electron–hole pairs in the semiconductor. If the Schottky barrier is sufficiently thin, the long-wavelength radiation ($h\nu_0 < h\nu < E_g$) can be incident on the metal either from the semiconductor or the air side. Pure hole injection can be obtained by shining strongly absorbed radiation ($h\nu \gg E_g$) through a transparent Ohmic back-contact. Corrections must be made to the zero multiplication current to allow for image-force lowering of the barrier for electron injection and to allow for widening of the depletion region with reverse bias for hole injection. Other effects discussed by Woods et al.[49] must also be taken into account to obtain accurate multiplication factors. Because of the short carrier diffusion lengths in GaAs compared to Si, it was necessary to fabricate special device structures to obtain sufficient hole injection current using this technique; that is, the distance from the edge of the depletion region to the Ohmic back-contact had to be approximately one diffusion length. The details of the fabrication of the devices and the experimental technique are described

TABLE II

Ionization Rate Measurements in GaAs (300°K)

$$\alpha = a \exp[-(b/|E|)^m]$$

Reference	a	b	m	Method[a]
Logan et al.[62]	1.34×10^6	2.03×10^6	2	Diffused $p^+ - n$ junctions, assumed $\alpha = \beta$
Logan and Sze[60]	3.5×10^5	6.85×10^5	2	Diffused $p^+ - n$ and $n^+ - p$ junctions $M_{np} \approx M_n \approx M_p$ implied $\alpha \approx \beta$
Kressel and Kupsky[63]	1.0×10^6	1.72×10^6	1	Vapor-phase epitaxial $p - n$ junctions, edge illuminated assumed $\alpha = \beta$
Hall and Leck[65]	2.0×10^5	5.5×10^5	2	Diffused $n^+ - p$ junctions, assumed $\alpha = \beta$
Chang and Sze[67]	4.0×10^6	2.1×10^6	1	Diffused $p^+ - n$ junctions, assumed $\alpha = \beta$
Shabde and Yeh[68]	3.7×10^6	7.2×10^5	2	Liquid-phase epitaxial $p - n$ junctions on $Al_{1-x}Ga_xAs$ $x = 1$, assumed $\alpha = \beta$
Salmer et al.[69]	1.18×10^5	5.55×10^5	2	From avalanche voltage measurements on Schottky barrier IMPATT devices, assumed $\alpha = \beta$
Glover[70]	$\sim 3.5 \times 10^5$	$\sim 6.85 \times 10^5$	2	Schottky barrier, assumed $\alpha = \beta$
McCarthy and Lee[70a]	$\sim 3.5 \times 10^5$	$\sim 6.85 \times 10^5$	2	Schottky barrier, assumed $\alpha = \beta$
Stillman et al.[73]	1.2×10^7	2.3×10^6	1	Electron ionization coefficient, α_n ⎱ Schottky barrier
	3.6×10^8	2.9×10^6	1	Hole ionization coefficient, β_p ⎰

[a] All methods (except for Salmer et al.[69]) used measurements of photocurrent versus bias voltage on the type device described.

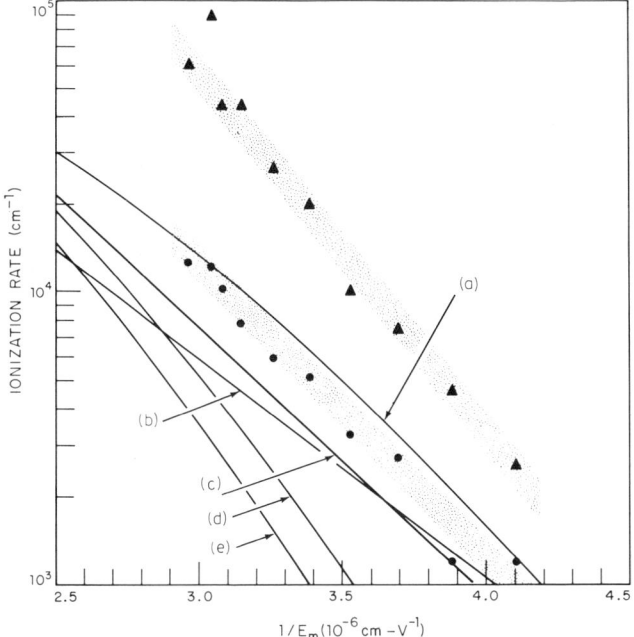

FIG. 21. Experimental ionization coefficients for GaAs at room temperature. The labeled solid curves from the indicated references were calculated with the assumption $\alpha_n = \beta_p$. Data: (a) Hall and Leck,[65] (b) Kressel and Kupsky,[63] (c) Chang and Sze,[67] (d) Logan and Sze,[60] and (e) Shabde and Yeh.[68] The data points for the electron (●) and hole (▲) ionization coefficients are from Stillman et al.[77]

by Stillman et al.[77] The results of these measurements as well as previous results are summarized in Table II and Fig. 21.

There are several sources of possible contamination of the injected photocurrents which could affect the experimental values of M_n and M_p and thus the calculated values of α_n and β_p obtained by Stillman et al.[77]:

(1) *Hole contamination of the injected electron current.* During the measurement of $M_n(V)$ using radiation with $h\nu_0 \leqslant h\nu \leqslant E_g$, the electron injection current could be contaminated by holes (and electrons) generated in the space-charge region of the diode due to a small electroabsorption (Franz–Keldysh effect) at long wavelengths. If $\beta_p \gg \alpha_n$, even a small amount of this contamination would cause the measured values of $M_n(V)$ to be much too large compared to the true value which would be measured with pure electron injection. The effect of this would be to increase the experimental values of α_n and decrease the experimental values of β_p. If the true electron and hole ionization rates are equal or even if $\alpha_n \gg \beta_p$, this source

of containation would have a negligible effect on the experimental results.

Another effect that would result in additional electron injection rather than hole contamination of the injected current is the excitation of electrons from deep impurity states or traps during the measurement of M_n. This could affect the results because where the electrons are injected can change the electric field profile in the device[35,78] and thus the multiplication. However, the results of CV measurements with and without illumination eliminated this as a complicating factor.

(2) *Electron contamination of the injected hole current.* There are two possible sources for this type of contamination: (i) During measurements of $M_p(V)$ with short wavelength ($hv > E_g$) radiation, the experimental arrangement is similar to optical pumping or photoluminescence experiments used to study radiative recombination in GaAs. The recombination radiation resulting from this optical pumping could be reabsorbed in the space-charge region and thus generate electron–hole pairs which would contaminate the injected hole current. (ii) The injected hole current could be contaminated by the "secondary" emission of electrons from the metal Schottky barrier due to the impact of high-energy holes which have been accelerated through the depletion region, but have acquired an energy slightly less than the ionization threshold energy.[79] This effect would be especially important if $\alpha_n \gg \beta_p$ (as in Si).

Although the other types of contamination could have an effect on the absolute values of α_n and β_p, the results demonstrated that β_p must be greater than α_n for GaAs.

Other more recent measurements, which will be discussed in Part VI, indicate that β_p is significantly higher at lower fields than in the experimental results just described and that β_p is much greater than α_n in GaAs. It then follows that a small amount of hole contamination of the electron injection current due to Franz–Keldysh absorption, as described above, could have a large effect on the experimental results of Stillman *et al.*[77] Thus, although the measurements reported so far do not give reliable values for the magnitudes and field variation of the electron and hole ionization coefficients in GaAs, they do show that the ionization coefficients are not equal, as has been widely assumed, and imply that they are significantly different. More work needs to be done to accurately determine these coefficients over a large electric-field range and at different temperatures to be able to evaluate the ultimate performance of GaAs avalanche photodiodes and microwave IMPATT diodes.

[78] R. H. Haitz, *J. Appl. Phys.* **35**, 1370 (1964).
[79] This mechanism was first suggested to us by H. Kroemer, and we thank him for several helpful discussions.

5. AVALANCHE PHOTODIODES

d. Ionization Rates in GaP

The ionization coefficients of electrons and holes in GaP have been measured by Logan and Chynoweth[80] and Logan and White[81] and have been reviewed by Chynoweth.[30] For these measurements the method indicated in Fig. 16 was used. The measured values of multiplication with strongly absorbed light were the same as the values with penetrating light. This implied that the electron and hole ionization rates are equal in GaP. The ionization rates were linear in $1/E^2$ and were in good agreement with Baraff's theory.

e. Ionization Rates in InSb

Baertsch[82] has studied the noise and multiplication for InSb avalanche photodiodes. From the variation of multiplication with wavelength for P^+N diffused photodiodes, he concluded that holes are much less effective in causing impact ionization than electrons. The electron ionization rate was calculated from the multiplication data under the assumption that the hole multiplication was unity. The resulting electron ionization rate was nearly independent of electric field for $E \gtrsim 5 \times 10^3$ V/cm and increased with increasing temperature in contrast to the results on other materials. The value of α_n in this field range was a little less than 2×10^3 cm^{-1}. More recent measurements of the wavelength dependence of multiplication in InSb avalanche photodiodes,[83] however, did not substantiate the large wavelength dependence observed by Baertsch.[82] Thus, more work remains to be done before a final conclusion can be made regarding the relative values of the electron and hole ionization coefficients in InSb.

f. Ionization Rates in InAs

The avalanche multiplication and electron and hole ionization coefficients in InAs have been studied by Mikhailova *et al.*[84] and Kim.[85] These workers conclude that in this material the ionization coefficient of electrons is much greater than that of holes. The electron ionization coefficient α_n was estimated to be within the limits of 10^3 to 10^4 cm^{-1} and the hole ionization coefficient β_p was estimated to be about 10^2 cm^{-1}, at least an order of magnitude smaller than α_n.[84] The difference was attributed to the lower probability of ionization by holes because of the much greater hole effective mass.

[80] R. A. Logan and A. G. Chynoweth, *J. Appl. Phys.* **33**, 1649 (1962).
[81] R. A. Logan and H. G. White, *J. Appl. Phys.* **36**, 3945 (1965).
[82] R. D. Baertsch, *J. Appl. Phys.* **38**, 4267 (1967).
[83] H. A. Protschka, unpublished, 1970.
[84] M. P. Mikhailova, D. N. Nasledov, and S. V. Slobodchikov, *Fiz. Tekh. Poluprov.* **1**, 123 (1967) [*English Transl.: Sov. Phys.—Semicond.* **1**, 94 (1967)].
[85] C. W. Kim, unpublished data, 1973.

g. Ionization Rates in Other Materials

Measurements of the electron and hole ionization coefficients have recently been reported for $In_xGa_{1-x}As$ and $GaAs_{1-y}Sb_y$ alloys suitable for far-infrared detectors in the 1.06 μm wavelength range.[85a] These measurements were made on abrupt liquid-phase epitaxial PN junctions which were grown on N-type GaAs substrates using step type intermediate grading layers to reduce the strain due to lattice mismatch. Pure hole and pure electron injection currents were obtained in the devices using a configuration similar to that shown in Fig. 15b except that a well was etched in the GaAs substrate and intermediate grading layers to expose the N-type alloy layer of the final composition. The wavelength of the incident radiation was 0.6328 μm. The doping of the P side of the junction was $\gtrsim 10^{18}$ cm^{-3} and the doping on the N side ranges from 10^{15} to 10^{16} cm^{-3}. The electron and hole multiplication factors were determined from the photoresponse measurements using a linear extrapolation of the photocurrents at low bias voltages. The analysis of the multiplication data showed that the hole ionization coefficient was greater than the electron coefficient for $In_{0.14}Ga_{0.86}As$ alloys and that the ratio $k = \beta_p/\alpha_n$ increased as the electric field increased, similar to the data of Stillman *et al.*[77] for GaAs in Fig. 21. In contrast, the results for $GaAs_{0.88}Sb_{0.12}$ indicate that the electron ionization coefficient is larger than the hole ionization coefficient and that $k = \alpha_n/\beta_p$ decreases with increasing field. It is surprising that the ionization rates are so sensitive to alloy composition. More work remains to be done on the variation of the ionization coefficients in these semiconductor alloy systems.

V. Avalanche Photodiode Detectors

In the previous sections it was shown that, to obtain low-noise, high gain-bandwidth-product avalanche photodiodes, it is important to use a semiconductor which has a large difference in the electron and hole impact-ionization coefficients. In addition, the device structure should be designed so that multiplication is initiated by the carrier with the higher ionization coefficient. To obtain the minimum detectable power [Eqs. (11) and (18)], the quantum efficiency should also be as large as possible for both avalanche and nonavalanche photodiodes. The semiconductor material must therefore be chosen to have a high optical-absorption constant at the desired wavelength of operation as well as a large difference between α_n and β_p. The quantum efficiency also depends on the device structure and can be a function

---

[85a] T. P. Pearsall, M. A. Pollack, R. E. Nahory, and C. A. Lee, *Bull. Amer. Phys. Soc. Ser. II* **20**, 273 (1975); T. P. Pearsall, R. E. Nahory, and M. A. Pollack, *Appl. Phys. Lett.* **27**, 330 (1975); T. P. Pearsall, R. E. Nahory, and M. A. Pollack, *Appl. Phys. Lett.* **20**, 403 (1976).

of the modulation frequency and the wavelength of the incident radiation.

In this part we will first discuss the influence of the material properties and device structure on the wavelength and frequency dependence of the efficiency η. We will then discuss some practical avalanche photodiode structures and the results obtained for these structures. Although most of the discussion applies to Si detectors, with slight modification the same considerations are appropriate for avalanche photodiodes in other materials.

11. DEVICE STRUCTURE

a. Wavelength and Frequency Dependence of η

The influence of the device structure on the wavelength and frequency dependence of the quantum efficiency of reverse-biased photodiodes has been considered by several authors.[53,86–90] The external quantum efficiency is defined as the ratio of the electron–hole pairs that are collected to the number of photons incident on the detector. Since for most semiconductor materials the reflectivity alone limits the quantum efficiency to about 70% or less, the device reflectivity must be minimized for high quantum efficiency. By suitable coating procedures[91,92] the reflectivity can be reduced to values close to zero over a small wavelength range. Other factors that limit the quantum efficiency of a photodiode are the absorption constant for the incident photons, which determines the number of electron–hole pairs that are generated, and the efficiency with which these electron–hole pairs are collected.

The absorption coefficients for several semiconductor materials useful for photodetectors are shown in Fig. 22.[93–100] The values shown are only

[86] R. H. Rediker and D. E. Sawyer, *Proc. IRE* **45**, 944 (1957).
[87] D. E. Sawyer and R. H. Rediker, *Proc. IRE* **46**, 1122 (1958).
[88] W. Gärtner, *Phys. Rev.* **116**, 84 (1959).
[89] R. P. Riesz, *Rev. Sci. Instrum.* **33**, 994 (1962).
[90] J. R. Biard, *IEEE Trans. Electron. Devices* **ED-14**, 233 (1967).
[91] M. V. Schneider, *Bell Syst. Tech. J.* **45**, 1611 (1966).
[92] R. A. Laff, *Appl. Opt.* **10**, 968 (1971).
[93] W. C. Dash and R. Newman, *Phys. Rev.* **99**, 1151 (1955).
[94] R. Newman, *Phys. Rev.* **111**, 1518 (1958).
[95] W. G. Spitzer, M. Gershenzon, C. J. Frosch, and D. F. Gibbs, *J. Phys. Chem. Solids* **11**, 339 (1959).
[96] W. G. Spitzer and H. Y. Fan, *Phys. Rev.* **106**, 882 (1957).
[97] M. D. Sturge, *Phys. Rev.* **127**, 768 (1962).
[98] D. D. Sell and H. C. Casey, Jr., *J. Appl. Phys.* **45**, 800 (1974).
[99] D. E. Hill, *Phys. Rev.* **133**, A866 (1964).
[100] T. S. Moss, "Optical Properties of Semiconductors." Butterworths, London and Washington, D.C., 1959.

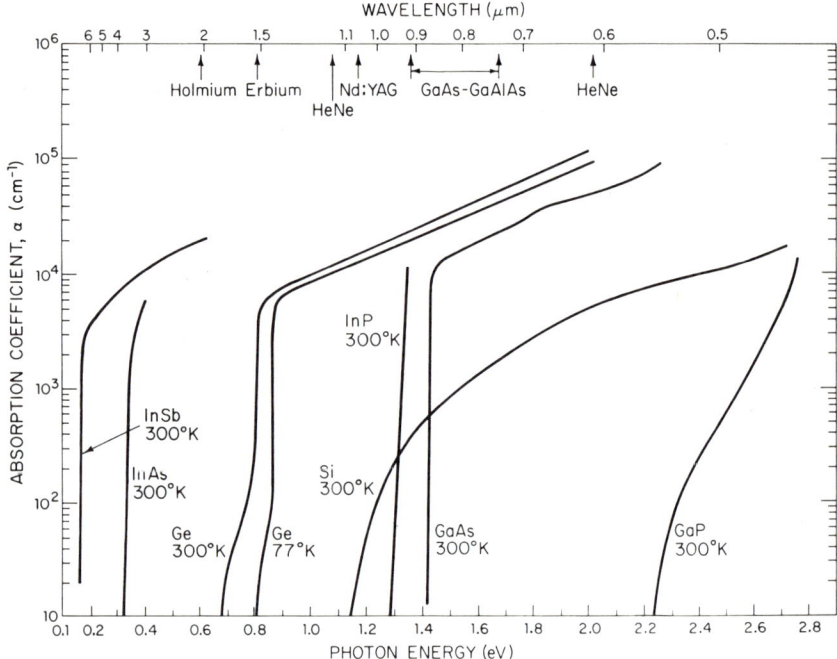

FIG. 22. Absorption coefficients for selected semiconductors in the visible and near-infrared spectral region: InSb,[96,100] InAs,[96,100] Ge,[93] Si,[93] InP,[94,100] GaAs,[97,99] and GaP.[95] Also shown are the emission energies of some promising lasers in the same spectral region.

typical values; the actual absorption coefficients in a given material can vary with carrier concentration, purity of the material, applied electric field, and temperature. If all the electron–hole pairs generated by the absorbed photons are collected, and we neglect multiple reflections from the back contact, the external quantum efficiency can be written as

$$\eta = (1 - R)(1 - e^{-\alpha x}).$$

where R is the reflectivity, α is the absorption coefficient, and x is the thickness of the semiconductor material where the radiation is absorbed. However, bulk and surface recombination can reduce the number of electron–hole pairs which are collected.

Figure 23 shows a schematic diagram of a photodetector illustrating the effects of the device structure on the quantum efficiency. The P^+ and N^+ regions serve only as Ohmic contacts to the active P and N regions of the diode. The depletion region of the reverse-biased diode extends a distance W_p into the P-type material and a distance W_n into the N-type material. Electron–hole pairs generated in the depletion region, represented by the

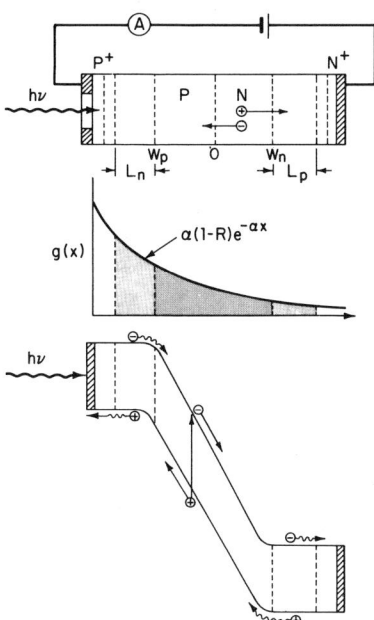

FIG. 23. Schematic diagram illustrating effects of device structure on quantum efficiency.

darkly shaded area of the pair-generation curve, are separated by the electric field. Because of this, these electrons and holes have little opportunity to recombine before they are collected. Of the minority carriers which are generated outside of the depletion region, some recombine with the majority carriers while others diffuse to the depletion region where they are collected. Carriers within the diffusion lengths L_n and L_p of the depletion region (the lightly shaded areas) can be considered to be collected, so that these distances as well as the depletion width contribute to the effective absorption.[88] The carriers generated in the P^+ and N^+ contacts and in the bulk material between the contacts and the edges of the diffusion region do not contribute to the quantum efficiency.

Thus, it is apparent that if the absorption coefficient α is high at the wavelength of interest, it is important to minimize the thickness of the P^+ contact region and the distance between the edge of the contact and the diffusion region. On the other hand, if α is very small at the desired wavelength, it is important to maximize the total width of the depletion region and the diffusion regions. However, the depletion width can only be increased so far before the transit time of the carriers across the depletion width will limit the frequency response of the device. In addition, the diffusion of minority carriers to the high-field region is a relatively slow process which will cause

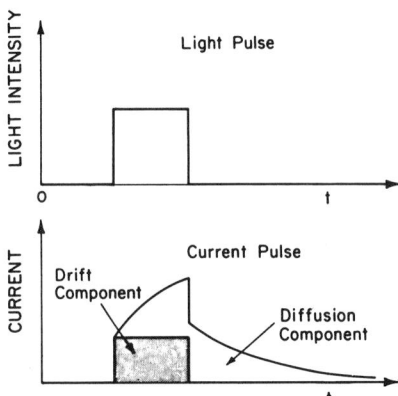

FIG. 24. Detector current pulse resulting from a rectangular light pulse showing drift and diffusion components of the photo-generated current.

the external quantum efficiency to be modulation-frequency as well as wavelength dependent.

Figure 24 shows the response of a detector, with the structure indicated in Fig. 23, to a rectangular light pulse. The wavelength of the light is assumed to be such that the radiation is completely absorbed before it reaches the bulk N-type material beyond the depletion region. The resulting detector pulse consists of a fast and a slow component and is considerably distorted from the incident light pulse. The fast component, shown by the shaded area in Fig. 24, is due to the carriers that are generated in the depletion region. The rise time of this component can be limited by the transit time, by $R_{eq}C$ effects, or (if the device is operated with avalanche gain) by the gain-bandwidth-product. The slow component is due to the diffusion of electrons that are absorbed within a distance L_n of the depletion region. This slow component would build up to some steady-state value if the light pulse were sufficiently long.

At the end of the light pulse, the carriers in the depletion region are collected quickly so that, neglecting $M(\omega)$ and $R_{eq}C$ effects, the detector response decreases by the same amount as the original increase due to the space-charge-generated current in a time comparable to the transit time. This is the origin of the fast component of the detector response at the end of the light pulse. The diffusion of the carriers which are within a distance L_n of the edge of the depletion region at the end of the light pulse produces the slow response at the end of the pulse. Although the time constant describing the diffusion process is the same for both the rising and falling part of the pulse, the shapes of these two parts of the response can appear to be drastically different, depending on how close the diffusion current in the rising part of the pulse is to the steady-state value. The actual shape of the pulse will depend on the device structure, the absorption coefficient, the minority-carrier diffusion-lengths, and the conditions under which the detector is operated (wavelength and avalanche gain).

The contribution of the diffusion current to the detector response can be reduced by increasing the depletion-layer width (for weakly absorbed radiation) and/or by reducing the diffusion length. Increasing the depletion width will also increase the drift component for weakly absorbed radiation, but in addition will decrease the transit-time-limited bandwidth. Since the minority-carrier diffusion-length generally decreases with increasing carrier concentration, the contribution of the diffusion current to the total detector current can be reduced by using epitaxial layers on substrates with higher carrier concentration.[101] For avalanche photodiodes the relative contribution of the diffusion current to that of the drift current can be further reduced by arranging the device structure so that the diffusion component of the current consists mainly of carriers with the lower ionization coefficient. In this manner the effective multiplication for the diffusion current is less than that for the drift current.

A study has been made of the detection properties of various types of Si avalanche photodiode structures by Takamiya.[102] In Si α_n is much greater than β_p,[7] and the minority-carrier diffusion length is significantly longer for electrons than for holes.[103] From these considerations Takamiya has shown that the diffusion component is smallest for a PN^+ structure with the radiation incident on the P region. This structure is thus suitable for high-speed detectors and is also capable of higher gain than the N^+P structure with the radiation incident on the N^+ region. With the optimum device structure it is possible to obtain an external quantum efficiency that is essentially independent of both wavelength and modulation frequency. For most device structures, however, there is some wavelength and/or frequency variation which must be considered in the evaluation of their performance.

As discussed above, for radiation where the absorption coefficient is small, not only can the frequency dependence of the efficiency be reduced by increasing the width of the depletion region, but the actual value of the efficiency can also be increased. Associated with these advantages is the disadvantage of an increase in the transit time of the carriers through the wider depletion region. Thus, there must be some compromise between the

---

[101] A. D. Lucas, *Opto-Electronics* **6**, 153 (1974).
[102] S. Takamiya and A. Kondo, Inst. Electron. and Commun. Eng. of Japan, Paper 1974-5.
[103] The minority carrier diffusion length can be calculated from the corresponding carrier mobility using the Einstein relation and the minority carrier lifetime.

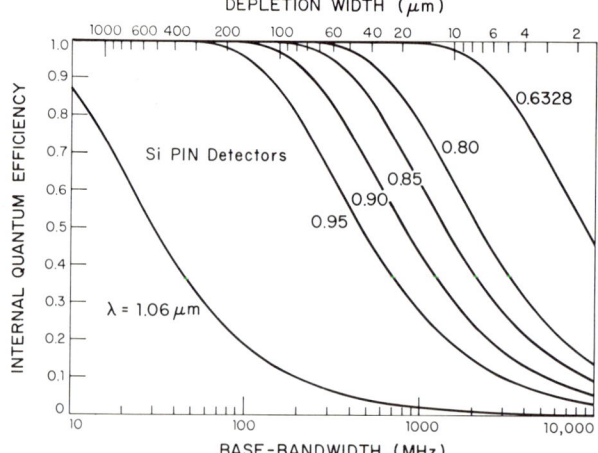

FIG. 25. Variation of internal quantum efficiency of Si PIN detectors with depletion width and transit-time-limited bandwidth for several different wavelengths, assuming a saturated drift velocity of 10^7 cm/sec.

quantum efficiency and bandwidth of such a detector. This is particularly true for Si devices since the absorption coefficient is relatively small for many wavelengths of interest. As shown in Fig. 22 the absorption coefficient for Si varies from about 10^4 cm^{-1} at 0.6 μm to less than 10 cm^{-1} at 1.1 μm.[104]

To further illustrate the compromise, Fig. 25 shows the calculated internal quantum efficiency $[R = 0, \eta = (1 - e^{-\alpha W})]$, with depletion width W or transit-time-limited base bandwidth at several different wavelengths for Si PIN detectors at room temperature assuming zero diffusion current. A saturated drift velocity $v_s = 10^{-1}$ cm/sec and the absorption coefficient data of Fig. 22 were used for these calculations. From these curves, it can be seen that at the longer wavelengths the response time or bandwidth will be limited by the required quantum efficiency or the quantum efficiency will be limited by the required bandwidth.

A means of obtaining high quantum efficiency for wavelengths where the absorption coefficient is small and at the same time avoiding the transit-time limitation is to use a narrow depletion width and illuminate the device from the side, transverse to the junction.[105-108] However, this procedure uses

[104] At the longer wavelengths the values of the absorption coefficient for Si can be slightly influenced by the electric field in the device through the Franz–Keldysh effect. Values of 10.6 cm^{-1} at zero electric field and 13 cm^{-1} under "appropriate" high-field conditions have been measured at room temperature.
[105] O. Krumpholz and S. Maslowski, Z. Angew. Phys. 25, 156 (1968).
[106] O. Krumpholz and S. Maslowski, Wiss. Ber. AEG-Telefunken 44, 73 (1971).
[107] O. Krumpholz, Wiss. Ber. AEG-Telefunken 44, 80 (1971).
[108] D. P. Matheu, R. J. McIntyre, and P. P. Webb, Appl. Opt. 9, 1842 (1970).

a very small active area and thus requires focusing of the incident light. Because of this, these detectors will probably only find application in special situations. The quantum-efficiency-bandwidth considerations we have discussed here can be extended to materials other than Si, particularly those with indirect bandgaps. For materials with direct bandgaps the absorption coefficient increases to large values in a narrow wavelength range (cf. Fig. 22), so it is not generally feasible to increase the quantum efficiency by increasing the depletion width. The frequency and wavelength dependence of the quantum efficiency due to the diffusion component of the current must, however, still be considered when designing photodiodes utilizing direct bandgap materials.

b. Practical Considerations

Besides the effects of the device structure on the frequency and wavelength response discussed above, which must be considered for both nonavalanche and avalanche photodiodes, the effects of the device structure on the type of injected carrier and the uniformity of the multiplication are especially important for avalanche photodiodes. This is because of the sensitivity of the gain and noise performance of these detectors to the device structure.

In order to achieve high uniform avalanche gain, microplasmas must be eliminated since these defects limit the gain which can be attained across the rest of the device.[109] The noise due to microplasmas is an additional complication.[110] The elimination of microplasmas is by no means simple, particularly in materials other than Si and Ge such as InSb, GaAs, In_xGa_{1-x} As, etc. Although a discussion of these areas is beyond the scope of this chapter, with sufficient development of materials and device technology, it should be possible to achieve reasonable yields of high-gain, uniform avalanche photodiodes in most semiconductor materials.

Even with material of sufficient quality (i.e., microplasma-free), in planar devices the electric field will usually be higher at the edge of the device than in the center. Therefore, some method of increasing the breakdown voltage and/or reducing the electric field at the edge is necessary. For nonplanar devices it is necessary to choose the shape of the mesa or edge of the structure so as to minimize the electric field at the surface. Some of the structures which have been used for avalanche photodiodes are shown in Fig. 26.

Figure 26a illustrates an N^+P guard-ring structure.[111] (Complementary P^+N and Schottky barrier guard-ring structures can also be envisioned from this figure.) This structure is easily fabricated using two separate

---

[109] H. Kressel, *RCA Rev.* **28**, 175 (1967).
[110] T. Tokuyama, *J. Appl. Phys.* **7**, 324 (1962).
[111] L. K. Anderson, P. G. McMullin, L. A. D'Asaro, and A. Goetzberger *Appl. Phys. Lett.* **6**, 62 (1966).

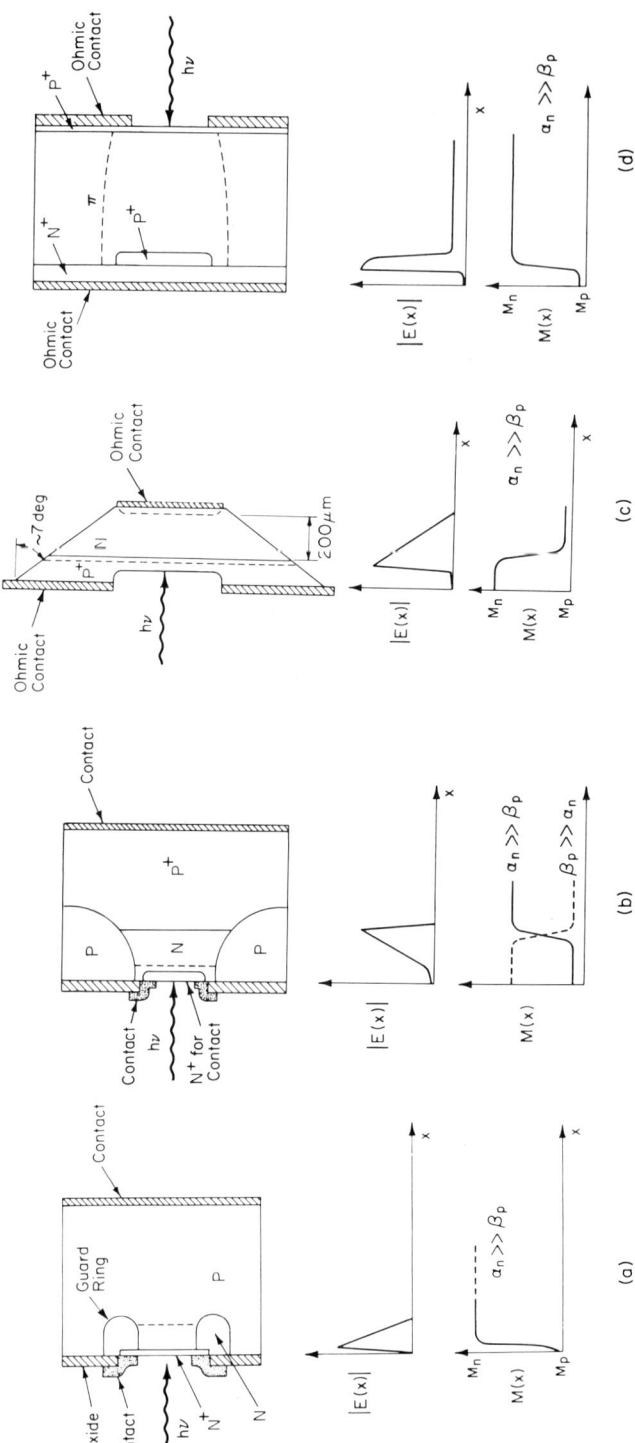

FIG. 26. Various avalanche photodiode devices; (a) N^+P guard-ring structure; (b) inverted, no guard ring, N^+P structure; (c) beveled-mesa or contoured-surface structure; (d) $N^+P\pi P$ reach-through structure.

diffusions to form the N-type guard ring and the N^+P active junction. The electric-field profile and multiplication for an electron–hole pair injected at position x (for $\alpha_n \gg \beta_p$) are shown in the lower portion of the figure. The depletion width for this type of device at breakdown is typically less than the depth of the guard ring, and is thus restricted to a few microns. Recently, however, devices of this type with wider depletion regions have also been fabricated.[112] The junction depth is usually kept as small as possible, and most of the incident radiation is absorbed beyond the metallurgical junction. Thus, for narrow depletion widths in Si this structure is best suited for relatively short wavelengths, $0.4 \lesssim \lambda \lesssim 0.8$ μm.

Although a guard-ring structure increases the breakdown voltage of the device, it also increases the capacitance, which can degrade high-speed performance. To overcome this limitation a buried or inverted structure[113] can be used to eliminate the guard ring and still prevent edge breakdown.[114] Figure 26b shows one example of this technique. The device was formed by starting with an epitaxial NP^+ junction, and then forming the P region and the thin N^+ contact by diffusion. The variation of $M(x)$ shown indicates that this structure would not be suitable for strongly absorbed radiation in a semiconductor with $\alpha_n \gg \beta_p$. However, by adjusting the doping and thickness of the N region so that it is fully depleted at the operating voltage, it would be close to the optimum structure for a material with $\beta_p \gg \alpha_n$.

Another method of eliminating the guard ring and its associated capacitance (and extra dark current) is to shape the device structure so that the electric field where the PN junction reaches the surface is less than that in the active region of the detector.[115,116] Similar procedures have also been used for Schottky barrier devices.[117,118] Figure 26c shows one method for obtaining this type of electric-field profile.[119] The structure is fabricated by diffusing Ga into high-resistivity Si to form the P^+N junction. The 6 to 10 degree angles are then lapped and etched to reduce the surface field. These devices have high breakdown voltages (~ 1800 V) and wide depletion widths. Part of the P^+ region is etched away to reduce the losses in the relatively thick diffused P^+ layer.

The variation of the multiplication with position shown in Fig. 26c indicates that this structure is most suitable for strongly absorbed radiation

---

[112] J. A. Raines, in "Solid State Devices" *(Proc. Solid State Device Res. Conf., 2nd, 1972)*, p. 225. Inst. Phys., London, 1973.
[113] D. P. Kennedy and R. R. O'Brien, *IBM J. Res. Dev.* **10**, 213 (1966).
[114] W. 1. Lynch, *IEEE Trans. Electron. Devices* **ED 15**, 735 (1968).
[115] R. L. Davies and F. E. Gentry, *IEEE Trans. Electron. Devices* **ED-11**, 313 (1964).
[116] G. C. Huth, H. E. Bergeson, and J. B. Trice, *Rev. Sci. Instrum.* **34**, 1283 (1963).
[117] J. A. Lewis and E. Wasserstrom, *Bell Syst. Tech. J.* **49**, 1183 (1970).
[118] D. J. Coleman, Jr., J. C. Irvin, and S. M. Sze, *Proc. IEEE* **59**, 1121 (1971).
[119] R. J. Locker and G. C. Huth, *Appl. Phys. Lett.* **9**, 227 (1966).

when $\alpha_n \gg \beta_p$ as in Si, but even then the frequency and wavelength dependence of the quantum efficiency, as discussed above, should be taken into account. For short-wavelength radiation the multiplication will be high and the excess noise will be low. In addition, this structure has the advantage that the dark current which is generated in the wide N-type depletion region is multiplied less than the photocurrent and thus contributes less to the noise. However, for long wavelengths where the absorption coefficient is small, a large part of the radiation can be absorbed in the N-type depletion region. This produces a higher quantum efficiency, but the multiplication will be low and the excess-noise factor will be high. Therefore, this structure is not as suitable for the detection of longer wavelengths (when $\alpha_n \gg \beta_p$) as some of the other structures. Nevertheless, reasonably good performance has been achieved with this structure.[120]

A somewhat more complicated device is shown in Fig. 26d. This structure is referred to as the $N^+P\pi P^+$ or "reach-through" avalanche photodiode.[53,121,122] The N^+ region acts as an Ohmic contact and as a guard ring for the N^+P junction since it extends out over the lightly doped π-region. For proper operation the carrier concentrations and thicknesses of the N^+ and P regions are adjusted so that the depletion region in the P-type material reaches through to the lightly doped π-region when the multiplication has attained a value of about 10 to 20. Further increase in the reverse-bias voltage quickly depletes the π-region, and the multiplication then varies much more slowly with reverse-bias voltage since the peak field at the N^+P junction increases slowly. This is a significant advantage in obtaining stable avalanche gain in practical devices. The same reach-through concept has also been used for large-area metal–oxide–semiconductor avalanche photodiodes.[123]

The electric-field profile and the variation of the multiplication with position for this device are also shown in Fig. 26d. The structure is fabricated so that the electric field in the π-region is high enough to cause the carriers in this region to move at their saturated drift velocity. By designing the structure so that the π-region is fully depleted, the quantum efficiency for radiation through the P^+ contact can be essentially independent of frequency over the entire useful wavelength range. The multiplication and noise can also be independent of wavelength. The trade-off between the quantum efficiency and transit time limitations discussed above can be accurately

[120] E. J. Schiel, R. R. Gammarino, and E. J. Savitsky, *Proc. Tech. Program Electro-Opt. Syst. Design Conf., 1971 East*, p. 45. Ind. Sci. Conf. Management, Inc., Chicago, Illinois, 1971.
[121] H. W. Ruegg, *IEEE Int. Solid-State Circuits Conf., Digest of Papers*, p. 56, 1966.
[122] J. R. Biard and W. N. Shaunfield, Jr., *IEEE Trans. Electron. Devices* **ED-14**, 233 (1967).
[123] N. A. Foss and S. A. Ward, *J. Appl. Phys.* **44**, 728 (1973).

controlled in this structure by controlling the thickness of the π-region. These advantages justify the additional effort required to fabricate this more complex structure.

In the next three sections we will discuss the results obtained using these structures for avalanche photodiodes in Si, Ge, and other materials.

12. Si Avalanche Photodiodes

The device structures discussed in the previous section are applicable to avalanche photodiodes in any material, but they were all developed specifically for use in the fabrication of Si avalanche photodiodes. The N^+P guard-ring devices were developed first because of their relative simplicity. The first devices which provided microplasma-free multiplication were described by Anderson et al.[111] and the breakdown voltage of these devices was about 30 V. The performance of these detectors was evaluated with a microwave-modulated HeNe laser (0.6328 μm). Power gains as high as 10^4 were observed at a modulation frequency of 3 GHz, corresponding to current gains of 100. For moderate values of multiplication these authors determined that the excess-noise factor was approximately unity. Later estimates have given the excess-noise factor for these devices as $F = M^x$ with $x = 0.5$ [2] and $x = 0.3$.[124] Even at the wavelength of 0.6328 μm most of the light in these devices was absorbed in the bulk P-type material beyond the edge of the depletion region. The multiplication was thus initiated primarily by electrons which diffused back to the high-field region. This is consistent with the relatively small excess-noise factor. Although the cutoff frequency and quantum efficiency of these devices were not measured, the diffusion component of the photocurrent would probably have caused the efficiency to be frequency and wavelength dependent, with η decreasing both with increasing frequency and increasing wavelength.

As discussed in the previous section the frequency dependence of the efficiency can be reduced by decreasing the diffusion component and increasing the drift component of the photocurrent. By using epitaxial P-type Si on a P^+ substrate to reduce the effective size of the zero-field absorption region (L_n is much smaller in the P^+ substrate material), Lucas[125] was able to fabricate devices which had essentially frequency-independent quantum efficiency and could accurately reproduce the light pulse from a GaAs light-emitting diode ($\lambda \approx 0.998$ μm). His devices were fabricated by diffusing both the guard ring and the active junction. The P-type epitaxial layer had a resistivity of 1.8–2.2 ohm-cm and the breakdown voltage of the resulting devices was about 80 V, corresponding to a depletion width of about 4 μm.

[124] S. Takamiya, A. Kondo, and K. Shirahata, *IEEE Int. Quantum Electron. Conf.* Digest of Tech. Papers, p. 85, 1974.
[125] A. D. Lucas, *Opto-Electronics* **6**, 153 (1974).

The measured quantum efficiency of the epitaxial devices was 10% and was independent of pulse length. This compares to a dc quantum efficiency of 35% and a fast component of the quantum efficiency of 10% for devices fabricated from bulk material. Multiplications of greater than 100 were attained. Lucas also compared the noise performance of the equivalent epitaxial and bulk devices. The dependence of the excess-noise factor on multiplication was about the same for both types of devices, but the noise current was smaller for the epitaxial devices because the slow component of the current contributes to the shot noise, but not to the multiplication at high frequencies.

To obtain higher quantum efficiencies in the 1 μm wavelength region, it is necessary to have wider depletion regions. Raines[126] has fabricated N^+P Si devices with depletion-layer widths of about 8 μm and breakdown voltages of 170–180 V. The dc quantum efficiency measured for these devices at 1.06 μm was 15%, but the high-frequency quantum efficiency was only about 6%. The difference between these quantum efficiency values and those obtained by Lucas[125] on lower breakdown-voltage devices is probably due to the slightly longer wavelength used by Raines ($\lambda = 1.06$ μm). The excess-noise factor for these devices, expressed as $F = M^x$, gave $x = 0.3$. Also, avalanche gains between 100 and 300 were obtained. By making careful measurements of the gain, dark current, and multiplied shot noise as a function of bias voltage, Raines was able to show that the main source of noise was due to multiplication of the thermally generated bulk-leakage current, even though the bulk-leakage current was about two orders-of-magnitude smaller than the total leakage current (approximately 5×10^{-11} A bulk leakage compared to about 1×10^{-8} A total leakage). This is in qualitative agreement with Eq. (19) where it is evident that the bulk dark current is much more detrimental to performance than the unmultiplied surface-leakage current.

In order to obtain higher quantum efficiencies at 1.06 μm, N^+P guard-ring devices were fabricated on higher-resistivity P-type Si.[127] Because the depletion width and breakdown voltage were much larger in this material, the guard ring had to be much deeper to prevent edge breakdown. The resulting devices had 90 μm deep guard rings with breakdown voltages greater than 700 V and the 2 μm N^+P junctions had 50 μm depletion widths with breakdown voltages of about 600 V. The contact was an aluminum ring and field plate to further reduce edge effects. The field plate produced sufficient spreading of the depletion region at the surface to significantly reduce surface-leakage current. Figure 27 shows a schematic cross section of the finished device. The back surface and contact were roughened by

[126] J. A. Raines, unpublished [see *SERL Tech. J.* **21**, 4.1 (1971)].
[127] J. A. Raines, unpublished [see *SERL Tech. J.* **23**, 13.1 (1973)].

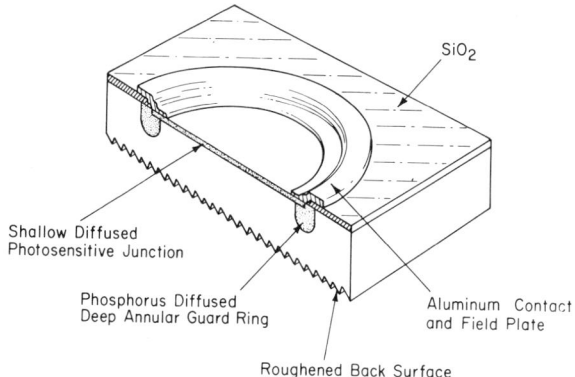

FIG. 27. Schematic cross section of a deep-diffused guard-ring Si avalanche photodiode structure designed for 1.06 μm. (After Raines.[127])

cutting shallow 75 μm wide grooves in an attempt to increase the quantum efficiency by scattering unabsorbed light back into the depletion region. With this structure a high-frequency quantum efficiency of 15% at 1.06 μm was obtained with multiplication values greater than 100 and response times less than 1×10^{-9} sec. The excess-noise factor was the same as for the lower voltage devices ($x = 0.3$).

Raines compared the noise performance of these devices with McIntyre's theory[22] [Eq. (53)] and obtained good agreement with a value of $k = 1/20$ for the ratio of the hole and electron ionization coefficients at a peak field between 2.1 and 2.3×10^5 V/cm. Approximately the same value of k was obtained for both the 170 and 600 V devices because the breakdown field does not change much in this range of voltage. This is in good agreement with the value determined for k at the same electric fields by van Overstraeten and DeMan.[48] The thermally generated bulk-leakage current was no larger than that for the lower voltage device. This was attributed to an increase in the minority carrier lifetime in the higher-resistivity P-type material used for the higher-voltage devices.

A P^+N detector, which was formed by diffusion, has recently been described by Nishida et al.[128] This structure, which was called a laterally diffused (LAD) planar diode had a very elongated P^+ region that prevented edge breakdown. The pulse response to a GaAs laser indicated an RC-limited rise time of 430 psec at a gain of 60. For gains greater than 50 the excess-noise factor was $F = M^x$ with $x = 0.4$ for 0.6370 μm radiation and with $x = 0.6$ for 0.8000 μm radiation. The excess-noise factor was larger

[128] K. Nishida, T. Takekawa, and M. Nakajima, *Appl. Phys. Lett.* **25**, 669 (1974).

for the longer wavelength because of the larger amount of injected hole current ($\alpha_n \gg \beta_p$).

The beveled-edge or contoured-surface devices described above (Fig. 26c) were first used for the detection of low energy protons, soft x rays, and 0.05–1.0 MeV electrons.[129] Because of the wide depletion region in these high-voltage devices, it was natural to extend the use of this type of device to the detection of 0.9 and 1.06 μm radiation.[120,130] The quantum efficiencies obtained were 70 to 75% at 0.9 μm and 25 to 35% at 1.06 μm. Multiplication values of 400 at 0.9 μm and 200 at 1.06 μm could be obtained reliably. The lower multiplication at the longer wavelength was produced by the relatively larger amount of hole injection and smaller amount of electron injection in this structure. This results in less avalanche gain since $\beta_p \ll \alpha_n$. Injection of the carrier with the lower ionization coefficient should also cause the excess-noise factor to be high at 0.9 μm and to be higher still at longer wavelengths. However, it was reported that noise measurements on detectors of this type resulted in excess-noise factors that were described by $F = M^x$ with $x = 0.3$ to 0.4,[120] although the wavelength used for these measurements was not specified. Response times as short as 4 nsec have been observed with large area (0.23 cm diameter) detectors of this type.[129]

The beveled-edge detector developed by Savitsky et al.[120,130] for Si does not satisfy the requirement that for optimum avalanche photodiode performance the injected carrier should be of the more highly ionizing type: electrons for Si. This requirement could be met with the same type of structure by interchanging the N and P regions. It could also be met by bringing the light in through the back contact at the edge of the depletion region. This procedure has been followed by Takamiya et al.[102,124] They fabricated a P^+PN^+ beveled-mesa type of structure, as shown in Fig. 28, in which the radiation was incident on the P^+ region. This structure not only reduced

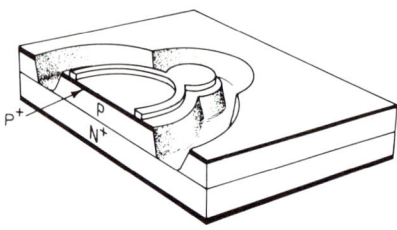

FIG. 28. Schematic cross section of beveled-mesa P^+PN^+ Si avalanche photodiode structure.

[129] G. C. Huth, J. B. Trice, and R. A. McKinney, *Rev. Sci. Instrum.* **35**, 1220 (1964).
[130] E. J. Savitsky and R. G. Trapani, *Proc. Electro-Opt. Syst. Design Conf., 1971 West*, p. 393. Ind. and Sci. Conf. Management, Inc., Chicago, Illinois, 1971.

the diffusion component of the current because of the shorter diffusion length in the heavily doped N^+ substrate, as discussed earlier, but it also provided pure electron injection at short wavelengths and nearly pure electron injection at longer wavelengths. The relative contributions of the small diffusion component of the current from the N^+ substrate (hole current) was further minimized by the multiplication process since the holes were not multiplied as much as the electrons. The depletion width of the device at operating-bias levels was about 8–9 μm and the breakdown voltage was about 160 V.

Takamiya et al.[131] made careful measurements of the excess-noise factor of these devices, as well as of N^+P devices, similar to the one used by Anderson

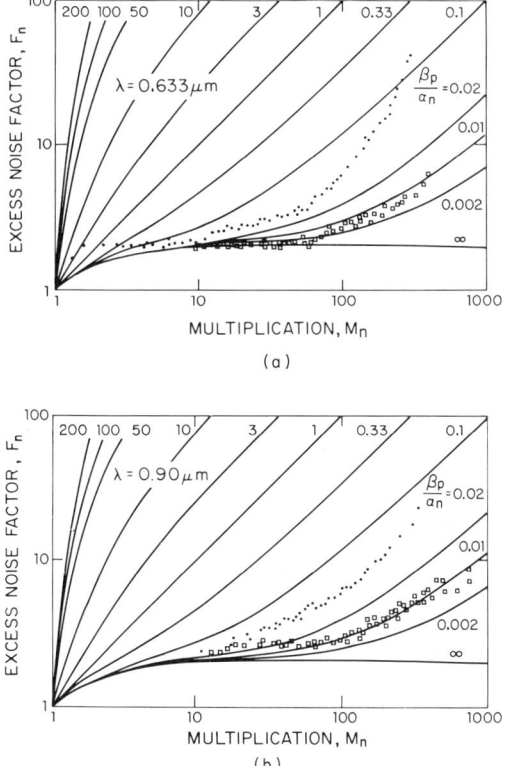

FIG. 29. Experimental excess-noise factors for Si P^+PN^+ (□) and N^+P (●) guard ring avalanche photodiodes at (a) 0.6328 μm and (b) 0.9000 μm. (Data from Takamiya.[131]). The solid curves in these figures were calculated using Eq. (53).

[131] S. Takamiya, A. Kondo, and K. Shirahata, Tech. Digest Mon. Meeting Opto- and Quantum Electron. (Japan), Paper 1974–8.

et al.,[111] but with breakdown voltages of about 40 V and depletion widths of about 1.2 μm. The maximum field in the P^+PN^+ device was about 3×10^5 V/cm while that in the N^+P device was about 5×10^5 V/cm. The results of these measurements at 0.6328 are shown in Fig. 29a where they are compared with the excess-noise factors calculated from Eq. (53) for pure electron injection. The excess-noise factor is significantly larger for the N^+P device than for the P^+PN^+ device. For both the P^+PN^+ and N^+P cases the injected current at this wavelength consists almost entirely of electrons, and a comparison with the calculated curves indicates that the ratio β_p/α_n is less than 0.01 for the P^+PN^+ case ($E_m = 3 \times 10^5$ V/cm) and greater than 0.02 for the N^+P case ($E_m = 5 \times 10^5$ V/cm).

Figure 29b shows the excess-noise factors for the same two devices at 0.9000 μm. The excess-noise factor for the P^+PN^+ device at this wavelength was slightly larger than that for 0.6328 μm, and the difference could be explained by the mixed injection (about 66% electron injection) which was estimated for a wavelength of 0.9000 μm (Fig. 11). For the N^+P device the injected current at 0.9000 μm consisted almost entirely of electrons. (The electron injection for the N^+P device at both wavelengths resulted primarily from the diffusion of electrons from the P-type substrate where the minority-carrier diffusion-length was estimated to be about 300 μm.) This is consistent with the results shown in Fig. 29 which show that the excess-noise factor for the N^+P device is essentially the same at both wavelengths. The fact that the fraction of the injection by electrons is slightly larger at 0.9000 μm than at 0.6328 μm is evident in the slightly lower excess-noise factor for multiplication greater than 100 at the longer wavelength.

The quantum efficiency of the beveled-mesa P^+PN^+ avalanche photodiode was about 30% at 0.9 μm, with no apparent variation of the quantum efficiency with frequency. Multiplication factors greater than 300 were obtained. Cutoff frequencies of 1 GHz for a 200 μm diameter device and 400 MHz for a 500 μm diameter device were measured, the difference being due to different junction capacitances. The excess-noise factors for these devices are the lowest yet reported and are close to the theoretical limit. Within the multiplication range from 50 to 100, the excess-noise factor is essentially independent of the multiplication. The ratio of α_n/β_p estimated from the noise measurements for the P^+PN^+ and N^+P devices was about 150 and 25, respectively. However, the same ratio calculated from the multiplication rate measurements was only about 36 for the P^+PN^+ device and 9 for the N^+P device. These values can be compared with the ratio α_n/β_p obtained from the ionization-rate data of Lee et al.[7] of about 25 and 4, respectively, for the peak electric fields in these devices. These discrepancies may be related to the difference between the effective k-value for the excess-noise factor k_{eff} and the effective k-value for the multiplication k_1 [see

Eqs. (55) and (57)] which occurs when the ratio of α_n and β_n is not constant. This will be discussed in more detail below.

The reach-through avalanche photodiode structure (Fig. 26d), first described by Ruegg,[53,121] has several desirable characteristics not present in the other device structures. The trade off between the quantum efficiency at long wavelengths and the speed of response can be accurately controlled by changing the thickness of the intrinsic region. Since multiplication occurs only in the narrow high-field region, the avalanche gain characteristics are not affected. The wide low-field region in these devices produces a much more gradual change in the avalanche gain with bias voltage, which is desirable in practical applications of these devices. In addition, the structure provides nearly pure electron injection regardless of wavelength (assuming the N^+ region is sufficiently thin), so that the noise performance of these devices should be essentially independent of wavelength and as good as can be obtained with Si. Because of these advantages McIntyre, Webb, Conradi, and co-workers at RCA Limited have expended considerable effort developing and characterizing this type of device,[9,29,58,132-132b] several versions of which are now available commercially.

The reach-through avalanche photodiode is fabricated by starting with very high quality π-type substrates with typical resistivities of about 5000 ohm-cm. The active N^+P junction is formed by two diffusions: the P-type using boron and the N-type using phosphorus. A thin P^+ diffusion is used to form the window and Ohmic contact to the π-region. The diffusion times and temperatures are adjusted so that, when a reverse-bias voltage is applied, the depletion region in the P-layer reaches through to the π-region just before the NP junction breaks down. Further increases in reverse-bias voltage then rapidly deplete the high-resistivity π-region and increase the uniform field in this region while only increasing slightly the peak field at the NP junction. This slower increase in the peak electric field results in the much slower variation of multiplication with bias voltage which is characteristic of this structure. The diffusion steps are critical to the performance of the device. If the phosphorus N-type diffusion is not deep enough, the NP junction will break down before reach-through occurs. In this case the gain will vary rapidly with voltage and the quantum efficiency for radiation incident on the P^+ contact will be low. If the phosphorus diffusion is too deep, an NP^- or $N\pi$ junction will be formed and sufficient gain will be obtained only at extremely high voltages. Recently, to achieve better

[132] P. P. Webb and R. J. McIntyre, *1970 Solid State Sensors Symp.*, Minneapolis, Conf. Record, p. 82, Inst. of Electron. and Electron. Eng., New York, 1970.
[132a] P. P. Webb and R. J. McIntyre, *Proc. Electro-Opt. Design Conf., 1971 East*, p. 51. Ind. and Sci. Conf. Management, Inc., Chicago, Illinois, 1971.
[132b] P. P. Webb and A. R. Jones, *IEEE Trans. Nucl. Sci.* **NS-21**, 151 (1974).

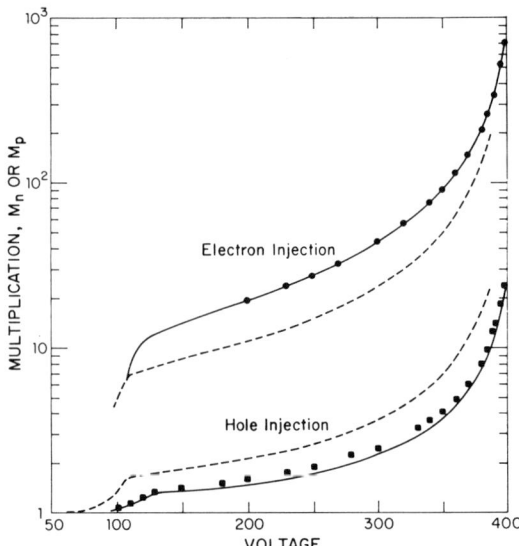

FIG. 30. Experimental (data points) and calculated (curves) variation of the electron and hole multiplication with bias voltage for a reach-through avalanche photodiode. The curves were calculated using the ionization coefficient data from Lee et al.[7] (solid curves) and van Overstraeten and DeMan[48] (dashed curves). (After Conradi.[58])

uniformity, ion implantation has been used to deposit the boron for the P-type diffusion.[133]

The experimental variation of the electron and hole multiplication with bias voltage for a reach-through avalanche photodiode is shown by the data points in Fig. 30.[58] Pure hole injection was obtained by using the 4047 Å line of a mercury lamp with a narrow line filter. This radiation was incident on the N side of the device where the junction was 8–10 μm deep. Pure electron injection was obtained by using 0.8 μm radiation incident through the P^+ window. As expected, the electron multiplication M_n was significantly larger than the hole multiplication M_p. The curves shown in Fig. 30 were calculated by Conradi[58] for the electric field distribution estimated for the device from the ionization-coefficient data of Lee et al.[7] (solid curves) and van Overstraeten and DeMan[48] (dashed curves). It is clear that much better agreement with experimental data is obtained by using the ionization-rate data of Lee et al.[7] The calculated values of M_n and M_p for various bias voltages are shown in Table III, together with the corresponding values of the maximum electric field, E_m, $k_m = \beta_p(E_m)/\alpha_n(E_m)$, k_1 [Eq. (55)], and k_{eff} [Eq. (59)] for both sets of ionization–coefficient data. It can be seen that

[133] W. N. Shaunfield and D. W. Boone, *1968 Solid State Sensors Symp., Minneapolis, Conf. Record*, p. 1281. Inst. of Electron. and Electron. Eng., New York, 1968.

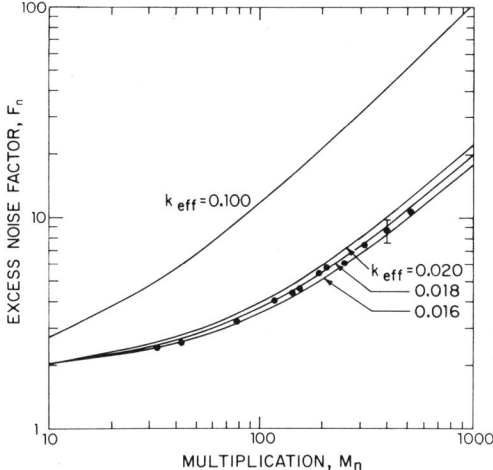

FIG. 31. Experimental excess-noise factor for electron injection in a reach-through avalanche photodiode. Curves are calculated using Eq. (57) and the indicated values of k_{eff}. (After Conradi.[58])

there is a significant difference between k_m and k_1, particularly when the ionization-rate data of Lee et al.[7] are used.

Low frequency room temperature noise measurements were made on these devices, and the experimental excess-noise factor for electron injection in one device is shown in Fig. 31. Also shown are curves calculated using Eq. (57) and the (constant) values of k_{eff} indicated. Good agreement is only achieved with the k_{eff} obtained from the data of Lee et al.[7] shown in Table III. The values of both k_{eff} and k_1 in Table III are nearly constant with voltage. Noise measurements on similar devices at 77 K were previously analyzed using Eq. (53) and, in the constant k approximation,[29a] a best fit was obtained with $k = 0.028$. This k value is somewhat higher than those for pure electron injection obtained by Takamiya et al.[131] (Fig. 29).

As discussed above, the quantum efficiency and speed of response for the reach-through avalanche photodiode are closely related; that is, an increase in the width of the high-resistance π-region increases the quantum efficiency for weakly absorbed radiation, but decreases the response time. The impulse response depends somewhat on where the radiation is absorbed.[9] For strongly absorbed radiation there is very little photocurrent until the injected electrons reach the high-field region (the electron transit time) and are multiplied. Then the multiplied current persists for essentially the transit time of the holes back to the P^+ contact. For weakly absorbed radiation some of the injected carriers are multiplied immediately and the resulting current pulse has a trapezoidal shape as shown in Fig. 32a. The

TABLE III

COMPARISON OF THE CONSTANTS k_m, k_{eff}, AND k_1 FOR VARIOUS VALUES OF GAIN[a]

	Volts	E_m (10^5 V/cm)	M_n	M_p	k_m	k_{eff}	k_1
(a)							
	107	3.06	6.1	1.10	0.032	0.017	0.020
	143	3.19	13.4	1.32	0.042	0.020	0.025
	218	3.25	21.5	1.56	0.047	0.020	0.027
	318	3.33	55	2.62	0.053	0.021	0.030
	393	3.37	402	13.6	0.059	0.021	0.031
	418	3.39	−553	−14.8	0.060	0.022	0.032
(b)							
	100	2.86	4.49	1.35	0.13	0.090	0.10
	135	2.98	7.9	1.76	0.15	0.096	0.11
	210	3.04	11.5	2.2	0.16	0.098	0.11
	310	3.12	26.1	4.0	0.17	0.10	0.12
	385	3.18	170	22	0.18	0.10	0.12
	410	3.20	−239	−29	0.18	0.10	0.13

[a] Using the ionization coefficient data of (a) Lee et al.[7] and (b) van Overstraeten and De Man.[48] The differences in E_m between the two sets of data result from slightly different field profiles which were adjusted to give a calculated breakdown voltage for the applicable ionization coefficient data that was approximately equal to the experimental breakdown voltage. A negative value of M indicates the breakdown voltage has been exceeded.

response times (pulse widths at 10% points) of the reach-through avalanche photodiodes, as a function of the width of the drift region and as a function of the operating voltage (the electron and hole transit times decrease with higher operating voltages) calculated for wavelengths of 0.9 and 1.06 μm, are shown in Fig. 32b. The corresponding calculated quantum efficiencies are shown in Table IV. Using these results the response time can be determined for a given quantum efficiency or vice versa. The quantum efficiencies obtained experimentally are not quite as high as indicated in Table IV, because of reflection and/or absorption losses in the window of the device package. However, typical values for the RCA Type C30817 avalanche photodiode are 85% at 0.9 μm and 15% at 1.06 μm with rise and fall times of about 2 nsec at 0.9 μm for an operating voltage of about 375 V.

The performance characteristics of Si avalanche photodiodes from the literature and from manufacturer's data sheets are summarized in Table V. The best designed structures are the reach-through avalanche photodiode and the beveled-mesa avalanche photodiode described by Takamiya et al.[124] However, in the wavelength range of GaAs and Nd:YAG lasers, 0.8–1.06 μm, the response time of Si avalanche photodiodes with reasonable quantum efficiencies will be greater than about 1 nsec. For detection at longer wave-

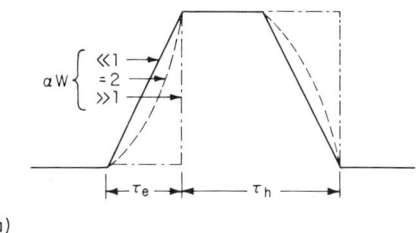

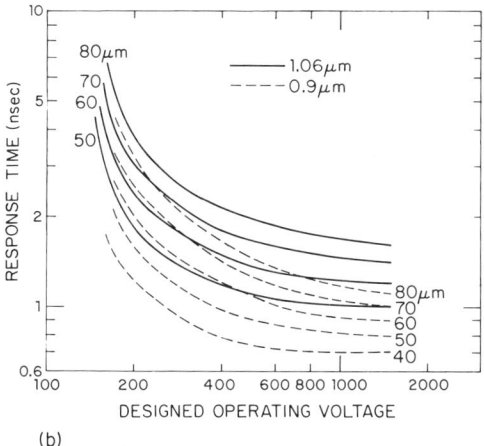

FIG. 32. Impulse response of Si reach-through avalanche photodiodes in which the transit time of holes is about twice that of electrons. (a) Pulse shape for $\alpha W \ll 1$ ($\lambda = 1.06$ μm), $\alpha W = 2$ ($\lambda = 0.9$ μm), and $\alpha W \gg 1$ (visible radiation), where α is the absorption coefficient and W is the width of the drift region. (b) Pulse widths versus bias voltage for $\lambda = 0.9$ μm and $\lambda = 1.06$ μm for the values of W shown. (After Webb et al.[9])

lengths or at faster response times, semiconductor materials with smaller bandgaps and/or higher absorption coefficients must be used.

13. Ge Avalanche Photodiodes

From the absorption coefficient data in Fig. 22 it is clear that at room temperature the absorption in Ge is high enough to provide sensitive photodetectors out to about 1.6 μm. Because of the smaller bandgap, the dark current of Ge devices at room temperature is relatively high. This limits the sensitivity and avalanche gain that can be obtained. The high dark current can be reduced by cooling, but then the shift of the absorption edge with temperature must be considered for operation at longer wavelengths.

TABLE IV

CALCULATED QUANTUM EFFICIENCY OF Si REACH-THROUGH AVALANCHE PHOTODIODES[a]

Depletion layer width (μm)	Quantum efficiency			
	0.9 μm	1.06 μm		
	300°K	300°K	325°K	350°K
40	0.766	0.094	0.118	0.167
50	0.812	0.116	0.146	0.206
60	0.841	0.137	0.172	0.240
70	0.861	0.158	0.199	0.273
80	0.873	0.177	0.223	0.305

[a] The reflectivities of the front and back surfaces were assumed to be 10% and 90%, respectively. (After Webb et al.[9])

The larger dark current is not peculiar to Ge, but is a consequence of the smaller bandgap. Thus, cooling will be required to obtain devices with high sensitivity and avalanche gain at wavelengths longer than about 1.5 μm. A fundamental limitation on the performance of Ge avalanche photodiodes, however, is the ratio of the electron and hole ionization coefficients. Although, as discussed in Section 10, there is some uncertainty in the actual values of α_n and β_p for Ge, there is general agreement that the ratio is close to unity. This leads to an excess-noise factor that varies approximately as the multiplication. Nevertheless, Ge avalanche photodiodes can significantly increase the sensitivity of a wide-bandwidth system where the main limitation is the noise of the first amplifier stage. Because the absorption coefficient is high for wavelengths shorter than about 1.4 μm, high-speed detectors can be fabricated in this range.

Since the ionization coefficients for electrons and holes are approximately equal, the multiplication and noise of Ge avalanche photodiodes are not as sensitive to the device structure as in Si devices. Melchior and Lynch[16] fabricated diffused N^+P devices with depletion width of about 1.2 μm and breakdown voltages of about 16 V. Edge breakdown was prevented by a diffused N-type guard ring, and to reduce the leakage current, mesas were also etched. Current gains as high as 200 at low frequencies and 10 at 6 GHz were obtained at room temperature for $\lambda = 1.15$ μm. Higher gains were achieved by cooling the devices to reduce the dark current [see Eq. (46)].

Ge avalanche photodiodes have been fabricated specifically for high-speed detection of the 1.54 μm wavelength of Er:glass lasers.[133] Because of the low absorption of Ge at 1.54 μm, wide depletion widths are required to obtain reasonable quantum efficiency. N^+P guard-ring devices were fabricated starting with about 5 ohm-cm P-type material. The breakdown voltages of the finished devices were between 100 and 200 V and avalanche gains between 30 and 50 were achieved. The depletion widths were about 25 μm. Because of the temperature dependence of the bandgap, the quantum efficiency was very temperature sensitive, but it was calculated that quantum efficiencies of about 68% at 1.54 μm could be attained with room temperature devices.

Thus, although $\alpha_n \approx \beta_p$ in Ge and the excess-noise factor for avalanche photodiodes varies approximately as the multiplication M, these devices can significantly increase the sensitivity of wide-bandwidth, long-wavelength systems where the main source of noise is from the amplifier following the detector. For these applications the maximum sensitivity is obtained when the multiplied shot noise is about equal to the amplifier noise. For a low-noise, 50-MHz amplifier this occurs with a gain of about 10 to 30.

14. Avalanche Photodiodes in Other Materials

Although only Si and Ge avalanche photodiodes are commercially available, there have been promising results obtained in other, less well-developed materials.

Avalanche gain in InAs photodiodes at 300°K was first reported by Lucovsky and Emmons,[134] and gains of ~50 were measured at a frequency of 125 MHz. They found that the noise increased faster than the signal for multiplication values higher than unity. Subsequent measurements at 77°K by Kim indicated that α_n is much greater than β_p,[85] and that the noise increases in the same manner as the photosignal. This implies that the excess-noise factor is approximately unity.[135]

Avalanche photodiodes have also been fabricated in InSb.[82] Multiplication measurements in these devices indicated that α_n is much greater than β_p, and analyses of noise measurements using McIntyre's theory [Eq. (53)] agreed with this conclusion; that is, the noise increased only slightly faster than the signal. Avalanche gain values as high as 10 were attained. In later work, Protschka[83] observed multiplication factors as high as 70 to 80 on devices with breakdown voltages of 19 V and long-wavelength cutoffs of ~5.4 μm at 77°K.

Uniform avalanche gain and low excess-noise factors have been reported for Schottky barrier photodiodes in GaAs.[76] A schematic cross section of the structure used for these devices is shown in Fig. 33. The Schottky barrier was a 100-Å-thick semitransparent Pt layer. A guard ring of high-resistivity

---

[134] G. Lucovsky and R. B. Emmons, *Proc. IEEE* **53**, 180 (1965).
[135] C. W. Kim, unpublished, 1971.

TABLE V

PERFORMANCE OF AVALANCHE PHOTODIODES

Material and structure		Wavelength (μm)	Quantum efficiency (%)	Response[a] time or frequency	Excess noise factor[b]	Gain	Depletion width (μm)	Bias voltage (V)	Reference
Si guard ring	N^+P	0.6328	—	~3 GHz	$M^{0.5}$	~100	—	~30	Anderson et al.[5]
	N^+P	0.998	10	<15 nsec	—	≳300	~4	~80	Lucas[101]
	N^+P	1.06	15	<30 nsec			50	500–600	
		0.925	52	≲1 nsec	$M^{0.3}$	≳200	—		Raines[126]
		0.800	75	≲1 nsec					
	N^+P	1.06	6	(<30 nsec)	$M^{0.3}$, $k \approx 1/20$	≳100	—	170	Raines[112,127]
	N^+P	0.694	—	~0.5 nsec	$M^{0.3}$	≳100	—	160–220	EMI types S30500–S3054
		0.900	—	≲10 nsec					
		1.065	—	—	—	—	—		
	N^+P	0.630	20	—	—	≳100	—	≳200	Texas Instruments type TIXL 55, 56, 69
		0.900	30–40						
	P^+N	0.637	—	~0.43 nsec	$M^{0.4}$	≳60	—	~100	Nishida et al.[128]
		0.800	—		$M^{0.6}$				
	P^+N	0.6328	70	~3 GHz	$M^{0.6}$	≳80	—	40–160	Nippon Electric Co. APD 200A, APD 200B
	P^+N	0.540	~9	~0.2 nsec	—	~100	—	150	Spectra Physics Model 403

Device	λ (μm)	η (%)	Response time	k_{eff} / M	Gain		Bias (V)	Reference
Si beveled P^+N	0.900	75	≤10 nsec	—	200–400	~150	1500–2000	Savitsky and Trapani[130]
mesa P^+N	1.06	30	≤10 nsec	—	100–200	~150	1500–2000	Schiel et al.[120]
P^+N	0.900	75	$\tau_R = 3$ nsec	—	200	—	1800–2000	GE type 50 EHS
	1.060	30	$\tau_F \approx 12$ nsec	—	100	—		
PN^+	0.900	~22	>1 GHz	$k_{eff} = 1/150$	≤100	8	150	Takamiya et al.[124]
Si reach- $N^-P\pi P^+$	1.06	25–30	~3–6 nsec	$k_{eff} \approx 1/55 = 0.018$	≤500	—	500	Webb et al.[9]
through $N^+P\pi P^+$	0.90	85						
device	1.06	15	$\tau_R = 2$ nsec	$k_{eff} \approx 1/50 = 0.020$	≤100	—	300–475	RCA type C30817
Ge guard N^-P ring	0.4–1.55	~48 at 1.45 μm	≥6 GHz	$\sim M^1$	250 at 300°K; >10^4 at 80 K	—	~16	Melchior et al.[2,16]
N^+P	1.54	~68 at 300 K	$\tau_R \lesssim 1$ nsec; $\tau_F \approx 3$ nsec	$\sim M^1$	30–50 at 300 K	~25	100–200	Shaunfield et al.[133]
N^+P	1.5	60 at 300 K	$\tau_R = 0.5$ nsec	$\sim M^1$	40 at 300 K	—	40	Optitron Inc. type Ga–1
N^+P	1.06	60	—	—	≳15 at 300°K	—	30–150	Texas Instruments type TIXL[57,68]
	1.54	25						

TABLE V (Cont.)

Material and structure	Wavelength (μm)	Quantum efficiency (%)	Response[a] time or frequency	Excess noise factor[b]	Gain	Depletion width (μm)	Bias voltage (V)	Reference
InAs PN etched mesa	0.6328	—	$\gtrsim$ 0.125 GHz	$\gtrsim M^1$	$\lesssim 12$ at 300°K	—	~10	Lucovsky and Emmons[134]
PN	0.5–3.5 at 77°K	High	$\lesssim$ 1 nsec	—	$\gtrsim 100$ at 77°K	—	~30	Kim[135]
InSb PN	0.5–5.5 at 77°K	—	—	~1	~10 at 77°K	—	~7	Baertsch[82]
PN	1–5.4 at 77°K	—	—	—	$\lesssim 80$ at 77°K	—	12–19	Protschka[83]
GaAs Pt-Schottky N GaAs barrier	0.4–0.87	30	$\lesssim$ 1 nsec	$\approx M^{0.1}$	$\gtrsim 100$	~3	~60	Lindley et al.[76]
Pt-N GaAs	0.86–0.95	~50	< 1 nsec	$k < 1/50$	$\lesssim 400$	20	~200	Stillman et al.[77]
GaInAs Pt-Schottky N GaInAs barrier	0.5–1.10	$\lesssim$ 50	$\lesssim$ 0.17 nsec	low	$\lesssim 250$	4	40–60	Stillman et al.[142–144]
GaAsSb P–N inverted mesa	1.06	$\approx$ 95	6 GHz	—	$\lesssim 11$	—	—	Eden and Nakano[141a]

[a] Response time is given either as a bandwidth, exponential time constant, or 10–90% rise time.
[b] Excess noise factor is given either as a power of M or in terms of k for Eq. (53) or (54).

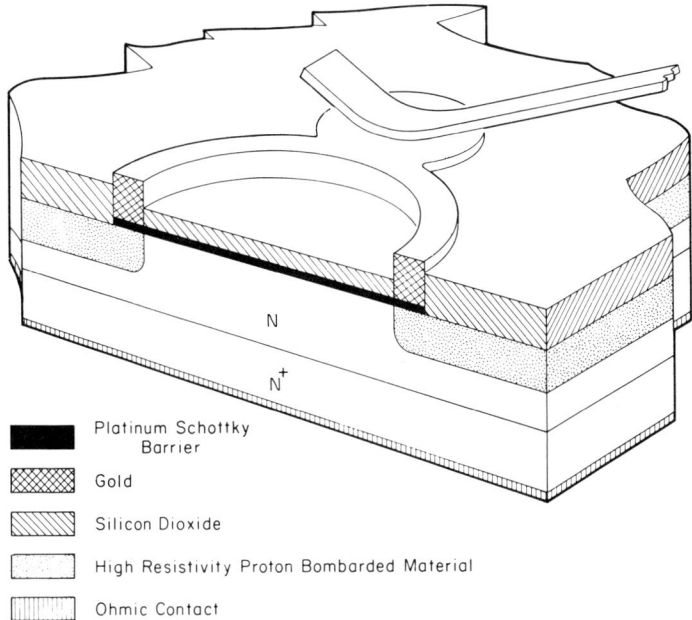

FIG. 33. Pictorial cross section of GaAs Schottky-barrier device. (After Stillman et al.[144])

GaAs (indicated by the shaded area) was formed by proton bombardment[136] or by vapor-epitaxial growth of high-purity GaAs around etched mesas.[137] The quantum efficiency of these devices was typically about 30% with no intentional antireflection coating. Response times were estimated to be less than 0.1 nsec from rise-time measurements of the response to a GaAsP[138] injection laser. Noise measurements on these devices indicated that the noise power increased as $M^{2.1}$ for multiplication values between 5 and 120. This corresponds to an excess-noise factor $F = M^x$ with $x = 0.1$. The slow variation of the excess-noise factor with multiplication was taken as an indication that there was a substantial difference in the hole and electron impact-ionization rates. Since for these devices most of the radiation was absorbed in the semiconductor beyond the high electric field region where avalanche occurs, this result implies, in contrast to earlier conclusions,[2] that the ionization coefficient for holes is much larger than that for electrons.

Other measurements[139] of noise in GaAs avalanche photodiodes have

[136] A. G. Foyt, W. T. Lindley, C. M. Wolfe, and J. P. Donnelly, *Solid-State Electron.* **12**, 209 (1969).
[137] C. M. Wolfe and W. T. Lindley, *J. Electrochem. Soc.* **116**, 276 (1970).
[138] N. Holonyak, Jr., and S. F. Bevacqua, *Appl. Phys. Lett.* **1**, 82 (1962).
[139] T. Igo and Y. Toyoshima, *Jap. J. Appl. Phys.* **9**, 1286 (1970).

indicated that the noise power increases as M^3. The diodes used for these measurements were diffused P^+N and epitaxial N^+P devices with junctions which were about 1 to 4 μm deep. The reasons for this disagreement with the previous results are not known. However, since the wavelength of the radiation used for the noise measurements by Igo and Toyoshima[139] was not given, it is possible that the higher excess-noise factor is the result of mixed injection or injection of the carrier type with the lower ionization coefficient (see Figs. 9 and 11). This explanation is consistent with the observation that the excess noise was higher in the P^+N diffused devices.[139] In general, further work is needed to characterize the noise performance of GaAs avalanche photodiodes.

To obtain a fast and sensitive high-gain avalanche photodiode for the Nd:YAG laser at 1.06 μm, a material with a high absorption coefficient at this wavelength is required. It is also desirable that the material have a wide bandgap to minimize the dark current at room temperature. In addition, the material should have electron and hole ionization coefficients that are significantly different. Promising results have been obtained from Schottky barriers on $In_xGa_{1-x}As$ alloys.[140] The room temperature bandgap of this alloy can be varied from about 0.35 eV for $x = 1$ to 1.41 eV for $x = 0$. However, there is also a large change in the lattice constant of the alloy over this composition range. For an alloy composition of $x = 0.20$, which is approximately that required for maximum detector response at 1.06 μm, the lattice constant of the alloy is about 1.4% larger than that of GaAs, which is generally used as a substrate for the epitaxial growth of the $In_xGa_{1-x}As$ layer. This lattice mismatch is important because it is the source of defects or dislocations which can affect the performance of a device. However, by using grading techniques during the growth to minimize the lattice-mismatch-induced dislocations,[141] epitaxial layers have been grown from which high-gain, uniform avalanche photodiodes could be selected. The device structure was the same as that shown in Fig. 33, except that the epitaxial material was $In_xGa_{1-x}As$.

The spectral responsivity at unity gain of two $In_xGa_{1-x}As$ avalanche photodiodes, along with the spectral responsivity curves for GaAs and Si avalanche photodiodes at unity gain, are compared in Fig. 34. Uniform avalanche multiplication factors of over 250 with 175 psec rise times were obtained in selected $In_xGa_{1-x}As$ devices. The response of one of these detectors (operating at a gain of greater than 250) to mode-locked pulses from a Nd:YAG laser is shown in Fig. 35. There was a high yield of devices with high quantum efficiency. However, only a few devices on each wafer could be selected which provided uniformly high values of avalanche gain;

[140] G. E. Stillman, C. M. Wolfe, A. G. Foyt, and W. T. Lindley, *Appl. Phys. Lett.* **24**, 8 (1974).
[141] C. M. Wolfe, G. E. Stillman, and I. Melngailis, *J. Electrochem. Soc.* **121**, 1506 (1974).

5. AVALANCHE PHOTODIODES

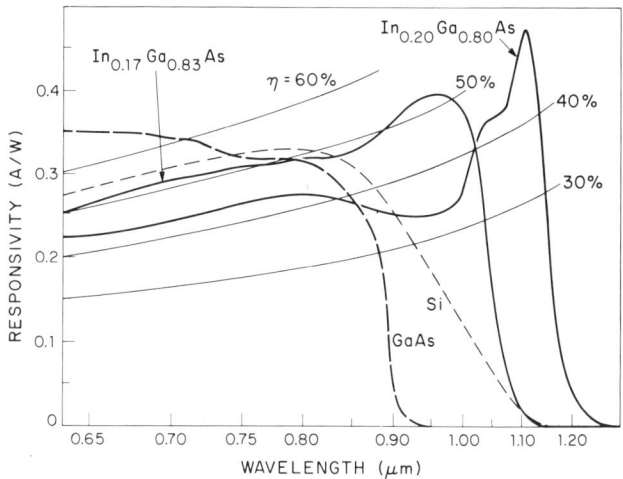

FIG. 34. Spectral responsivity of two $In_xGa_{1-x}As$ avalanche photodiodes at unity gain. The spectral response of GaAs and Si avalanche photodiodes are also shown for comparison. (After Stillman et al.[140])

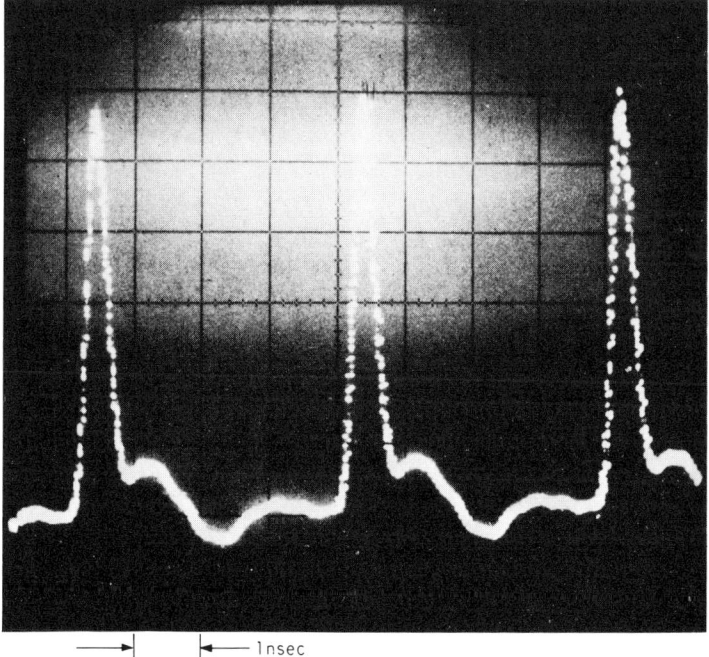

FIG. 35. Response of an $In_{0.17}Ga_{0.83}As$ avalanche photodiode to 1.06 μm mode-locked Nd:YAG laser pulses. The detector was operating into a 50-ohm sampling oscilloscope and the avalanche gain shown in the photograph was greater than 250.

the gain of most devices was limited by microplasmas. The origin of these microplasmas was not determined, but they were probably related to defects or dislocations generated by the lattice mismatch. Thus, although the most sensitive avalanche photodiodes for 1.06 μm have been fabricated using $In_xGa_{1-x}As$ alloys, considerable materials development remains to be done before it will be feasible to manufacture these devices with reproducibly high gain. Other alloy systems, such as $GaAs_{1-y}Sb_y$ may also be suitable for avalanche photodiodes[85a,141a] in the 1.06 μm wavelength range, but they are even less well developed than $In_xGa_{1-x}As$ alloys.

VI. Electroabsorption Avalanche Photodiode Detectors

In further work on GaAs Schottky barrier avalanche photodiodes,[142-144] the spectral responsivity of devices fabricated with low carrier concentration ($< 10^{15}$ cm^{-3}) GaAs was found to be very different from that of devices fabricated with higher carrier concentration ($\sim 10^{16}$ cm^{-3}) material. Since these results have important implications for integrated optical circuits and for the ultimate performance of GaAs avalanche photodiodes, they will be discussed in detail.

15. EXPERIMENTAL RESULTS

The spectral responsivity of a GaAs Schottky barrier avalanche photodiode fabricated on epitaxial material with $n \approx 7 \times 10^{15}$ cm^{-3} is shown in Fig. 36a for several bias voltages. At low bias, the long wavelength cutoff occurs at about 0.866 μm and is quite sharp. At higher bias voltages the response at wavelengths shorter than 0.86 μm increases with increasing voltage, and a long tail develops for wavelengths longer than about 0.87 μm. Similar behavior was observed in earlier work.[76] At low voltages, the increase in response at the shorter wavelengths is due mainly to a widening of the depletion region with increasing bias. At intermediate and high voltages the increase at shorter wavelengths is due to avalanche multiplication. At high bias voltages the increase in response at the longer wavelengths is due primarily to increased absorption caused by the Franz–Keldysh shift of the absorption edge. The long wavelength threshold is shifted significantly

---

[141a] R. C. Eden and K. Nakano, unpublished. [See *IEEE Trans. Electron. Devices* **ED-21**, 742 (1974).]

[142] G. E. Stillman, C. M. Wolfe, J. A. Rossi, and J. L. Ryan, *Proc. Symp. Opt. Acoust. Microelectron. New York, 1974*, p. 543. Polytechnic Press, Brooklyn, New York, 1975.

[143] G. E. Stillman, C. M. Wolfe, J. A. Rossi, and J. P. Donnelly, *Appl. Phys. Lett.* **25**, 671 (1974).

[144] G. E. Stillman, C. M. Wolfe, J. A. Rossi, and J. L. Ryan, in "Gallium Arsenide and Related Compounds" *(Proc. 5th Int. Symp. GaAs, Deauville, 1974)*, p. 210. Inst. Phys. and Phys. Soc., London, 1975.

by this effect, but the response decreases monotonically with wavelength at any bias voltage for wavelengths longer than 0.8 μm. Avalanche gains of over 100 at wavelengths shorter than 0.86 μm were obtained with bias voltages close to breakdown.

The variation of the spectral responsivity of GaAs avalanche photodiodes on material with $n < 10^{15}$ cm^{-3}, shown in Figs. 36b and 36c, is quite different from that for devices with $n \gtrsim 7 \times 10^{15}$ shown in Fig. 36a. Even at very low bias-voltages (~ 30 V) there is a significant increase in responsivity and quantum efficiency at wavelengths close to and beyond the usual low-electric-field absorption edge in GaAs, and at higher voltages the avalanche gain in this region is much higher than that at shorter wavelengths. In some devices the shoulder visible at about 0.9 μm in Fig. 36b is more significant so that the peak in the spectral response extends beyond 0.91 μm.[142] The absorption responsible for the photoresponse at these long wavelengths is due to the Franz–Keldysh effect. The dominant peak in the spectral response at these wavelengths can be qualitatively explained by highly asymmetric values of the electron and hole impact-ionization coefficients ($\beta_p/\alpha_n \gg 1$). Diodes designed to exploit the Franz–Keldysh shift of the absorption edge are referred to as electroabsorption avalanche photodiode (EAP) detectors.

16. Franz–Keldysh Effect and Quantum Efficiency

The absorption coefficient of a direct-bandgap semiconductor in a uniform electric field E, for radiation of wavelength λ, can be expressed in mks units as[145-147b]

$$\alpha(\lambda, E) = \frac{2^{7/3}\mu^{4/3}e^{7/3}|\mathscr{E} \cdot \mathbf{P}_{nn'}|^2 E^{1/3}}{\hbar^{8/3}\omega m^2 n\varepsilon_0 c} \int_{\beta}^{\infty} |\text{Ai}(z)|^2 \, dz. \quad (73)$$

In this equation, $\beta = (2\mu)^{1/3}(E_g - \hbar\omega)/(\hbar eE)^{2/3}$, μ is the electron–hole reduced mass, e the electronic charge, $\mathscr{E}$ the radiation polarization vector, $\mathbf{P}_{nn'}$ the zero-field interband matrix element, $\hbar$ Planck's constant/2π, ω the frequency of the incident radiation of wavelength λ, m the free electron mass, n the index of refraction, ε_0 the free-space dielectric constant, c the speed of light, and E_g the energy gap at zero field. The Airy functions Ai(z) are defined by

$$\text{Ai}(z) = (1/\pi) \int_0^{\infty} \cos(sz + \tfrac{1}{3}s^3) \, ds. \quad (74)$$

---

[145] J. Callaway, *Phys. Rev.* **130**, 549 (1963).
[146] J. Callaway, *Phys. Rev.* **134**, A998 (1964).
[147] K. Tharmalingham, *Phys. Rev.* **130**, 2204 (1963).
[147a] D. F. Blossey and P. Handler, in "Semiconductors and Semimetals" (R. K. Willardson and A. C. Beer, eds.), Vol. 9, p. 257. Academic Press, New York, 1972.
[147b] D. E. Aspnes and N. Bottka, in "Semiconductors and Semimetals" (R. K. Willardson and A. C. Beer, eds.), Vol. 9, p. 457. Academic Press, New York, 1972.

The interband matrix element can be estimated using either the f-sum rule[148] or the $\mathbf{k}\cdot\mathbf{p}$ band model,[149,150] or it can be used as an adjustable parameter

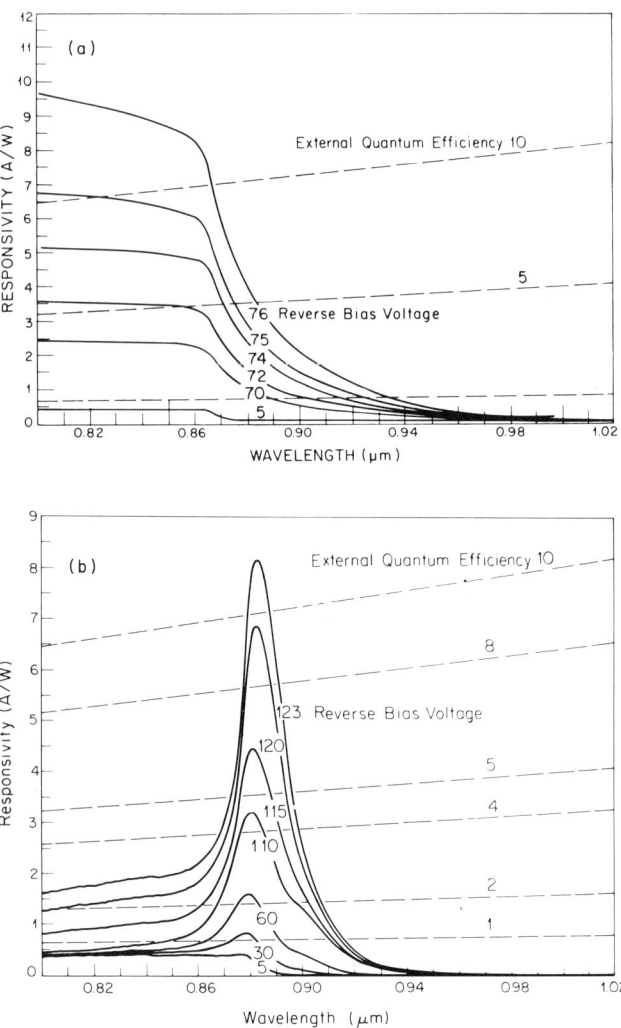

FIG. 36. Experimental spectral response of (a) GaAs Schottky barrier avalanche photodiode on material with $N_D - N_A = 7 \times 10^{15}$ cm^{-3}; and (b) and (c) GaAs EAP detector on material with $N_D - N_A = 5 \times 10^{14}$ cm^{-3} for low and high bias voltages. (After Stillman et al.[143,144])

[148] J. Bardeen, F. J. Blatt, and L. H. Hall, in Photoconduct. Conf. Atlantic City, 1954 (R. G. Breckenridge, B. R. Russell, and E. E. Hahn, eds.), p. 146. Wiley, New York, 1956.
[149] E. O. Kane, J. Phys. Chem. Solids 1, 249 (1957).
[150] J. Houghton and S. D. Smith "Infrared Physics," p. 131. Oxford Univ. Press, London and New York, 1966.

to obtain agreement with experimental measurements above the bandgap. Using the f-sum rule, the interband matrix element can be written as

$$|P_{nn'}|^2 = \tfrac{1}{2}mh\omega f_{nn'}, \tag{75}$$

and the sum rule of the f's is

$$\sum f_{nn'} = 1 + (m/m_v), \tag{76}$$

where the sum is over direct transitions to all higher bands and m_v/m is the effective mass ratio for the valence band. Using Eqs. (75) and (76) and $f \approx 1 + m/m_v$ the Franz–Keldysh absorption coefficient for wavelengths close to the absorption edge can be written as

$$\alpha(\lambda, E) = 1.0 \times 10^4 (f/n)(2\mu/m)^{4/3} E^{1/3} \int_\beta^\infty |\text{Ai}(z)|^2 \, dz, \tag{77}$$

where $\beta = 1.1 \times 10^5 \, (E_g - \hbar\omega)(2\mu/m)^{1/3} E^{-2/3}$ for α in centimeters^{-1}, E in volts/centimeter, and $(E_g - \hbar\omega)$ in electron volts. The integral of the square of the Airy function can be evaluated in terms of the Airy function and its derivative as[146]

$$\int_\beta^\infty |\text{Ai}(z)|^2 \, dz = [|(d \, \text{Ai}(z)/dz)_\beta|^2 + \beta |\text{Ai}(\beta)|^2]. \tag{78}$$

The calculated absorption for GaAs at low electric fields and energies greater than the bandgap was in good agreement with experimental measurements[151] when the parameters $n = \text{constant} = 3.63$, $m_v/m = 0.087$ for the

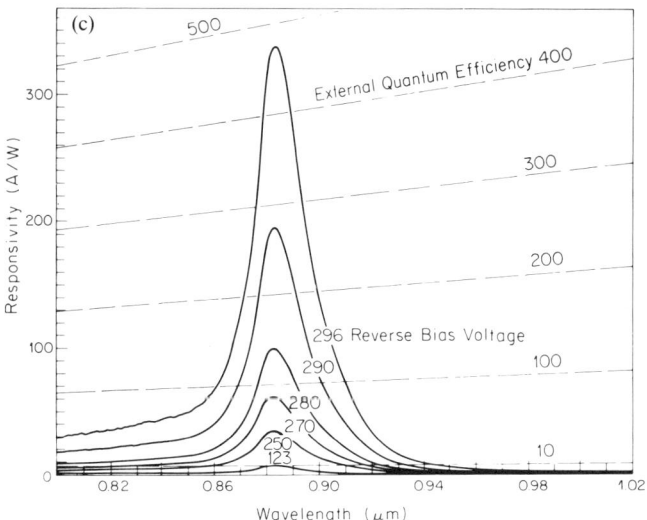

FIG. 36. (Cont.)

[151] D. D. Sell and H. C. Casey Jr., J. Appl. Phys. **45**, 800 (1974).

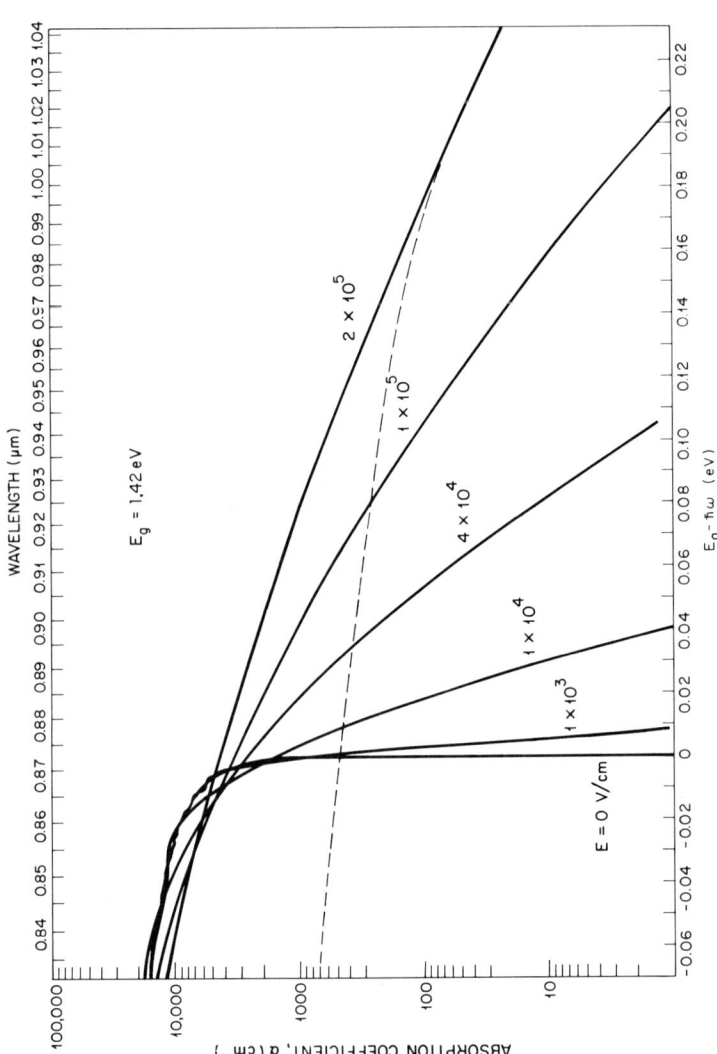

Fig. 37. Absorption coefficients of GaAs in uniform electric fields from 0 to 2×10^5 V/cm as calculated from Eq. (77). The dashed curve indicates the low field absorption coefficient for Si.

light-hole band, and $\mu/m = 0.0377$ were used. Alternatively, the matrix element $\mathbf{P}_{nn'}$ in Eq. (73) could be treated as an adjustable parameter to obtain agreement with experimental absorption values. The calculated absorption coefficients for GaAs in uniform electric fields from 0 to 2×10^5 V/cm are shown in Fig. 37. The wavelengths shown were determined from $(E_g - \hbar\omega)$ using a bandgap energy of 1.42 eV. There is a significant increase in the absorption at the longer wavelengths with increasing electric field, and at a wavelength of 1 μm the calculated absorption coefficient for an electric field of 2×10^5 V/cm is about the same as that for Si at the same wavelength. (The absorption coefficient for Si is shown in Fig. 37 by the dashed line.)

Using Eq. (77) we can express the absorption coefficient at position x in the depletion region of a GaAs Schottky barrier or P^+N junction diode as $\alpha(\lambda, x)$, where the electric field at position x, $E(x)$ is determined by the doping profile and applied bias voltage of the device. Then, the generation rate of electron–hole pairs at x for an incident flux of photons of wavelength λ, $\varphi(\lambda)$, and sample reflectivity R, can be written as

$$G(\lambda, x) = \varphi(\lambda)(1 - R)g(\lambda, x) = \varphi(\lambda)(1 - R)\alpha(\lambda, x) \exp\left[-\int_0^x \alpha(\lambda, x')\, dx'\right], \quad (79)$$

assuming that each absorbed photon creates one electron–hole pair. The internal quantum efficiency of the detector for a given wavelength and applied voltage V, without avalanche gain, is then given by

$$\eta_0(\lambda, V) = \frac{\int_0^W G(\lambda, x)\, dx}{\varphi(\lambda)(1 - R)} = \int_0^W g(\lambda, x)\, dx = \int_0^W \alpha(\lambda, x) \exp\left[-\int_0^x \alpha(\lambda, x')\, dx'\right] dx, \quad (80)$$

where W is the depletion width corresponding to the applied voltage V.

The calculated quantum efficiency without avalanche gain for two different GaAs Schottky barrier devices with uniform doping levels of 5×10^{14} and 2×10^{16} cm^{-3} is shown in Fig. 38. For each device the internal quantum efficiency was evaluated at 0.05 and 0.90 of the corresponding breakdown voltage. There is a significant increase in the quantum efficiency of the more heavily doped device at the shorter wavelengths for the larger bias, $V = 0.90\ V_B$. This increase in quantum efficiency is due to the widening of the depletion region (as the bias voltage is increased) which results in an increased collection efficiency for the photogenerated carriers. At both bias levels, the long-wavelength cutoff is more gradual for the device on 2×10^{16} cm^{-3} material, and this is due to the higher fields and larger Franz–Keldysh shift of the absorption edge in this material. However, the quantum efficiency is still low because for this doping level the depletion

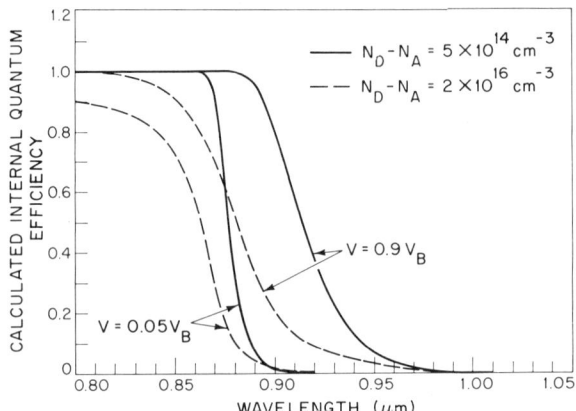

FIG. 38. Calculated internal quantum efficiency for two Schottky barrier diodes with $N_D - N_A = 5 \times 10^{14}$ and 2×10^{16} cm^{-3}, respectively. The calculations were done at applied voltages of 0.05 and 0.90 of the corresponding breakdown voltages for the case of no multiplication or avalanche gain, but including the Franz–Keldysh electroabsorption. (After Stillman et al.[144])

width, even at avalanche breakdown, is less than 2 μm. The gradual long-wavelength cutoff is similar to that observed experimentally on devices of this doping level.[76] In contrast, at low bias voltages the device on high-purity material has a higher quantum efficiency at the shorter wavelengths and a much sharper long-wavelength cutoff due to the wider depletion region and lower electric fields in this device, respectively. At high reverse-bias voltages, the Franz–Keldysh effect becomes important, and, along with the wider depletion region (about 30 μm at the breakdown voltage), results in a significant extension of the long wavelength cutoff for this device.

17. Avalanche Gain

The results presented in the previous section show that the Franz–Keldysh effect in GaAs Schottky barrier detectors in high-purity material can result in an increase of the quantum efficiency to nearly unity at wavelengths which are beyond the long-wavelength cutoff at low bias voltages. Any increase in the quantum efficiency above unity must be due to avalanche multiplication.

Using Eqs. (24) and (25) and considering only space-charge photogeneration of carriers as described by Eq. (79), the internal quantum efficiency *with avalanche gain* $\eta_1(\lambda, V)$ can be written as

5. AVALANCHE PHOTODIODES

$$\eta_1(\lambda, V) = \frac{\int_0^W g(\lambda, x) \exp\left[-\int_0^x (\alpha_n - \beta_p) dx'\right] dx}{\left\{1 - \int_0^W \alpha_n \exp\left[-\int_0^x (\alpha_n - \beta_p) dx'\right] dx\right\}}$$

$$= \frac{\exp\left[\int_0^W (\alpha_n - \beta_p) dx\right] \cdot \left\{\int_0^W g(\lambda, x) \exp\left[-\int_0^x (\alpha_n - \beta_p) dx'\right] dx\right\}}{\left\{1 - \int_0^W \beta_p \exp\left[\int_x^W (\alpha_n - \beta_p) dx'\right] dx\right\}}.$$

(81)

The device multiplication for a given wavelength and applied voltage is given by $M_0(\lambda, V) = \eta_1(\lambda, V)/\eta_0(\lambda, V)$. From Eqs. (80) and (81) it can be seen that for $\alpha_n = \beta_p$, M_0 is independent of the wavelength of the incident radiation and the position of the photogeneration of carriers.

The internal quantum efficiency and avalanche gain calculated from Eq. (81) using the ionization coefficients determined by Hall and Leck[65] are shown in Fig. 39a. As expected, for the case where $\alpha_n = \beta_p$, the quantum efficiency is independent of wavelength out to the long wavelength cutoff, and the multiplication is constant for all wavelengths. The results are shown for applied voltages of 0.90, 0.95, 0.98, and 0.99 of the breakdown voltage of 496 V determined using this ionization-coefficient data. The value of the multiplication or internal quantum efficiency with gain is quite small and the shapes of the curves have no resemblance to the experimental results given in Fig. 36b and 36c. Figure 39b shows the calculated internal quantum efficiency with gain for $\beta_p > \alpha_n$ using the data of Stillman et al.[73] The calculated breakdown voltage for this ionization-coefficient data was 468 V, and the internal quantum efficiency curves are shown for the same fractions of the breakdown voltage as in Fig. 39a. These curves show a peak beginning to develop at the absorption edge, similar to that observed experimentally, but the overall multiplication values obtained are much too low (cf. Fig. 36c). Also, compared to the experimental results, the quantum efficiency at the peak is too low relative to that at shorter wavelengths.

The breakdown voltages calculated from the ionization-coefficient data of Hall and Leck[65] and Stillman et al.,[73] 496 and 468 V, respectively, are considerably higher than the breakdown voltages observed experimentally on uniformly doped material with $N_D - N_A = 5 \times 10^{14}$ cm^{-3}. The breakdown voltages and corresponding maximum electric fields at breakdown calculated from several sets of ionization rate data are shown in Fig. 40 as a function of the net donor concentration of uniformly doped devices. Also shown in this figure are experimental breakdown voltage data on

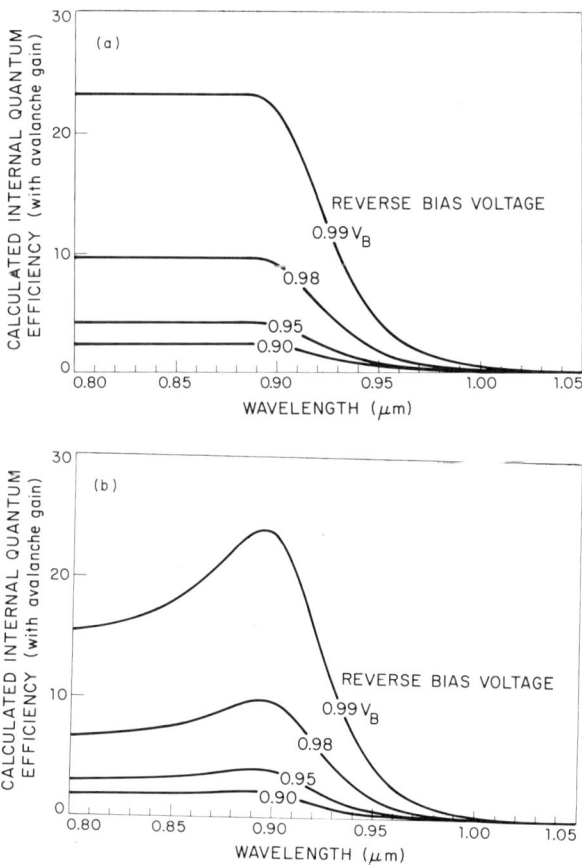

FIG. 39. Calculated internal quantum efficiency with gain for a GaAs EAP detector on uniformly doped material with $N_D - N_A = 5 \times 10^{14}$ cm^{-3} for two different sets of ionization coefficient data and the given fraction of the corresponding calculated breakdown voltage. (a) Ionization coefficients of Hall and Leck,[65] $\alpha_n = \beta_p = 2 \times 10^5 \exp[(-5.5 \times 10^5/E)^2]$, $V_B = 496$ V; (b) Ionization coefficients of Stillman et al.,[73] $\alpha_n = 1.2 \times 10^7 \exp[-2.3 \times 10^6/E]$, $\beta_p = 3.6 \times 10^8 \exp[-2.9 \times 10^6/E]$, $V_B = 468$ V.

abrupt junctions from the literature. The calculated breakdown voltages determined from the data of Hall and Leck[65] and Stillman et al.[73] are considerably lower than those determined from the data of Logan and Sze.[60] In addition, it is clear that a linear extrapolation of the ionization coefficient data of Stillman et al.[73] to higher fields results in calculated breakdown voltages that are much too low for doping concentrations higher than about 5×10^{15} cm^{-3}.

It should be pointed out, however, that none of these calculations takes

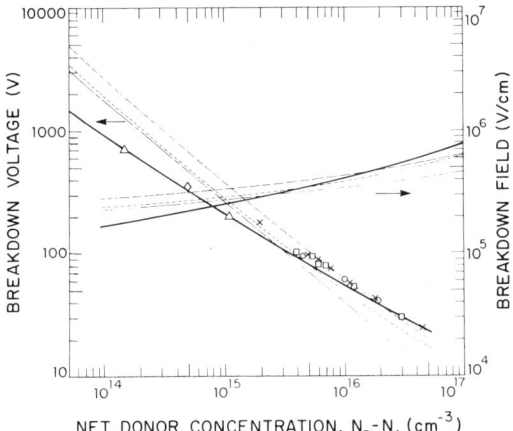

FIG. 40. Breakdown voltage and corresponding breakdown field for abrupt P^+N junctions and Schottky barrier diodes on N-type GaAs. The broken curves were calculated using the ionization coefficient data noted:—Logan and Sze,[60] $\alpha_n = \beta_p$; --- Hall and Leck,[65] $\alpha_n = \beta_p$; --- Stillman et al.,[73] $\beta_p > \alpha_n$. The solid curves were calculated using ionization coefficients which were adjusted to obtain agreement with the breakdown voltages at the lower doping levels from Stillman et al.[144] The data points are experimental results from the literature: △, Weinstein and Mlavsky[153]; ○, Kuno et al.[154]; □, Salmer et al.[69]; ×, Miller and Casey[155]; ◇, Stillman et al.[144])

into account the ionization threshold energies of electrons and holes. This would cause the calculated breakdown voltages to be too high at high doping levels if the correct ionization coefficients were used.[152] At low doping levels ($N_D - N_A < 10^{15}$ cm^{-3}) all of these data give calculated breakdown voltages that are too high when compared with the few experimental breakdown voltages available in this doping range.[153] This is not too surprising, however, since the older experimental values may have been limited by microplasmas. The experimental point for $N_D - N_A = 5 \times 10^{14}$ cm^{-3}, on the other hand, was obtained from a Schottky barrier device on material with a uniform doping level (determined by C–V measurements) and the multiplication at voltages close to breakdown was quite uniform. The solid curves in Fig. 40 were obtained with electron and hole ionization coefficients which were independently adjusted to give calculated breakdown voltages in agreement with the experimental breakdown voltage data over the lower doping range. When ionization coefficients of the form $\alpha_n = a_n \exp[-(b_n/E)^m]$ and $\beta_p = a_p \exp[-(b_p/E)^m]$ were used, the best agreement was obtained with $m = 1$. With $m = 2$ the curvature of the calculated curve was too large to give good

[152] Y. Okuto and C. R. Crowell, Solid-State Electron. **18**, 161 (1975).
[153] M. Weinstein and A. I. Mlavsky, Appl. Phys. Lett. **2**, 97 (1963).

agreement with the experimental data. The solid curves in Fig. 40 were calculated using

$$\alpha_n = 2 \times 10^6 \exp\left[-\left(\frac{2 \times 10^6}{E}\right)^1\right]$$

and

$$\beta_p = 1 \times 10^5 \exp\left[-\left(\frac{5 \times 10^5}{E}\right)^1\right].$$

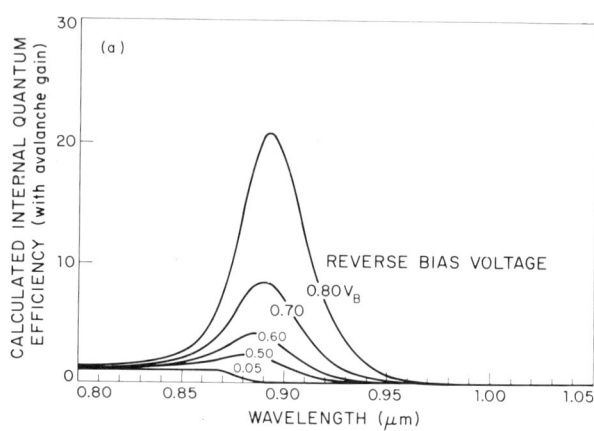

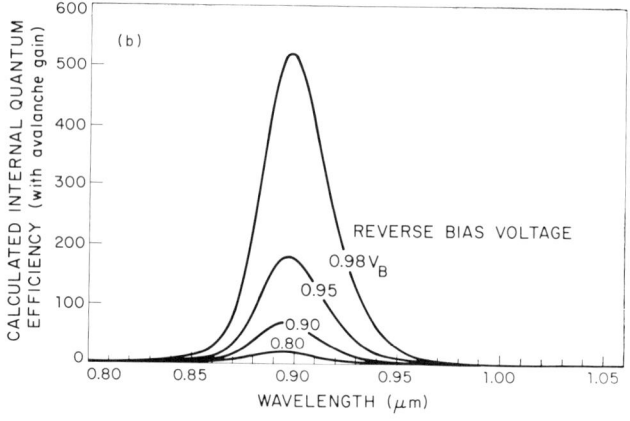

FIG. 41. Calculated internal quantum efficiency with gain for a GaAs EAP detector as in Fig. 39, but using the ionization coefficients which were adjusted to give agreement with the reported breakdown voltages: $\alpha_n = 2 \times 10^6 \exp[-2 \times 10^6/E]$, $\beta_p = 1 \times 10^5 \exp[-5 \times 10^5/E]$. The calculated breakdown voltage for this case (with $N_D - N_A = 5 \times 10^{14}$ cm^{-3}) was 316 V.

The use of ionization coefficients of the form $\alpha_n = \alpha_0(E/E_0)^m$ as given by Kuno et al.[154] may give better agreement over the entire doping range.

The internal quantum efficiency with gain calculated with these ionization coefficients is shown in Fig. 41a for low applied voltages and in Fig. 41b for higher applied voltages. The magnitude of the calculated quantum efficiency with gain and the variation of the quantum efficiency with both wavelength and bias voltage are similar to what is observed experimentally. Attempts to obtain more quantitative agreement are not warranted because of the accuracy of the absorption calculation and the lack of more detailed breakdown-voltage data at low net donor concentrations.[155] However, these results show that the experimental effect observed in the GaAs EAP detectors can be explained by electron and hole ionization coefficients which are consistent with the available breakdown voltage data and the previous conclusion that $\beta_p > \alpha_n$. Before further quantitative comparisons can be made, more work must be done to accurately determine the electron and hole ionization coefficients over a wider range of electric fields.

VII. Summary and Conclusions

Solid-state photodiodes have gained favor as radiation detectors because they are smaller, more durable, more convenient and less expensive to use than photomultipliers. The minimum required optical power for a given signal-to-noise ratio with a given bandwidth [Eqs. (11) and (12)] can be limited by noise due to background radiation, dark current, or thermal noise from the load resistor and excess noise from the amplifier. If the background is the limiting noise source, the lowest minimum required power can be obtained by making the quantum efficiency as close to unity as possible. Simple *PN* or *PIN* junction photodiodes can attain close to the theoretical minimum optical power required in most background-limited situations, and avalanche gain will not reduce the minimum required power. If the dark current is the limiting noise source, the minimum detectable power can be decreased by operating the detector at a lower temperature to reduce the dark current, provided that the decreased temperature does not adversely affect the quantum efficiency. Avalanche gain will decrease the minimum required power for this case only if the dark current consists mainly of unmultiplied surface leakage current. If the load resistor and/or preamplifier are the major sources of noise, the minimum required power can be reduced by using a photodiode which provides avalanche gain. This is because the multiplication reduces the relative contribution of the

---

[154] H. J. Kuno, J. R. Collard, and A. R. Gobat, *Appl. Phys. Lett.* **14**, 343 (1969).

[155] B. I. Miller and H. C. Casey, Jr., *in* "Gallium Arsenide and Related Compounds" *(Proc. 4th Int. Symp. GaAs, Denver, 1972)*, p. 23. Inst. Phys. and Phys. Soc., London, 1973.

thermal-noise terms [Eqs. (18) and (19)] (although it also adds excess noise). The multiplication also decreases the relative importance of the surface-leakage current. This is especially significant since the surface-leakage current in avalanche photodiodes is typically more than two orders-of-magnitude larger than the bulk-leakage current.

For wide-bandwidth applications the thermal-noise contributions of the load circuit are larger because small values of equivalent resistance R_{eq} are required to avoid the time-constant limitations of $R_{eq}C_j$, where C_j is the junction capacitance. In addition, the wide-bandwidth amplifiers required for these applications add considerable extra noise to the amplified signal ($T_{eff} > T_L$). Thus, unless the signal-to-noise ratio is extremely high in these wide-bandwidth applications, a considerable reduction in the minimum required power can be obtained by using avalanche photodiodes compared to unity-gain *PIN* detectors with the same quantum efficiency. Equations (11) and (12) and Eqs. (18) and (19) can be used to compare the performance of two detectors, with and without internal gain, for a particular application to determine which detector will have the minimum required power. If the performance can be increased with an avalanche device, the added cost and more stringent operating conditions (constant temperature and/or gain control) of avalanche photodiodes must be considered in the final determination of which detector to use.

The only commercially available avalanche photodiodes utilize the elemental semiconductors, Si and Ge. The Si devices have been highly developed and, for wavelengths and/or bandwidths where high quantum efficiencies can be obtained, these devices can provide very sensitive and rugged detectors. The Ge devices do not provide the same low-noise performance as the Si detectors since the electron and hole ionization coefficients of Ge are nearly equal. However, for long wavelengths and/or wide bandwidths where the Si detectors are not satisfactory, the Ge detectors can still significantly improve the system signal-to-noise ratio when the limiting noise source is the equivalent load resistance–amplifier combination.

Avalanche photodiodes in GaAs seem particularly promising for low-noise wide-bandwidth applications at wavelengths where high quantum efficiency can be obtained. This is because the difference between the electron and hole ionization coefficients in GaAs is quite large. Thus, although GaAs detectors are not as highly developed as Si devices, with additional work they may ultimately replace Si avalanche photodiodes in wide-bandwidth applications. For longer wavelengths, below the normal absorption edge of GaAs, the quantum efficiency of GaAs devices can be greatly increased by utilizing the Franz–Keldysh effect. These GaAs electroabsorption avalanche photodiode (EAP) detectors should be of considerable importance in applications involving integrated optical circuits where fast response is required at long wavelengths.

For applications at wavelengths beyond the range of Si avalanche photodiodes and GaAs EAP detectors ($\lambda \gtrsim 1.1$ μm), lower bandgap compound semiconductors should be explored to determine which are most suitable for use as avalanche devices. Clearly, this exploration should be first directed toward finding a semiconductor with the appropriate energy gap which also has a large asymmetry of the electron and hole ionization coefficients. Only after this has been done should additional effort in materials and device technology, which will be necessary to obtain high performance avalanche photodiodes, be considered.

ACKNOWLEDGMENT

The authors would like to thank Miss Peggy Southard for her careful typing of the manuscript.

NOTE ADDED IN PROOF

Since this chapter was finished many papers giving new fabrication methods, further refined device structures, and improved performance data, mainly on Si avalanche photodiodes, have appeared in the literature. Time does not permit the inclusion of this recent work; therefore the reader should refer to the literature for developments since 1975. However, two recent papers contradict previous results for InAs discussed on pp. 349 and 373 so they will be mentioned. Mikhailova et al.[156] have studied avalanche multiplication in diffused P–N junctions in InAs and In-rich $In_xGa_{1-x}As$, and their new results show that the hole ionization coefficient is over an order of magnitude greater than the electron ionization coefficient. This result was confirmed by measurements of the multiplication dependence of the noise in InAs and $In_xGa_{1-x}As$ avalanche photodiodes,[157] and is in general agreement with the implications of the gain-bandwidth-product observed in Ga-rich $In_xGa_{1-x}As$ avalanche photodiodes.[140]

[156] M. P. Mikhailova, N. N. Smirnova, and S. V. Slobodchikov, Fiz. Tekh. Poluprov. **10**, 860 (1976) [Engl. Transl.: Sov. Phys.—Semicond. **10**, 509 (1976)].

[157] M. P. Mikhailova, S. V. Slobodchikov, N. N. Smirnova, and G. M. Filaretova, Fiz. Tekh. Poluprov. **10**, 978 (1976) [Engl. Transl.: Sov. Phys.—Semicond. **10**, 578 (1976)].

CHAPTER 6

The Josephson Junction as a Detector of Microwave and Far-Infrared Radiation

P. L. Richards

I.	INTRODUCTION	395
II.	THE ALTERNATING CURRENT JOSEPHSON EFFECT	396
III.	REAL JUNCTIONS AND EQUIVALENT CIRCUITS	399
IV.	THE JUNCTION IMPEDANCE	405
V.	OPTIMUM JUNCTIONS FOR DETECTOR APPLICATIONS	408
VI.	NOISE	410
VII.	BOLOMETER	414
VIII.	VIDEO DETECTOR	417
IX.	HETERODYNE DETECTOR WITH EXTERNAL LOCAL OSCILLATOR	421
X.	PARAMETRIC AMPLIFIER	429
XI.	OTHER DEVICES OF INTEREST	435
	1. *Heterodyne Detector with Internal Local Oscillator.*	436
	2. *Regenerative Detector.*	438
	3. *Injection Locking*	438
XII.	CONCLUSIONS.	439

I. Introduction

The macroscopic quantum phenomena associated with the superconducting state provide a unique bridge between the microscopic quantum "world" and the macroscopic world of everyday experience. They are, therefore, much studied because of their intrinsic interest. For the purposes of this volume, however, we will consider only those aspects of the phenomena which involve the interaction of electromagnetic waves with weak links between superconductors.

Most of those effects discussed here were predicted theoretically by Josephson[1] from the quantum theory of tunneling through a thin oxide barrier between two plane superconducting films. The tunneling current through such a junction has been studied extensively. It can be divided into two parts: single particle tunneling and Josephson tunneling. The first to be discovered was the tunneling of individual electrons or quasi particles, which usually dominates the dc current when voltages comparable to, or

---

[1] B. D. Josephson, *Phys. Lett.* **1**, 251 (1962); *Advan. Phys.* **14**, 419 (1965).

greater than the energy gap are applied. This effect has been widely used for the study of the energy gap in the density of states for excitations from the superconducting ground state.[2] Josephson showed that correlated "superconducting" pairs of electrons will also tunnel and should allow current flow with no associated voltage drop. This Josephson current is extremely sensitive to electromagnetic fields in a way which permits the construction of a variety of sensitive detectors of radiation at frequencies up to the far-infrared. Regenerative and parametric as well as video and heterodyne modes of operation are possible.

An elementary introduction to the ac Josephson effect is presented in Part II. The discussion is extended to include the properties of real junctions in Part III. The junction impedance and the problem of junction optimization are treated in Parts IV and V. The general review of the Josephson effect concludes with a discussion of noise in Part VI.

The types of devices that appear to the author to have the most potential practical value are reviewed as follows: The bolometer in Part VII, the square law video detector in Part VIII, the heterodyne detector with external local oscillator in Part IX, and the parametric amplifier in Part X. Finally, Part XI contains a partial list of a number of other devices which have been suggested or developed. Many of these rely on unusual, or even unique properties of the Josephson junction and thus have substantial intrinsic interest. The emphasis in this review, however, is on those devices which appear to have the most potential applications.

The number of papers written which are relevant to this review is very large. Fortunately, there are books and review articles which can be used as secondary references. An attempt will be made to refer to those original papers which remain of fundamental importance, especially when they are well written. Frequent use will be made of secondary references, however, when they contain material in a more useful form.

Of particular general value for this subject are the books on tunneling by Solymar,[2] the review of detection from the point of view of the resistively shunted junction model by Richards et al.[3] and the Proceedings of the 1973 Perros–Guirec Conference.[4]

II. The Alternating Current Josephson Effect

Any discussion of the Josephson effect must be quantum mechanical since no valid classical description is known. The order parameter for the

---

[2] L. Solymar, "Superconductive Tunneling and Applications." Chapman & Hall, London, 1972.

[3] P. L. Richards, F. Auracher, and T. Van Duzer, *Proc. IEEE* **61**, 36 (1973).

[4] *Proc. Int. Conf. Detect. Emission Electromagn. Radiat. with Josephson Junctions; Rev. Phys. Appl.* **9**, 1–312 (1974).

superconducting state can be written in the simple form $\psi = \psi_0 e^{i\phi}$, where the phase factor ϕ is a function of both position and time. Josephson's calculations showed that the lossless pair tunneling current depends on the phase difference between the order parameters for the superconductors on the two sides of the tunnel junction,[1,2]

$$I = I_c \sin \phi. \tag{1}$$

(We follow the usual practice of using ϕ for this phase difference.) Here I_c is the maximum zero-voltage current which can be carried by the junction.

We now consider the effect of the finite potential drop V which occurs across the oxide barrier when I_c is exceeded. The quantum theory of tunneling[1] then predicts a time dependence of the phase difference

$$\hbar \, d\phi/dt = 2eV. \tag{2}$$

This result for superconductors is analogous to that obtained from the elementary quantum mechanics of a single particle system. If we assume a wave function of the above form and use Schrödinger's equation for a single particle, $-i\hbar \, \partial\psi/\partial t = \mathcal{H}\psi$, the time dependence of the phase difference ϕ between two states is $\hbar \, d\phi/dt = \Delta E$. In Josephson's result [Eq. (2)] for the superconducting tunnel junction the chemical potential difference, or change in system energy $2\,eV$ when a superconducting pair moves from one side of the barrier to the other plays the role of the energy difference ΔE between the two single particle states.

If the voltage is constant in time $V(t) = V_0$, Eq. (2) can be integrated to give $\phi = 2eV_0 t/\hbar + \phi_0$. Equation (1) then predicts an alternating current flow across the barrier at the frequency $\omega_0 = 2eV_0/\hbar$. The Josephson junction is thus a quantum mechanical oscillator. Its frequency can be high since $\omega_0/2\pi = 484$ GHz/mV. The existence of this Josephson alternating current has been verified by observing[5,6] microwave radiation emitted at the frequency ω_0. The ac Josephson currents were first detected, however, by an easier experiment.[7] An alternating voltage $V_{rf} \cos \omega_{rf}$ at a microwave frequency ω_{rf} was induced across the barrier in addition to the steady voltage

---

[5] I. K. Yanson, V. M. Svistunov, and I. M. Dmitrenko, *Zh. Eksper. Teor. Fiz.* **48**, 976 (1965) [*English Transl.: Sov. Phys.—JETP* **21**, 650 (1965)]; I. M. Dmitrenko, I. K. Yanson, and V. M. Svistunov, *Zh. Eksper. Teor. Fiz. Pis'ma Red.* **2**, 17 (1965) [*English Transl.: JETP Lett.* **2**, 10 (1965)]; I. M. Dmitrenko and I. K. Yanson, *Zh. Eksper. Teor. Fiz. Pis'ma Red.* **2**, 242 (1965) [*English Transl.: JETP Lett.* **2**, 154 (1965)].

[6] D. N. Langenberg, D. J. Scalapino, B. N. Taylor, and R. E. Eck, *Phys. Rev. Lett.* **15**, 294 (1965).

[7] S. Shapiro, *Phys. Rev. Lett.* **11**, 80 (1963); S. Shapiro, A. R. Janus, and S. Holly, *Rev. Mod. Phys.* **36**, 223 (1964).

V_0. Equation (2) then yields

$$\phi = (2eV_0 t/\hbar) + (2eV_{rf}/\hbar\omega_{rf}) \sin \omega_{rf} t + \phi_0,$$

so from Eq. (1)

$$I(t) = I_c \sin[(2eV_0 t/\hbar) + (2eV_{rf}/\hbar\omega_{rf}) \sin \omega_{rf} t + \phi_0]. \quad (3)$$

This equation shows that the frequency of $I(t)$ is modulated by the rf frequency. It is usual to expand such expressions using the relations

$$\cos(X \sin \alpha) = \sum_{n=-\infty}^{\infty} J_n(X) \cos(n\alpha)$$

and

$$\sin(X \sin \alpha) = \sum_{n=-\infty}^{\infty} J_n(X) \sin(n\alpha),$$

where $J_n(X) = (-1)^n J_{-n}(X)$ is Bessel's function of order n. Choosing the value $\phi_0 = \pi/2$ which corresponds to a maximum zero-voltage current at $V_0 = \omega_0 = 0$, and using standard trigonometric identities, we obtain

$$I(t) = I_c \sum_{n=-\infty}^{\infty} J_n(2eV_{rf}/\hbar\omega_{rf}) \cos(\omega_0 + n\omega_{rf})t. \quad (4)$$

The junction is thus a nonlinear device in which the Josephson currents beat with the induced ac signal. A measurement of the dc junction properties reveals the presence of the zero frequency beat currents,

$$I_0 = I_c J_n(2eV_{rf}/\hbar\omega_{rf}). \quad (5)$$

These beats are a lossless contribution to the dc current through the junction. They appear whenever the voltage is adjusted so that $\omega_0 + n\omega_{rf} = 0$ for any given value of n. That is, whenever the ac Josephson frequency $\omega_0 = 2eV_0/\hbar$ equals a harmonic of the rf frequency. The amplitude of this dc beat current is given by Bessel's functions $J_n(2eV_{rf}/\hbar\omega_{rf})$, where n is the order of the harmonic.

The observation of constant voltage steps on the I_0–V_0 characteristic of a Josephson junction placed in a microwave field thus verifies the existence of the ac Josephson currents. Examples of such I_0–V_0 characteristics are shown in Fig. 1. Careful measurements of the applied microwave frequency and the voltage at which the steps occur have yielded the best numerical value for the ratio of fundamental constants e/h,[8] and are used to provide a voltage standard.[9]

[8] W. H. Parker, B. N. Taylor, and D. N. Langenberg, *Phys. Rev. Lett.* **18**, 287 (1967).
[9] B. F. Field, T. F. Finnegan, and J. Toots, *Metrologia* **9**, 155 (1973).

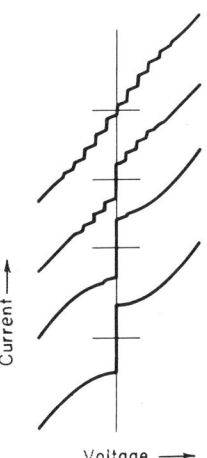

FIG. 1. Current—Voltage curves for a Nb point contact without rf bias (lowest curve) and for increasing values of 24 GHz rf bias (higher curves).

III. Real Junctions and Equivalent Circuits

The theory of the Josephson effect outlined in Part II was developed for tunnel junctions in which the barrier between the two superconductors is 10–30 Å of insulating oxide. The complete description of such structures from the microscopic theory includes contributions to the total current from quasi-particle tunneling[2] as well as from the junction shunt capacitance.

$$I(t) = I_c \sin \phi + [G_0(V) + G_1(V) \cos \phi] V + C \, dV/dt. \qquad (6)$$

The first term on the right is the phase dependent Josephson pair tunneling current from Eq. (1). The second term $G_0(V)V$ is the dissipative quasi-particle current neglecting coherence. The third term, which is dissipative and phase-dependent, appears when the influence of coherence on the quasi-particle distribution is included.[1] Since high frequencies are of interest in this review, a displacement current term which arises from the junction capacitance C has been included in Eq. (6).

The theory outlined here is valid if all of the junction currents flow at frequencies and voltage below the superconducting energy gap. This is a useful approximation for microwave detectors, but is not even strictly valid for them because of the efficiency with which harmonics are generated in the junction. A more correct form of the theory for high frequencies has been given by Werthamer[10] for zero temperature. It is much more

[10] N. R. Werthamer, *Phys. Rev.* **147**, 255 (1966).

complicated and has rarely been used for device calculations. The most striking results of the complete theory are step discontinuities in $G_0(V)$ and $G_1(V)$ at the energy gap voltage, and a peak (the Riedel singularity) in the pair current amplitude for frequencies equal to the full superconducting energy gap 2Δ. The discontinuities in $G_0(V)$ are frequently observed. The effect of the Riedel peak on the steps produced in the ac Josephson experiment can be seen under the proper circumstances.[11,12]

The tunneling theory [Eq. (6)] has been used with remarkable success to explain the measured properties of other types of weak links between two

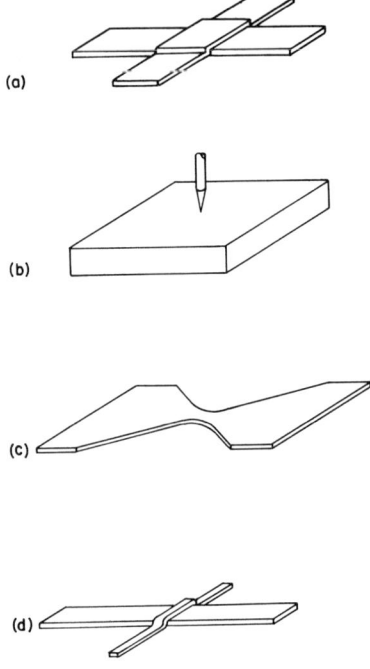

FIG. 2. Various forms of weak link between two superconductors: (a) Sandwich junction in which the barrier may be an insulating oxide, a semiconductor, or a normal metal. (b) Point contact junction. (c) Constricted film junction. (d) Superconducting film with narrow overlay of normal metal to produce a region of weak superconductivity.

[11] C. A. Hamilton and S. Shapiro, *Phys. Rev. Lett.* **26**, 426 (1971); C. A. Hamilton, *Phys. Rev. B* **5**, 912 (1972).
[12] S. A. Buckner, T. F. Finnegan, and D. N. Langenberg, *Phys. Rev. Lett.* **28**, 150 (1972).

superconductors. Figure 2 shows some of the physical arrangements which have been used to make weak links. The following are included:

(a) A sandwich arrangement of evaporated films in which the barrier between the two superconductors may be 10–30 Å of insulating oxide[13] (SIS), several hundred angstroms of semiconductor[14] (SSS), or several thousand angstroms of normal metal[15] (SNS);

(b) A point contact where one superconductor is sharpened to a point of a few microns radius[16];

(c) A constricted thin superconducting film where the length and width of the narrow region are of the order of 1 μm (often called a Dayem bridge)[17];

(d) A region of a superconducting thin film which is weakened by overlaying a narrow cross strip of normal metal (often called a proximity effect, or Mercereau, bridge).[18]

The voltage biased Josephson element model outlined in Part II does not include many important features of real junctions in detector configurations. The simplified resistively shunted junction (RSJ) model shown in Fig. 3a consists of an "ideal Josephson element" with current $I_c \sin \phi$ in parallel with a voltage independent shunt resistance R.[19-21] It contains many of these important features without the full complexity of Eq. (6). The shunt resistance R does not of course play an important role if the junction is driven from an ideal voltage source at all frequencies as was assumed in Part II. In practice, however, the impedance seen by the junction is frequency-dependent and, because of the low junction impedance, is often more nearly a current than a voltage source.

The RSJ model neglects the voltage dependence of $R = 1/G_0(V)$, the cos ϕ term, and the displacement current. The voltage dependence of R is known to be complicated only for the case of SIS tunnel junctions. We will postpone the discussion of such junctions to Part X. The effect of the cos ϕ term on the dc $I_0 - V_0$ curves of junctions with dc and/or microwave frequency bias have been computed for the simplified case of constant G

[13] P. W. Anderson and J. M. Rowell, *Phys. Rev. Lett.* **10**, 230 (1963).
[14] I. Giaever, *Phys. Rev. Lett.* **20**, 1286 (1968).
[15] J. Clarke, *Proc. Roy. Soc. Ser. A* **308**, 447 (1969).
[16] J. E. Zimmerman, *Proc. 1972 Appl. Superconduct. Conf., Annapolis*, IEEE Conf. Rec. 72CHO 682-5-TABSC, p. 544, 1972.
[17] P. W. Anderson and A. H. Dayem, *Phys. Rev. Lett.* **13**, 195 (1964); A. H. Dayem and J. J. Wiegand, *Phys. Rev.* **155**, 419 (1967).
[18] H. A. Notary and J. E. Mercereau, *Proc. 1969 Stanford Conf. Sci. Superconduct.*, p. 424. North-Holland Publ., Amsterdam, 1971.
[19] L. G. Aslamazov and A. I. Larkin, *Zh. Eksp. Teor. Fiz. Pis'ma Red.* **9**, 150 (1969) [*English Transl.: JETP Lett.* **9**, 87 (1969)].
[20] W. C. Stewart, *Appl. Phys. Lett.* **12**, 277 (1968).
[21] D. E. McCumber, *J. Appl. Phys.* **39**, 3113 (1968).

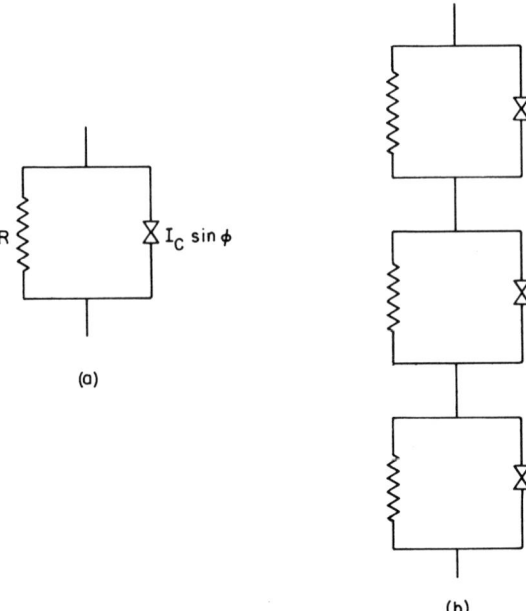

FIG. 3. (a) Equivalent circuit of a resistively shunted junction. (b) A series array of such junctions.

and have been found to be subtle and difficult to observe.[22,23] The displacement current term can be neglected for junctions which have $I_0 - V_0$ curves with no negative resistance or hysteresis, such as those shown in Fig. 1.

Hysteresis on junction I_0–V_0 curves can arise either from capacitance or from bias current heating effects,[21,24,25] neither of which is desirable for detector applications. Point contacts can be made which are in good agreement with calculations from the RSJ model.[26] Typical I_0–V_0 curves for such a point contact are shown in Fig. 1. The agreement is generally less good with Dayem bridges[27] and other types of weak links, but the RSJ model is usually the best available compromise between simplicity and detailed agreement with experiment.

Unfortunately, there are only a limited number of analytical solutions to the RSJ model. In the absence of an rf bias, the finite voltage portions

[22] F. Auracher, P. L. Richards, and G. I. Rochlin, *Phys. Rev. B* **8**, 4182 (1973).
[23] D. N. Langenberg, *Rev. Phys. Appl.* **9**, 35 (1974).
[24] W. J. Skocpol, M. R. Beasley, and M. Tinkham, *Rev. Phys. Appl.* **9**, 19 (1974).
[25] T. A. Fulton and L. N. Dunkleberger, *J. Appl. Phys.* **45**, 2283 (1974).
[26] Y. Taur, P. L. Richards, and F. Auracher, *Proc. XIII Int. Conf. Low Temp. Phys.*, p. 276. Plenum Press, New York, 1974.
[27] M. T. Levinsen, *Rev. Phys. Appl.* **9**, 135 (1974).

of the I_0-V_0 curve are hyperbolas that asymptotically approach the I_0-V_0 curve of the shunt resistor alone at high voltages. Perturbation theory can often be used to obtain solutions for small rf bias.[28-30] Other approximate analytical models can be used for frequencies $\hbar\omega/2e$ and voltages $V \gg I_c R$.[30,31]

For the general case, however, computer calculations are required. Fortunately, ways have been found to simulate the RSJ model with relatively simple electronic circuits.[32,33] As a result, solutions to the RSJ model for a complicated detector configuration can be obtained by an analogous low frequency computer "experiment." References to both analytical and computer solutions will be given throughout this review.

Calculations from the microscopic theory of SIS tunnel junctions show that the product of the zero voltage current times the normal state resistance $I_c R$ is given by[1]

$$I_c R = \pi \Delta / 2e, \qquad (7)$$

for the case $T \ll T_c$. Here Δ is the energy gap parameter and $\Delta \approx 2kT_c$ for the strong coupling[2] superconductors which are most useful for detector applications. This relation appears to also give an upper limit for $I_c R$ in point contacts and thin film bridges.[16] The product $I_c R$, which is a property of the material used to make the junction and of its operating temperature, is used in the definition of a normalized frequency

$$\Omega = \hbar\omega_{rf}/2eI_c R, \qquad (8)$$

which plays an important role in RSJ theory. When $T \ll T_c$, $\Omega \lesssim \hbar\omega/\pi\Delta$. For small values of Ω, the heights of the steps induced by a microwave current source of frequency ω_{rf} differ markedly from the Bessel's function behavior obtained from the theory of the voltage biased junction in Part II.[34] This dependence is shown in Fig. 4. For $\Omega > 1$, the step heights approach a Bessel's function dependence

$$J_n(I_{rf}/I_c\Omega), \qquad (9)$$

where I_{rf} is the rf current. The argument of J_n is identical to that in Eq. (5) if we assume that, for large Ω, all of the rf current flows through the shunt resistance so that $I_{rf} R = V_{rf}$. This result can be understood by computing

---

[28] H. Kanter and F. L. Vernon, Jr., *J. Appl. Phys.* **43**, 3174 (1972).
[29] F. D. Thompson, *J. Appl. Phys.* **44**, 5587 (1973).
[30] A. N. Vystavkin, V. N. Gubankov, L. S. Kuzmin, K. K. Likharev, V. V. Migulin, and V. K. Semenov, *Rev. Phys. Appl.* **9**, 79 (1974).
[31] M. J. Renne and D. Polder, *Rev. Phys. Appl.* **9**, 25 (1974).
[32] C. A. Hamilton, *Rev. Sci. Instrum.* **43**, 445 (1972).
[33] C. K. Bak, *Rev. Phys. Appl.* **9**, 15 (1974).
[34] P. Russer, *J. Appl. Phys.* **43**, 2008 (1972).

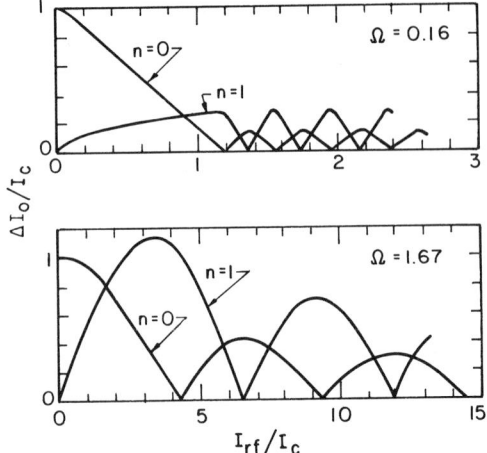

FIG. 4. Dependence of the (half) height of the $n=0$ step and the (full) height of the $n=1$ step on the rf current for two values of normalized frequency Ω.

the inductive reactance of the Josephson currents from Eqs. (1) and (2),

$$\omega L = \omega V/(dI/dt) = \Omega R/\cos\phi. \tag{10}$$

For $I_0 = 0$ we have $\cos\phi \simeq 1$, so Ω is the fraction of the rf current which flows through the resistive path. For finite V, the situation is much more complicated, but the qualitative conclusion, that for large Ω the resistive path carries most of the rf current, seems to remain valid.[19]

This conclusion that the pair currents do not carry an imposed current at high frequencies ($\Omega \gg 1$) is the primary limitation on the high frequency performance of Josephson detectors. Generally speaking, detectors function best with $\Omega < 1$. Good high frequency performance is therefore only obtained from junctions with largest usable $I_c R$ product. An ideal Nb junction at $T = 0$ has $\Omega = 1$ for $\omega_{rf}/2\pi \approx 10^{12}$ Hz. In most junctions, however, hysteresis arising from bias current heating prevents operation at $T \ll T_c$ where I_c reaches its full value. Experiments with point contacts are often limited to less than half the theoretical value of $I_c R$. Thin film structures appear to be even more sensitive to heating effects,[24] but some progress is being made in reducing them.[35]

If all applied frequencies $\hbar\omega/2e$ and voltages V are $\gg I_c R$, then the ideal Josephson element in the RSJ model is voltage biased and parts of the voltage biased model of Part II can be used. The range of validity of this model is limited for the nearly ideal junctions which are useful for devices since for them $\Omega \gg 1$ implies $\hbar\omega \gg 2\Delta$. Close to or above the energy gap

[35] T. M. Klapwijk and T. B. Veenstra, *Phys. Lett.* **47A**, 351 (1974).

voltage, the details of the theory differ significantly from the RSJ model. The term in Eq. (6) which varies as cos ϕ, for example, becomes important. The results of these changes in the theory are only beginning to be appreciated.[36] At high frequencies and voltages, therefore, we should use the Werthamer theory[10] which assumes a voltage bias, but includes the energy gap effects in the proper way. Roughly speaking, the RSJ model is valid for optimized millimeter wave devices, and the Werthamer theory for submillimeter wave devices.

IV. The Junction Impedance

The rf impedance of a junction which obeys the RSJ model is a complicated function of all junction parameters. On the zero voltage step, it can be computed from the parallel combination of the inductive reactance of Eq. (10) and the shunt resistance. On the resistive portion of the curve, perturbation theory[28-30] can be used to obtain results for small rf signals. In the limit of $\Omega \gg 1$, the impedance on the first induced step can be computed by modeling the junction as an rf current source of magnitude I_c shunted by a resistance R. For the general case, however, computer calculations are required.

The results of computer calculations[37,38] of the rf impedance V_{rf}/I_{rf} for two values of Ω and with large enough I_{rf} to induce finite constant voltage steps are shown in Fig. 5. The reactive components arise from supercurrent flow synchronized to the applied rf, so only appear on the steps. Between the steps, the rf resistance is complicated, but generally has the magnitude of the shunt resistance R. Note that for certain values of Ω, R_{rf} becomes negative for bias currents immediately below the first step. Experimental evidence for this effect has been found.[30,39]

In general, the radiation source can be modeled by an equivalent circuit consisting of a current source with amplitude I_s and a shunt resistor R_s. Depending on the coupling technique chosen, R_s usually lies in the range from 10 to 400 ohm. The maximum power from such a source, $I_s^2 R_s/8$, is available to a matched load $R_{rf} = R_s$. Many experiments on Josephson junctions are done in the limit $R_s \gg R_{rf}$, so that the coupled power is only $I_s^2 R_{rf}/2$, where R_{rf} is in general a complicated function of the junction parameters. For the purpose of computing a detector performance parameter

---

[36] R. E. Harris, *Phys. Rev. B* **10**, 84 (1974).
[37] F. Auracher and T. Van Duzer, *J. Appl. Phys.* **44**, 848 (1973).
[38] Strictly speaking, the rf impedance seen by a small signal source is not the ratio V_{rf}/I_{rf} plotted in Fig. (5). Since both a small signal and a strong local oscillator (I_{lo}) are involved, the rf impedance seen by the signal is the derivative dV_1/dI_1 evaluated at $I_1 = I_{lo}$.
[39] H. Kanter, *Appl. Phys. Lett.* **23**, 350 (1973).

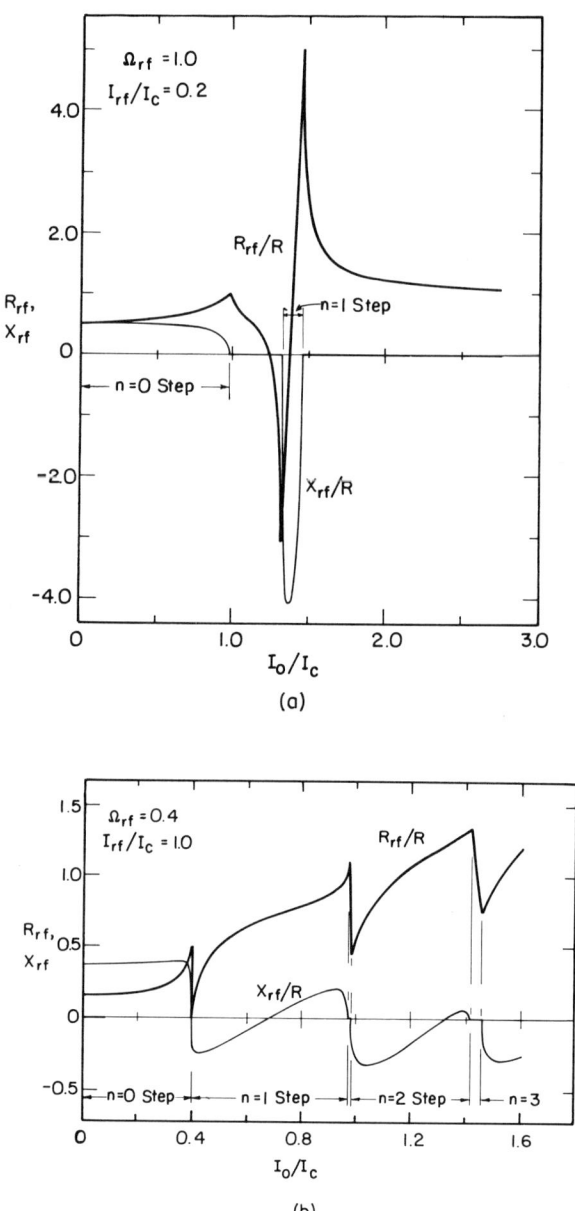

FIG. 5. Dependence of the rf impedance on dc bias. (a) For small rf current $I_{rf}/I_0 = 0.2$ and $\Omega_{rf} = 1$. (b) For large rf current $I_{rf}/I_0 = 1$ and $\Omega_{rf} = 0.4$. These cases correspond roughly to a submillimeter wave video detector (a) and a millimeter wave heterodyne detector (b). (After Auracher and Van Duzer.[37])

such as the signal power required to equal the detector noise (NEP), the responsivity S should properly be defined as the voltage output divided by the maximum available input power. To compare the performance of detector with theory, however, the output voltage is often compared with the input power available under the matching conditions actually employed (the coupled power).

The proper choice of rf source impedance is that value which minimizes the noise temperature or NEP of the detection system. In the absence of detailed information about the sources of system noise, one usually chooses the impedances that maximize the detector responsivity or conversion efficiency. The general rule that the rf source impedance should match the rf impedance of the junction must be applied with caution because, in the RSJ model, the junction impedance R_{rf} depends on the rf source impedance R_s. An analysis of this problem has been given[40] for the heterodyne mixer. It will be outlined here since it illustrates the complexity associated with choosing the optimum rf source impedance in any device.

The heterodyne mixer is operated about halfway between steps so that the rf junction impedance is real and is similar to the shunt resistance R. If the rf source impedance R_s is tightly coupled over a broad band of frequencies, then the source resistance is in parallel with the junction and the effective normalized frequency of the device is

$$\Omega' = (\hbar\omega_{rf}/2eI_c)[(1/R) + (1/R_s)]. \quad (11)$$

Since the junction impedance and the device responsivity depend on Ω', a self-consistent solution must be found for the optimum R_s. In practice, it proved simpler to optimize the device responsivity directly without reference to its rf impedance.[40] Analogous procedures should be followed to determine the optimum R_s for other detectors.

The output impedance of Josephson effect down converters (such as bolometers, video detectors, and heterodyne detectors) is different from the rf impedance. The inductance of the output circuit is generally so large that the applied rf and Josephson frequency currents do not flow in it. Consequently, the output load resistance does not modify Ω for the rf frequency and the output impedance (between steps) is simply the dynamic resistance $R_D = dV_0/dI_0$ of the dc I_0-V_0 curve at the bias point.

As can be seen from Fig. 1, R_D is frequently as large as $10R$, so that the output impedance can be substantially larger than the rf input impedance. As will be discussed below, this impedance gain is intimately related to the responsivity or conversion gain of several types of detectors. It can also influence the contribution to the system noise which arises from the first

[40] Y. Taur, J. H. Claassen, and P. L. Richards, *Rev. Phys. Appl.* **9**, 263 (1974); *Appl. Phys. Lett.* **24**, 101 (1974).

stage of amplification. Since audio frequency FET amplifiers have low noise temperature for high impedance sources, this impedance gain can be of advantage. In the megahertz frequency range, however, amplifiers are generally designed for 50 ohm impedance, so that a Josephson junction which matches the ~400 ohm impedance of a free space rf wave could have an inconveniently high impedance at the intermediate frequency.

V. Optimum Junctions for Detector Applications

Several junction properties can be listed as being important for most detector applications. The properties required for bolometers and for parametric amplifiers are sufficiently different that they will be considered separately in Parts VII and X. The junctions should, of course, be rugged, reliable, and reproducible. They should be small compared with an rf wavelength to avoid the complication of internal resonance effects. Their rf impedance should be high enough to allow proper matching to the incident power. This usually implies $R \gtrsim 10$ ohm and small C. The low frequency impedance must be in a range which can be efficiently coupled to the output circuit. The critical current I_c should be as large as possible to minimize the fractional effects of current noise in the junctions. Since R and I_c should both be large, and since many detectors function best when $\Omega < 1$, it is important to use superconductors, such as Pb or Nb with large energy gaps and use junctions which have nearly ideal values of the product $I_c R$.

In order to obtain large values of $I_c R$ and low junction thermal noise, it must be possible to operate at low temperatures $T \ll T_c$ without hysteresis due to bias current heating. The heat generated varies as $I_c^2 R$ and, at least for some types of junctions, the thermal conductance which dissipates the heat is a decreasing function of R. Heating effects can, therefore, scale roughly as $(I_c R)^2$. They represent the most serious deviation from ideal behavior of most types of junctions.

Arrays of junctions in series are potentially useful for detector applications. The properties of such arrays have been very little studied. For the purpose of estimating their usefulness we will assume that the junctions in the array have values of I_c and R which differ by only a few percent. There is considerable interest in the possibility that the junctions in an array are coupled sufficiently strongly[41,42] through electromagnetic fields or through the injection of nonequilibrium densities of pairs and quasi particles, that the Josephson currents flow coherently in the array. Except for the internally pumped parametric amplifier of Part X, none of the detectors selected for detailed description in this review require such coherence for their successful

[41] T. D. Clark, *Phys. Lett.* **27A**, 585 (1968); *Rev. Phys. Appl.* **9**, 207 (1974).
[42] J. Rosenblatt, *Rev. Phys. Appl.* **9**, 217 (1974).

operation. Unless otherwise noted, our model of array performance will assume simple superposition of junction properties as is given by the equivalent circuit in Fig. 3b.

First, we consider the case in which the array length is small compared with an rf wavelength. If we compare a single junction with an array of p identical junctions, then $R \to pR$, $I_c \to I_c$, and $2e/\hbar \to 2e/p\hbar$ (since the Josephson frequency is determined by V_0/p, where V_0 is the voltage across the array). Consequently, several important junction parameters, such as Ω, R_D/R, and the *normalized* I_0-V_0 curve are the same for the array as for the single junction. In these latter respects, the array behaves somewhat like an impedance transformation[43] with a turns ratio of $p^{1/2}$. The primary value of the array is thus to raise the junction impedance into the desired range. One other favorable property of an array is that the power required to saturate[43] the device is proportional to $RI_c^2 \to pRI_c^2$. Detectors made from arrays may therefore be able to avoid the bandwidth limitations of resonant impedance transformations and also have larger dynamic range than the corresponding detectors made from single junctions.

Arrays should be especially useful at infrared frequencies where resonant coupling is less efficient. Usually, it will be desirable to keep the overall array length small compared with λ_{rf}, as will be discussed later, larger arrays are of value for multimode video detectors, but of doubtful value for heterodyne or parametric devices.

Because the best combinations of properties have been available in the past from point contacts, most high frequency detection experiments to date have been done using them. The primary disadvantages of point contacts are their limited reliability and the difficulty of making series arrays. The availability of thin film arrays,[44,45] which should in principle be stable and reproducible, may shift the emphasis in detector development away from point contacts. The undesirably low impedance of the usual Dayem or proximity effect bridges can be avoided by using series arrays. Bias current heating effects, however, remain a serious problem for many applications. Thin film SIS junctions have not been used for high frequency detection because of their large capacitance, and the internal strip line resonances[46] which generally complicate their behavior. The development of small area SIS junctions and the possibility of making arrays of them will help to alleviate the impedance problem, but will not generally reduce the junction cutoff frequency $1/RC$ which is too low for many high frequency applications. Thin film SIS junctions, however, will be mentioned again in Part X.

[43] J. H. Claassen, private communication, 1973.
[44] P. T. Parrish and R. Y. Chiao, *Appl. Phys. Lett.* **25**, 627 (1974).
[45] P. W. Palmer and J. E. Mercereau *IEEE Trans. Mag.* **11**, 667 (1975).
[46] D. D. Coon and M. D. Fiske, *Phys. Rev.* **138**, A744 (1965).

VI. Noise

The nonlinearity of the Josephson junction is so strong that, under certain circumstances, it can be saturated by thermal noise at 1°K. Because of this nonlinearity, the effects of noise on measurable junction properties is quite complicated. In spite of these complications, however, much is known about the fundamental noise mechanisms and also about the effects of noise on detector performance.

The first studies of noise in Josephson junctions were oriented toward SIS tunnel junctions.[47] For such junctions, the large capacitance insures that the voltage bias model can be used. In the limit $eV \ll kT$ this theory can be interpreted as leading to a shot noise contribution from the Josephson pair current plus the thermal (Johnson) noise associated with the quasi particle resistance. The former is often neglected in the case of point contacts and other types of weak links for which the pair current fluctuations are not expected to be as important.[48] Some measurements on point contacts[40,49] are in reasonable agreement with the RSJ model if the only noise source is the thermal (Johnson) noise in the shunt resistance. Other experiments are interpreted as showing pair current fluctuations as well.[50]

Noise experiments on low capacitance junctions are generally current biased at all important frequencies. This is the easiest well-defined experimental situation to achieve and has a particularly simple equivalent noise circuit. The noise currents flow around the loop in Fig. 3a. Noise experiments should, of course, also be done in detector configurations. Since detectors are tightly coupled to input and output circuits, more complicated equivalent noise circuits are required for calculations. The results of calculations of this type will be given in Part IX.

At some low frequency, noise arising from junction thermal fluctuations will dominate certain measurements. In practice, this noise is of particular importance to bolometers and video detectors whose output occurs at low frequencies. The thermodynamic equilibrium value of the thermal fluctuations in any system of heat capacity C_V in contact with a thermal reservoir is

$$<(\Delta T)^2> = kT^2/C_V. \qquad (12)$$

The amount of this fluctuation which appears within a given bandwidth

---

[47] A. J. Dahm A. Denenstein, D. N. Langenberg, W. H. Parker, D. Rogovin, and D. J. Scalapino, *Phys. Rev. Lett.* **22**, 1416 (1969).
[48] K. K. Likharev and V. K. Semenov, *Zh. Eksper. Teor. Fiz. Pis'ma Red.* **15**, 625 (1972) [*English Transl.: JETP Lett.* **15**, 442 (1972)].
[49] J. H. Claassen, Y. Taur, and P. L. Richards, *Appl. Phys. Lett.* **25**, 759 (1974).
[50] G. Vernet and R. Adde, *Appl. Phys. Lett.* **19**, 195 (1971).

depends on its power spectrum. This has been shown to vary inversely with frequency for certain thermal geometries.[51] A contribution to the power spectrum which varies as the inverse of the frequency to the second power arises from a constant rate of temperature drift.

Whenever the junction properties are temperature-dependent, these low frequency thermal fluctuations can appear as electrical noise in a circuit which carries a current. Since this type of noise commonly occurs in all measuring circuits, reports on the presence of such noise in junctions must be treated with caution. The frequency at which inverse frequency junction noise becomes important has been observed to be as low as ~ 0.1 to 10 Hz for SIS and SNS junctions.[52]

Unlike most other devices, the static I_0-V_0 characteristic of a Josephson junction is modified by the junction noise. The effects of noise depend on the magnitude of the noise current compared with the critical current. A convenient dimensionless noise parameter is $\Gamma = 2ekT/\hbar I_c$. If the system is well-shielded from external rf disturbances T is the ambient temperature. If not, T is an effective noise temperature which can be many hundreds of degrees Kelvin.

One property which is important for a number of detectors is the dynamic resistance $R_D = dV_0/dI_0$ between constant voltage steps. The conversion efficiency or responsivity of detectors biased between steps is proportional to R_D at the bias point. For $\Omega \ll 1$, the RSJ model predicts a very large R_D between steps. In practice, as is seen in Fig. 1, a finite slope which is due to noise is generally observed. Although no simple theoretical relationship is available for the dependence of R_D on the junction parameters Γ, Ω, and V_0, calculations for a single set of values were found to be in reasonable agreement with experiments on point contacts.[26] Generally speaking, R_D/R is large for low Ω and for large I_c (that is, for low impedance junctions).

Noise in the junction also causes a rounding of sharp corners on the I_0-V_0 curve. The importance of this rounding for detectors is that R_D remains finite as V_0 approaches the voltage of a step. In the absence of noise the hyperbolic I_0-V_0 curve gives $R_D \to \infty$ at the corner. This noise rounding has been examined in detail and is also in agreement with theory.[53]

One important effect of low frequency noise currents in the junction is that the junction voltage, and therefore the Josephson frequency $2eV_0/\hbar$ is modulated. In this way, low frequency noise can frequency modulate the ac Josephson currents.[47,50] The constant voltage steps, which are dc beats between the internal Josephson oscillator and the external rf source, can occur only if the rf signal power P_s is strong enough to overcome the noise

[51] J. Clarke and R. T. Voss, *Phys. Rev. Lett.* **33**, 24 (1974).
[52] J. Clarke, private communication, 1973.
[53] W. H. Henkels and W. W. Webb, *Phys. Rev. Lett.* **26**, 1164 (1971).

and synchronize the junction. The power required to synchronize a junction with noise temperature T and resistance R is[54]

$$P \gtrsim (2ekT/\hbar)^2 R, \qquad (13)$$

or $\sim 10^{-14}$ W for $R = 10$ ohm, $T = 1°$K. This effect places some limitations on the usefulness of the ac Josephson currents as the internal local oscillator of a heterodyne receiver for the detection of small signals.

Junction noise contributes in the usual way to detector noise if it appears in the output bandwidth of the detector, or if it is converted into the output bandwidth by mixing with ac signals in the junction. The results[49] of direct measurements of the noise at 50 MHz in various point contacts are shown

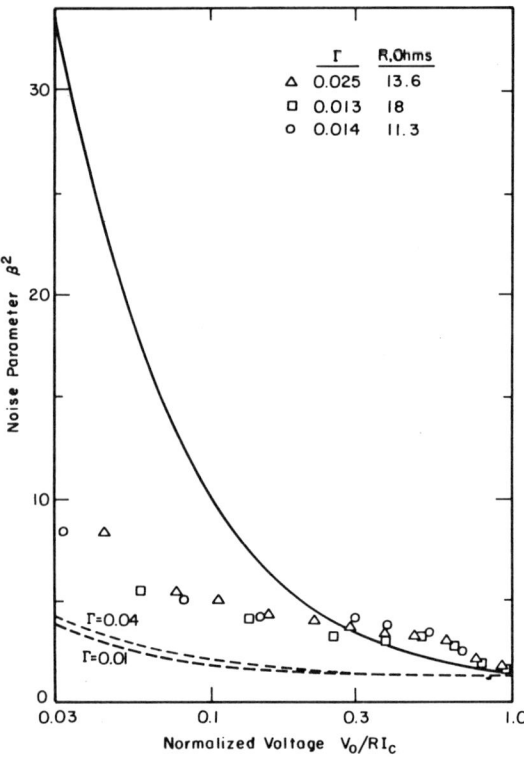

FIG. 6. Noise parameter β^2 measured at 50 MHz on several Nb point contact junctions with different values of the temperature parameter Γ. The data are compared with the thermal noise model for two values of Γ (---) and a simplified underestimate of the noise if pair fluctuations are included (—). (After Claassen et al.[49])

[54] M. J. Stephen, *Phys. Rev.* **186**, 393 (1969).

in Fig. 6. In order to make comparisons with the theory which assumes thermal noise in the RSJ model, it is important to choose an output frequency which is high enough to avoid low frequency thermal fluctuation noise. A noise parameter β^2 is defined in terms of a low frequency equivalent circuit model which contains a noise current source I_N in parallel with the junction dynamic resistance R_D. Then β^2 is defined as the ratio of $\langle I_N \rangle^2$ to $4kTB/R$, the thermal noise current in the junction shunt resistance. The noise parameter β^2 is thus a measure of the contribution to the noise of the nonlinear processes in the junction. The junction amplifies the low frequency noise power by an amount proportional to R_D whenever $V_0 > 0$. The complexity of the process is indicated by the fact that R_D itself depends on the noise. The experimental data in Fig. 6 lie about a factor two above the theoretical curves obtained both from analytical calculations[48] and from an analog simulator.[49] They lie well below the predictions of the tunnel junction theory which includes additional terms due to fluctuations in the pair current.

When a strong rf bias is applied to the junction (as is the case for some heterodyne detectors and parametric amplifiers) constant voltage steps appear in the I_0-V_0 curve which indicate that current is flowing in the junction at many harmonics of ω_{rf}. Noise in the neighborhood of each of these harmonic frequencies is down-converted into the output bandwidth. Values of β^2 measured on point contacts with a 36 GHz rf bias are shown in Fig. 7. Again, the experimental values lie somewhat above the results

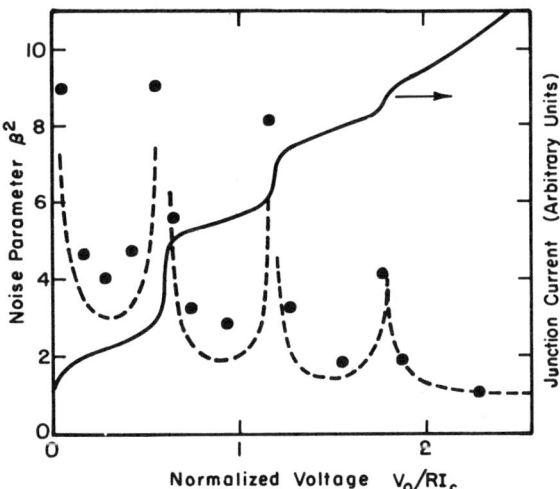

FIG. 7. I_0-V_0 curve of a Nb point contact biased with a 36 GHz rf current (—) and the corresponding values of the noise parameter β^2 measured at 50 MHz. The data points are compared with calculations from the thermal noise model (---). (After Claassen et al.[49])

of the simulator calculations.[49] These noise measurements will be discussed further in Part IX since they are of particular interest in the study of heterodyne mixers.

The noise in a series array of junctions has not been studied. In order to guess the result, we can assume that the dynamic resistance of a junction in an array is the same as for the junction alone, and then compute the noise delivered by the array to a low frequency output circuit. Since the individual junctions are characterized by an equivalent noise circuit made up from a noise current source $\langle I_N^2 \rangle$ and a shunt resistance R_D, then the array will be characterized by a noise current source $\langle I_N^2 \rangle / p$ and a shunt resistance pR_D.

Our assumption that the dynamic resistance of a junction is not decreased by extra noise when the junction is put into an array is based on the equivalent circuit in Fig. 3b. The thermal noise current from a shunt resistor R flows primarily around its own loop, which loop has an impedance that is lower, by a factor of $\sim p$, than that of any loop connecting the whole array. Consequently, each junction sees its own noise, but not the amplified noise of its neighbors.

VII. Bolometer

A bolometer consists of a radiation absorbing element and a resistance thermometer with total heat capacity C, which are attached to a thermal bath through a thermal conductance G. The most sensitive broadband detectors of infrared radiation available today are bolometers operated at ^{3}He and ^{4}He temperatures. There is a well-developed technology of heavily doped semiconducting bolometers (usually Si or Ge). Using them, it is possible to approach theoretically optimum performance for those infrared experiments which are limited by fluctuations in the room temperature blackbody background radiation. There exists, however, an important class of experiments (in the far-infrared or in cooled or space environments) for which better bolometers are needed.

Josephson junctions have temperature dependent critical currents and so can be used as the thermometric element of a bolometer. They can have several advantages over the now conventional Ge or Si bolometers: The noise and the thermal response of Josephson junctions is better understood and more easily controlled than for heavily doped semiconductors at low temperatures. Since they can be evaporated film structures, Josephson thermometers can be easily duplicated, can have very low heat capacity, and can be used with any convenient low heat capacity substrate. The radiation absorbing function can be provided by a thin film structure on a transparent substrate so that it can be optimized essentially independently of the thermal properties of the bolometer. Although only one Josephson effect bolometer has been made thus far,[55] its properties are sufficiently attractive to stimulate widespread interest.

As shown in Fig. 8, this bolometer consists of a sapphire substrate which is transparent in the far-infrared and has an index of refraction $n \simeq 3$. Electrical and thermal contact is made to the bolometer through nylon threads coated with superconducting Pb. The radiation is absorbed in a thin Bi film with surface impedance $Z_0/(n-1) \approx 200$ ohm/□ where Z_0 is the impedance of free space. The theoretical value of the absorption is $4(n-1)/(n+1)^2 \approx 1/2$, independent of frequency. This has been experimentally verified over the frequency range from 5 to 50 cm^{-1}. Alternatively, selective absorbers can be used if desired. The temperature is measured using the temperature-dependent critical current of an SNS Josephson junction made from evaporated films of Pb/Cu–Al/Pb. The bolometer is biased with a constant current slightly greater than I_c, and the voltage is measured with a superconducting amplifier. Since the normal metal is a Cu–Al alloy in the dirty limit, the critical supercurrent I_c of the junction is of the form[15]

$$I_c = I_0 \exp[-(T/T_0)^{1/2}]. \qquad (14)$$

The minimum detectable amount of absorbed radiation power is

$$\text{NEP} = G\delta T = G \frac{\partial T}{\partial I_c} \delta I = \frac{2G\sqrt{TT_0}}{R_D I_c} V_N. \qquad (15)$$

If we assume that the noise arises from Johnson noise in the shunt resistance, then, from the noise theory discussed in Part VI, $V_N = R_D(4kT/R)^{1/2}$. Typical operating parameters and results for the (dark electrical) NEP are given in Table I. Low frequency thermal fluctuation noise (inverse frequency noise) is negligible down to frequencies below ≈ 1 Hz.

Perhaps the most unusual feature of this SNS bolometer is that this low

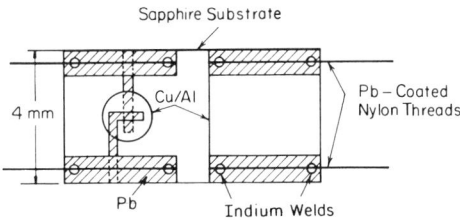

FIG. 8. Version of the SNS bolometer used to make electrical NEP measurements. Since bias current heating is negligible, a Cu/Al alloy heater was evaporated on the right side. The SNS junction is centered on the left side. (After Clarke et al.[55])

[55] J. Clarke, G. I. Hoffer, and P. L. Richards, *Rev. Phys. Appl.* **9**, 69 (1974).

TABLE I

Typical Numerical Parameters for an SNS Bolometer[a]

$T = 1.5°K$	$G = 10^{-8}$ W/°K
$T_0 = 0.1°K$	$\tau = 0.1$ sec
$R_D \approx R = 3 \times 10^{-6}$ ohm	NEP = 2×10^{-15} W/Hz$^{1/2}$ (calculated)
$I_c = 2 \times 10^{-5}$ A	NEP = 5×10^{-15} W/Hz$^{1/2}$ (measured)
$C = 10^{-9}$ J/°K	NEP = 10^{-15} W/Hz$^{1/2}$ (thermal fluctuation limit)

[a] Clarke et al.[55]

value of NEP was obtained with a large area $A \simeq 0.3$ cm^2. The resulting specific detectivity $D^* = A^{1/2}/\text{NEP} \approx 10^{14}$ cm Hz$^{1/2}$/W is the largest reported for a ^{4}He bolometer. This means that the bolometer can accept a large number of spatial modes, $A\Omega \approx 0.3$ sr cm^2 compared with a single mode detector for which $A\Omega \approx \lambda^2$. In this case, the symbol Ω denotes a solid angle. The least desirable feature of this bolometer is its very low output impedance $R_D = 3 \times 10^{-6}$ ohm. A superconducting quantum interference amplifier (SQUID)[56] must be used to measure the voltage output at this low impedance level. Such amplifiers have adequately low noise temperatures, but are less convenient to use than conventional room temperature FET electronics.

The NEP of a bolometer with no "excess noise" can be written in the form[57] of a sum of squares of noise powers referred to the input,

$$(\text{NEP})^2 = (4kTR/S^2) + 4kT^2G + (8\alpha\sigma kT_b^5 A\Omega/\pi), \tag{16}$$

for $\tau \approx \ll 1$ and for unit bandwidth. The first term is the square of the Johnson noise voltage divided by the voltage responsivity $S = dV/dP_s$. The second term arises from thermal fluctuations in the bolometer element as a whole. The third term arises from the fluctuations in the blackbody background assuming a constant absorptivity α, a background temperature T_b, a throughput $A\Omega$, and a Stefan–Boltzmann constant σ. In low background applications the third term, which dominates most room temperature infrared measurements, is made small by reducing T_b, or by reducing the bandwidth with cooled filters. The first term can be made negligible if S is sufficiently large.

The ideal bolometer has high absorptivity and the lowest possible heat capacity. The thermal conductance G is the smallest consistent with the required thermal time constant and with an acceptable temperature rise due to radiation heating. The responsivity should be high enough that the noise is limited by the thermal fluctuation term $\text{NEP} = (4kT^2G)^{1/2}$ without excessive bias current heating.

[56] J. Clarke, *Proc. IEEE* **61**, 8 (1973); *Science* **184**, 1235 (1974).
[57] F. J. Low and A. R. Hoffman, *Appl. Opt.* **2**, 649 (1963).

The SNS bolometer meets this variety of exacting conditions better than any other bolometer. In its present form the responsivity is slightly too low. More careful optimization of junction parameters, or the use of arrays, should make it possible to reach the thermal fluctuation limit of 10^{-15} W/Hz$^{1/2}$.

It is perhaps useful to mention at this point that bolometers can also be constructed using the temperature dependent quasi particle tunneling current in an SIS or SIN evaporated film tunnel junction. If ideal junctions made from high T_c superconductors were available, such bolometers would have very favorable properties.[55] Unfortunately, however, the current in most junctions is dominated by a temperature independent leakage current in the useful operating range of $T \ll T_c$. If a series array of junctions were used, it might be possible to obtain adequate responsivity with lower T_c superconductors operated with T close to T_c.

VIII. Video Detector

The first way in which a Josephson junction was used to detect far-infrared radiation was as a video square law detector.[58] In this application, a point contact is biased with a constant current slightly larger than I_c as is shown in Fig. 9, and a modulated rf signal is focused onto it. Figure 4 shows that a small signal current at any rf frequency depresses the zero voltage current

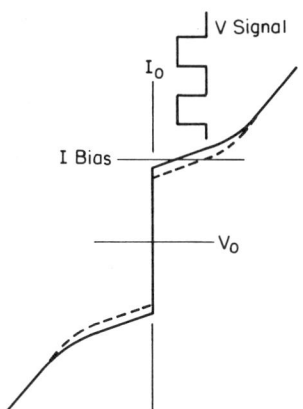

FIG. 9 $I_0 - V_0$ curves for a chopped rf signal. The video square law signal voltage is obtained when bias current is applied.

---

[58] C. C. Grimes, P. L. Richards, and S. Shapiro, *Phys. Rev. Lett.* **17**, 431 (1966); *J. Appl. Phys.* **39**, 3905 (1968).

I_0 (the $n = 0$ step) quadratically. This can be written as $I_0 = I_c(1 - \gamma I_{rf}^2)$. The voltage output at the modulation frequency is then

$$\delta V = \delta I_0 R_D = -\gamma I_c I_{rf}^2 R_D. \quad (17)$$

If the rf source impedance $R_s \gg R_{rf}$, then I_{rf} can be computed from the available rf power and the source impedance, so Eq. (17) can be tested without reference to R_{rf}. We will give results in the form of a coupled power responsivity S, and a value for R_{rf}, so that they can be compared with experiments for any value of R_s/R_{rf}. From Eq. (17) we obtain

$$S = -2\gamma I_c R_D/R_{rf}, \quad (18)$$

and $R_{rf} \simeq R$ at this bias point.[28,37]

The parameter γ which is a measure of the curvature of the $n = 0$ curve at $I_{rf} = 0$ in Fig. 4, clearly decreases with Ω. When $\Omega > 1$, the Bessel function model of Eq. (9) can be used to obtain

$$S = -R_D/2I_c R\Omega^2. \quad (19)$$

This result turns out to be valid even for $\Omega \lesssim 1$.[19,28] The responsivity of the video detector is thus proportional to Ω^{-2}.

A more thorough analysis of the video detector has been carried out using second order perturbation theory.[28-30,59] The results divide naturally into three classes depending on the Josephson frequency $\omega_0 = 2eV_0/\hbar$ compared with the signal frequency ω_{rf}. For $\omega_0 \ll \omega$, we have the case discussed above. For $\omega_0 \gg \omega$, we have classical detection which depends on the *curvature* of the I_0–V_0 curve. The rf impedance approaches the dynamic resistance R_D, and the responsivity is[28]

$$S \approx (2R_D)^{-1} d^2V_0/dI_0^2. \quad (20)$$

When $\omega_0 \approx \omega$ there is a resonant response with a change in sign as is shown in Fig. 10. The perturbation theory without noise predicts a singular responsivity at $\omega_0 = \omega$,

$$S = (R_D \omega^2)/[4I_0 R_{rf} \Omega^2 \omega_0(\omega_0 - \omega)]. \quad (21)$$

The rf resistance is given by

$$R_{rf} = R_D - [\omega^2/(\omega^2 - \omega_0^2)](R_D - R), \quad (22)$$

for all values of ω_0. A singularity in the responsivity is also expected from Eq. (18) because, for a hyperbolic I_0–V_0 curve, $R_D = \infty$ at $V = 0$.

Noise in the junction has the effect of rounding the corner on the I_0–V_0 curve so that R_D is everywhere finite, and desynchronizing the $n = 1$ step which is the source of the singularity at $\omega_0 = \omega$. In the data shown in Fig. 10c,

---

[59] H. Ohta, M. J. Feldman, P. T. Parrish, and R. Y. Chiao, *Rev. Phys. Appl.* **9**, 61 (1974).

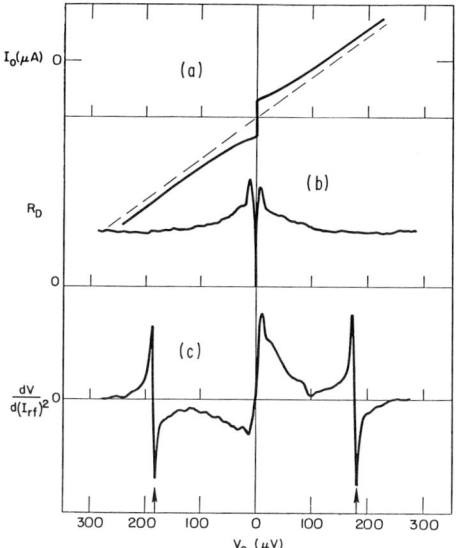

FIG. 10. I_0–V_0 curve of a Nb point contact junction (a), dynamic resistance (b), and video response (c) for an applied 90 GHz rf signal too weak to synchronize the junction. The response peaks near $V_0 = \hbar\omega_{rf}/2e$ are marked by arrows. (After Kanter and Vernon.[28])

TABLE II

THEORETICAL PERFORMANCE OF A VIDEO DETECTOR BIASED ON THE FIRST STEP[a]

$R = 12.5$ ohm	$T = 2°$K
$R_D = 24$ ohm	$\lambda = 2$ mm
$R_s = 50$ ohm	$S = 1.4 \times 10^5$ W/ω
$I_c = 20$ μA	NEP = 4.8×10^{-16} W/Hz$^{1/2}$

[a] Ohta et al.[59]

peaks in the response are seen for small bias voltages and also at $\omega_0 = \omega$ as is expected from the theory. An analysis of the square law response of a matched junction biased at $\omega_0 \approx \omega$ in the presence of noise leads to the predicted detector performance given by the values in Table II.[59]

This video detector has a very great speed advantage over the bolometer of Part VII. In applications where speed is not required, however, it has few advantages. The theoretical performance for $\lambda = 2$ mm in Table II is only a factor 2 better than the theoretical limit for the bolometer in Table I. Since the NEP is expected to increase as $\sim \Omega^2$, the bolometer is superior in principle for wavelengths $\lambda \lesssim 1$ mm.

In practice it is much more difficult to obtain efficient coupling to a point contact video detector than to the bolometer. The point contact is limited to ~ 1 spatial mode, whereas the bolometer in Table 1 can collect up to 30 modes for $\lambda = 1$ mm. Even for single mode application the coupling efficiency is usually less for the point contact. If the junction impedance is $\gtrsim 100$ ohm then I_c becomes so small compared with the noise current that the responsivity suffers. If $R \lesssim 10$ ohm, high Q resonant coupling is required and broadband operation cannot be obtained. Although many factors must be considered in an accurate comparison, the experimental data generally support the expected superiority of bolometer detectors for $\lambda \lesssim 1$ mm.[28,60]

The availability of series arrays of junctions with $pR > 100$ ohm and RI_c approaching the theoretical maximum should permit broadband matching to the video detector and thus improve its performance. Let us consider the case $\omega_0 \ll \omega$ and compare a single junction with parameters I_c, R, and S, to a p-junction series array with the same I_c, with shunt resistance pR, and with dynamic resistance pR_D. We first consider the case of detection of a single spatial mode with an array whose length is less than λ_{rf}. In the limit in which the array impedance pR is $\ll R_s$, the responsivity varies in proportion to p. If $pR \approx R_s$, then the available power responsivity becomes independent of p. The output noise voltage discussed in Part VI increases as $p^{1/2}$. Consequently, under the above conditions, the NEP decreases as $p^{-1/2}$ as long as $pR < R_s$ and increases as $p^{1/2}$ after $pR \approx R_s$. If the array becomes longer than λ_{rf}, then the mode must be defocused to illuminate all of it. Consequently, the responsivity will not increase in proportion to p, even when $pR \ll R_s$.

In the case of a multimode source, each mode has a minimum focal spot of area $\simeq \lambda_{rf}^2$. The optimum detector array is then a series connection of optimized single mode detector arrays, one for each available mode. The signal will then increase as p and the noise as $p^{1/2}$.

Square arrays of junctions have been constructed with superconducting spheres and proposed for infrared detection.[41] Since the impedance of a square array is independent of its size, the only advantage of this arrangement appears to be as a detector for multimode sources which cannot be coupled to a single junction. In the approximation used to guess the properties of series arrays, the noise from a square array is the same as for a single junction. The signal is the same if the power per junction is independent of p. This argument suggests that square arrays will have an NEP independent of p. A single junction would perform as well.

This brief description is only sufficient to illustrate the complexity of the problem of array detector optimization. It does suggest, however, that

---

[60] B. T. Ulrich, *Rev. Phys. Appl.* **9**, 111 (1974).

series arrays may be useful at infrared frequencies where matching is difficult, and for detecting multimode sources.

IX. Heterodyne Detector with External Local Oscillator

The heterodyne mixer with external local oscillator has appeared to many to be the most promising of Josephson detectors, and consequently has been most extensively developed. Most of the work done to date can be divided into two categories: low noise receivers, primarily for millimeter waves ($\Omega < 1$), and far-infrared ($\Omega > 1$) harmonic mixers for frequency comparison. We will first consider the low noise application.

Unlike infrared mixer receivers which can be shot noise limited, the noise in millimeter-wave heterodyne receivers is generally dominated by thermal noise from the mixer, or the intermediate frequency (i.f.) amplifier. Consequently, the useful figure of merit is the receiver noise temperature T_R, which contains contributions from the mixer T_M, and from the i.f. amplifier $T_{i.f.}$:

$$T_R = T_M + T_{i.f.}/\eta. \tag{23}$$

In order to refer $T_{i.f.}$ to the mixer input, it is scaled with the conversion efficiency $\eta = P_{i.f.}/P_s$, the ratio of the output power coupled into the i.f. amplifier to the available signal power. The best performance possible from a conventional single sideband mixer using a nonlinear resistor is $\eta T_M \gtrsim$ half the ambient temperature T and $\eta \lesssim 0.5$. Josephson effect mixers with $\Omega < 1$ are characterized by $\eta T_M > T$ and $\eta > 1$ (conversion gain).

Since a heterodyne down converter must have a reasonably narrow band at the signal frequency, a resonant coupling circuit such as that shown in Fig. 11 can be used to provide the optimum R_s. The equivalent circuit for such a resonant coupling scheme is shown in Fig. 12. The reactances are such that the dc, i.f., and rf currents flow primarily in their respective circuits.

If an rf local oscillator of sufficient strength to depress the height of the

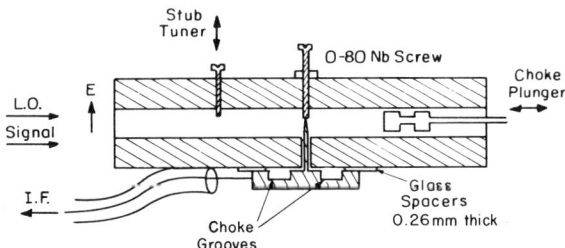

FIG. 11. Waveguide matching assembly for 36 GHz point contact mixer. Except for the glass insulators the entire structure was made from Nb so that junctions could be preset at room temperature. (After Taur et al.[40])

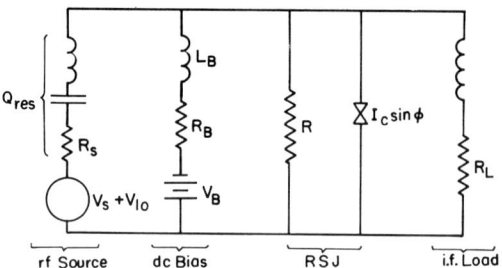

FIG. 12. Equivalent circuit of matching structure in Fig. 11.

zero voltage step I_0 to $\sim I_c/2$ is coupled to the junction then (from Fig. 4) I_0 decreases linearly with I_{rf}. The rf current arises from the sum of the signal current I_s and local oscillator current I_{lo}. The effect of a small I_s is to amplitude modulate I_{lo} at the frequency $\omega_{i.f.} = \omega_s - \omega_{lo}$, and thus to modulate I_0 at the same frequency. The phase modulation of I_{rf} does not play an important role.[61] The modulation of I_0 can be measured if the junction is current biased at a point of high differential resistance close to the $\eta = 0$ step, and the junction voltage coupled to an i.f. amplifier. The I_0–V_0 curve for such a mixer without and with P_{lo}, as well as the conversion efficiency η are shown in Fig. 13. The peak value of η appears halfway between the $n = 0$ and $n = 1$ steps where the maximum value of R_D occurs.

It is convenient to define a dimensionless parameter α which is a measure of the coupling of the available signal power P_s to the junction,

$$\alpha = \partial(I_0/I_c)/\partial[(8P_s/RI_c^2)^{1/2}]. \tag{24}$$

In terms of this parameter, the conversion efficiency is

$$\eta = C_{i.f.}\alpha^2 R_D/R, \tag{25}$$

where $C_{i.f.}$ is the output coupling efficiency. Experimental values for α can thus be obtained from measurements of η. Unlike η, which contains R_D, α is not sensitive to noise and so can be computed from simple RSJ theory. The source resistance R_s can be chosen to maximize α. As was discussed in Part IV, the optimum R_s/R may then be different from unity since the response α is a function of Ω, which itself depends on R_s. The optimum source resistance and the maximum value of α^2 are shown in Fig. 14.

The mixer theory outlined here assumes broadband coupling to the rf source so that R_s is effectively in parallel with R. Calculations with an analog junction simulator using the equivalent circuit in Fig. 12, however, show

[61] F. Auracher and T. Van Duzer, *Rev. Phys. Appl.* **9**, 233 (1974); *Proc. 1972 Appl. Superconduct. Conf., Annapolis*, IEEE Conf. Rec. 72CO682–5–TABSC, p. 603, 1972.

6. JOSEPHSON JUNCTION DETECTORS

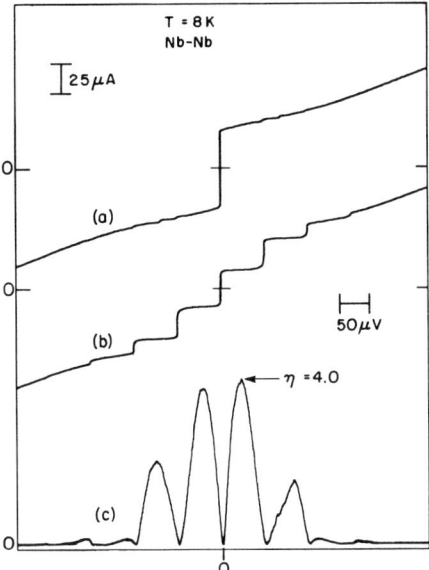

FIG. 13. Junction current without (a) and with (b) 36 GHz rf local oscillator and conversion efficiency (c) for a Nb point contact mixer with conversion gain. (After Taur.[65])

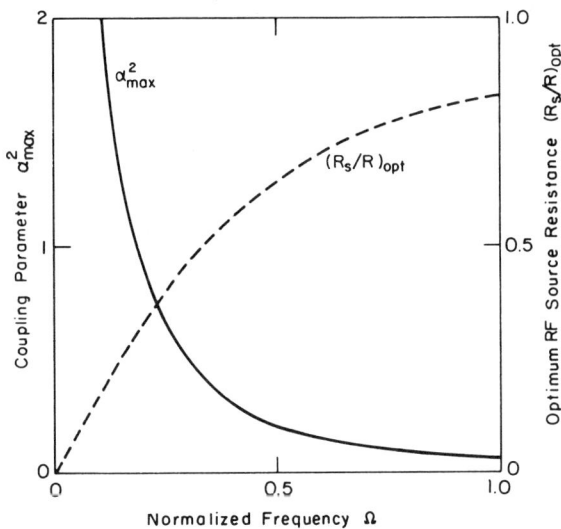

FIG. 14. The maximum value of the mixer coupling parameter α and the optimum rf source impedance R_s computed from the RSJ model as a function of the normalized frequency Ω. (After Taur et al.[40])

TABLE III

Performance of a 36 GHz Josephson Effect Mixer[a] Compared with a Cooled Schottky Barrier Diode Mixer[b]

	η	T_M (°K)	P_{lo} (W)
Josephson	1.3	53	$\sim 10^{-9}$
Schottky	0.25	200	$\sim 10^{-3}$

[a] Claassen et al.[49]
[b] Weinreb and Kerr.[62]

that it is valid for resonant coupling provided that R_s is not $\ll R$.[40] Measurements of α^2 in the neighborhood of $\Omega = 0.5$ using the 36 GHz apparatus in Fig. 11 often yield 90 to 100% of α^2_{max} for $Q_{res} \approx 50$. Measured values of η are 80 to 100% of the values predicted from Eq. (25) using the experimental R_D.[40] Typical results for 36 GHz mixers are compared in Table III with results for a cooled resistive mixer.[62] Mixing experiments at higher frequencies have thus far been much less successful. This has been generally due to the fact that the junctions have not been properly matched.[63,64]

The noise parameter β^2 has been measured for a variety of current biased junctions at various temperatures.[49] The results obtained at the values of dc and rf (36 GHz) bias required for efficient mixer operation are shown in Fig. 15. The experimental values for β^2_{min} scale with T and Ω in a way which is similar to the prediction of the thermal noise theory computed using the junction simulator, but are a factor ~ 2 higher.

Unfortunately, the noise is seen to increase by another factor of ~ 1.5 to 2 when the junction is resonantly coupled to the optimum R_s. This effect is not fully understood, but it is also seen in the simulation of resonantly coupled junctions.[40]

The mixer contribution to the system noise [Eq. (23)] can be obtained from[65]

$$T_M = \beta^2_{min} T / \alpha^2_{max}. \qquad (26)$$

Both α^2_{max} and β^2_{min} are decreasing functions of Ω as shown in Figs. 14 and 15. The excess of the experimental β^2_{min} over the theory in Fig. 15, and the excess noise due to resonant coupling both decrease with increasing Ω. Taking these effects into account, it seems possible that T_M will not be a

[62] S. Weinreb and A. R. Kerr, *Proc. IEEE* **SC-8**, 58 (1973).
[63] H. Kanter, *Rev. Phys. Appl.* **9**, 255 (1974).
[64] T. G. Blaney, *Rev. Phys. Appl.* **9**, 279 (1974).
[65] Y. Taur, Ph.D. thesis, Univ. of California, Berkeley, 1974 (unpublished).

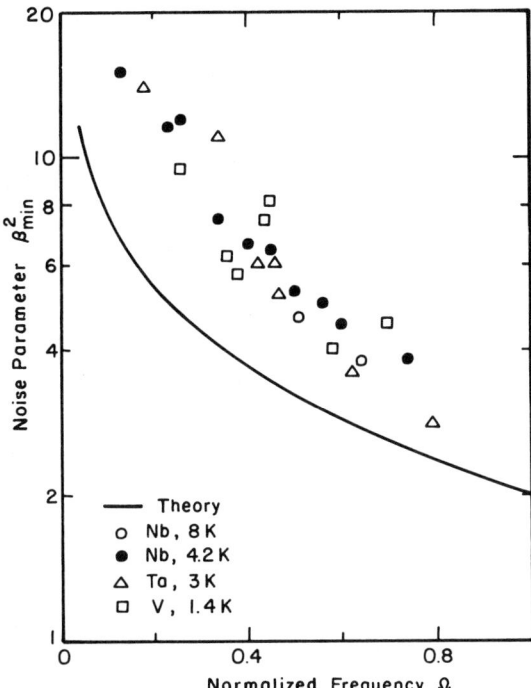

FIG. 15. Values of the noise parameter β^2_{min} which are measured under conditions of dc and rf bias appropriate to the operation of an efficient mixer. The data obtained at 50 MHz with a 36 GHz local oscillator from a variety of point contact junctions are compared with simulator calculations (—) as a function of normalized frequency Ω. (After Claassen et al.[49])

sensitive function of frequency in the millimeter range. The results for η and T_M at 36 GHz in Table III suggest that Josephson effect mixers will be very useful for millimeter-waves. Both parameters are presently about a factor 5 better than has been done by cooling conventional Schottky barrier diode resistive mixers.

The strong nonlinearity of the Josephson junction reduces the required P_{lo} (and thus the sensitivity to noise from the local oscillator) and makes possible efficient harmonic mixing. Figure 16 shows the results for a 4th order harmonic mixing experiment at $\omega_s/2\pi = 36$ GHz. In this case, the maximum value of the conversion efficiency was obtained at $V_0 = 0$ for a value of I_{lo} which completely suppressed I_0.

There is considerable interest in low noise mixers for the submillimeter frequency region. As the frequency is increased, photon shot noise limits the detectable signal. If a noise temperature is still used to characterize the mixer performance, it cannot be less than $\sim h\nu/k = 14°$K at 300 GHz, and

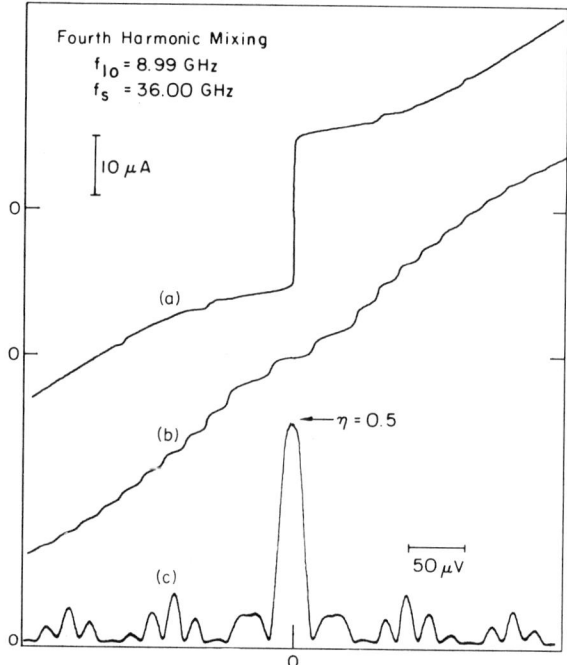

FIG. 16. Junction current without (a) and with (b) an ~9 GHz local oscillator, and conversion efficiency (c) for a fourth harmonic Nb point contact mixer. (After Taur et al.[40])

47°K at 1 THz. In practice, the noise temperatures of submillimeter mixers are far above these shot noise limits.

The mixer theory presented above is simplified somewhat if $\Omega \gg 1$. Then $R_s = R$ so that for optimum coupling, $P_s = I_{rf}^2 R/8$. The effective normalized frequency is $\Omega' = 2\Omega$, so from Eq. (9)

$$\alpha = \left(\frac{\partial J_0(I_{rf}/I_c 2\Omega)}{\partial (I_{rf}/I_c)}\right)_{max} \approx \frac{1}{3.5\Omega}. \quad (27)$$

Using this result and the fact that $\beta_{min}^2 \to 1$ for high frequencies we find

$$T_M = 12\Omega^2 T, \quad \Omega \gg 1. \quad (28)$$

The contribution to T_R from the i.f. amplifier can be similarly estimated from Eq. (25). For perfect i.f. coupling $\eta = \alpha^2 R_D/R$, so

$$T_{i.f.}/\eta = 12\Omega^2 R T_{i.f.}/R_D, \quad (\Omega \gg 1). \quad (29)$$

If the i.f. amplifier noise is not to dominate T_R, then we require $T_{i.f.} < TR_D/R$. This is a stringent requirement which can only be approached by the best cooled parametric i.f. amplifiers.

The best high frequency fundamental mixing experiments[64] reported thus far were at 891 GHz using a Nb contact with $RI_c = 2.2$ mV, so that $\Omega = 0.81$. The experimental conversion efficiency of $\eta = 6 \times 10^{-4}$ is much smaller than our theoretical estimate of $\eta \approx 0.2$. Most of the discrepancy was certainly due to poor input coupling which is a serious problem at these frequencies. The system noise was dominated by the i.f. amplifier since $T_{\text{i.f.}} = 500°K \gg TR_D/R \approx 30°K$. In spite of these difficulties, the minimum detectable power obtained in these experiments compares favorably with that obtained using other types of submillimeter wave mixers. If the matching can be improved substantially by better techniques with point contacts,[66] or by the use of arrays, Josephson junctions should be useful wide-band low-noise mixers for frequencies beyond 1 THz.

The highest frequency fundamental mixing reported in a Josephson junction made use of two CO_2 lasers at $\sim 9.6\mu$ (32 THz).[67] The Werthamer theory was used to compute the supression of the zero voltage current by rf signals with $\hbar\omega_{rf} \gg 2\Delta$. The results are quantitatively similar to those of the Bessel's function model. Although rf heating becomes increasingly severe at high frequencies and can cause thermal i.f. response, it appears that fundamental Josephson mixing at 32 THz may have been observed.

Harmonic mixing of infrared laser signals with microwave local oscillators has been explored extensively. Most such experiments have frequency comparison as their goal. Consequently, large signal powers have been used and questions of conversion efficiency and noise temperature have not generally been investigated.

There exists no complete theory of harmonic mixing comparable to that given above for fundamental mixing. A perturbation calculation[68] carried out to high enough order to be valid for $I_{rf} \lesssim I_c/3$ will be useful for some small signal detection experiments, but not for frequency comparison experiments done at high power. It is tempting to use the voltage source approximation of Part II with two applied rf voltages,

$$V = V_0 + V_{lo}\cos(\omega_{lo}t + \theta_{lo}) + V_s\cos(\omega_s t + \theta_s). \quad (30)$$

Then Eq. (4) becomes

$$I = I_c \sum_{n,m=-\infty}^{\infty} J_n\left(\frac{2eV_{lo}}{\hbar\omega_{lo}}\right) J_m\left(\frac{2eV_s}{\hbar\omega_s}\right) \sin[\omega_0 t + \theta_0 + n(\omega_{lo}t + \theta_{lo}) + m(\omega_s t + \theta_s)]. \quad (31)$$

---

[66] L. M. Matarrese and K. M. Evenson, *Appl. Phys. Lett.* **17**, 8 (1970).

[67] D. G. McDonald, F. R. Petersen, J. D. Cupp, B. L. Danielson, and E. G. Johnson, *Appl. Phys. Lett.* **24**, 335 (1974).

[68] M. T. Levinsen and B. T. Ulrich, *IEEE Trans. Mag.* **11**, 807 (1974).

Equation (31) is the general expression for the Josephson currents at all frequencies and voltages in the presence of two applied rf voltages. It predicts the appearance of dc (current) steps whenever $\omega_0 + n\omega_{lo} + m\omega_s = 0$, where n and m are integers with all positive and negative values. These include the familiar steps expected for each frequency acting separately, as well as steps spaced by multiples of the difference frequency $\omega_{lo} - \omega_s$ from the steps associated with each frequency acting alone.[69] When the junction is used as a small signal harmonic mixer these steps are too small to be synchronized in the presence of noise. When used for frequency comparison, both rf signals are large, but steps arising from mixing with ω_0 are commonly absent. This may be due to the low value of ω_{lo} and the large noise environment of laser sources of ω_s and ω_{lo}.[70]

As was the case for fundamental mixing, both types of harmonic mixing experiments are usually dominated by mixing processes in which ω_0 does not appear explicitly. The problem with the model is that in a real experiment the junction is not voltage biased, so the Josephson currents $I(t)$ in the loop of Fig. 3a impose a voltage $V(t) = I(t)R$ on the junction. The additional voltages at ω_0 and its harmonics give mixing terms of the form $\omega_{i.f.} = n\omega_{lo} + m\omega_s$, which do not include ω_0.[71]

Since experiments are usually done with $V_0 \ll \hbar\omega_{rf}/2e$, it is common practice to patch up Eq. (31) by setting $V_0 = 0$.[64,47,49,70,72] The i.f. current amplitudes are then expected to vary as the zero voltage current, $J_n(2eV_{lo}/\hbar\omega_{lo})J_m(2eV_s/\hbar\omega_s)$. Although this approximation is undoubtedly convenient, its validity has not been tested in detail. It is possible to speculate that for $\Omega < 1$, the Bessel's functions should be replaced by the appropriate step height dependences from Fig. 4. The Werthamer[10] theory could be used if the signal source is well above the gap frequency.

Frequency comparison experiments are limited either by sidebands arising from phase noise on P_{lo} or by i.f. amplifier noise. Since low frequency crystal oscillators have relatively better phase stability than microwave oscillators, the highest order mixing (825th harmonic) has been observed using them.[72] The maximum i.f. voltage swing available from the mixer is set by the separation of the steps induced by the lo: $V_{i.f.} \lesssim \hbar\omega_{lo}/2e$. The maximum power which can be coupled into a matched i.f. amplifier with impedance R_D is then

$$P_{i.f.} \lesssim (V_{i.f.}^2)/(8R_D) = (\hbar^2\omega_{lo}^2)/(32e^2R_D). \tag{32}$$

[69] C. C. Grimes and S. Shapiro, *Phys. Rev.* **169**, 397 (1968).

[70] D. G. McDonald A. S. Risley, J. D. Cupp, and K. M. Evenson, *Appl. Phys. Lett.* **18**, 162 (1971); D. G. McDonald, A. S. Risley, J. D. Cupp, K. M. Evenson, and J. R. Ashley, *Appl. Phys. Lett.* **20**, 296 (1972).

[71] E. D. Thompson, *J. Appl. Phys.* **44**, 3310 (1973).

[72] T. G. Blaney and D. J. E. Knight, *J. Phys. D* **6**, 936 (1973); **7**, 1882 (1974).

For a signal to be seen, ω_{lo} must be high enough for P_{if} to exceed the i.f. amplifier noise.[72]

For purposes of infrared frequency comparison Josephson junctions compete with room temperature metal–oxide–metal point contacts and Schottky barrier diodes. The Josephson junction has the disadvantage of low temperature operation, but can be used for much higher order mixing.

Some comments about the usefulness of arrays for heterodyne mixers are appropriate here. If a series array is used with length $< \lambda_{rf}$, the power conversion efficiency $\eta \propto (dI_0/dR_{rf})^2 R_D/R$ will not depend on the number of junctions in the array. The noise will be similar to that in a single junction with the same R_D as the array. The principal benefit of using an array is that is provides higher junction impedances. This effective impedance transformation may be of great value in increasing the bandwidth of the mixer and/or making possible the use of thin film structures which are too low impedance to be used individually. The greatest improvement in mixer performance is to be expected at infrared frequencies where resonant impedance matching is difficult. Arrays with lengths $> \lambda_{rf}$ must be planar and must be coupled to radiation with wave fronts perpendicular to the plane in order to preserve the phase coherence of the i.f. currents. It is doubtful that arrays with length $\gg \lambda_{rf}$ will be of value. Square arrays also appear not to have useful properties.

An entirely different type of superconducting junction called the "super-Schottky" diode is being developed for low noise millimeter-wave detection.[73] This is a junction whose operation depends on the nonlinear quasi particle tunneling between a superconductor and a semiconductor. It is like a classical Schottky diode, except that the curvature of the I_0–V_0 curve is controlled by the superconducting energy gap ~ 1 meV rather than by the semiconducting gap ~ 1 eV. The super-Schottky diode can be used as a video detector, or as a classical resistive mixer in a heterodyne receiver. As is the case with the Josephson junction, the super-Schottky has a much stronger nonlinearity than a conventional Schottky diode. This leads to large video responsivity, and low local oscillator power requirements in the heterodyne application. The response in both modes should be independent of frequency up to the capacitive cutoff $(RC)^{-1}$, in contrast to the Josephson response which varies as Ω^{-2}. Although it is too early to compare the two devices critically, the super-Schottky diode may well compete favorably with the Josephson junction in certain applications.

X. Parametric Amplifier

Since the Josephson current is reactive and nonlinear, a variety of

[73] M. McColl, M. F. Millea, and A. H. Silver, *Appl. Phys. Lett.* **23**, 263 (1973).

parametric effects can be observed in Josephson junctions. In a simple frequency converter based on a nonlinear reactance, there is one output photon for each input photon, so that the power conversion efficiency is proportional to the ratio of the final to the initial frequency.[74] Consequently, nonlinear reactors are useful for up-conversion and for parametric amplification, but generally not for down-conversion. This is the reason that the Josephson heterodyne mixer of Part IX is, for example, operated in an essentially resistive mode, and not as a nonlinear reactor.[75]

We are concerned here with the properties of Josephson effect parametric amplifiers. These could be useful as preamplifiers at the rf signal frequency, or as low noise i.f. amplifiers for heterodyne receivers. As was the case with heterodyne mixers, parametric amplifiers can be operated with an external pump, or by using the ac Josephson currents as pump. The study of Josephson effect parametric amplifiers has not developed sufficiently to make it clear which type of amplifier will prove more useful. The arbitrary choice has been made to emphasize in this section the parametric amplifier with external pump.

A Josephson junction which obeys the RSJ model and which is biased with $I_0 < I_c$ behaves as a nonlinear inductor with reactance given by Eq. (10) in parallel with a resistor R. The admittance of the junction is thus

$$Y_{RSJ} = 1/R - j/[\omega L(I)]. \tag{33}$$

As a circuit element, the RSJ at $V_0 = 0$ is closely analogous to a varactor diode whose equivalent circuit is a nonlinear capacitor in series with a resistor, giving an impedance

$$Z = R - j/[\omega C(V)]. \tag{34}$$

The well-developed theory of varactor devices[76] can be extended to the case of parametric Josephson devices biased at $V_0 = 0$ by means of the simple transformation $R \to R^{-1}$, $C \to L$, and $Z \to Y$.

In order to introduce some of the relevant ideas, we first consider the case of a degenerate parametric oscillator. If a parallel capacitance C is added to the RSJ model, then the result is the resonant circuit in Fig. 17, with $\omega_R = (\cos \phi_0/L_J C)^{1/2}$, where $L_J = \hbar/2eI_c$. The capacitance can be supplied by an external circuit if the junction obeys the RSJ model, or it can be internal to the junction as in the case of an SIS tunnel junction. In the latter case, the circuit resonance is called the Josephson plasma resonance. In the case of the degenerate oscillator, the circuit is driven by an external pump

---

[74] J. M. Manley and H. E. Rowe, *Proc. IRE* **44**, 904 (1956).
[75] H. A. Combet, *Ann. Telecommun.* **27**, 507 (1972).
[76] P. Pennfield, Jr., and R. P. Rafuse, "Varactor Applications." MIT Press, Cambridge, Massachusetts, 1962.

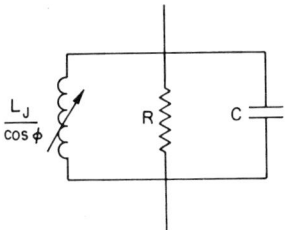

FIG. 17. Equivalent circuit for the Josephson parametric oscillator.

current at $\omega_p = 2\omega_R$. Since the inductance is a function of the Josephson current, there is a threshold current beyond which oscillations will occur at the frequency ω_R. Beyond the threshold, a signal and an idler photon ($\omega_s = \omega_i = \omega_R$) are emitted for each pump photon absorbed. The threshold for this degenerate parametric oscillation can be computed from the criterion for the pumping of a child's swing.[77] If the pump is strong enough to change the resonant frequency by $\Delta\omega$, then oscillations will be supported when $\Delta\omega/\omega_R = \Delta L/2L \geqslant Q^{-1}$. In terms of the phase, the criterion can be written

$$\alpha_p \tan \phi_0 \geqslant 2/Q, \qquad (35)$$

where $\alpha_p = \Delta\phi = 2eV_p/\hbar\omega_p$. For the parallel resonant circuit $Q = R \cos \phi_0/\omega L_J$ so that the criterion for oscillation becomes $\alpha_p \sin \phi_0 \geqslant 2\Omega$. To obtain $\sin \phi_0 \approx 1$, the bias point is selected to be close to the top or bottom of the zero voltage step. The criterion for oscillation is easily satisfied only if $\Omega < 1$, and if there are no other circuit losses comparable to R. This parametric excitation of the Josephson plasma resonance was first observed on an analog junction simulator and then Eq. (35) derived from a detailed solution of the pumped Josephson equations in the limit $\alpha_p \ll 1$.[78] Indirect experimental evidence for the effect has been observed.[79]

Although these simple considerations give the oscillation criterion, the design of a parametric amplifier requires a more detailed treatment to obtain such important properties as the gain and the bandwidth.[76] The usual procedure is to assume that the circuit of Fig. 17 sees a short circuit at all frequencies except ω_p, ω_s, and ω_i. If we assume that the voltage across the circuit is $V_p \cos(\omega_p t + \phi_p) + V_s \cos(\omega_s t + \phi_s) + V_i \cos(\omega_i + \phi_i)$, then the currents in the ideal Josephson element $I = I_c \sin(2e/\hbar) \int V \, dt$ can be

---

[77] C. Kittel, W. D. Knight, and M. A. Ruderman, "Berkeley Physics Course, Machanics," Vol. 1, p. 229. McGraw-Hill, New York, 1965.
[78] N. F. Pedersen, M. R. Samuelsen, and K. Saermark, *Rev. Phys. Appl.* **9**, 223 (1974); *J. Appl. Phys.* **44**, 5120 (1973).
[79] C. K. Bak, B. Kofoed, N. F. Pedersen, and K. Saermark, *J. Appl. Phys.* **46**, 886 (1965).

written in terms of an admittance matrix equation,

$$\begin{bmatrix} I(\omega_s) \\ I^*(\omega_i) \end{bmatrix} = \begin{bmatrix} Y_{ss} & Y_{si}^* \\ Y_{is} & Y_{ii}^* \end{bmatrix} \begin{bmatrix} V_s \\ V_i^* \end{bmatrix}. \qquad (36)$$

If a signal source with an internal shunt conductance G_0 at ω_s and ω_i is connected to the circuit, then the idler is terminated so that $I_2(\omega_i) = -V_i(G + G_0 + j\omega_i C)$, where $G = R^{-1}$. The admittance seen by this signal source can then be computed in terms of the matrix elements in Eq. (36),

$$Y_T(\omega_s) = G + j\omega_s C + Y_{ss} - \frac{Y_{si}^* Y_{is}(G + G_0 + j\omega_i C + Y_{ii})}{|G + G_0 + j\omega_i C + Y_{ii}|^2}. \qquad (37)$$

The first three terms give the circuit admittance of the parallel resistor, capacitor, and averaged inductor. The last term includes the effects of the pump and provides the negative resistance.

The gain $G(\omega_s)$ for an amplifier which is operated with a circulator can be computed from Y_T since it is just the reflection coefficient for a signal with source admittance G_0,

$$G(\omega_s) = \left| \frac{Y_T(\omega_s) - G_0}{Y_T(\omega_s) + G_0} \right|^2. \qquad (38)$$

A few collected results for the three photon parametric amplifier[80] $\omega_p = \omega_s + \omega_i$ are given in Table IV. The corresponding results are also given for the doubly degenerate four photon parametric amplifier[81] in which two pump photons are employed, $2\omega_p = \omega_s + \omega_i$, and $\omega_s \approx \omega_L$. If we consider the limit of small pump amplitude $\alpha_p \ll 1$, and set $G_0 = 0$, then for the three

TABLE IV

THEORETICAL PARAMETERS OF EXTERNALLY PUMPED PARAMETRIC AMPLIFIERS

Quantity	Three photon[a]	Four photon[b]
Resonant frequency	$\omega_R[J_0(\alpha_p)\cos\phi_0]^{1/2}$	$\omega_R[J_0(\alpha_p)]^{1/2}$
Gain condition	$J_1(\alpha_p)\sin\phi_0 \approx \Omega(1+G_0/G)$	$J_2(\alpha_p) \approx \Omega(1+G_0/G)$
"Gain-bandwidth" product $\sqrt{G_{max}}\,\Delta\omega/\omega$	$\dfrac{\Omega(1+G_0/G)}{2J_0(\alpha_p)\cos\phi_0}$	$\dfrac{\Omega(1+G_0/G)}{2J_0(\alpha_p)}$

[a] Taur.[80]
[b] Parrish et al.[81]

[80] Y. Taur, private communication, 1974.
[81] P. T. Parrish, M. J. Feldman, H. Ohta, and R. Y. Chiao, *Rev. Phys. Appl.* **9**, 229 (1974).

photon case, the resonant frequency and the condition for high gain (essentially the oscillation condition) are the same as was derived from the simpler approach presented earlier.

In any attempt to build a parametric amplifier (or oscillator), the details of the actual microwave circuit must be carefully considered. These differ from the simple circuit in Fig. 17 because they have high order resonant response at a number of frequencies. Since parametric devices can generate harmonics, real circuits have potential losses which are not included in the model calculation. In practice, practical microwave parametric amplifier cavities are resonant at ω_s and ω_i. Every effort is made to prevent power dissipation at any other combination (or harmonic) frequency. The microwave circuit can sometimes be simplified by operation in a degenerate or doubly degenerate mode.

Evaporated film SIS tunnel junctions can probably be used for parametric devices since they can be fabricated with the Josephson plasma resonance in the microwave frequency region.[82] Although the quasi particle current is a complicated function of junction voltage, the observed damping of ω_R is comparable to that expected from the normal state junction resistance.[83] Consequently, $\Omega \gtrsim \hbar\omega/\pi\Delta$ as is the case for other junctions. Care must be taken to avoid harmonic excitation of the stripline resonances in the tunnel junction. Junctions which obey the RSJ theory can also be used if the external rf circuit supplies the required capacitance.

The only successful externally pumped parametric amplifier built thus far is of the doubly degenerate type.[44,81] It is operated at 10 GHz and makes use of a series array of ~ 80 Dayem bridges made of Sn with the array length much less than a guide wavelength λ_g. The microwave circuit is a parallel plate stripline which has an open circuit at a distance $\lambda_g/4$ from the junction array so that the junctions effectively terminate the transmission line. As shown in Fig. 18, there is a tuning stub consisting of an open ended $\lambda_g/12$ orthogonal strip immediately before the junctions. This stub is intended to short out the third harmonic of the pump and the undesirable up-conversion products $2\omega_p + \omega_s$ and $2\omega_p + \omega_i$.

The observed gain was 12 dB and the bandwidth ~ 1 GHz. The gain bandwidth product $G_{max}^{1/2}\Delta\omega/\omega = 0.3$. This result is roughly consistent with the theoretical product in Table IV if we assume that the gain condition is met and use the array resistance $R = 100$ ohm, the stripline impedance $R_0 = 50$ ohm and the normalized frequency $\Omega = 0.15$. The noise temperature of this amplifier including the room temperature circulator is 26°K. The noise temperature of the low temperature portions was less than 20°K.

[82] A. J. Dahm, A. Denenstein, T. F. Finnegan, D. N. Langenberg, and D. J. Scalapino, *Phys. Rev. Lett.* **20**, 859, 1020E (1968).

[83] N. F. Pedersen, T. F. Finnegan, and D. N. Langenberg, *Phys. Rev. B* **6**, 4151 (1972).

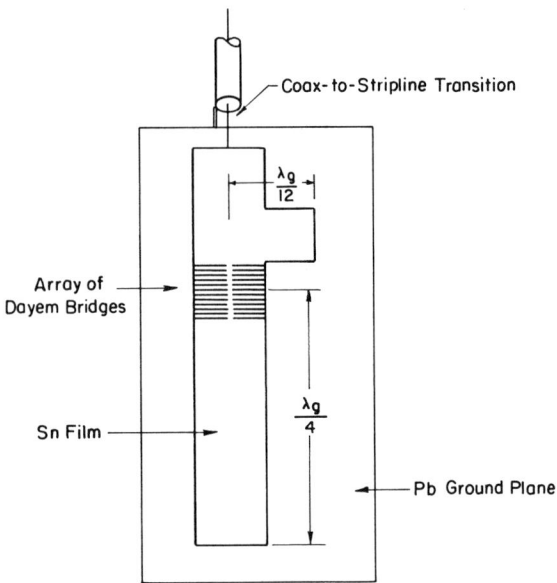

FIG. 18. Strip line 9 GHz parametric amplifier.

Modern cooled varactor paramps with a similar center frequency but smaller bandwidth have noise temperatures of $\sim 20°K$.

The performance of this Josephson parametric amplifier makes it appear likely that such devices will ultimately compete favorably with other cooled paramps in the microwave frequency range. At its present frequency, it would be particularly useful as an i.f. amplifier for a Josephson mixer operating in the high frequency region, where we require bandwidths approaching 1 GHz and, from the analysis of Part IX, $T_{i.f.} \lesssim R_D T/R$. With an ideal Nb junction array and the same value of Ω, the operating frequency could exceed 100 GHz. Although it may prove very difficult to produce arrays with length $\ll \lambda_{rf}$ at such frequencies, the resulting amplifier would be of great value.

From Table IV, we see that the gain-bandwidth product increases with Ω. Unfortunately, however, much larger values of Ω cannot be used since $\Omega \lesssim J_1(\alpha_p)/(1 + G_0/G) \lesssim 0.5$ is required for significant gain. Consequently, these Josephson parametric amplifiers will probably never be used in the submillimeter band.

A line of development parallel to that described above has been pursued on self pumped Josephson parametric amplifiers. The Manly–Rowe relations have been generalized to include the case in which a bias voltage V_0 can be a source of rf power at frequency ω_0,[84,85] and amplifier performance has been computed for particular choices of equivalent circuit.[30,84,86,87] A parametric gain of 11 dB has been obtained at 30 MHz with a Josephson frequency pump at 60 MHz.[88] Similar gain has also been reported at ~ 9 GHz with an ~ 18 GHz internal pump.[89] Both of these amplifiers make use of point contacts in a voltage bias circuit which has low impedance at the pump frequency. For historical reasons this configuration is known as a resistive SQUID.

A third approach is related to the negative rf resistance of a resistively shunted junction biased below the first step. A point contact junction is voltage biased so that ω_0 is slightly below $\omega_s \simeq 9$ GHz. In addition to signal gain, considerable noise is seen which can be shown to have been up-converted from a low frequency idler at $\omega_i = \omega_s - \omega_0$.[39,89] Because of the extra noise this amplifier is less attractive than the ones pumped at $\omega_0 \simeq 2\omega_s$. It is expected, however, to operate at a larger value of Ω.[89] This device also functioned well as an up converter with gain from $(\omega_s - \omega_0)/2\pi = 115$ MHz to ~ 9 GHz.[89]

Except under special circumstances, the admittance of a pumped Josephson junction is rich in harmonic frequencies. When a signal is present, current will flow through the junction shunt resistance at many combination frequencies. The resulting dissipation is prejudicial to the operation of a parametric amplifier. The internally pumped amplifier described above is short circuited at all frequencies, except the signal frequency. The externally pumped amplifier is arranged to short particular combination frequencies. To short a larger number of these "idler" frequencies is a formidable task.

These recent results are sufficient to show that many types of Josephson effect parametric amplifier are sufficiently attractive to deserve further development. The relative importance of self pumped and externally pumped amplifiers is not clear at this time.

XI. Other Devices of Interest

Because of the complexity of the response of Josephson junctions, there are a very large number of ways in which they can be used as high frequency detectors. In this review, four devices have been emphasized which are closely analogous to useful conventional devices. This selection has been

[84] P. Russer, *Proc. IEEE (Lett.)* **59**, 282 (1971).
[85] E. D. Thompson, *IEEE Trans* **ED-20**, 680 (1973).
[86] P. Russer, *Arch. Elektrisch. Übertrag.* **23**, 417 (1969).
[87] A. N. Vystavkin, V. N. Gubankov, G. F. Leshchenko, K. K. Likharev, and V. V. Migulin, *Radiotekh. Elektron.* **15**, 2404 (1970) [*English Transl.: Radio Eng. Electron. Phys.* **15**, 2121 (1970)].
[88] H. Kanter and A. H. Silver, *Appl. Phys. Lett.* **19**, 515 (1971).
[89] H. Kanter, *IEEE Trans. Mag.* **11**, 789 (1975).

made on the basis of estimated or demonstrated performance characteristics such as responsivity, noise, bandwidth, gain, linearity, etc. The selection made seems reasonable at the present time, but may not prove valid in the future. It should not, therefore, influence active workers in the field to ignore other possibilities.

In this part a description will be given of a number of other high frequency devices whose properties have been explored. Where necessary, an introduction will be given to Josephson phenomena which have not been previously described.

1. Heterodyne Detector with Internal Local Oscillator

We first consider the possibility of a useful heterodyne detector with an internal local oscillator. The constant voltage steps on the I_0–V_0 curve of a Josephson junction appear as the result of mixing between the ac Josephson currents and harmonics of an applied rf signal. A dc output appears because the Josephson currents are synchronized to the external signal over a range of bias currents. Linear detection is obtained if the height of the $n = 1$ step is measured. This can be done in practice[65] by voltage biasing the junction at the step and imposing a small modulation so that the current is switched between the values at the top and bottom of the step. The result of measuring this ac current is a linear heterodyne detector for rf signals large enough to synchronize the junction. For smaller signals, square law detection is obtained in a way which is related to the video detector of Part VIII.

Unfortunately, the power levels at which synchronization occurs are $\gtrsim 10^{-14}$ W from Eq. (13) for $R = 10$ ohm. These are too large to be of interest for small signal detection unless R is made too small for good coupling over a useful bandwidth. It is possible that this mode of detection could be employed in measuring small changes in the amplitude of a strong carrier wave.[65,90] The phase locked loop, which is an exact analog of the RSJ model[33] is used for automatic frequency control receivers at lower frequencies.

Heterodyne down conversion with the Josephson frequency currents as local oscillator has also been studied for the case of a finite $\omega_{i.f.}$.[91-93] For current biased junctions, it has been reported that true mixing signals are seen only for very large intermediate frequencies $\omega_{i.f.}/2\pi \gtrsim 10$ GHz.[93] Among the factors which contribute to this behavior are the frequency modulation of the Josephson currents by low frequency noise, the tendency of the

---

[90] P. L. Richards and Y. Taur, *Bull. Amer Phys. Soc. Ser. II* **17**, 45 (1972).
[91] A. Longacre, *Proc. 1972 Appl. Superconduct. Conf., Annapolis* IEEE Conf. Rec. 72CO682-5 TABSC, p. 712, 1972.
[92] A. Longacre, Ph.D. thesis, Univ. of Rochester, 1972 (unpublished).
[93] G. Vernet and R. Adde, *Rev. Phys. Appl.* **9**, 275 (1974); *J. Appl. Phys.* **45**, 2678 (1974).

junction to synchronize so that $\omega_{i.f.} \to 0$, and parametric amplification due to the negative rf resistance close to the induced step.

Down conversion with small $\omega_{i.f.}$ has been reported with a point contact junction that is voltage biased in a resistive SQUID.[94,95] As long as the inductive reactance of the bias circuit $\omega_s L$ is small ($< R_B$), the noise in the junction is controlled not by the junction resistance, but by the much smaller bias resistor R_B, which is in parallel with it. The bandwidth of the Josephson currents is thereby less than for a current biased junction.

If a Josephson junction is coupled to a resonant circuit such that the junction resistance dominates the losses, then cavity mode steps are observed on the I_0-V_0 curve at $V_1 = \hbar\omega_R/2e$. A theory of this effect exists for voltage biased junctions, which is useful for the case of strip line resonances in SIS tunnel junctions.[10] For current biased junctions various approximate theories can be used for large Ω,[30] but analog computers are necessary for $\Omega \lesssim 1$.[92]

Mixing experiments have been carried out with the junction synchronized to a cavity mode at the Josephson frequency[40] and also for the case of a cavity mode at the i.f. frequency.[91,92] The inherent complexity of these experiments must be considered a disadvantage when applications are contemplated. The possibility also exists of cavity mode effects at the signal frequency whenever a junction is resonantly coupled to the rf source. These effects need not be strong if the junction is properly matched, because the Q will be loaded by the rf source resistance. Under these circumstances, the junction noise can be sufficient to prevent synchronization of the junction to the resonant mode.

In general, the reason for studying mixers which make use of the Josephson currents for the local oscillator is the difficulty of providing local oscillators for submillimeter wavelengths. This difficulty is severe for conventional Schottky barrier diode mixers because of the high power requirement ($P_{lo} \approx 10^{-3}$ W). It is much less important for the Josephson mixers of Part IX because of the lower power requirement ($P_{lo} \approx 10^{-9}$ W at 36 GHz). Submillimeter power at low levels can be provided by external harmonic generators, by internal harmonic mixing, or perhaps even from tightly coupled external Josephson effect oscillators.[96,97] Heterodyne mixers which make use of the ac Josephson currents will not be able to compete successfully with the mixers of Part IX unless they show comparable values of noise temperature and conversion efficiency. This is not the case in their present state of development.

[94] A. A. Fife and S. J. Gygax, *J. Appl. Phys.* **43**, 2391 (1972); S. Gygax and W. Zingg, *Rev. Phys. Appl.* **9**, 269 (1974).
[95] S. N. Erné, *Rev. Phys. Appl.* **9**, 243 (1974).
[96] D. N. Langenberg, D. J. Scalapino, and B. N. Taylor, *Proc. IEEE* **54**, 560 (1966).
[97] R. K. Elsley and A. J. Sievers, *Rev. Phys. Appl.* **9**, 295 (1974).

2. REGENERATIVE DETECTOR

A junction with a cavity mode step on its I_0-V_0 curve can be operated as a narrow band square law detector.[98] This is a regenerative effect which occurs when the rf current in the junction excites the cavity mode and the cavity excitation is coupled back to the junction with a loop gain approaching unity. The square law detection is observed when the junction is biased at a point of high differential resistance near the cavity mode step. The observed bandwidth is much narrower than the video detection at the same frequency. This is interpreted as evidence for line narrowing due to positive feedback.

The best results obtained with this regenerative detector were NEP $\simeq 10^{-14}$ W/Hz$^{1/2}$ in a 300 MHz bandwidth at a frequency of 180 GHz using an unmatched Nb point contact. This corresponds to a heterodyne detector with system noise temperature $T_s \approx 40{,}000°$K. Although this detector could in principle be useful, the effect is sufficiently sensitive to the junction and cavity parameters that it is difficult to observe, and has never been properly optimized. The only published theoretical description of the detector is given[98] in terms of the voltage bias model. No proper description in terms of the RSJ model is available.

The regenerative mode of detection has apparently been observed in a square array of point contacts made by pressing superconducting spheres together.[41] These arrays show cavity mode resonances for wavelengths comparable to the diameter of the spheres, and narrow band square law detection effects. For the reasons given in Part VIII, the square array is not expected to have fundamental advantages as a detector over a single junction operated in an optimum way.

In an experiment related to the two above, the radiation from a point contact biased near a cavity mode was extracted from the cryostat and coupled to a Fabry–Perot interferometer. As the interferometer separation was varied, the voltage across the junction responded in a way which was interpreted as demonstrating a very narrow linewidth for the radiation from the synchronized junction ($Q > 10^3$) and also very sensitive self detection.[99]

3. INJECTION LOCKING

Injection locking of the Josephson oscillation has been observed when a ~ 9 GHz signal with power $\gtrsim 10^{-17}$ W was coupled through a circulator to a junction in a coaxial cavity which was current biased to a nearby frequency.[100] The output in a narrow bandwidth around ω_s showed gain

---

[98] P. L. Richards and S. A. Sterling, *Appl. Phys. Lett.* **14**, 394 (1969).
[99] B. T. Ulrich and E. O. Kluth, *Proc. 1972 Appl. Superconduct. Conf., Annapolis.* IEEE Conf. Rec. 72CO682-5 TABSC, p. 709, 1972.
[100] C. V. Stancampiano and S. Shapiro, *Appl. Phys. Lett.* **25**, 315 (1974).

as large as 51 dB resulting from the fact that the Josephson oscillation is pulled in frequency and phase locked by the signal power. In practice the usefulness of this effect as an amplifier will be limited by the nonlinearity of the gain. It could, however, be used to increase the level of a weak frequency modulated carrier.

Another example of gain arising from injection locking has been reported.[40,101] In this case, the oscillation which is locked is the relaxation oscillation of a voltage-biased junction which shows hysteresis in its current biased I_0–V_0 curve. The frequency of this relaxation oscillation is controlled by the junction shunt resistance R and the inductance L_B of the voltage bias circuit. The source of the hysteresis, whether capacitance or heating, is not important as long as the relaxation oscillation frequency $\omega = R/L_B$ is slow compared with the junction response. When a signal with frequency $\omega_s = R/L_B \simeq 5$ MHz was coupled to the junction, a reflected signal was observed with gain up to 20 dB. When used as a fundamental mixer at 36 GHz, a conversion gain of $\eta = 50$ was observed when $\omega_{i.f.} = R/L_B \simeq 5$ MHz These devices are not particularly useful because of their nonlinear gain, narrow bandwidth, and large noise temperature. In the case of the mixer, $T_M \approx 1000°$K compared with $< 100°$K for nonhysteretic mixers at the same frequency.

XII. Conclusions

In this review, four types of Josephson effect devices have been described which appear to have useful properties for the detection of millimeter and submillimeter waves. In each case there is strong competition from other He temperature devices, most of which are made from semiconductors. It is difficult to predict which detectors will prove superior in the future. The vast technology of semiconducting devices does not generally extend to their low temperature properties. At present, more is known about noise mechanisms in superconductors than is known about heavily doped semiconductors at He temperature.

The primary disadvantages of Josephson effect devices arise from the undesirably low junction impedance and the variation of high frequency response as Ω^{-2}. The use of series arrays may overcome the impedance deficiency. The high frequency rolloff will always be present. With ideal resistively shunted junctions, $\Omega \approx \hbar\omega/\pi\Delta$, so with present superconductors it need not be serious for frequencies below ~ 1 THz. New superconductors could increase this limit by more than a factor of two. It is an open question at this time whether devices can be made using tunnel junctions which

[101] Y. Taur and P. L. Richards, *J. Appl. Phys.* **46**, 1793 (1975).

have $\Omega < \hbar\omega/\pi\Delta$ because of the small quasi particle conductance below the gap voltage.

Acknowledgments

The author is greatly indebted to Drs. J. H. Claassen, P. W. Forder, and Y. Taur for numerous discussions about all aspects of Josephson effect detectors. This is especially the case with the properties of arrays and of parametric amplifiers. A portion of this review was written while the author was on leave at the Cavendish Laboratory, Cambridge, England. The support of the J. S. Guggenheim Memorial Foundation and the Office of Naval Research is gratefully acknowledged.

CHAPTER 7

The Pyroelectric Detector—An Update

E. H. Putley

The pyroelectric detector was discussed in Chapter 6 of Volume 5 of this treatise. This short note has been written for inclusion in this present volume on Infrared Detectors to bring the pyroelectric story up to date. Since the publication of Volume 5 there has been steady development of the pyroelectric detector. While there is little to add to the basic principles discussed in Chapter 6 of Volume 5, the introduction of new materials and the refinement of manufacturing techniques has lead to the increased use of pyroelectric detectors for a wider range of applications.

The most sensitive pyroelectric detectors still use what is essentially TGS[1] (triglycine sulfate). However one of the disadvantages of the earlier TGS detectors was that if heated above the Curie point (49°C) the material became depoled. Although the detector was not permanently damaged, the necessity to repole the detector could be a considerable nuisance. This fault is possessed not only by pure TGS, but also to a greater or lesser extent by any ferroelectric with a Curie temperature close to ambient. These materials have the most attractive pyroelectric properties, but because of this problem, they must be used with care. A solution to this problem has been discovered by Lock[2] who found that the introduction into TGS of a few percent of the related amino acid L-alanine introduced a degree of asymmetry into the crystal structure which inhibited domain switching. Comparison of the molecular formulas for glycine and alanine indicates the origin of the asym-

---

[1] E. Schwarz and R. R. Poole, *Appl. Opt.* **9**, 1940–1 (1970); E. H. Putley, *Opt. Laser Tech.* **3**, 150–156 (1971); H. P. Beerman, F. Schwarz, and S. Weiner, *Proc. Electro-Optic 1972 Tech. Sem.*, 12–14 September, New York, pp. 343–350.

[2] P. J. Lock, *Appl. Phys. Lett.* **19**, 390–391 (1971); E. T. Keve, K. L. Bye, P. W. Whipps, and A. D. Annis, *Ferroelectrics* **3**, 39–48, (1971). K. L. Bye, P.W. Whipps, and E. T. Keve, *Ferroelectrics* **4**, 253–256 (1972); E. T. Keve, *Philips Tech. Rev.* **35**, 247–257 (1975).

metry.[2a] Figure 1 shows how the presence of a small amount of alanine displaces the hysteresis loop of TGS.[3] It was also found that in alanine doped TGS, the pyroelectric coefficient was slightly higher and the dielectric constant and the dielectric loss were slightly lower than for pure TGS. Thus the performance of detectors made from this material is somewhat better than that of ones made from pure TGS. These differences in properties come about because in normal pure TGS it is not possible to secure 100% uniform orientation of the domains, while this is automatically obtained in the permanently poled material. Figure 2[3-3b] shows the performance obtained from a detector made with this material (LATGS) while Fig. 3[3c-3l] shows an up-to-date comparison of the best types of uncooled thermal detectors.

Although LATGS is very suitable for use at normal room temperature, for some purposes its relatively low Curie temperature is a disadvantage. Some improvement can be achieved by using deuterated TGS which has a higher Curie temperature ($\sim 60°C$). Triglycine fluorberyllate (TGFB) also has a higher Curie temperature (75°C), but there are two objections to its use. First, at lower temperatures it is not as satisfactory as TGS[4] and second

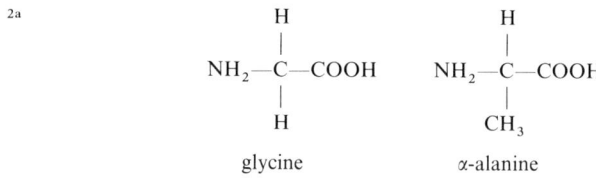

[2a]

Due to the presence of the CH_3 radical, alanine is an optically active compound. It is essential to use material of the same symmetry when adding it as a dopant to TGS. The laevo form is normally used, since this is the most readily available, but the dextro would be equally effective.

[3] S. E. Stokowski, J. D. Venables, N. E. Byer, and T. C. Ensign, *Infrared Phys.* **16**, 331–334 (1976).
[3a] G. Baker, D. E. Charlton, and P. J. Lock, *Radio Electron. Eng.* **42**, 260–264 (1972).
[3b] H. P. Beerman, *Infrared Phys.* **15**, 225–231 (1975).
[3c] S. T. Liu and R. B. Maciolek, *J. Electron. Mater.* **4**, 91–100 (1975).
[3d] P. B. Fellgett, *Radio Electron. Eng.* **42**, 476 (1972).
[3e] D. A. H. Brown, R. P. Chasmar, and P. B. Fellgett, *J. Sci. Instrum.* **30**, 195–199 (1953).
[3f] E. Schwarz, *Research* **5**, 407–411 (1952).
[3g] J. R. Hickey and D. B. Daniels, *Rev. Sci. Instrum.* **40**, 732–733 (1969).
[3h] O. Stafsudd and N. B. Stevens, *Appl. Opt.* **7**, 2320–2322 (1968).
[3i] J. C. Gill, private communication, 1958.
[3j] M. Chatanier and G. Gauffre, *IEEE Trans. Instrum. Meas.* **IM22**, 179–181 (1973).
[3k] N. B. Stevens, *in* "Semiconductors and Semimetals" (R. K. Willardson and A. C. Beer, eds.), Vol. 5, Chap. 7, Table I and Fig. 15. Academic Press, New York, 1970.
[3l] R. De Waard and S. Weiner, *Appl. Opt.* **6**, 1327–1331 (1967).
[4] E. Leiba and A. Hadni, *Nouv. Rev. Opt. Appl.* **3**, 263–266 (1972).

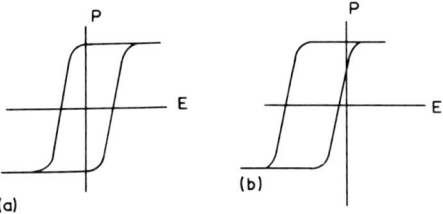

FIG. 1. TGS hysteresis loops: (a) normal material; (b) alanine doped material with displaced loop.

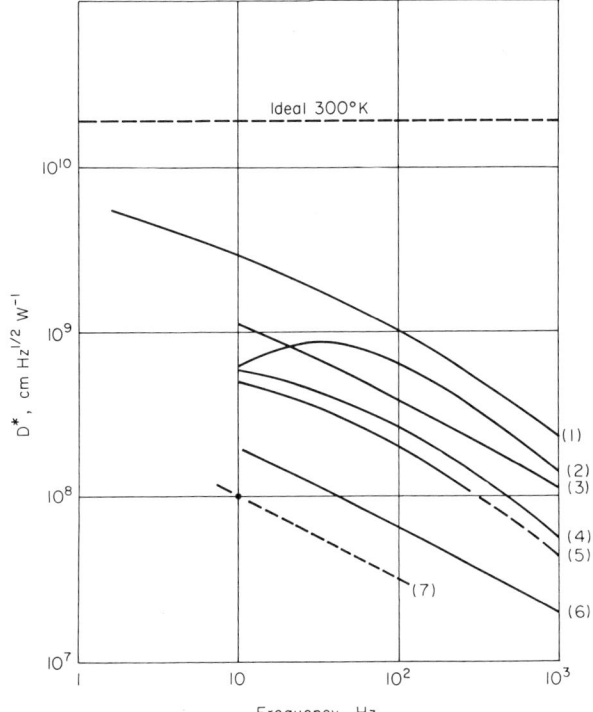

FIG. 2. Performance of some pyroelectric detectors: (1) Mullard research sample using alanine doped TGS 10 μm thick and area 1.5 × 1.5 mm². This detector was mounted in a space-qualified encapsulation and its performance was independently verified. (2) Research sample LiTao₃ detector produced by Martin Marietta. (Stokowski et al.[3]) The thickness of the sensitive element was reduced to less than 10 μm by ion-beam milling. The sensitive area was 1.0 × 1.0 mm². The fall off in performance at low frequencies is thought to be due to airborne vibrations, which better encapsulation should eliminate. (3) Mullard production TGS detector in ruggedized encapsulation. Detector area 0.5 × 0.5 mm². (Baker et al.[3a]) (4) Barnes LiTaO₃ detector. (Beerman.[3b]) Element thickness 13 μm and area 1.0 × 1.0 mm². (5) Honeywell rare earth doped SBN detector. (Liu and Maciolek[3c] and Honeywell data sheet.) (6) Plessey modified PZT ceramic detector. Thickness 60 μm, area 2 × 2 mm². Performance measured at RRE (now RSRE), using RRE low noise amplifier. (7) A. D. Little PVF₂ polymer film detector of 1-cm diameter. Plotted point taken from ADL data sheet. Broken line shows frequency dependence expected from measurements on similar polymer film detectors.

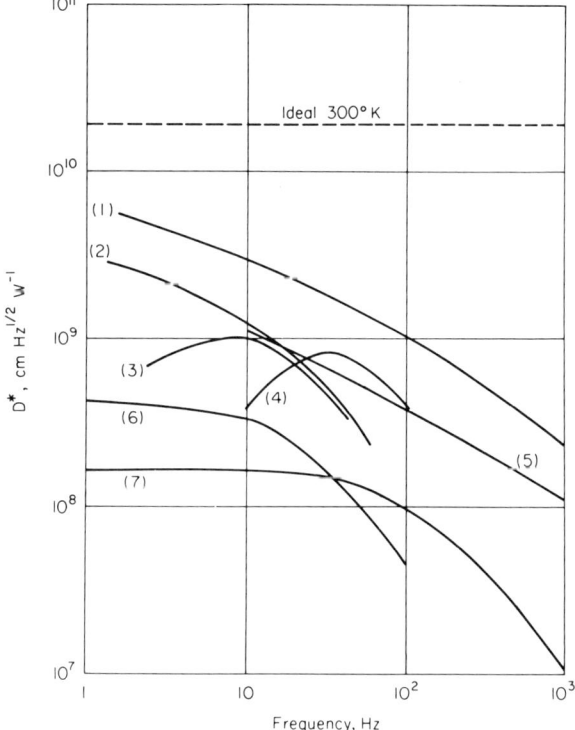

FIG. 3. Performance of some uncooled thermal detectors: (1) Mullard research sample using alanine doped TGS 10 μm thick and area 1.5 × 1.5 mm². Mounted in space qualified encapsulation and performance independently verified. (2) Spectroscopic thermopile. (Based on Fellgett,[3d] Brown et al.,[3e] and Schwarz.[3f]) Typical receiver area 0.4 mm², time constant 40 msec. (3) Golay cell. (After Hickey and Daniels,[3g] Stafsudd and Stevens[3h] and Gill.[3i]) (4) TRIAS cell. Space qualified encapsulation. (Chatenier and Gauffre.[3j]) (5) Mullard production TGS detector in ruggedized encapsulation. Area 0.5 × 0.5 mm². (Baker et al.[3a]) (6) Evaporated film thermopile. (Stevens.[3k]) Receiver 0.12 × 0.12 mm², time constant 13 msec. (7) Immersed thermistor. (De Waard and Weiner.[3l]) Area of flake 0.1 × 0.1 mm², time constant 2 msec.

the toxic hazards of working with beryllium raise objections to the use of this material. For some applications triglycine selenate (TGSe) seems an attractive material, but thermostatting would be required for operation under normal ambient conditions. Recently workers at Mullard[5] have found that alanine doped mixed TGS–Se crystals have a superior performance to the LATGS material at room temperature. The Curie temperature of the mixed crystals is about 40°C, but the transition becomes very broad when enough alanine is present so that the maximum operating temperature is about 60°C. Typical properties of LATGS are pyroelectric coefficient

[5] K. L. Bye, P. W. Whipps, E. T. Keve, and M. R. Josey, *Ferroelectrics* 7, 179–181 (1974).

$p \sim 3 \times 10^{-8}$ C cm^2 deg^{-1} volume specific heat c 2.5 J cm^{-3} deg^{-1}, dielectric coefficient 25–30, loss tangent tan δ 0.005–0.01.

Apart from the TGS group of materials some useful refractory pyroelectrics are now available. These materials are useful where greater robustness at high power levels, operation at higher temperatures, or very high long term stability are required, or when a cheap reliable detector of somewhat lower performance is adequate. Strontium barium niobate,[6] lithium tantalate,[7,7a] lead germanate,[8] various ceramic derivatives of PZT[9] (lead zirconate titanate) including PLZT[10] (lathanum doped PZT), lead titanate,[11] and some other materials[12] are being used for these detectors.

A very interesting recent development was the discovery that thin films of certain polymers, chiefly polyvinyl flouride (PVF)[13] and polyvinylidene fluoride (PVF$_2$),[7a,14] can after suitable heat treatment show piezoelectric and pyroelectric properties. Their pyroelectric properties are sufficiently good to make them very attractive where a simple cheap detector is required. They are particularly useful when a very large area is needed (as for a simple far infrared or submillimeter laser detector) and have also been used in absolute radiometers[15] and other instruments.[7a]

Further studies have been made of pyroelectric heterodyne techniques.[16] The best results obtained so far, at 10.6 μm of an NEP of 2×10^{-15} W Hz^{-1} with a 20 KHz i.f. and 2×10^{-14} W Hz^{-1} with 1.5 MHz i.f. are presently the most sensitive results obtained with a room temperature receiver. The

---

[6] A. M. Glass and R. L. Abrams, *Polytech. Inst. Brooklyn Symp. Proc.* **20**, 281–294 (1971); W. M. Doyle, *Laser Focus*, pp. 34–37 (July 1970); S. Lavi and M. Simhony, *J. Appl. Phys.* **44**, 5187–5189 (1973).

[7] A. M. Glass and R. Abrams, *J. Appl. Phys.* **41**, 4455–4459 (1970). C. B. Roundy and R. L. Byer, *J. Appl. Phys.* **44**, 929–931 (1973).

[7a] J. Geist and W. R. Blevin, *Appl. Opt.* **12**, 2532–2535 (1973); R. L. Peterson, G. W. Day, P. M. Gruzensky, and R. J. Phelan, Jr., *J. Appl. Phys.* **45**, 3296–3303 (1974); W. R. Blevin and J. Geist, *Appl. Opt.* **13**, 1171–1178 (1974); **13**, 2212–2217 (1974).

[8] G. R. Jones, N. Shaw, and A. W. Vere, *Electron Lett.* **8**, 345–346 (1972).

[9] R. J. Mahler, R. J. Phelan, Jr., and A. R. Cook, *Infrared Phys.* **12**, 57–59 (1972).

[10] S. T. Liu, J. D. Heaps, and O. N. Tufte, *Ferroelectrics* **3**, 281–285 (1972).

[11] E. Yamaka, T. Hayashi, and M. Matsumoto, *Infrared Phys.* **11**, 247–248 (1971).

[12] H. Iwasaki, *J. Radio Res. Lab. (Japan)* **17**, 147–151 (1970); F. G. Ullman, B. N. Ganguly, and J. R. Zeidler, *J. Electron. Metals* **1**, 425–534 (1972); O. M. Stafsudd and M. Y. Pines, *J. Opt. Soc. Amer.* **62**, 1153–1155 (1972).

[13] R. J. Phelan, Jr., R. J. Mahler, and A. R. Cook, *Appl. Phys. Lett.* **19**, 337–338 (1971).

[14] A. M. Glass, J. H. McFee, and J. G. Bergman, Jr., *J. Appl. Phys.* **42**, 5219–5222 (1971); J. H. McFee, J. G. Bergman, Jr., and G. R. Crane, *Ferroelectrics* **3**, 305–313 (1972).

[15] R. J. Phelan, Jr., and A. R. Cook, *Appl. Opt.* **12**, 2494–2500 (1973).

[16] S. T. Eng and R. A. Gudmundsen, *Appl. Opt.* **9**, 161–166 (1970); S. J. Lee and A. van der Ziel, *Physica* **67**, 119–124 (1973); A. C. Baynham, C. T. Elliott, N. Shaw, and D. J. Wilson, *Proc. Electro-Opt. Syst. Design Conf. Brighton, 1974*; *Int. Conf. Submillimeter Waves Appl., June 5–7, 1974. Conf. Dig.* (IEEE Cat. No. 74 CHO 856–5MTT), pp. 153–155.

performance is probably limited partly by the heat sinking of the detector element, which limits the amount of local oscillator power which can be used, but mainly by instabilities in the local oscillator laser itself.

The performance of a pyroelectric detector depends critically on that of the amplifier used in conjunction with it, as discussed in Section 3(a) of Chapter 6, Volume 5. Superior amplifiers to those described in Fig. 5 of that chapter are now available. These are described in the present volume in Chapter 4 on the InSb submillimeter detectors where Fig. 12 is particularly relevant.

With the best pyroelectric detectors the dominant source of noise is Johnson noise associated with the dielectric loss of the material. Even so, the performance at low frequencies is within an order of the theoretical limit (see Fig. 2). To obtain further improvement the dielectric loss of the pyroelectric must be reduced to reduce the Johnson noise. If this were possible, however, temperature fluctuation noise would become dominant. Thermal fluctuations occur not only via radiative exchanges with the detector's surroundings, but in detectors as normally constructed via conduction in to surrounding air or gas filling the encapsulation and via thermal conduction laterally into the material surrounds of the pyroelectric element. Logan[17] has calculated the magnitude of both these sources of noise. It turns out that if Johnson and amplifier noise sources could be completely eliminated the performance of a pyroelectric detector at low frequencies (~ 10 Hz) would still be a factor of 2 or 3 worse than that of an ideal detector and as the frequency increased the degree of degradation would increase as $f^{1/4}$.

If the Johnson noise could be reduced to even lower values than at present, then further improvements in the amplifier would be required to obtain full advantage of the Johnson noise reduction. Not only would it be necessary to reduce the amplifier noise sources, but also its leakage current. The leakage current sets the maximum value of the effective input resistance across the amplifier (see Fig. 9, p. 276, Volume 5) for stable operation. If this is smaller than the effective resistance of the element, then the Johnson noise in this component will be dominant. If it were practicable to carry out these improvements, then the gaseous thermal conduction noise could be eliminated by evacuating the encapsulation. It would, in principle, at least, be possible to reduce the lateral thermal conduction noise by using a more complex constructional technique such as forming the active element into a mesa configuration or by using as the active element a small flake supported on threads. These techniques would produce a delicate and expensive detector and may not be worthwhile. It would be more valuable to seek ways of reducing the dielectric loss at higher frequencies (1 kHz or above) since

[17] R. M. Logan and K. Moore, *Infrared Phys.* **13**, 37–47 (1973); R. M. Logan, *Infrared Phys.* **13**, 91–98 (1973).

Johnson noise is dominant up to 10–100 kHz, but at these frequencies the performance is two or three orders of magnitude worse than that of an ideal thermal detector.

There have been a number of other theoretical studies on the operation of pyroelectric detectors.[18] These have been mainly concerned with study of the transient response and their results have given detailed support to the broad picture presented in Volume 5, Chapter 6.

Although strictly outside the terms of reference of the article, a related application of the pyroelectric effect is the pyroelectric vidicon in which a sheet of pyroelectric material (usually TGS) is used as the target in an image tube. The charge distribution produced by the thermal image is read out by a scanning electron beam in a similar way to a photoconductive vidicon. This is the first successful camera tube working in the 8–13 μm band. The performance attainable from it is still somewhat inferior to that of the best thermal imaging systems using cooled photon detectors, but the pyroelectric camera tube has a number of advantages. It operates at room temperature, is compatible with standard television techniques, and is comparatively cheap. In the latest pyroelectric vidicons, the target is manufactured from deuterated TGS or deuterated TGFB and techniques for reticulating the target at about 50 μm pitch are being developed to reduce the sideways thermal diffusion of the picture and hence improve the spatial resolution. These tubes are capable of resolving a thermal pattern with a minimum resolvable temperature not greater than 0.5°C and at a spatial frequency not less than 300 TV lines per picture height and are thus capable of producing thermal infrared pictures comparable in quality to visible TV.[19]

[18] A. Shaulov and M. Simhony, *Appl. Phys. Lett.* **20**, 6–7 (1972); M. Simhony and A. Shaulov, *Appl. Phys. Lett.* **21**, 375–377 (1972); M. Simhony, A. Shaulov, and S. Lavi, *Appl. Phys. Lett.* **22**, 99–100 (1973); M. Simhony, A. Shaulov, and A. Maman, *J. Appl. Phys.* **44**, 2464–2469 (1973); A. van der Ziel and S. T. Liu, *Physica* **61**, 589–593 (1972); A. van der Ziel and S. T. Liu, *J. Appl. Phys.* **43**, 4260–4261 (1972); A. van der Ziel, *J. Appl. Phys.* **44**, 546–549 (1973); B. R. Holeman, *Infrared Phys.* **12**, 125–135 (1972).

[19] S. D. Pel'ta, L. M. Dun, V. K. Novik, and I. S. Rez, *Radio Eng. Electron. Phys.* **13**, 157–158 (1968); F. Le Carvennec, *Advan. Electron. Electron Phys.* **28A**, 265–272 (1969); M. F. Tompsett, *IEEE Trans. Electron. Devices* **ED-18**, 1070–1074 (1971); E. H. Putley, R. Watton, and J. H. Ludlow, *Ferroelectrics* **3**, 263–268 (1972); D. R. Charles and F. Le Carvennec, *Advan. Electron. Electron Phys.* **33A**, 279–284 (1972); E. H. Putley, R. Watton, W. M. Wreathall, and S. D. Savage, *Advan. Electron. Electron Phys.* **33A**, 285–292 (1972); R. G. F. Taylor and H. A. H. Boot, *Contemp. Phys.* **14**, 55–87 (1973); R. M. Logan and T. P. McLean, *Infrared Phys.* **13**, 15–24 (1973); R. Watton, C. Smith, B. Harper, and W. M. Wreathall, *Electron Lett.* **9**, 534–535 (1973); R. M. Logan and R. Watton, *Infrared Phys.* **12**, 17–28 (1972); R. Watton, C. Smith, B. Harper, and W. M. Wreathall, *IEEE Trans. Electron. Devices* **ED-21**, 462–469 (1974); B. Singer, M. H. Crowell, and T. Conklin, *Proc. Electron. Devices Meeting, Washington, 1973*, pp. 293–295; L. E. Garn and F. C. Petito, *Proc. Electron Devices Meeting, Washington 1973*, pp. 296–297; H. A. H. Boot and J. G. Castledine, *Electron. Lett.* **10**, 452 (1974); L. E. Garn and E. J. Sharp, *IEEE Trans.*

What is now becoming clear is that for those applications for which uncooled thermal detectors are generally used, pyroelectric detectors are gradually replacing the earlier types, because they offer an attractive combination of performance, reliability, and cheapness. The use of pyroelectric detectors in burglar alarms and similar security systems is growing rapidly and represents one of the largest applications for the rugged ceramic detector. One unsolved problem which has arisen with the more widespread use of pyroelectric devices is the discovery of a new noise source, giving rise to random pulses from the detector some time after it has undergone a small temperature change. This thermally induced transient (TIT) acts as a false alarm in applications such as satellite horizon sensors or burglar alarms. Fundamental research is required to establish its precise mechanism but empirical studies have enabled the effect to be reduced to a tolerably low enough level for most applications. Examples of recent applications include radiometers for use in the Nimbus satellites[20] and for absolute measurements[7] and rugged arrays for monitoring high power laser beam profiles.[21] The use of pyroelectric devices for certain specialized microwave applications such as the analysis of very short pulses is also being developed.[22]

Parts Hybrids Packag. **PHP-10**, 208–221 (1974); B. Singer, M. H. Crowell, and T. Conklin, *IEEE Trans. Electron Devices* **ED-21**, 744 (1974); R. Watton, G. R. Jones, and C. Smith, *Electron. Lett.* **10**, 469 (1974); R. M. Logan, *Infrared Phys.* **15**, 51–64 (1975); W. M. Wreathall and A. L. Harmer, *IEE Conf. Publ. London* No. 124, 62–63 (1975); C. N. Helmick, Jr., and W. H. Woodworth, *Ferroelectrics* **11**, 309–313 (1976); S. G. Porter, D. Appleby, and F. W. Ainger, *Ferroelectrics* **11**, 351–354 (1976); B. Singer and J. Lalak, *Ferroelectrics* **10**, 103–107 (1976); P. W. Steinhage and R. R. Zeyfang, *Ferroelectrics* **11**, 301–304 (1976); R. Watton, *Ferroelectrics* **10**, 91–98 (1976); E. Yamaka, A. Teranishi, K. Nakamura, and T. Nagashima, *Ferroelectrics* **11**, 305–308 (1976); T. Conklin, B. Singer, M. H. Crowell, and R. Kurczewski, *Int. Electron Devices Meeting, Washington, D.C., 1974*, pp. 451–454; T. Conklin and B. Singer, *Int. Electron Devices Meeting, Washington, D.C., 1975*, pp. 66–69; H. A. H. Boot, J. G. Castledine, P. G. R. King, K. E. Trezise, and B. Turner, *J. Phys. D* **9**, 679–695 (1976); J. Mangin, G. Morlot, P. Strimer, R. Thomas, J. Weber, A. Hadni, P. Félix, S. Véron, and G. Moirou, *Nouv. Rev. Opt.* **5**, 305–311 (1974); C. R. W. Richardson, *IEE Conf. Publ. London* No. 124, 194–200 (1975); R. M. Logan, *Infrared Phys.* **16**, 75–79 (1976); B. Singer, *Electro-Opt. Syst. Des.* 30–34 (1975); J. P. Klozenberg, *Infrared Phys.* **16**, 487–488 (1976); B. Turner and H. A. H. Boot, *Infrared Phys.* **16**, 367–374 (1976); A. A. Turnbull, *Proc. Electro Opt. Laser Int.* 76/UK, 9–11 March 1976, Brighton, England, in press.

[20] P. Ellis, et al., *Proc. Roy. Soc. London* **A334**, 149–170 (1973).

[21] B. C. McIntosh, D. W. Sypek, *Laser Focus*, pp. 38–40 (December 1972).

[22] R. A. Lawton, *IEEE Trans. Instrum. Measurements* **IM-22**, 299–306 (1973).

7. THE PYROELECTRIC DETECTOR—AN UPDATE

The use of pyroelectric arrays in thermal imaging systems is being actively pursued[23] and looks to offering a useful alternative to the use of cooled photon detectors, possibly in conjunction with a CCD read-out structure.[24] Some recent references to pyroelectric detectors[25] and their applications[26] are given below.

---

[23] P. A. Schlosser and D. D. Glower, *Ferroelectrics* **3**, 257–262 (1972); H. Blackburn, H. C. Wright, R. Eddington, and R. S. King, *Radio Electron. Eng.* **42**, 369–372 (1972).

[24] A. J. Steckl, R. D. Nelson, B. T. French, R. A. Gudmundsen, and D. Schechter, *Proc. IEEE* **63**, 67–74 (1975).

[25] W. R. Blevin and J. Geist, *Appl. Opt.* **13**, 1171–1178 (1974); C. Chi-Shu, *Acta Phys. Sin.* **23**, 429–436 (1974); G. W. Day, C. A. Hamilton, R. L. Peterson, R. J. Phelan, Jr., and L. O. Mullen, *Appl. Phys. Lett.* **24**, 456–458 (1974); F. C. Gabriel, *Appl. Opt.* **3**, 1294–1295 (1974); A. Hadni, *IEEE Trans. Microwave Theory Tech.* **MTT-22**, 1016–1018 (1974); P. A. Jansson, *Appl. Opt.* **13**, 1293–1294 (1974); S. T. Liu and J. D. Zook, *Ferroelectrics* **7**, 171–173 (1974); C. B. Roundy, R. L. Byer, D. W. Phillion, and D. J. Kuizenga, *Opt. Commun.* **10**, 374–377 (1974); A. van der Ziel, *J. Appl. Phys.* **45**, 4128 (1974); E. L. Dereniak and F. G. Brown, *Infrared Phys.* **15**, 39–43 (1975); M. W. Geis, K. A. Smith, and R. D. Rundel, *J. Phys.* **E8**, 1011–1014 (1975); A. S. Martynyuk, E. P. Nikolaev, and V. K. Novik, *Sov. J. Opt. Technol.* **42**, 35–37 (1975); G. W. Day, C. A. Hamilton, P. M. Gruzensky, and R. J. Phelan, Jr., *Ferroelectrics* **10**, 99–102 (1976); S. T. Liu, *Ferroelectrics* **10**, 83–89 (1976); K. L. Bye, P. W. Whipps, E. T. Keve, and M. R. Josey, *Ferroelectrics* **11**, 525–534 (1976); M. Petersen, *Infrared Phys.* **16**, 465–473 (1976).

[26] G. S. Bowen, *ISA Trans.* **13**, 101–119 (1974); C. A. Hamilton, R. J. Phelan, Jr., and G. W. Day, *Opt. Spectra*, **9**, 37–38 (1975); R. C. Milward, *IEEE Trans. Microwave Theory Tech.* **MTT-22**, 1018–1023 (1974). See also *Proc. Soc. Photo-Optical Instrumen. Eng.* **62**, 151–213 (1975).

Author Index

Numbers in parentheses are footnote numbers and are inserted to enable the reader to locate those cross references where the author's name does not appear at the point of reference in the text.

A

Abaulina-Zavaritzkaya, E. I., 63
Abrahams, E., 87
Abrams, R. L., 34, 445, 448(7)
Adams, A. R., 290
Adams, E. N., 245
Adams, H. D., 104
Adde, R., 410, 411(50), 436
Afinogenov, V. M., 167
Afonchenkov, N. G., 166
Aggarwal, R. L., 50, 130, 179
Aigrain, P., 80, 231
Ainger, F. W., 447(19), 448
Akasaki, I., 204
Allan, C., 163
Anderson, C. L., 329, 330(43), 344
Anderson, L. K., 292, 293, 297, 298, 299(1), 311, 357, 361, 366(111), 374
Anderson, P. W., 401
Annis, A. D., 441
Antell, G. R., 189, 191(50)
Appleby, D., 447(19), 448
Arams, F. R., 138, 163
Asai, S., 204
Ascarelli, G., 60, 63, 119(43, 44), 211, 213
Ashley, J. R., 428
Aslamazov, L. G., 401, 404(19), 418(19)
Aspnes, D. E., 290, 381
Assour, J., 339, 340(55)
Auracher, F., 396, 402, 403(3), 405, 406, 411 (26), 418(37), 422

B

Baertsch, R. D., 339, 349, 373(82), 376
Bak, C. K., 403, 431, 436(33)
Baker, G., 442, 443, 444
Balkanski, M., 80

Ballantyne, J. M., 167, 287
Balliager, R. A., 329
Balslev, I., 189, 191(45)
Baltensperger, W., 230, 231, 235
Baraff, G. A., 327, 328, 330, 342
Bardeen, J., 231, 382
Baron, R., 110, 113
Bartelink, D. J., 326
Bastin, J. A., 164
Batdorf, R. L., 303, 306(7), 331(7), 335(7), 338 (7), 339(7), 340(7), 355(7), 366(7), 368(7), 369(7), 370(7)
Bate, R. T., 207
Bates, R. L., 11, 13(13)
Baukus, J. P., 167, 287
Baynham, A. C., 445
Beasley, M. R., 402
Bebb, H. B., 47, 48, 195
Bechara, M. R. E., 189
Beckman, J. E., 166, 167
Beer, A. C., 207
Beerman, H. P., 441–443
Belanger, L. J., 288
Beleznay, F., 213, 217(120), 227
Bell, E. E., 41, 104(3), 105(3), 130, 175, 231(9)
Bell, R. J., 168, 224
Belyaev, Yu. I., 86
Belyantsev, A. M., 154, 167
Bergeson, H. E., 359
Bergman, J. G., Jr., 445
Berman, L. V., 167
Bernard, W., 99
Besson, J., 150
Bethe, H. A., 210
Bevacqua, S. F., 377
Beyen, W. J., 101, 104
Biard, J. R., 351, 360
Biskupski, G., 167
Blackburn, H., 449
Blakemore, J. S., 50, 59, 60, 99, 100, 110, 196

451

Blaney, T. G., 424, 427(64), 428, 429(72)
Blatt, F. J., 382
Blevin, W. R., 445, 449
Blossey, D. F., 381
Blouke, M. M., 127–129
Bode, D. E., 104
Bolger, D. E., 197, 198, 229, 235
Bonch-Bruevich, V. L., 233
Boone, D. W., 368, 372(133), 375(133)
Boot, H. A. H., 447, 448
Borrello, S. R., 41, 104(8)
Bosomworth, D. R., 208, 209, 213, 227(106), 236, 238, 271(106)
Bottka, N., 381
Bowen, G. S., 449
Boyle, W. S., 130, 245, 246
Bradley, C. C., 166, 182(37), 184
Braggins, T., 99, 100(107)
Brandt, R. C., 190, 191(57), 237(57), 238(57), 251(57), 252(57), 253(57), 254(57)
Braslau, N., 189, 191
Bratt, P. R., 41, 64, 79, 101, 103, 104, 114, 120, 135, 263
Bray, R., 66
Breene, R. G., 257
Bridges, T. J., 138
Brooks, H., 204
Brown, D. A. H., 442, 444
Brown, F. G., 449
Brown, J. M., 99, 100
Brown, M. A. C. S., 146, 162
Brown, R. A., 60, 119(45), 213
Brown, R. D., III, 63, 64(49), 65(49), 99(49), 211, 264
Brown, S. C., 211
Bube, R. H., 53
Buckner, S. A., 400
Buczek, C. J., 116, 117, 119(141), 138, 139
Bulucea, C. D., 330
Burger, R. M., 86
Burstein, E., 41, 42, 50, 52, 53, 57, 58, 60, 63, 64, 66, 74, 98, 103–105, 130, 175, 189, 191(45), 202, 209, 231(9), 264(107)
Bye, K. L., 441, 444, 449
Byer, N. E., 442
Byer, R. L., 445, 448(7), 449

C

Cairns, B., 193
Callaway, J., 381, 383(146)
Camphausen, D. L., 329
Cano, R., 150
Carballes, J. C., 201
Cardona, M., 181, 184(30)
Caringella, P. C., 9
Carlson, R. O., 109
Casey, H. C., Jr., 351, 383, 389, 391
Castellan, G. W., 232
Castledine, J. G., 447, 448
Chamberlain, J. M., 182–184, 186
Champlin, K. S., 189, 190(53), 191
Chang, J. J., 308–310
Chang, R. K., 190
Chang, T. Y., 138
Chang, Y. J., 342, 343, 346, 347
Chapman, R. A., 47, 48, 50, 83, 130, 133(173)
Charles, D. R., 447
Charlton, D. E., 442, 443(3a), 444(3a)
Chasmar, R. P., 6, 268, 442, 444
Chatanier, M., 442, 444
Cheban, A. G., 233
Cheo, P. K., 138
Chiao, R. Y., 409, 418, 419(59), 432, 433(44, 81)
Chi-Shu, C., 449
Cholet, P., 99
Christian, S. M., 113
Chusov, I. I., 167
Chynoweth, A. G., 325, 331(30), 332(30), 340(30), 342, 346(62), 349
Claassen, J. H., 407, 409, 410(40), 412(49), 413(49), 414(49), 421(40), 423(40), 424, 425, 426(40), 428(49), 437(40), 439(40)
Clark, T. D., 408, 420(41), 438(41)
Clarke, J., 401, 411, 415, 416, 417(55)
Clegg, P. E., 166
Cohen, G. G., 342, 346(62)
Cohen, M. E., 65
Colbow, K., 233
Cole, R., 79
Coleman, D. J., Jr., 359
Collard, J. R., 389(154), 391

AUTHOR INDEX

Collins, A. T., 220
Collins, C. B., 109
Combet, H. A., 430
Conklin, T., 447, 448
Connes, J., 224
Conradi, J., 306, 313(9), 318(9), 321(9), 322, 323(9), 324(9), 340, 367–369, 371(9), 372(9), 375(9)
Conwell, E. M., 87, 118, 231, 343
Cook, A. R., 445
Coon, D. D., 409
Coriell, A. S., 179
Craford, M. G., 179
Crandall, R. S., 203, 208, 209(106), 213(106), 227(106), 236(106), 238(106), 271(106)
Crane, G. R., 445
Cronin, G., 83
Crowell, C. R., 328–330, 344, 389
Crowell, M. H., 447, 448
Cupp, J. D., 427, 428

D

Daehler, M., 168
Dahm, A. J., 410, 411(47), 428(47), 433
Dalven, R., 42(16), 43
Daniels, D. B., 442, 444
Danielson, B. L., 427
Darviot, Y., 79, 83, 84
D'Asaro, L. A., 297, 357, 361(111), 366(111), 374(5)
Dash, W. C., 351, 352(93)
Davies, R. L., 339, 340(54), 359
Davis, E. A., 233
Davis, H., 101
Davis, W. D., 99, 100
Davisson, J. W., 41, 63, 104(3), 105(3), 130, 175, 231(9)
Day, G. W., 444, 445(7a), 449
Dayem, A. H., 401
Debye, P. P., 104, 231
Decker, D. R., 340, 341, 375(57)
DeMan, H., 332, 338–340, 363, 368, 370
DeMeis, W. M., 181–183, 185, 187, 188
Denenstein, A., 410, 411(47), 428(47), 433

Denisova, A. D., 99, 100(109)
Dereniak, E. L., 449
De Waard, R., 442, 444
Di Domenico, M., Jr., 292, 293(1), 298(1), 299(1), 311(1)
Diguet, D., 201
DiLorenzo, J. V., 193
Dimmock, J. O., 35, 182(38), 184, 196, 197, 198(66), 199(66), 200(66), 201(70), 203, 205, 206(100), 208(84), 212, 222, 227(66), 229(66), 233(66, 132), 235(66), 241(38), 242(38, 66), 243(66), 244(66), 246(38), 247(66), 272(133), 273(133), 276(133), 279(113, 132), 280(132), 284(84), 288(84)
Dingle, R., 234, 236(160), 237(160), 238(160), 239(160), 248(160), 254(160)
Dmitrenko, I. M., 397
Donnelly, J. P., 376(143), 377, 380, 382(143)
Donovan, R. P., 86
Doyle, W. M., 445
Dubois, H., 167
Dumke, W. P., 329, 330
Dun, L. M., 447
Duncan, W., 180
Dunkleberger, L. N., 402
Dunlap, W. C., Jr., 42(13), 43, 97
Dunn, C. N., 340, 341, 375(57)
Dunn, D., 233
Durban, J., 233

E

Eastman, P. C., 118
Eck, R. E., 397
Eddington, R., 449
Eddolls, D. V., 199, 201, 205(74), 229, 235
Eden, R. C., 376, 380
Effer, D., 191
Efremova, G. D., 167
Ehrenreich, H., 178, 180, 182, 189
Eisenman, W. L., 9, 11, 13(13), 23, 32
Elad, E., 164
Eldumiati, I. I., 166
Elliott, C. T., 445
Elliott, R. J., 245, 246

AUTHOR INDEX

Ellis, P., 448
Elsley, R. K., 437
Emelyanenko, O. V., 181, 182
Emmons, R. B., 308–311, 373, 376
Eng, S. T., 445
Engeler, W., 41, 103(114), 104
Ensign, T. C., 442
Enstrom, R. E., 208, 209(106), 213(106), 227 (106), 236(106), 238(106), 271(106)
Erginsoy, C., 230, 231
Erlandson, R. J., 189, 191(47)
Erné, S. N., 437
Estle, T. L., 130, 133(173)
Evans, P. R., 191
Evenson, K. M., 427, 428

F

Fairman, R., 193
Fan, H. Y., 41, 50, 104, 130, 211, 351, 352(96)
Farrayre, A., 342, 346(69), 389(69)
Fassett, J. R., 67, 69(66)
Faulkner, R. A., 130, 175
Fawcett, W., 290
Feldman, M. J., 418, 419(59), 432, 433(81)
Feldman, P. D., 278
Félix, P., 447(19), 448
Fellgett, P. B., 442, 444
Fenton, E. W., 233
Ferre, D., 167
Fetterman, H., 182, 184, 248, 249, 252(178), 254(36), 284–286, 288
Field, B. F., 398
Fife, A. A., 437
Filaretova, G. M., 393
Finnegan, T. F., 398, 400, 433
Fisher, M. B., 292, 293(1), 298(1), 299(1), 311(1)
Fisher, P., 50, 130–132, 218, 226(129)
Fiske, M. D., 409
Fistul', V. I., 181, 182(29)
Forbes, L., 50, 135(29), 220
Foss, N. A., 360
Foyt, A. G., 343(77), 344, 346(73), 347(77), 348(77), 350(77), 373(76), 376(76, 77), 377, 378, 379(77), 380(76), 386(76), 387(73), 388 (73), 389(73)
Franks, J., 197(71), 198, 229(71), 235(71)
Franz, W., 344

Fray, S. J., 105
French, B. T., 449
Fritzche, H., 87
Frosch, C. J., 351, 352(95)
Fuller, C. S., 86
Fulton, T. A., 402

G

Gabriel, F. C., 449
Gärtner, W., 351, 353(88)
Galagher, C. J., 109
Galipern, Yu. S., 167
Gammarino, R. R., 360, 364(120), 375(120)
Gamo, T., 167
Ganguly, B. N., 445
Garbuny, M., 99, 257
Gardrud, W. B., 34
Garn, L. E., 447
Gauffre, G., 442, 444
Geballe, T. H., 42(14), 43
Gebbie, H. A., 163
Geis, M. W., 449
Geist, J., 445, 449
Genkin, V. N., 154, 167
Gentry, F. E., 359
Gershenzon, M., 351, 352(95)
Gershenzon, Ye. M., 167
Giaever, I., 401
Gibbs, D. F., 351, 352(95)
Gill, J. C., 442, 444
Glasko, V. B., 233
Glass, A. M., 445, 448(7)
Glinchuk, K. D., 99, 100
Glodeanu, A., 45
Glover, G. H., 189, 190(53), 191, 342, 344, 346
Glower, D. D., 449
Gobat, A. R., 389(154), 391
Godik, E. E., 111
Goetzberger, A., 297, 357, 361(111), 366(111), 374(5)
Gol'tsman, G. N., 167
Goodwin, D., 5, 16(6)
Gordon, J., 197(71), 198, 229(71), 235(71)
Gornik, E., 167
Graham, H. A., 104
Grant, W. N., 337, 338, 340(50a)
Greene, P. E., 191
Grimes, C. C., 417, 428

Grube, R. H., 268, 270(191)
Gruzensky, P. M., 445, 449
Gubankov, V. N., 153, 166, 167, 403, 405(30), 418(30), 435, 437(30)
Gudmundsen, R. A., 445, 449
Gulyaev, Yu. V., 167
Gunther-Mohr, G. R., 63, 64, 202
Gurvich, Yu. A., 167
Gygax, S. J., 437

H

Haddad, G. I., 166
Hadni, A., 442, 447(19), 448, 449
Haering, R. R., 233
Haitz, R. H., 348
Halbo, L., 204
Hall, L. H., 382
Hall, R., 342, 343, 346, 347, 387–389
Hamann, D. R., 212
Hambleton, K. G., 186–191
Hamilton, C. A., 400, 403, 449
Handler, P., 381
Hange, P. S., 189, 191(47)
Hanley, P. E., 233
Hansen, J. R., 99
Hara, T., 204
Harman, T. C., 27
Harmer, A. L., 447, 448
Harp, E. E., 127, 128(154), 129(154)
Harper, B., 447
Harris, R. E., 405
Harty, W. E., 41, 114(7), 271
Harwit, M. O., 150, 164(14)
Hasegawa, H., 243
Hauser, J. R., 329, 330
Hayashi, T., 445
Heaps, J. D., 445
Hearn, C. J., 329
Helmick, C. N., Jr., 447(19), 448
Henkels, W. H., 411
Henvis, B., 50, 130
Herman, F., 113
Herman, J. M., III, 50
Hicinbothem, W. A., Jr., 189
Hickey, J. R., 442, 444
Hill, D. E., 234, 236(160), 237(160), 238(160), 239(160), 248(160), 254(160), 351, 352(99)

Hilsum, C., 179, 186, 187(41), 188, 189(41), 190(41), 191(41)
Hirao, M., 204
Hoffer, G. I., 415, 416(55), 417(55)
Hoffman, A. R., 416
Holcomb, T., 288
Holeman, B. R., 186, 187(41), 188, 189(41), 190(41), 191(41), 447
Holloway, H., 35, 36(24)
Holly, S., 397
Holonyak, N., Jr., 179, 377
Holter, M., 3, 5(4), 7(4)
Holtsmark, J., 257
Houck, J. R., 150, 164(14)
Houghton, J., 382
Hoversten, E. V., 322
Howard, N. R., 303
Howard, R. E., 243, 245, 246
Hrostowski, H. J., 130
Hudson, R. D., Jr., 141
Huizinga, J. S., 166
Hutchinson, W. G., 50, 83, 130, 133(173)
Huth, G. C., 359, 364
Hutson, A. R., 179

I

Igo, T., 377, 378
Irvin, J. C., 196, 359
Iskhakov, I. A., 167, 168
Iwasa, S., 189, 191
Iwasaki, H., 445

J

Jackson, J. K., 12
Jacobs, S. F., 41, 57(5), 103(5), 104(5)
Jamieson, J. A., 9, 268, 270(191)
Jansson, P. A., 449
Janus, A. R., 397
Javan, A., 34
Jayaraman, A., 179
Jefferts, K. B., 166, 167
Jeffus, C. R., 127, 128(154), 129(154)
Johnson, C. J., 189, 191
Johnson, E. G., 427
Johnson, E. R., 113
Johnson, K. M., 297

Johnson, L. F., 50, 99–101
Johnson, W. C., 334, 339(49), 345(49)
Joly, B., 79, 83(80), 84(80)
Jones, A. R., 367
Jones, F. E., 6, 268
Jones, G. R., 445, 447(19), 448
Jones, R. C., 2, 5, 7(1–3), 16(6), 72
Jones, R. L., 50, 132
Jones, S., 189, 191
Jordan, A. G., 99, 100
Jordan, A. S., 239
Josephson, B. D., 395, 397, 399, 403(1)
Josey, M. R., 444, 449
Jostad, L. L., 80, 86

K

Kaiser, R. H., 130
Kaiser, W., 41, 50, 104
Kanai, Y., 154
Kane, E. O., 179, 382
Kang, C. S., 191
Kanter, H., 403, 405, 418(28), 419, 420(28), 424, 435
Kaplan, R., 150, 162, 236, 245, 247
Katana, P. K., 233
Kaufman, S. A., 129
Keating, D. E., 190
Keldysh, L. V., 329, 344
Kennedy, D. P., 359
Kennedy, R. S., 322
Kerr, A. R., 424
Keve, E. T., 441, 444, 449
Keyes, R. J., 121, 138
Keyes, R. W., 245
Khaikin, N. Sh., 129
Kikuchi, S., 198, 229(73), 235(73)
Kim, C. W., 349, 373, 376
Kimmitt, M. F., 146, 162, 163, 208, 224(105), 289, 290
Kinch, M. A., 150, 153, 161, 163, 245, 247(171)
King, P. G. R., 447(19), 448
King, R. S., 449
Kingston, F. E., 164
Kingston, R. H., 138
Kittel, C., 174, 431
Klapwijk, T. M., 404

Kleimack, J. J., 303, 306(7), 331(7), 335(7), 338(7), 339(7), 340(7), 355(7), 366(7), 368(7), 369(7), 370(7)
Kleiner, W. H., 130, 245, 246
Klozenberg, J. P., 447(19), 448
Kluth, E. O., 438
Knight, D. J. E., 428, 429(72)
Knight, J. R., 191, 199, 229(75)
Knight, W. D., 431
Kobayashi, S., 167
Koenig, S. H., 63, 64, 65(49), 65(55), 100, 202, 211, 264
Kofoed, B., 431
Kogan, Sh. M., 131, 145, 215, 227(121)
Kohl, F., 167
Kohn, W., 130, 172–174, 245
Kokosa, R. A., 339, 340(54)
Kon, S., 162
Kondo, A., 355, 361, 364(102), 365, 369, 370(124)
Kopec, Z., 181
Korn, D. M., 224
Kornilov, B. V., 50
Kosenko, V. E., 86
Kozlovskaya, V. M., 80
Krag, W. E., 130
Kramer, B., 342, 346(69), 389(69)
Kressel, H., 312, 342, 346, 347, 357
Krivopolenova, M. M., 79
Kroemer, H., 348
Krumpholz, O., 356
Kruse, P. W., 5, 7(5), 8(5), 23, 67, 76, 268
Kuizenga, D. J., 449
Kuno, H. J., 389, 391
Kupsky, G., 342, 346, 347
Kurczewski, R., 447(19), 448
Kurosawa, L., 203
Kuvas, R., 319
Kuzmin, L. S., 403, 405(30), 418(30), 434(30), 437(30)

L

Laff, R. A., 351
Lalak, J., 447(19), 448
Lambert, L. M., 65, 203
Lampert, M. A., 174, 334, 339(49), 345(49)

AUTHOR INDEX

Landsberg, P. T., 65, 199, 213
Langenberg, D. N., 397, 398, 400, 402, 410, 411(47), 428(47), 433, 437
Lao, B. Y., 287
Larkin, A. I., 401, 404(19), 418(19)
Larrabee, R. D., 189
Larsen, D. M., 182(36), 184, 190, 191(57), 237 (57), 238(57), 245, 246, 248, 249(178), 250, 251(57), 252(57), 178), 253(57), 254, 255 (179), 257–261
Lasser, M. E., 99
Lauritzen, P. O., 340
Lavi, S., 445, 447
Lavin, J. M., 167, 287
Lawton, R. A., 448
Lax, B., 179
Lax, M., 46, 60, 119(41, 42), 130, 175, 210, 212, 231(9)
Lazarev, S. D., 167
Lebaily, J., 201
Le Carvennec, F., 447
Leck, J. H., 342, 343, 346, 347, 387–389
Lee, C. A., 303, 306, 319, 331, 335, 338–340, 342, 346, 350, 355(7), 366, 368–370, 380 (85a)
Lee, K., 164
Lee, S. J., 445
Lees, J., 179
Leguerre, J. R., 313
Leiba, E., 442
Leonov, A. M., 154, 167
Leshchenko, G. F., 435
Levinsen, M. T., 402, 427
Levinstein, H., 23, 32(15), 41, 49, 50, 63, 68, 99–101, 103, 104
Lewis, J. A., 359
Lewis, W. B., 70
Li, S. P., 233
Lichtenberg, A. J., 153
Lifshits (Lifshitz), T. M., 105, 107, 130–132, 135, 137(122), 176, 216–218, 221, 249(11)
Lightowlers, E. C., 220
Likharev, K. K., 403, 405(30), 410, 413(48), 418(30), 435, 437(30)
Likhtman (Lichtman), N. P., 132, 135(179), 216, 217, 218(122, 126), 221(122, 126)
Lile, D. L., 35, 37
Limperis, T., 32, 33

Lindley, W. T., 343(77), 344, 347(77), 350(77), 373(76), 376–378, 379(77), 380(76), 386(76)
Lipson, H. G., 41, 104(3), 105(3)
Listvin, V. N., 166–168
Litovchenko, N. M., 99, 100(109)
Litvak, M. M., 287
Liu, S. T., 442, 443, 445, 447, 449
Lock, P. J., 441, 442, 443(3a), 444(a)
Locker, R. J., 359
Logan, R. A., 303, 306(7), 331(7), 335(7), 338 (7), 339(7), 340(7), 341–344, 346, 347, 349, 355(7), 366(7), 368(7), 369(7), 370(7), 388, 389
Logan, R. M., 446–448
Logothetis, E. M., 35, 36
Loh, E., 135, 217, 264(123)
Long, D., 26
Longacre, A., 436, 437(91, 92)
Loudon, R., 245, 246
Love, W. F., 233
Low, F. J., 416
Lowenstein, E. V., 224
Lu, T., 189–191
Lucas, A. D., 355, 361, 362, 374
Lucovsky, G., 47, 308, 373, 376
Ludlow, J. H., 447
Luttinger, J. M., 172, 173, 245
Lyddane, R. H., 190, 191(54)
Lynch, D. W., 290
Lynch, W. T., 312, 313(16), 359, 372, 375

M

McAfee, K. B., 297, 330(3)
McCarthy, D., 342, 346
McColl, M., 429
McCumber, D. E., 401, 402(21)
McDonald, D. G., 427, 428
McFee, J. H., 445
McFee, R. H., 268, 270(191)
McGlauchlin, L. D., 5, 7(5), 8(5), 23(5), 67, 76(68), 268
Machala, A. E., 193
McIntosh, B. C., 448
McIntyre, R. J., 306, 313(9), 314, 315, 317, 318, 320, 321(9), 322, 323(9, 29), 324(9), 356, 363, 367, 369(9), 371(9), 372(9), 375(9)

Maciolek, R. B., 442, 443
McKay, K. G., 297, 330, 335, 338(46), 339(50)
McKinney, R. A., 364
McLean, T. P., 447
McMullin, P. G., 297, 357, 361(111), 366(111), 374(5)
McNurtry, B. J., 292, 293(1), 298(1), 299(1), 311(1)
McNutt, D. P., 168, 278
McQuistan, R. B., 5, 7(5), 8(5), 23(5), 67, 76(68), 268
MacRae, A. U., 41, 68, 101, 103(114), 104
McWhorter, A. L., 63, 64, 203, 212
Madelung, O., 179
Mahler, R. J., 445
Major, K. G., 329
Mallenson, J. R., 329
Maman, A., 447
Mangin, J., 447(19), 448
Manley, J. M., 430
Mao, S., 189, 191
March, N. H., 233
Margolin, N. W., 167
Marple, D. T. F., 186–188
Martin, A. E., 11
Martin, D. H., 222, 224(134)
Martynyuk, A. S., 449
Maruyama, M., 198, 229, 235
Maslowski, S., 356
Mason, H. J., Jr., 110
Matarrese, L. M., 427
Matheu, D. P., 356
Matsumoto, M., 445
Matteoli, M., 150
Mead, C. A., 138
Melchior, H., 292, 312, 313(16), 361(2), 372, 375, 377(2)
Melngailis, I., 27, 203, 208(84), 212, 279, 284(84), 288, 378
Mercereau, J. E., 401, 409
Merriam, J. D., 11
Mertz, L., 224
Messinger, R. A., 50, 110(25)
Meyer, N., 326, 327
Migulin. V. V., 166, 167, 403, 405(30), 418(30), 435, 437(30)
Mikhailova, M. P., 349, 393
Millea, M. F., 429
Miller, A., 87
Miller, B. I., 389, 391

Miller, S. C., 233
Miller, S. M., 312, 331, 338(18), 340, 341
Milton, A. F., 127, 128
Mil'vidskii, M. G., 181, 182(29)
Milward, R. C., 449
Mitchell, A. H., 174
Miyao, M., 182, 243, 245–247
Mizuno, O., 198, 229(73), 235(73)
Mlavsky, A. I., 389
Mönch, W., 325, 329, 330(31)
Moirou, G., 447(19), 448
Moll, J. L., 303, 326, 327, 332(33), 338, 340
Moller, K. D., 224
Mooradian, A., 190, 191
Moore, C. B., 138
Moore, G. E., Jr., 193
Moore, K., 446
Moore, T. G., 80, 86
Moore, W. J., 42, 103, 105–107, 130, 176, 217(10), 268
Morenkov, A. D., 166
Morlot, G., 447(19), 448
Morton, G. A., 41, 114, 271
Moss, T. S., 181–183, 351, 352(100)
Mott, N. F., 87, 232–234
Mourzine, V., 130
Müller, W., 167
Mullen, L. O., 449
Murotani, T., 167

N

Nad', F. Ya., 105, 107, 130, 131, 137(122), 176, 217, 249(11)
Nagasaka, K., 105, 107, 218–220, 249(127)
Nagashima, T., 447(19), 448
Nahory, R. E., 350, 380(85a)
Nakajima, M., 363
Nakamura, K., 447(19), 448
Nakamura, M., 164
Nakano, K., 376, 380
Naqvi, I. M., 318, 320
Narita, S., 105, 107, 182, 218–220, 243, 245–247, 249(127)
Nasledov, D. N., 181, 182(28), 349
Naugle, A. B., 10
Navrotskiy, V. I., 166, 167
Nelson, R. D., 449

AUTHOR INDEX

Nesmeyanov, A. N., 84
Neuberger, M., 196
Neuringer, L. J., 99
Newman, R., 49, 53, 79, 85, 96, 98, 100, 101, 351, 352(93, 94)
Niblett, G. B. F., 146, 162
Nikolaev, E. P., 449
Ning, T. H., 50, 135(29), 220
Nishida, K., 363
Nisida, Y., 167
Norton, P., 63, 97, 99, 100, 120, 137
Notary, H. A., 401
Novak, R., 144
Novik, V. K., 447, 449
Nudelman, S., 3, 5(4), 7(4)

O

Oberly, J. J., 41
O'Brien, R. R., 359
Ogawa, T., 338, 339
Ohta, H., 418, 419, 432, 433(81)
Oka, Y., 105, 218, 249(127)
Okuto, Y., 389
Oliver, B. M., 314
Oliver, D. J., 202
Oliver, J. F. C., 105
Olsen, C. G., 290
Orlova, S. L., 167
Osgood, R. M., 34
Ovrebo, P. J., 141
Owens, E. B., 241

P

Pace, F. P., 138
Paige, E. G. S., 66
Palik, E. D., 181, 182
Palmer, P. W., 409
Papoular, R., 150
Parker, C. D., 203, 208(84), 284, 285(197), 286(197), 288
Parker, J. V., 138
Parker, W. H., 398, 410, 411(47), 428(47)
Parrish, P. T., 409, 418, 419(59), 432, 433(44, 81)
Pataki, G., 213, 217(120), 227
Pauling, L., 79, 80, 84(78)

Pearsall, T. P., 350, 380(85a)
Pearson, G. L., 231
Pedersen, N. F., 431, 433
Pehek, J., 41, 103(114), 104
Pel'ta, S. D., 447
Pennfield, P., Jr., 430, 431(76)
Pernett, J. M., 10
Personick, S. D., 322, 324, 325
Peskett, G. D., 153
Petersen, F. R., 427
Petersen, M., 449
Peterson, R. L., 444, 445(7a), 449
Petito, F. C., 447
Petritz, R. L., 104
Peyton, B. J., 138, 163
Pfann, W. G., 77, 78(76)
Phelan, R. J., Jr., 35, 344, 373(76), 376(76), 380(76), 386(76), 445, 449
Philippeau, B., 150
Phillion, D. W., 449
Phillips, T. G., 166, 167
Pickering, C., 290
Picus, G. S., 41, 42, 50, 52, 53(12), 57(5), 58(12), 66, 74, 98, 103(5), 104(5), 116, 117, 119(141), 120, 130, 135, 139, 209, 217, 264(107, 123)
Piller, H., 182, 183
Pines, M. Y., 110, 113, 445
Pinson, W. E., 66
Pipher, J., 236, 239(162), 271(162)
Pitt, G. D., 179, 290
Plass, G. N., 268, 270(191)
Platt, C. M. R., 153, 161
Poehler, T. O., 182, 184, 186, 204
Pokrovskii, Ya. E., 111
Polder, D., 403
Pollack, M. A., 350, 380(85a)
Poole, R. R., 441
Porter, S. G., 447(19), 448
Potter, R. F., 10, 23, 32
Powell, R. L., 155
Praddaude, H. C., 246
Pratt, W. K., 322
Preier, H., 111
Pribetich, J., 342, 346(69), 389(69)
Prior, A. C., 163
Prisecaru, D. C., 330
Proklav, V. V., 111
Protschka, H. A., 349, 373, 376
Ptitsina, N. G., 167

Pullan, G., 5, 16(6)
Putley, E. H., 32, 53, 57, 58, 143, 146, 151, 153, 154, 161(24), 163, 170–172, 262, 266, 268(1), 269, 441, 447

Q

Quist, T. M., 121–123

R

Rafuse, R. P., 430, 431(76)
Raines, J. A., 359, 362, 363, 374
Ralston, J. M., 190
Ramdas, A. K., 50, 130
Rashevskaya, E. P., 181, 182
Read, W. T., Jr., 309
Redfield, D., 256
Rediker, R. H., 63, 64, 203, 351
Reine, M., 179
Reiss, H., 86
Renne, M. J., 403
Reuszer, J. H., 50, 130(31), 131, 218, 226(129)
Reynolds, R. A., 83, 116, 202, 203
Rez, I. S., 447
Rhoderick, E. H., 233
Richards, P. L., 144, 162, 224, 396, 402, 403(3), 407, 410, 411(26), 412(49), 413(49), 414(49), 415, 416(55), 417, 421(40), 423(40), 424(40, 49), 425(49), 426(40), 428(49), 436, 437(40), 438, 439
Richards, R. G., 268, 270(191)
Richardson, C. R. W., 447(19), 448
Riesz, R. P., 351
Risley, A. S., 428
Rittner, E. H., 54
Roberts, V., 163
Robinson, L. C., 167
Rochlin, G. I., 402
Rodriquez, S., 60, 63, 119(43–45), 213
Rogers, C. B., 189, 191
Rogers, C. G., 164
Rogers, W. H., 155
Rogovin, D., 410, 411(47), 428(47)
Rollin, B. V., 41, 145, 146, 153
Rose, A., 55
Rose-Innes, A. C., 179
Rosenblatt, J., 408
Rosier, L. L., 50, 135(29), 220

Rossi, J. A., 179, 207, 344, 346(73), 376(142–144), 377(144), 380, 381(142), 382(142–144), 386(144), 387(73), 388(73), 389(73, 144)
Rothschild, W. G., 224
Roundy, C. B., 445, 448(7), 449
Rowe, H. E., 430
Rowell, J. M., 401
Rubenshtein, R. N., 80
Ruderman, M. A., 431
Ruegg, H. W., 338, 339, 351(53), 360, 367
Rundel, R. D., 449
Russer, P., 403, 434, 435
Ryan, J. L., 376(142, 144), 377(144), 380, 381(142), 382(142, 144), 386(144), 389(144)
Ryder, E. J., 118

S

Sachs, R. G., 190, 191(54)
Saermark, K., 431
Sah, C. T., 50, 135(29), 220
Salmer, G., 342, 346, 389
Salomon, S. N., 211
Salpeter, E. E., 210
Samuelsen, M. R., 431
Sanchez, A., 34
Sard, E. W., 138, 163
Sarver, C. E., 110
Savage, S. D., 447
Savich, N. A., 167, 168
Savitsky, E. J., 360, 364, 375
Sawyer, D. E., 351
Sayle, W. E., 340
Scalapino, D. J., 397, 410, 411(47), 428(47), 433, 437
Schechter, D., 449
Schiel, E. J., 360, 364(120), 375
Schillinger, W., 264
Schlosser, P. A., 449
Schmit, J. L., 26
Schneider, E. E., 180
Schneider, M. V., 351
Schneider, W. E., 12
Schreiber, P. J., 135
Schultz, M. L., 41, 42(15), 43, 114(7), 271
Schwarz, E., 441, 442, 444
Sclar, N., 42, 52, 53(12), 58(12), 60, 63, 64, 66, 98, 113, 202, 209, 264(107)

AUTHOR INDEX

Scott, W. C., 245, 247(171)
Sedunov, B. I., 131, 215, 227(121)
Seitz, F., 232
Sekiguchi, T., 167
Sell, D. D., 351, 383
Semenov, V. K., 403, 405(30), 410, 413(48), 418(30), 434(30), 437(30)
Sesnic, S., 153
Shabde, S. N., 342, 344, 346, 347, 375(68)
Shapiro, S., 397, 400, 417, 428, 438
Sharp, E. J., 447
Shaulov, A., 447
Shaunfield, W. N.. Jr., 360, 368, 372(133), 375
Shaw, J. A., 166, 167
Shaw, N., 153, 163, 445
Shenker, H., 42, 57, 103–107, 154, 176, 217, 264, 266, 268
Sherman, G. H., 189, 191(52)
Shifrin, K. S., 199, 224, 229, 232
Shillinger, W., 63, 64(49), 65(49), 99(49)
Shirahata, K., 361, 364(124), 365, 369(131), 370(124)
Shivanandan, K., 150, 164(14), 168
Shockley, W., 118, 177, 312, 326, 329, 348(35)
Shotov, A. P., 331, 338, 340, 341
Sidorov, V. I., 105, 107, 132, 135(179), 137(122), 181, 182(28), 216, 217, 218(122, 126), 221(122, 126)
Sievers, A. J., 437
Silver, A. H., 429, 435
Simhony, M., 445, 447
Simmonds, P. E., 182(37), 184
Simmons, E. L., 41
Singer, B., 447, 448
Shirpkin, V. A., 181, 182(28)
Skocpol, W. J., 402
Sladek, R. J., 154, 204
Slater, J. C., 156, 172
Slobodchikov, S. V., 349, 393
Smirnova, N. N., 393
Smith, C., 447, 448
Smith, E. F., 213
Smith, K. A., 449
Smith, R. A., 6, 156, 268
Smith, S. D., 382
Sodha, M. S., 118
Sokoloff, D. R., 34
Sokolov, A. V., 167, 168
Soloman, R., 194
Solymar, L., 396, 397(2), 399(2), 403(2)

Soref, R. A., 42, 112
Sorrentino, A., 79, 83(80), 84(80)
Spirito, P., 330
Spitzer, W. G., 181–183, 351, 352(95, 96)
Stafsudd, O. M., 442, 444, 445
Stair, R., 12
Stancampiano, C. V., 438
Steckl, A. J., 449
Steinhage, P. W., 447(19), 448
Stephen, M. J., 412
Sterling, S. A., 438
Stern, F., 231
Stevens, N. B., 32, 442, 444
Stevenson, J. R., 181, 182(26)
Stewart, W. C., 401
Stillman, G. E., 179, 182, 184, 190–192, 193(61), 194, 196–201, 203, 205, 206(100), 207, 208(84, 104a), 212, 215, 222, 224, 227, 229, 233(66, 132), 235, 237, 238(57), 241, 242(38, 66), 243, 244, 246–248, 249(178), 251, 252(57, 178), 253–255, 261, 272, 273, 276, 279, 280(132), 284(84), 288(84), 343, 344, 346–348, 350, 376, 377, 378–380, 381, 382, 386–389
Stockton, J. R., 166
Stokowski, S. E., 442, 443
Stone, N. W. B., 163
Stradling, R. A., 182–184, 186, 236, 239(162)
Strimer, P., 447(19), 448
Sturge, M. D., 351, 352(97)
Suits, G. H., 3, 5(4), 7(4)
Summers, C. J., 234, 236–239, 248(160), 254(160)
Sunshine, R. A., 339, 340(55)
Svistunov, V. M., 397
Swiggard, E. M., 103, 105, 106(121), 107(121), 176, 217(10)
Sypek, D. W., 448
Sze, S. M., 196, 328, 341–344, 346, 347, 359, 388, 389

T

Tager, A. S., 314, 316
Takamiya, S., 355, 361, 364, 365, 369, 370
Takekawa, T., 363
Talalakin, G. N., 181, 182(28)
Talley, R. M., 231

Tanimoto, A., 167
Tannenwald, P. E., 182(36), 184, 203, 208(84), 248, 249(178), 252(178), 254(36), 284, 285 (197), 286(197), 288
Taur, Y., 402, 407, 410, 411(26), 412(49), 413 (49), 414(49), 421. 423, 424, 425(49), 426, 428(49), 432, 436, 437(40), 439
Taylor, B. N., 397, 398, 437
Taylor, R. G. F., 447
Teich, M. C., 138
Teller, E., 190, 191(54)
Teranishi, A., 447(19), 448
Tharmalingham, K., 381
Thomas, R., 447(19), 448
Thompson, E. D., 403, 405(29), 418(29), 428, 434
Thomson, G. H. B., 189, 191(50)
Tinkham, M., 402
Tiron, S. D., 233
Tokuyama, T., 357
Tompsett, M. F., 447
Toots, J., 398
Toyabe, T., 204
Toyoshima, Y., 377, 378
Trapani, R. G., 364, 375
Trezise, K. E., 447(19), 448
Trice, J. B., 359, 364
Trifonov, B. A., 154, 167
Trifonov, V. I., 167
Trumbore, F. A., 79, 80
Tsunawaki, Y., 167
Tufte, O. N., 445
Turnbull, A. A., 447(19), 448
Turner, B., 447(19), 448
Turner, W. J., 41, 63, 104(3), 105(3)
Twose, W. D., 87
Tyapkina, N. D., 79
Tyler, W. W., 49, 53, 79, 84–86, 96, 98, 100, 101

U

Ukhanov, Yu. I., 182, 183
Ullman, F. G., 445
Ulrich, B. T., 420, 427, 438
Urgell, J., 313

V

Valov, V. A., 154, 167
van der Ziel, A., 67, 445, 447, 449
Van Duzer, T., 396, 403(3), 405, 406, 418(37), 422
van Overstraeten, R., 326, 332, 338–340, 363, 368, 370
van Vliet, K. M., 67, 69, 70, 76, 268, 269(189)
Vardanyan, A. S., 167
Varga, A. J., 35, 36(24)
Vavilov, V. S., 79
Veenstra, T. B., 404
Venables, J. D., 442
Vere, A. W., 445
Vernet, G., 410, 411(50), 436
Vernon, F. L., Jr., 403, 405(28), 418(28), 419, 420(28)
Véron, S., 447(19), 448
Vinson, P. J., 290
Vogl, T. P., 99
Voronenko, V. P., 166, 167
Voss, R. T., 411
Vrehen, Q. H., 180, 182, 183
Vystavkin, A. N., 153, 166–168, 403, 405(30), 418(30), 435, 437(30)

W

Waldman, J., 182(36), 184, 248, 249(178), 252 (178), 254(36)
Wallis, R. F., 50, 57, 104, 106, 130(24), 181, 182(26), 264, 266
Walls, D. F., 167
Walters, W. R., 12
Walton, A. K., 181–183
Ward, S. A., 360
Wasserstrom, E., 359
Watton, R., 447, 448
Webb, P. P., 306, 313, 318(9), 321, 322(9), 323, 324, 356, 367, 369(9), 371, 372, 375
Webb, W. W., 411
Weber, J., 447(19), 448
Weil, R., 189, 191(52)
Weiner, M. E., 239
Weiner, S., 441, 442, 444

AUTHOR INDEX

Weinreb, S., 424
Weinstein, M., 389
Werthamer, N. R., 399, 405, 428, 437(10)
Wertheim, G. K., 99, 100
Westgate, C. R., 167, 284
Whalen, J. J., 167, 284
Whelan, J. M., 181–183
Whipps, P. W., 441, 444, 449
Whitaker, J., 197(71), 198, 229(71), 235(71)
Whitbourn, L. B., 167
White, H. G., 349
Wieder, H. H., 35, 37(26)
Wiegand, J. J., 401
Wiegmann, W., 303, 306(7), 331(7), 335(7), 338(7), 339(7), 340(7), 355(7), 366(7), 368(7), 369(7), 370(7)
Wilkes, E., 35, 36(24)
Williams, E. W., 195
Williams, R., 342
Williams, R. L., 124, 125, 127, 128(154), 129
Wilson, B. L. H., 199, 229(75)
Wilson, D. J., 445
Wolfe, C. M., 179, 182(38), 184, 190–192, 193(61), 194, 196, 197, 198(66), 199(66), 200(66), 201(70), 203, 205–207, 208(84, 104a), 212, 215, 222, 224, 227(66), 229, 233(66, 132), 235(66), 237(57), 238(57), 241, 242(38, 66), 243(66), 244(66), 246(38), 247(66), 248(61), 251(57), 252(57), 253(57), 254, 255(179), 261(179), 272(133), 273(133), 276(133), 279(113, 132), 280(132), 284(84), 288(84), 343(77), 344, 346(73), 347(77), 348(77), 350(77), 373(76), 376(76, 77, 142–144), 377, 378, 379(77), 380, 381(142), 382(143, 144), 386(76, 144), 387(73), 388(73), 389(73, 144)
Wolfe, W. L., 3, 5(4), 7(4)
Wolff, P. A., 326
Wood, R. A., 138, 139, 182, 183, 186
Woods, M. H., 334, 339, 345

Woodward, D., 271
Woodworth, W. H., 447(19), 448
Wreathall, W. M., 447, 448
Wright, G., 190, 191
Wright, H. C., 449
Wul, B. M., 331, 338, 340, 341
Wurst, E. C., Jr., 99

Y

Yafet, Y., 245
Yamada, K., 167
Yamaka, E. 445, 447(19), 448
Yamamoto, J., 162
Yamashita, J., 203
Yanson, I. K., 397
Yardley, J. T., 138
Yariv, A., 116, 117, 119, 138
Yeh, C., 342, 344, 346, 347, 375(68)
Yokavleva, G. T., 129
Yoshinaga, H., 146, 162, 167

Z

Zabolotnyi, V. F., 166
Zavadskii, Yu. I., 50
Zeidler, J. R., 445
Zeiger, H. J., 130
Zeyfang, R. R., 447(19), 448
Zhidkov, V. A., 86
Zhukov, A. G., 167
Zimmerman, J. E., 401, 403(16)
Zingg, W., 437
Zissis, G. J., 3, 5(4), 7(4)
Zook, J. D., 449
Zyabrev, B. G., 166, 167
Zylbersztejn, A., 65, 203

Subject Index

A

Absorption, 51–53, 222, 265, 351–356, 381–386
  device implications, 353
  electric field effects, *see* Electroabsorption, Franz–Keldysh effect
  excited states, 129, 130, 175, 176, 209, 210
  free carrier, 144
  GaAs, 209–211
  impurity atom, 45–52, 209–211
  lattice, 52
  magnetic field effects, 146
  measurement, 222–224
    Fourier transform spectroscopy, 222
  narrow-depletion-width device, side illumination, 356, 357
Absorption coefficient, 46, 51–53, 144, 265, 352, 381–385
  electric field effects, 381–385
  GaAs, GaP, Ge, InAs, InP, InSb, and Si, Si, 352
Applications, *see also* specific detectors
  hot electron photoconductive detectors, 163, 164, 166–168
  impurity Ge and Si detectors, 142
  pyroelectric detectors, 447, 448
Arrays
  impurity Ge and Si detectors, 92–95
  Josephson junctions, 402, 408, 409, 420, 429, 439
  pyroelectric detectors, 449
Auger process, 59–65, 211, 265, *see also* Recombination
Avalanche gain, *see* Alvanche photodiode, Multiplication
Avalanche gain mechanism, 300–314, *see also* Avalanche photodiode, Multiplication
  gain–bandwidth product limitations, 308–313
  gain saturation effects, 312–314
  general features, 301–308
    breakdown voltage, 303, 304

impact ionization, 325–350, *see also* Impact ionization
  measurement of rates, 330–337
impact ionization coefficients, 300, 308, 337–350, *see also* Impact ionization
  optimum performance, 313, 314, 350
  low frequencies, 301–308
  measured gains, 361–366, 368–370
  multiplication factor, 304–308
    field dependence, 306
    frequency dependence, 308–311
    maximum, 312, 313
  noise, 314–325
    effective, 319, 321
    excess, 314–317, 321, 361, 363–366, 369, 373–378
    spectral density, 315, 316, 318–320
  statistics, 321–325
    pulse code modulation, 321–325
Avalanche multiplication, *see* Multiplication
Avalanche photodiode, 291ff, *see also* Avalanche gain mechanism, Ge avalanche photodiodes, Multiplication, Si avalanche photodiodes
  advantages, 300, 391–393
  bandwidth considerations, 300, 392
  current gain, *see* Multiplication, Avalanche gain mechanism
  dark current, 298, 299
    bulk, 299
    surface leakage, 299
  device structure considerations, 350–380
    external quantum efficiency, 351–354
    internal quantum efficiency, 356, 357
    microplasma problems, 357, 361, 380
    narrow-depletion-width design, side illumination, 356, 357
    pulse response, 354, 355
    type of injected carrier, 357
    uniformity of multiplication, 357
  device structure designs
    beveled mesa, 358–360, 364–367, 375
    guard ring, 357, 358, 362, 363, 374, 375

SUBJECT INDEX

inverted arrangement, 358, 359, 376
large area MOS design, 360
laterally diffused planar diode (LAD), 363
"reach-through" design, 358, 360, 367, 375
Schottky barrier, 376
electric field effects, *see also* Electroabsorption avalanche photodiodes
Schottky barrier EAP detector, 380–382, 386, 389
error probability
Gaussian distribution, 322, 323
Poisson distribution, 321–323
error rate limits, 324, 325
excess noise factor, 314–317, 321, 361, 363–366, 369, 374–378
impact ionization, 325–350, *see also* Impact ionization
measurement of rates, 330–337
optimum detector performance, 313, 314, 350
rates (Baraff curves), 328–330
threshold energy, 329, 330
impact ionization noise, 315
Poisson statistics, 315, 321, 322
minimum detectable power, 300
quantum noise limit, 300
multiplication factor, 298, *see also* Multiplication
optimum, 299
multiplication noise, 314–325
noise, *see also* Avalanche gain mechanism, Multiplication, Noise
amplitude modulation, 299, 321
pulse-code modulation, 321–325
optimum materials, 350
performance limits, 297–300
signal current, 298
total noise current, 298, 299
pulse-code modulation, 321–325
error probability, 321–325
pulse-input response, 354, 355
signal-to-noise ratio, 299, 321
Schottky barrier, 343–345, 376
EAP detector, 380–382, 386, 389
surface leakage reduction, 357, 362
various materials, *see also* specific materials
GaAs, 373, 376–378

Ge, *see* Ge avalanche photodiodes
InAs, 373, 376
InSb, 373, 376
Si, *see* Si avalanche photodiodes

B

Background limited infrared photoconductor (BLIP), 74–77, 102, 103, 110–112, *see also* BLIP, Impurity detectors
Background radiation, 69–71, 121–129, 267, 276–284, *see also* Noise
current
PIN photodiode, 293, 294
reduced magnitude, 121–129, 276–284
V–I characteristics, GaAs, 281–283
responsivity, 283, 284
Baraff curves, 328–330, *see also* Impact ionization
Bias, 3, 13, 15, *see also* specific detectors
circuit, 263
effects
noise spectrum, 30, 31, 158, 275
signal-to-noise ratio, 275
hot electron photoconductive detector, 150–152, 159, 160
shot noise, 158
maximum responsivity, 282
BLIP, 74–77, 102, 103, 110–112, 123, *see also* Background limited infrared photoconductor; specific detectors
Bolometer, 262, 414–417, *see also* Thermal detectors
detectivity, 289
free electron (electronic), 145, 171, 172, 262
Josephson junction, 414–417
advantages, 414–417
NEP, 415, 416
Boson factor, 70, 76
Breakdown, *see* Impact ionization

C

Carrier drift length, 54
Carrier kinetics, *see* Charge carrier properties
Carrier lifetime, 54–56
electric field dependence, 117

Ge(Au) detectors, 99
  optimization, 55
  response time, 61, 62
Carrier scattering, *see also* Mobility
  acoustic mode, 118
  ionized impurity, 204
  neutral impurity, 55, 118
  optical mode, 118, 119, 326–328
  polar mode, 329
Carrier sweepout, 124–127
Charge carrier properties, *see also* entries beginning with "Carrier"
  kinetics of generation and recombination, 58–66, 264–267, *see also* Generation, Recombination
  response time, 61, 62
Compensation, *see* Impurity compensation
Conductivity, *see also* Mobility
  high frequency, 144
  hot electron regime, 145, 146, *see also* Electric field dependences
  low frequency, 143
  non-Ohmic, 145, *see also* Non-Ohmic behavior
Cooling systems, 25
  cryogenic liquids, 25
  thermoelectric systems, 25
Current–voltage characteristics, *see* Josephson junction, Non-Ohmic behavior; specific detectors

D

D^*, *see also* Detectivity, Performance; specific detectors
  bandwidth product, 129
  BLIP, 75, 76, 102, 103, 110–112
  field of view reduction, 76
  measurement, low background, 122, 123
  modulation-frequency dependence, 19, 29, 30, 110, 443, 444
  temperature dependence, 25, 73, 74
    InAs, InSb, PbS, PbSe detectors, 25
    (PbSn)Te detector, 26
    Si impurity detector, 110
    thermal detectors, 33

various detectors compared, 289
wavelength dependence, 289
  bolometers, 289
  GaAs impurity detectors, 289
  Ge impurity detector, 102, 103, 106, 289
  Ge–Si alloy impurity detector, 114
  Golay cell, 289
  (HgCd)Te detector, 24, 26
  InAs, Pbs, PbSe detectors, 24
  InSb detectors, 24, 25, 28–30, 289
  (PbSn)Te detector, 26
  PbTe Schottky diode, 36
  pyroelectric detector, 289
  Si impurity detector, 111, 112
  vacuum thermopile, 289
D-Star, *see* D^*, Detectivity
Debye Hückel screened potential, 232
Detectivity, 7, 17–21, 72–74, 289, *see also* D^*, Detector parameters, Performance; specific detectors
  calculation, 17–21, 23, 24
  low background conditions, 121, 122
  temperature dependence, 73, 74, *see also* D^*, temperature dependence
  various detectors, *see also* specific detectors
    (HgCd)Te, InAs, InSb, PbS, PbSe, 24
  wavelength dependence, *see also* D^*, wavelength dependence, Wavelength dependence
  various detectors, 289
Detector fabrication, 90–95, *see also* specific detectors
  contact metals, 91, 92
  etch solution
    Ge, 90
    Si, 90, 91
  multielement arrays, 92–95
Detector parameters, 2–8, *see also* Detectors, Performance, Photoconductive detectors, Spectral response; specific detectors, specific materials
  background conditions, 5, *see also* Background radiation
  bias, 3, *see also* Bias
  detectivity (D-star or D^*), 7, *see also* D^*, Detectivity
  figures of merit, 5–8, *see also* Figure of merit

frequency characteristics, 6, see also Frequency response
incident signal power, 3
measurement, 8–18, see also Measurement of detector parameters
noise, 4, 5, see also Noise
noise equivalent irradiance, 7
noise equivalent power, 7, see also NEP, Noise equivalent power
output signal, 3, 4
   factorability property, 4, 13
performance calculations, 16–23, see also Performance
response times, see also Response; specific detectors
   various detectors, 290
responsive time constant, 7, see also Responsive time constant
responsivity, see Responsivity
spectral power efficiency, 6
Detectors, see also Detector parameters, Photoconductive detectors, Photoconductor; specific detectors, specific materials
   circuit, 56–58, 263
   extrinsic, see Impurity detectors
   frequency response, see Frequency response
   GaAs, see GaAs far-infrared impurity detectors
   (HgCd)Te, InAs, InSb, PbS, PbSe, (PbSn)Te detectors D^*, 24–26
   hot electron InSb detectors, 143–168, see also Hot electron photoconductive detectors
   impurity, 39ff, 169ff, see also Impurity detectors
   Josephson junction devices, 395–440, see also Josephson junction, Josephson junction detectors
   MIS InSb photovoltaic detector, 35
      polycrystalline films, 35, 37
   MOM point contact diodes, 34
      NEP, 34
   photoconductive, see Photoconductive detectors
   photodioces, 291ff, see also Photodiodes
      avalanche, see Avalanche photodiode
      PIN, see PIN photodiodes
   photovoltaic, see Photovoltaic detector
   pyroelectric, 262, 289, 441–449, see also Pyroelectric detector
   Schottky barrier, see Schottky barrier detector
   short-circuit current, 54–58, see also specific detectors
   super-Schottky diode, 429
      compared with Josephson junction detectors, 429
   thermal, see Bolometer, Thermal detectors
Dielectric relaxation, 124–127

E

EAP detector, see Electroabsorption avalanche photodiodes
Electric field dependences, see also Franz–Keldysh effect, Non-Ohmic behavior
   impact ionization, 119, 120, see also Impact ionization
   internal fields, see Stark effect
   lifetime, 117
   mobility, 116
   photoconductivity, 135–137, 227, see also Electroabsorption avalanche photodiodes
   Stark effects, 257–262, see also Stark effect
Electroabsorption, 347, 380–386, see also Absorption, Electroabsorption avalanche photodiodes, Franz–Keldysh effect
   Airy functions, 381, 383
   coefficient, 381, 383–385
   impact ionization measurement, 347
Electroabsorption avalanche photodiodes, 380–391
   absorption coefficient, 381–385
      Airy functions, 381, 383
   advantages, 381, 392
   Franz–Keldysh shift, 344, 347, 348, 380–386
   internal quantum effect, 385–391
      avalanche gain, 386–391
      zero multiplication, 385, 386
Emitter
   GaAs, 278–281

Energy equivalence chart, 170
Excited states, 129, 130, 175, 176, 199–202, *see also* Impurity excitation; specific detectors
  cascade process, 60, 212, 215–217
  GaAs, 175, 176, 199–202, 213–227
  ionization probability, temperature dependence, 226
  Ge, 130, 217–221
  inclusion in donor statistics, 199–201
  magnetic field effects, 241–261
  negative donor ion, 137
  photoconductivity, 175, 176, 199–202, 213–227
    electric field effects, 227
    measurement, 222–224
    mechanisms, 214–217
    photothermal ionization, 222–227, 244–247
    residual, 218, 219
    spectra, 215
    temperature variation, 224–226
  Si, 129, 130, 220
Extrinsic photoconductive detectors, *see* Impurity detectors

F

Factorability property, 4, 13
Far-infrared photoconductivity in high purity GaAS, 169ff, *see also* GaAs, GaAs far-infrared impurity detectors, Performance, Photoconductivity
  bias circuit, 263
  detectivity (comparison, various detectors), 289
  excited states, 175, 176, 199–202, 213–217, 222–227, *see also* Excited states
    spectra, 215, 225, 228, 238, 242, 243, 248
  experimental arrangement, 223, 278
  long wavelength response, 239–241
    electron trap model, 240
  magnetic field effects, 241–261
  maximum responsivity, 263, 282, 283
  measurements, 222–224, 248–251, 278
    high resolution, 248–251
    reduced background conditions, 276–278
    use of far-infrared laser, 248
  millimeter and centimeter wavelengths, 284–288
    compared with InSb, 285
    reduced background conditions, 276–278
  response
    donor concentration dependence, 228, 238, 272
    relevant factors, 264–266
    speed, 288–290
    wavelength dependence, 215, 225, 228, 238, 242, 243, 248
Figure of merit, 5–8, *see also* Detector parameters
  responsivity, 5, 6, *see also* Responsivity
Fourier transform spectroscopy, 222
Franz–Keldysh effect, 344, 347, 348, 380–386, *see also* Electroabsorption
  Airy functions, 381, 383
  EAP detectors, *see* Electroabsorption avalanche photodiodes
  impact ionization measurements, 344, 347, 348
Frequency characteristic, 6, *see also* Frequency response, Wavelength dependence
Frequency response, 6, 7, 57, *see also* Responsive time constant, Wavelength dependence

G

GaAs, *see also* Far infrared photoconductivity in high purity GaAs, GaAs far infrared impurity detectors
  absorption coefficient, 209–211, 265, 352, 384
  avalanche photodiode, 373, 376–378
  avalanche gain, 376, 377
  bandwidth advantage, 392
  electron/hole ionization rates, 377
  excess noise factor, 376–378
  quantum efficiency, 376, 377
  response time, 376, 377
  band structure, 178–180, 290
  nonparabolic effects, 179–184
  carrier concentration
    temperature dependence, 197, 199

SUBJECT INDEX

vapor phase grown, 197–200
characterization, 196–208
  Hall coefficient analysis, 196–204
  mobility analysis, 198–200, 204–207
crystal structure, 177
current–voltage characteristics, 202, 203, 273, 274
dielectric constant, 189–191, 253
  improved determination, 253
  Lyddane–Sachs–Teller relation, 190
  tabulated values, 191
  temperature variation, 189, 190
donor ionization energies, 195, 196, 198, 200–202
  concentration dependence, 227–237
  high resolution measurements, 248–250
  screening effects, 232–234
effective mass, 180–186, 244, 256
  carrier concentration dependence, 183
  density of state mass, 181
  nonparabolic band effects, 181, 184
  optical, 184, 185
  tabulated values, 182, 186
effective Rydberg determination, 250–254
  static dielectric constant, improved value, 253
electroabsorption coefficient, 384
emitter, 278–281
excited states, 175, 176, 194–202, 213–227
far-infrared absorption, 209
Hall coefficient, 196–204
  factor, 197
inhomogeneity effects, 207, 208
impact ionization rates, 342–348, 377, 387–391
  EAP detector, 387–391
  temperature dependence, 343, 344
impurity density determinations, 195–207
  effective, 205
  total, 206
impurity energy levels, 173, 195, 196, 198, 200–202, 227–237, *see also* Impurity energy levels
  concentration dependence, 227–237
ionized impurity scattering, 204–207
long wavelength response, 239–241
  electron trap model, 240
material preparation, 191–196
  defects and purity control, 193, 195, 196

liquid phase epitaxy, 193–195
vapor phase epitaxy, 192, 193
non-Ohmic behavior, 202, 203, 273, 282, 286, 287
photoconductivity, 208–261, *see also* Photoconductivity
  high resolution measurements, 248–251
  magnetic field effects, 242–261
  measurement, 222–224, 248–251
  spectra, 215, 225, 228
  temperature dependence, 224–227
  use of far-infrared laser, 248
refractive index, 186–189
  wavelength dependence, 187, 188
Stark effect, 257–262
Zeeman effects, 242–261
GaAs EAP detector, *see* Electroabsorption avalanche photodiodes
GaAs far infrared detectors, 169ff, *see also* Far infrared photoconductivity in high purity GaAs, GaAs, Impurity detectors (GaAs), Photoconductive detectors, Photoionization
bias, 275
  effect on noise and signal-to-noise ratio, 275
current–voltage characteristics, 202, 203, 273, 274, 286, 287
  microwave radiation, 287
detectivity, 289
  various detectors compared, 289
fabrication information, 270, 271
  contacts, 271
  integrating sphere arrangement, 270, 271
material properties, *see* GaAs
millimeter and centimeter wavelengths, 284–288
  compared to InSb, 285
noise mechanisms, 267–269, 275–277
  background radiation, 267–269
  bias current effect, 275
  generation–recombination, 268, 269, 275, 277
  impact ionization, 267
  Johnson noise, 268, 275
  thermal carrier generation, 269
performance, 262–290
  background radiation, 267–269, 276–278
  bias circuit, 263

reduced background conditions, 276–278
temperature, 267
photoconductivity, 208–261
  donor concentration dependence, 228, 238
  electric field variation, 227
  excited states, 175, 176, 199–202, 213–227
  field-induced tunneling, 214
  high resolution measurements, 248–251
  impact ionization, 214
  magnetic field effects, 242–261
  photoionization, 214
  photothermal processes, 214–217, 222–227, 244–247
  spectra, 215, 225, 228, 242, 243, 248, 251, 255, 261
  temperature dependence, 224–227
  thermal ionization, 214–217
reduced background conditions, 276–284
  cold filters, 277
  experimental arrangement, 278
  responsivity, 283
  V–I characteristics, 281–283
response time, 288, 290
responsivity, 271–273, 276, 283
  donor concentration dependence, 271–273
  maximum, 263, 264
  V–I characteristics, 273, 274
  reduced background conditions, 281–283
Ga(AsSb)
  avalanche photodiode, 376, 380
  impact ionization rates, 350
Gain
  photoconductive, 55, 128
  saturation, 124–129
  system, 15
GaP
  absorption coefficient, 352
  impact ionization rates, 349
Ge, *see also* Impurity detectors, Photoconductivity, Photoionization
  absorption coefficient, 352
  avalanche photodiode, *see* Ge avalanche photodiodes
  excited-state photoconductivity, 130, 217–221
  impurity atom capture cross sections, 99, 100
  impurity detectors, 39ff, 263, *see also* Impurity detectors, Performance

D^*, 29, 74, 102, 103, 106, 289
  noise spectrum, 28, 31
  response, modulation frequency dependence, 27
  spectral response, 98–104, 218–221
  impact ionization rates, 340–342
  material preparation, 77–90, *see also* Material preparation—Ge and Si
  properties of various substitutional impurities, 79, 86, 99, 100
    capture cross section, 99, 100
    diffusion coefficient, 86
    distribution coefficient, 79
    solid solubility, 79
    tetrahedral radius, 79
Ge avalanche photodiodes, 371–373
  dark-current problem, 371, 372
  device structure considerations, 372
  excess-noise factor, 373, 375
  multiplication, 372, 373, 375
  quantum efficiency, 373, 375
  response speed, 375
  wide-band system application, 372
Ge–Si alloys
  impurity detectors, 113–116, *see also* Impurity detectors
Generation (charge carrier), 58–66, 264–267, *see also* Charge carrier properties, Noise
  cascade emission, 215, *see also* Excited states
  excited states, 129, 130, *see also* Excited states
  external photons, 59, 60, 264, 265
  impact ionization, 59, 60, 63, 202, 264, *see also* Impact ionization
  internal photons, 59
  phonons, 59
  quantum efficiency, 61, 265
Generation–recombination noise, *see* Noise
Golay cell, 262
G–R noise, *see* Noise, generation–recombination

H

Hall coefficient, *see also* GaAs
  factor, 196
  material characterization, 196–204
Harmonic mixing

SUBJECT INDEX

Josephson junction heterodyne detector, 425–428
   infrared, with microwave local oscillator, 427
Heterodyne detector, 138, 163, 167, 421–429, *see also* specific detectors
   high speed impurity Ge detector, 138
   InSb hot electron detector, 163, 167
   Josephson junction
      external local oscillator, 421–429
      internal local oscillator, 437, 438
   pyroelectric detector, 445
(HgCd)Te
   D^*, 24–26, 29
   noise spectrum, 28
   response
      modulation-frequency dependence, 27
High speed performance, 137–140, 419, *see also* specific detectors
Hot electron effects, 116–120, 135–137, *see also* Electric field effects, Hot electron photoconductive detectors
   GaAs, 167
   Ge, 167
   InSb detector, 143–168
   magnetic field enhancement, 146
Hot electron photoconductive detectors, 143ff, 262, 263, *see also* InSb, submillimeter photoconductive detectors
   amplifiers, 164–166
   applications, 163, 164, 166–168
   current–voltage characteristics, 150, 151
   cyclotron resonance regime, 162, 163
   description of detector, 147–150
   heterodyne detection, 163, 167
   impedance matching, 146, 147
   magnetic field effects, 146
      freeze-out, 146
      NEP, 159, 160
      responsivity, 151–153
      time constant, 152
   NEP, 154–161
      bias, 159, 160
      magnetic field, 159
      noise limited, 159–161
      wavelength dependence, 156
   noise, 156–160, 161, 164–166
   physical principles, 143–147
   response time, 152, 153, 262
   responsivity, 151–153
      current dependence, 151
      magnetic field dependence, 151, 152
      temperature dependence, 153
      wavelength dependence, 152, 263
   voltage responsivity, 145
Hydrogenic model, 172–175, 210, 250–254, *see also* Impurity ionization energies
   central cell correction, 173, 174, 237, 238, 253, 254
   chemical shift, *see* Hydrogenic model, central cell correction
   valley–orbit splitting, 174

I

Impact ionization, 63–68, 325–350, *see also* Avalanche gain mechanism, Generation, Impact ionization rates, Impurity ionization, Multiplication
   Baraff theory, 327–330, 338, 344
      curves, 328–330
   breakdown, 55, 64–67, 105, 120, 304, 329, 387–391
   critical field strengths, 64, 66, 67
   carrier generation, *see* Generation
   configurations for measurement, *see also* measurement of rates, *below*
      abrupt junction, 334
      abrupt or "punch through" structure, 334, 335
      linearly graded junction, 335
      PIN junction, 332–334
      other structures, 336, 337
   GaAs, 202, 203, 214, 264, 278, 342–348, 377, 387–391
   general discussion, 325–330
   lattice collisions, 326
   optical phonon collisions, 326–328
   optimum detector performance, 313, 314, 350
   polar mode scattering, 329
   impurities in Ge and Si, 119, 120, 202
   breakdown field strengths, 64, 66, 67, 119, 120
   measurement of rates, 330–337, *see also* Impact ionization, configurations for measurement, Impact ionization rates
   electric field effect, 347, 348, *see also* Franz–Keldysh effect
   electron contamination of hole injection current, 337, 348

experimental results, 337–350, see also Impact ionization rates
  hole contamination of injected electron current, 347
  noise, 63, 269, 315
  Poisson statistics, 315, 321, 322
  threshold energy, 329, 330
Impact ionization rates, see also Impact ionization; specific materials
  GaAs, 342–348, 377, 387–391
  EAP detector, 387–391
  temperature dependence, 343, 344
  Ga(AsSb), 350
  GaP, 349
  Ge, 340–342
  InAs, 349, 373
  (InGa)As, 350
  InSb, 349, 373
  Si, 337–340, 363, 366, 368
Impurities—substitutional, see also Impurity trapping centers, Material preparation
  absorption, 45ff, see also Absorption
  impact ionization, see Impact ionization
  impurity banding, 87–90, see also Impurity banding
  model, 43–45
  optimized concentrations, 85–90
  photoconductivity, see Impurity photoconductor
  photothermal ionization conductivity, 46, see also Photothermal ionization conductivity
  various impurities in Ge and Si
    capture cross section, 100
    diffusion coefficient, 86
    distribution coefficient, 79, 80
    solid solubility, 79, 80
    tetrahedral radius, 79, 80
Impurity banding, 87–90, 214, 229–237, see also Impurity conduction
Impurity compensation, 55
  high speed detector, 137, 138
  hot electron photoconductive detector, 154
Impurity conduction, 87–90, see also Impurity banding
  hopping mechanism, 87–90
  impurity density dependence, in Ge, 89
  limiting densities, in Ge and Si, 90
  temperature dependence, in Ge, 88

Impurity detectors, see also Impurity photoconductor, Photoconductive detectors, Photoconductivity; specific detectors
  applications, 142
  background limited operation, 74–77, 102, see also BLIP
  carrier sweep-out, 124–127
  D^*–bandwidth product, 129
  detectivity, 72, 73, 102, 103, 106, 289
  device fabrication, 77–94
    detector fabrication, 90–95, see also Detector fabrication
    material preparation, 77–90, see also Material preparation
  dielectric relaxation, 124–127
  electric-field-assisted photoconductivity, 135–137
  excited states, 130, see also Excited states
  field of view, reduced, 76
  GaAs, see GaAs far-infrared impurity detectors
  gain saturation, 124, 129
  Ge, see Impurity detectors, Ge
  Ge–Si alloys, see Impurity detectors, Ge–Si alloys
  heterodyne detector, 138, see also Heterodyne detector
  high speed performance, 137–140, see also Performance
  low background conditions, 121–129
  maximum performance, 73
  modulation-frequency variation, 129
  negative donor ion states, 137
  non-Ohmic, 57, 116–120, see also Non-Ohmic behavior
  photothermal effects, 130–135, 214–222
  responsivity, 71, 72, see also Responsivity
  Si, see Impurity detectors, Si
  space-charge effects, 124–127
Impurity detectors, Ge, 39ff, 263, see also Ge, Impurity detectors, Impurity photoconductor
  BLIP, 102, 103
  current–voltage characteristics, 120, see also Non-Ohmic behavior
  D^*
    temperature dependence, 74
    wavelength dependence, 102, 103, 106, 289

SUBJECT INDEX

dopant
  Al, 221
  As, 105, 217
  Au, 41, 42, 98, 102, 138
  B, 105–107, 221
  Be, 101, 104
  Cd, 41, 42, 101, 104, 129, 132, 135–138
  Co, 101
  Cu, 41, 42, 101, 104, 120, 123, 129, 135–137
  Fe, 101
  Ga, 42, 105–107
  Hg, 41, 42, 101, 114, 129, 132, 135–138
  In, 105–107, 132, 217
  Mn, 101
  Ni, 101, 102
  Sb, 105, 130, 131, 217–220
  Zn, 41, 42, 101, 104, 120
excited-state photoconductivity, 130, 217–221, see also Excited states
field-assisted photoconductivity, 135–137
heterodyne detection, 138
NEP versus background flux, 123
photoconductive spectral response, 98–101
response speed, 27, 263, 290
spectral response, 98–104, 218–221
Impurity detectors, Ge–Si alloys, 41, 113–116, see also Impurity photoconductor
  D^*, 114
  dopant
    Au, 41
    Hg, 114
    Zn, 41, 113
  Ge–Si(Hg) compared to Ge(Hg), 114
  gradual long wavelength fall-off, 115
  spectral response, 113, 114
Impurity detectors, Si, 39ff, 263, see also Impurity photoconductor
  BLIP, 110–112
  D^*, see also D^*, Detectivity
    modulation frequency dependence, 110
    temperature dependence, 110
    wavelength dependence, 111, 112
  dopant
    Al, 111, 113
    As, 112
    Au, 108, 109
    B, 111, 113
    Cd, 108

  Cu, 109
  Fe, 109
  Ga, 111, 113
  Hg, 108
  In, 110, 111
  P, 112
  Zn, 108, 109
  excited-state photoconductivity, 129, 130, 220
  photoconductive response, 109, 263
Impurity energy levels, see also Impurity ionization energies; specific materials
  broadening, 229–237
  in GaAs, 173, 195, 196, 198, 200–202, 227–237
    concentration dependence, 227–237
  in Ge, 44, 173
  high resolution measurements, 248–254
  hydrogenic model, 47, 51, 172–175, 250–254
  in InSb, 173
  magnetic field effects, 146, 204, 241–247
  quantum defect method of calculation, 47, 51
  screening effects, 232–234
  in Si, 45, 173
Impurity excitation, see also Absorption, Excited states, Impurity ionization
  electric field assisted, 135–137
  photothermal, 130–135, 214–222
  probability, in GaAs
    magnetic field effects, 244–247
    temperature dependence, 226
Impurity ionization, see also Impact ionization, Impurity energy levels, Impurity excitation, Photoionization
  cross section, see Photoionization, cross section
  impact ionization, see Impact ionization
  impurity density dependence, in GaAs, 227–235
  probability, in GaAs
    magnetic field effects, 244–247
    temperature dependence, 226
Impurity ionization energies, see also Hydrogenic model, Impurity energy levels, Photoionization

in GaAs, 173, 195, 196, 198, 200–202
in Ge, 40, 173
hydrogenic model, 172–175, 250–254
  central cell correction, 173, 174, 237, 238, 253, 254
  chemical shift, see Impurity ionization energies, central cell correction
  valley–orbit splitting, 174
impurity concentration effects, 229–237
in InSb, 173
magnetic field effects, 244–247
optical, 229
screening effects, 232–234
in Si, 40, 173
thermal, 229
Impurity photoconductor, see also GaAs far infrared detectors, Impurity detectors, Photoconductivity, Photoconductor
  noise, 66–71, see also Noise
  optimum condition, 73
  theory, 53–56
Impurity trapping centers, 60, 136, 212, 215–217, 240, see also Impurities—substitutional
  isoelectronic-type centers, 239
  long wavelength response, 239–241
  electron trap model, 240
InAs
  absorption coefficient, 352
  avalanche photodiode, 373
    avalanche gain, 373, 376
    electron/hole ionization rates, 373
    excess noise factor, 373, 376
    response time, 376
  D^*, 24–26
  impact ionization rates, 349, 373
(InGa)As
  avalanche Schottky barrier photodiode, 376, 378, 379
    microplasma, 380
    multiplication factor, 376, 378
    quantum efficiency, 376, 378
    response time, 376, 378
    responsivity, 378, 379
  impact ionization rates, 350
InP
  absorption coefficient, 352
InSb
  absorption coefficient, 352
  avalanche photodiode, 373, 376
    avalanche gain, 373, 376
    electron/hole ionization rates, 373
    excess noise factor, 376
  D^*, 24, 25, 28–30, 289
  detector performance, compared to GaAs, 285
  free electron bolometer, 145, 171, 172, 262, see also InSb, submillimeter photoconductive detectors
  impact ionization rates, 349, 373
  impurity ionization energy, 173
  noise spectrum, 28, 30
  response, 27, 150–154, 285, see also Hot electron photoconductive detectors
    modulation frequency dependence, 27, 285
  submillimeter photoconductive detectors, 143ff, 262, 263, see also Hot electron photoconductive detectors
Irradiance, 14
  distribution, 22
  noise equivalent, 21

J

Johnson–Nyquist, see Noise
Josephson effect (ac Josephson effect), 396–405, see also Josephson junction
  current steps, 398, 399, 403
  order parameter, 396, 397
  phase difference, 397
  quantum-mechanical oscillator, 397, 398
    determination, fundamental constants, 398
    voltage standard, 398
  tunneling current, 397–404
Josephson junction, 396–408, see also Josephson, effect, Josephson junction detectors, Josephson pair tunneling
  beats, 398
  detectors, see Josephson junction detectors
  dynamic resistance, 411
  heterodyne mixer, 407, 421, see also Heterodyne detector, Josephson junction detectors
  I_0–V_0 characteristic, 398, 399, 402, 403, 413, 417

cavity mode steps, 437
effects of noise, 411, 413
hysteresis, 402
steps, 398, 399, 403, 417
impedance, 405–408
  dc-bias dependence, 406
  matching, 407, 408
injection locking, 438, 439
noise, 410–414
  $1/f$ spectrum, 411
  parameter β^2, 412, 413
  power spectrum, 411
  series array, 414
  thermal fluctuations, 410
nonlinear characteristics, 429, 430
  parameter amplifier, 429–435
  up-frequency conversion, 430
normalized frequency Ω, 403, 404, 407
optimization for detector applications, 408, 409
  requirements, 408
  series arrays, 408, 409, 414, 439
Riedel singularity, 400
RSJ model (resistively shunted junction), 401–405, 407, 430
  phase locked loop analog, 436
  series array, 402, 409
SQUID (superconducting quantum interference amplifier), 416
validity of approximations
  RSJ model, 404, 405
  Werthamer theory, 405
weak-link arrangements, 400–402
  Dayem bridge, 400–402
  Mercereau bridge, 400, 401
  point contact, 400, 401
  sandwich type, 400, 401
Werthamer theory, 399, 400, 405, 427
Josephson junction detectors, 34, 35, 395–440, *see also* Josephson effect, Josephson junction
advantages, 439
bolometer, 414–417
  advantages, 414–417
  NEP, 415, 416
  performance, 416
compared with super-Schottky diode, 429
heterodyne detector (external local oscillator), 421–429

array, 429
conversion efficiency, 421–424, 427
design, 421
equivalent circuit, 422
harmonic mixing, 425–428
infrared frequencies comparisons, 429
mixer coupling parameter, 423
noise, 421, 424, 425
performance, 424
heterodyne detector (internal local oscillator), 437, 438
  cavity mode steps, 437
  down-conversion, 436, 437
  phase locked loop, 436
injection locking, 438, 439
NEP, 35
parametric amplifier (and oscillator), 429–435
  degenerate oscillator, 430–432
  fabrication, 433–435
  gain–bandwidth product, 433, 434
  theoretical performance parameters, 432
regenerative detector, 438
  NEP, 438
video detector, 417–421
  I_0–V_0 curves, 417, 419
  NEP, 419
  series arrays, 420
  speed advantage over bolometer, 419
  square arrays, 420
  theoretical analysis, 418, 419
  theoretical performance, 419
Josephson pair tunneling, 395–405, *see also* Josephson effect

L

Laser, far infrared, 248
  high resolution measurements, 248
Lifetime, *see* Carrier lifetime
LMB (light-modulated breakdown mode), 287
  NEP, 287
Lyddane–Sachs–Teller relation, 190

M

Magnetic field effects

energy levels, 242–247
freeze-out effects, 146, 204
  GaAs, 204
  InSb, 146
GaAs current–voltage behavior, 202
GaAs emitter, 279
GaAs excited-state photoconductivity, 241–261
high magnetic field quantum numbers, 245
hot electron photoconductor, 146
  NEP, 159, 160
  responsivity, 151–153
  time constant, 152
photothermal ionization probability, 244–247
resonant photoconductive effect, 146, 162, 163
Zeeman effect, 242–261
Zeeman effective mass, 244, 256
Material preparation—GaAs, see GaAs, material preparation
Material preparation—Ge and Si, 77–90
  diffusion doping, 84–86
    concentrations, 85
    diffusion coefficients, 86
  impurity atom properties, 79, 80, 86, 99, 100
    capture cross section, 99, 100
    diffusion coefficient, 86
    distribution coefficient, 79, 80
    solid solubility, 79, 80
    tetrahedral radius, 79, 80
  optimized impurity concentrations, 85–90
    impurity banding, 87–90
  vapor doping, 81–84
    concentrations, 83
    vapor pressures, 83, 84
  zone refining, 77–80
    floating zone method, 78–81
Measurement of detector parameters, 8–16, see also Detector parameters; specific detectors; specific parameters
  blackbody source, 9, 10
  electronics equipment, 8, 9
  modulation-frequency source, 10, 11
  monochromatic source, 11, 12
  noise, 8, 9, 15, 16
  reduced background conditions, 276–278
  reference detector, 11, 12
  responsivity, 12–14

Mobility, see also Carrier scattering, Conductivity
  analysis (material characterization), 198–200, 204–207, see also GaAs, characterization
  Brooks–Herring formula, 204
  electric field dependence (Ge), 116
  hot electron detector, 144–146
  inhomogeneity effects, 207, 208
  ionized impurity scattering, 204
Modulation frequency, 6, 7, 10, see also D^*, Response
  effect on performance
    D^*, 29, 30, 110
    external quantum efficiency, 354
    noise, 71
    response, 27, 129
  source, 10, 11
Monochromatic source, 11, 12
Multiplication, 298–314, see also Avalanche gain mechanism, Avalanche photodiode; specific detectors
  breakdown voltage expression, 303, 304
  electric field effects, 306, 387, see also Franz–Keldysh effect
  excess-noise factor, 314–317
  factor, 303–308
    field dependence, 306
    frequency dependence, 308–311
    time-dependent transport equations, 308, 309
  frequency response function, 309, 310
  gain–bandwidth product limitations, 308–313
  gain saturation effects, 312–314
  impact ionization, 325–350, see also Impact ionization
  impact ionization coefficients, 300, 308
    breakdown, 303, 304
    optimum performance, 313, 314
  maximum photomultiplication, 312, 313, see also Multiplication, observed values
  mechanism, 301–308
  microplasma-free, 361
  noise, 314–325
    amplitude modulation, 321
    excess, 314–317, 321, see also specific detectors

SUBJECT INDEX

pulse-code modulation, 321–325
spectral density, 315, 316, 318–320
observed values, 361–366, 368–370, 372, 373, 374–376
optimum, 299
Schottky barrier avalanche photodiode, 343–345, 376
temperature dependence, 328, 329
uniformity, 357

N

NEP, see also Noise equivalent power; specific detectors
GaAs far infrared impurity detector, 276, 287
heterodyne detector, 138, 163, 167, 421
hot electron photoconductive detector, 154–161
   bias dependence, 159, 160
   limiting noise, 161
   magnetic field, 159
   wavelength dependence, 152
Josephson junction bolometer, 415, 416
Josephson junction regenerative detector, 438
LMB mode (GaAs detector), 287
negative donor ion states-type detector, 137
PIN photodiode, 295, 296
pyroelectric detector, 445
Noise, 4, 5, 15, 28–31, 66–71, 267–269, 275–277, see also specific detectors
amplifier, 157, 164–166
avalanche gain, 314–325, see also Avalanche gain mechanism, Multiplication
background radiation, 69–71, 121, 267–269
   boson factor, 70
bandwidth, 16
bias-current dependence, 275
excess (avalanche process), 314–317, 321, 361, 363–366, 369, 373–378
$1/f$, 68, 277, 411
generation–recombination (g–r), 69, 121, 124, 268, 269, 275
hot electron photoconductive detector, 156–161
impact ionization, 63, 269, 277
Johnson–Nyquist, 68, 124, 157, 268, 275

Josephson junction, 410–414
   mixer, 421, 424, 425
low background conditions, 121–129
measurement, 8, 9, 15, 16
microplasma, 357, 361, 380
modulation-frequency dependence, 71
optimum condition, 73
PIN photodiode, 294, 295
pulse-code modulation, 321–325
   error probability, 321–325
pyroelectric detector, 445, 446
quantum noise limit, 300
relative contributions, 71
signal-to-noise (S/N) ratio, 5, see also specific detectors
spectrum, 5, 28–31, see also Noise spectrum
temperature fluctuation, 158, 161
thermal, see also Noise, Johnson–Nyquist
   Josephson junction, 410
   load resistor, 16
Noise equivalent power, 7, see also NEP
versus background (Ge:Cu detector), 123
calculations, 21
heterodyne detector, 138, 163
quantum noise limit, 300
Noise spectrum, 5, see also Noise; specific detectors
spectral density
   avalanche gain processes, 315, 316, 318–320
   various detectors, 28, 30, 31, see also specific detectors
   different bias currents, 30, 31
Non-Ohmic behavior, 57, 116–120, 202, 203, 266, 267, 273, 281–283, 286, 287, see also Electric field dependences
background temperature effect, 120
current–voltage behavior, see also specific detectors
   GaAs, 202, 203, 273, 281–283, 286, 287
   Ge:Zn and Ge:Cu detectors, 120
   hot electron photoconductive detector, 150, 151
reduced-background conditions, 281–283

O

Operational characteristics, 1ff, see also Performance; specific detectors detector parameters, 2–7, see also Detector parameters

P

PbS
  D^*, 24, 25, 29
  noise spectrum, 28
  response
    modulation frequency dependence, 27
PbSe
  D^*, 19–21, 23–25
  NEP, 21
  noise equivalent irradiance, 21
  responsivity, 18–22
(PbSn)Te
  D^*, 26
  noise spectrum, 28
PbTe
  D^* Schottky diode, 36
Performance, 16–23, 95–107, 262–290, see also Detector parameters, Spectral response; specific detectors
  avalanche photodiode, 297–300, 361–375, 391–393, see also Avalanche photodiode, Ge avalanche photodiodes, Si avalanche photodiodes
    optimum ionization coefficients, 313, 314, 350
  bolometer, 145, 262, see also Bolometer, Performance, thermal detector
  GaAs far infrared impurity detector, 215, 225, 228, 237–255, 261–290
    maximum responsivity, 263
    millimeter and centimeter wavelengths, 284–288
    reduced background conditions, 277–284
    response times, 288
    responsivity, 272
  gain saturation, 124, 128, 129
  Golay cell, 262
  high speed, 27, 137–140, 263, 290
  hot electron photoconductive detector, 150–163, 262
    resonance effect, 162, 163
  impurity GaAs detector, see Performance, GaAs far infrared impurity detector
  impurity Ge detectors, 95–107, 218–221, 263
    BLIP, 102, 103
    D^*, 102, 103
    long IR range, 104–108
    medium IR range, 101–104
    photoconductive spectral response, 98–101
    short IR range, 96–101
  impurity Ge–Si alloy detectors, 113–115
  impurity Si detectors, 108–113, 263
    D^*, 110–112
    medium IR range, 111–113
    photoconductive response, 109
    short IR range, 108–111
  InSb submillimeter photoconductive detectors, 150–163, 262
  Josephson junction detectors
    bolometer, 416
    heterodyne detector, 424
    video detector, 419
  low background conditions, 121–129, 277–284
  non-Ohmic behavior, 57, 116–120, 266, 267
  PIN photodiodes, 292–297
    bandwidth limitation, 296, 297
    NEP, 295, 296
  pyroelectric detector, 262, 443–447
  reduced background, 276–278
  thermal detector, 32, 33, see also Performance, bolometer
  various detectors, 23–36, see also specific detectors
Photoconductive detectors, see also Detector parameters, Detectors, GaAs far-infrared impurity detectors, Impurity photoconductor, Impurity detectors; specific materials
  background radiation, 69–71, 121–129, 267
  bias circuit, 56–58, 263, 275
  current–voltage characteristics, see Non-Ohmic behavior
  magnetic field effects, 242–247
  magnetic field enhancement, 146, 162, 163
  mechanisms, 170–172, 214–217
    extrinsic, 170, see also Impurity detector
    field-induced tunneling, 214
    free-carrier absorption, 171, 172, see also Hot electron effects
    impact ionization, 214
    intrinsic, 170
    photoionization, 214

SUBJECT INDEX

photothermal processes, 130–135, 214–222, 244–247
  thermal ionization, 214–222
noise mechanisms, 66–71, 267–269, see also Noise
noise spectrum
  photoconductive versus photovoltaic detector, 28
photodiodes, see Photodiodes
state of the art, 23–27
temperature effects, 74, 153, 267
Photoconductive gain, 55
Photoconductivity, 53–66, 170–172, 208–261, see also Far-infrared photoconductivity in high purity GaAs, Photoconductive detectors, Photoconductor, Photocurrent, Photothermal ionization conductivity; specific materials
  carrier concentration change (response time), 61, 62
  dynamic processes, 58–66
  electric-field assistance, 135–137
    experimental arrangement, 135
  excited states, 129, 130, 175, 176, 199–202, 213–227
  field-induced tunneling, 214
  Ge, 130, 217–221
  impact ionization, 214
  magnetic field effects, 241–261
  photoionization, 214
  photothermal processes, 214–222, 244–247
  Si, 129, 220
  spectra, 215, 225
  GaAs, 208ff, 261, see also GaAs
  hot electron (InSb), 143ff, see also Hot electron photoconductive detectors
    GaAs, 167
    Ge, 167
  magnetic field effects, 241–261
  measurement, high resolution, 248–251
    far-infrared laser, 248
  mechanisms, 170–172, 214–217
    extrinsic, 170, see also Impurity detectors
    free-carrier absorption, 171, 172, see also Hot electron effects
    intrinsic, 170
  negative donor ion, 137
  non-Ohmic effects, see Non-Ohmic behavior

  reduced-background conditions, 121–129, 277–284
  residual, 218, 219
  resonance region, 146, 162, 163
Photoconductor, see also Impurity photoconductor, Nonohmic behavior, Photoconductive detectors, Photoconductivity, Photocurrent
  bias circuit, 56, 58, 263
  non-Ohmic, 57, 116–120, 202, 203, 266, 267
  Ohmic, 57
Photocurrent, 54–56, see also Photoconductor
  frequency response, 57, 58, 288
  non-Ohmic, 57, 116–120, 202, 203
  optimization, 55
  short circuit, 56–58
Photodiodes, 291ff, see also Avalanche photodiodes, Electroabsorption avalanche photodiodes, PIN photodiodes
Photoionization, see also Impurity ionization energies, Photothermal ionization conductivity
  Au in Ge, 96–100
  cross sections, 46–51
    impurities in Ge and Si, 48–50
    impurities in GaAs, 210, 216
  hydrogenic model, 47, 51, 172–175, 210
  photothermal processes, 130–135, 214–222, 244–247
    cross section, 132, 216
    spectra, 131–134, 215
  quantum defect method, 47, 51
  quantum efficiency, 51–53, 265
Photothermal ionization conductivity, 46, 130–135, 214–222, see also Photoconductivity, Photoionization
  magnetic field effects, 244–247
  spectra, 131–134, 215
Photovoltaic detector
  MIS structure, InSb, 35–37
    array, 35, 37
    D^*, 35
    polycrystalline films, 35–37
  noise spectrum, versus photoconductive detector, 28
  Schottky barrier, 35, see also Schottky barrier detector
PIN phodiodes, 291
  bandwidth limitation, 296, 297, 300
  BLIP operation, 296

equivalent circuit, 293, 294
NEP, 295, 296
need for internal gain, 297
noise, 294, 295
performance limits, 292–297
background-radiation current, 293
"dark" current, 294
signal current, 293
total noise current, 294
signal-to-noise ratio, 295
Pyroelectric detector, 262, 441–449, *see also*
Pyroelectric detector materials
applications, 447–449
arrays, 449
detectivity, 289, 443, 444
heterodyne arrangement, 445, 446
NEP, 445
noise, 445–447
performance, 443–447
vidicon, 447
Pyroelectric detector materials, 441–447, *see also* Pyroelectric detector
LATGS (L-alanine doped TGS), 441–445
refractory materials, 443, 445
TGFB (triglycine fluorberyllate), 442, 447
TGS (triglycine sulfate), 441–444
alanine doped, 441–445
deuterated, 442, 447
thin-film polymers, 443, 445

Q

Quantum efficiency, 51–53, 265, 351–357, *see also* specific detectors
avalanche photodiode, 298, *see also* Ge avalanche photodiodes, Si avalanche photodiodes
carrier generation rate, 61
electric field effects, 385–391
external, 351–354, 361
modulated-frequency dependence, 354, 357
wavelength dependence, 351–354, 357
GaAs avalanche photodiode, 376, 377
Ge avalanche photodiode, 373, 375
(InGa)As avalanche Schottky barrier photodiode, 376, 378
internal, 52, 356, 385, 386
EAP detector, 385–391

narrow depletion width, side illumination, 356, 357
PIN photodiode, 292, 293
responsive, 52
Si avalanche photodiode, 361–372, 374, 375
frequency independence, 361

R

Radiant power, incident, 14
Recombination, 58–66, 264–267, *see also* Charge carrier properties, Noise
Auger process, 59, 60, 62, 63, 65, 211, 265
cascade process, 60, 130, 132, 215
GaAs far-infrared emitter, 278–281
phonon, 59, 60, 212, 213
cascade process, 60, 215–217
Lax giant trap model, 60, 212, 215–217
photo emission, 59, 211
radiative, 59, 211, 278–281
Reference detector, 11, 12
Reflectivity, 51–53, 265, 351
antireflection coatings, 52, 351
total internal reflection, 52, 53
Response, *see also* Modulation frequency, Responsivity, Spectral response; specific detectors
diffusion current contribution, 353–355
heterodyne detector, 138, 163, 167
impulse input, 371
Si avalanche photodiode, 370, 371, 374, 375
magnetic field effects, 146, 162, 163, 242, 243, 248, 251, 255, 261
measurement, 8, 9, 12–14
millimeter and centimeter wavelengths
GaAs detector, 284–288
InSb photoconductive detectors, 145, 156, 285
modulation-frequency dependence
Ge detector, 27, 129
(HgCd)Te, InSb, and PbS detectors, 27
pulse input, 354, 355
relative, 14
resonance effect, 146, 162, 163
speed, *see also* Response, time constant; specific detectors
hot electron photoconductive detector, 152–154

SUBJECT INDEX 481

Si avalanche photodiode, 370, 371, 374, 375
tabulation, various detectors, 288, 290
time constant, *see also* Responsive time constant, Response, speed
  GaAs detectors, 285, 288
  hot electron photoconductive detectors, 152
  various detectors, 374–376, 378
Responsive time constant, 7, *see also* Response, time constant
  hot electron photoconductive detector, 152
    bias dependence, 152
    magnetic field dependence, 152
  GaAs impurity detector, 288
  various detectors, 290
Responsivity, 5, 6, 71, 72, 145–147, 264–267, 271–273, 282, 283, *see also* Response; specific detectors
  background temperature effects, 283, 284
  calculation, 16–22
  electric field (Franz–Keldysh effect), 381–383
  GaAs far-infrared impurity detector, 272, 275–277
    bias-current dependence, 275, 282
    donor concentration dependence, 271–273
  high speed detectors, 138
  hot electron photoconductive detectors, 151–153
    current dependence, 151
    magnetic field dependence, 151
    optimization, 151, 153
    temperature dependence, 153
    wavelength dependence, 152
  impedance matching, 146
  (InGa)As avalanche Schottky barrier photodiode, 378, 379
  maximum, 263, 282, 283
    bias, 282, 283
  measurement, 12–14
  relevant factors, 145–147, 264–267

S

Scattering, *see* Carrier scattering
Schottky barrier detector, 35
  D^*, PbTe, 36

quantum efficiency, 35
superconducting elements, 429, *see also* Super-Schottky diode
Si, *see also* Impurity detectors, Photoconductivity, Photoionization, Si avalanche photodiodes
  absorption coefficient, 352
  avalanche photodiode, *see* Si avalanche photodiodes
  excited-state photoconductivity, 130, 220
  impact ionization rates, 337–340, 363, 366, 368
    hole/electron, 363, 365, 366
  impurity atom capture cross sections, 99, 100
  impurity detectors, 39ff, 263, *see also* Impurity detectors
  material preparation, 77–90, *see also* Material preparation—Ge and Si
  PIN photodiodes, 292–297
  properties of various substitutional impurities, 80, 86, 99, 100
    capture cross section, 100
    diffusion coefficient, 86
    distribution coefficient, 80
    solid solubility, 80
    tetrahedral radius, 80
Si avalanche photodiodes, 361–371, *see also* Avalanche photodiode
  bias voltage, 374, 375
  current gain, 361, *see also* Si avalanche photodiodes, multiplication
  excess-noise factor, 361, 363–366, 369, 374, 375
  impulse response, 371
  microplasma-free device, 361
  multiplication, 361–366, 368–370, 374, 375
  quantum efficiency, 361–372, 374, 375
  structures, *see also* Avalanche photodiode
    beveled edge, 364–367, 375
    guard ring, 362, 363, 374
    laterally diffused planar diode (LAD), 363
    "reach-through" design, 367, 375
    wavelength range, 374, 375
Signal-to-noise (S/N) ratio, 5, *see also* specific detectors
Space charge effects, 124–127
Spectral power efficiency, 6
Spectral response, *see also* D^*, Detector pa-

rameters, Performance, Response; specific detectors
Fourier transform spectroscopy, 222
GaAs EAP detector, 381–383
GaAs far-infrared impurity detectors, 215, 225, 228, 272
 donor concentration effect, 228, 238, 272
 electric field effects, 227
 high resolution measurements, 248–251
 long wavelength response, 239–241
 magnetic field effects, 242, 243, 248, 251, 255, 261
 temperature effects, 224–226
high resolution measurements, 248–252
hot electron photoconductive detector, 152–154
 resonance effect, 162
impurity Ge detector, 98–104, 218–221
impurity Ge–Si alloy detector, 113, 114
impurity Si detector, 109
(InGa)As avalanche Schottky barrier photodiode, 379
measurement, 12–14, see also Spectral response, high-resolution measurements
PbS detector, 27
photothermal ionization conductivity, 131–134, 214–222, 244–247
reference detector, 11, 12
relative response, 14
Stark effect, 256–262
 broadening, 257–262
  Holtsmark theory, 257, 258
 internal electric fields, 256
  quadrupole nature, 257–260
 linear shift, 257–262
 quadratic shift, 260–262
Super-Schottky diode, 429
 compared to Josephson junction detector, 429
System gain, 15

T

Temperature dependence, see also specific detectors
 avalanche multiplication, 328, 329

D^*, 25, 73, 74, see also D^*
 responsivity, 267
 background temperature effects, 283, 284
 hot electron photoconductive detector, 153
Thermal detectors, 32, 33, see also Bolometer
 D^*, 33
 differences from photon detectors, 32, 33
 pyroelectric, 32, see also Pyroelectric detector
 radiation thermocouple, 32
Time constant, 7, see also Frequency response, Modulation frequency, Responsive time constant
 detector response time, 62, 152
 GaAs detectors, 285–288
 compared with InSb, 285
Traps, see Impurity trapping centers
Tunneling
 Josephson pair, 395–398, see also Josephson effect, Josephson pair tunneling
 single particle, 395, 396
 super-Schottky diode, 429

W

Wavelength dependence, see also Spectral response; specific detectors
 D^*
  (HgCd)Te, InAs, InSb, PbS, PbSe detectors, 24
  (PbSn)Te detectors, 26
  various detectors compared, 289
 external quantum efficiency, 351–354
 free-carrier absorption, 144
 GaAs emitter, 278–281
 response, various detectors, 374–376
 responsivity, 152

Z

Zeeman effect
 GaAs, 241–261
  Zeeman effective mass, 244, 256
  high magnetic field quantum numbers, 245

QC
612
S4
W5
v.12

OCT 17 1978